exploring Mathematics

INVESTIGATIONS WITH FUNCTIONS

The Jones & Bartlett Learning Series in Mathematics

Geometry

Geometry with an Introduction to Cosmic Topology
Hitchman (978-0-7637-5457-0) © 2009

A Gateway to Modern Geometry: *The Poincaré Half-Plane*, *Second Edition*
Stahl (978-0-7637-5381-8) © 2008

Lebesgue Integration on Euclidean Space, *Revised Edition*
Jones (978-0-7637-1708-7) © 2001

Liberal Arts Mathematics

Exploring Mathematics: Investigations with Functions
Johnson (978-1-4496-8854-7) © 2015

Understanding Modern Mathematics
Stahl (978-0-7637-3401-5) © 2007

Precalculus

Precalculus: A Functional Approach to Graphing and Problem Solving, Sixth Edition
Smith (978-1-4496-4916-6) © 2013

Precalculus with Calculus Previews, Fifth Edition
Zill/Dewar (978-1-4496-4912-8) © 2013

Essentials of Precalculus with Calculus Previews, Fifth Edition
Zill/Dewar (978-1-4496-1497-3) © 2012

Algebra and Trigonometry, Third Edition
Zill/Dewar (978-0-7637-5461-7) © 2012

College Algebra, Third Edition
Zill/Dewar (978-1-4496-0602-2) © 2012

Trigonometry, Third Edition
Zill/Dewar (978-1-4496-0604-6) © 2012

Calculus

Multivariable Calculus
Damiano/Freije (978-0-7637-8247-4) © 2012

Single Variable Calculus: Early Transcendentals, Fourth Edition
Zill/Wright (978-0-7637-4965-1) © 2011

Multivariable Calculus, Fourth Edition
Zill/Wright (978-0-7637-4966-8) © 2011

Calculus: Early Transcendentals, Fourth Edition
Zill/Wright (978-0-7637-5995-7) © 2011

Calculus: The Language of Change
Cohen/Henle (978-0-7637-2947-9) © 2005

Applied Calculus for Scientists and Engineers
Blume (978-0-7637-2877-9) © 2005

Calculus: Labs for Mathematica
O'Connor (978-0-7637-3425-1) © 2005

Calculus: Labs for MATLAB®
O'Connor (978-0-7637-3426-8) © 2005

Linear Algebra

Linear Algebra: Theory and Applications, Second Edition
Cheney/Kincaid (978-1-4496-1352-5) © 2012

Linear Algebra with Applications, Eighth Edition
Williams (978-1-4496-7954-5) © 2014

Linear Algebra with Applications, Alternate Eighth Edition
Williams (978-1-4496-7956-9) © 2014

Advanced Engineering Mathematics

A Journey into Partial Differential Equations
Bray (978-0-7637-7256-7) © 2012

Advanced Engineering Mathematics, Fifth Edition
Zill/Wright (978-1-4496-9172-1) © 2014

An Elementary Course in Partial Differential Equations, Second Edition
Amaranath (978-0-7637-6244-5) © 2009

Complex Analysis

Complex Analysis, Third Edition
Zill/Shanahan (978-1-4496-9461-6) © 2015

Complex Analysis for Mathematics and Engineering, Sixth Edition
Mathews/Howell (978-1-4496-0445-5) © 2012

Classical Complex Analysis
Hahn (978-0-8672-0494-0) © 1996

Real Analysis

Elements of Real Analysis
Denlinger (978-0-7637-7947-4) © 2011

An Introduction to Analysis, Second Edition
Bilodeau/Thie/Keough (978-0-7637-7492-9) © 2010

Basic Real Analysis
Howland (978-0-7637-7318-2) © 2010

Closer and Closer: Introducing Real Analysis
Schumacher (978-0-7637-3593-7) © 2008

The Way of Analysis, Revised Edition
Strichartz (978-0-7637-1497-0) © 2000

Statistics

Essentials of Mathematical Statistics
Albright (978-1-4496-8534-8) ©2014

Topology

Foundations of Topology, Second Edition
Patty (978-0-7637-4234-8) © 2009

Discrete Mathematics and Logic

Essentials of Discrete Mathematics, Second Edition
Hunter (978-1-4496-0442-4) © 2012

Discrete Structures, Logic, and Computability, Third Edition
Hein (978-0-7637-7206-2) © 2010

Logic, Sets, and Recursion, Second Edition
Causey (978-0-7637-3784-9) © 2006

Numerical Methods

Numerical Mathematics
Grasselli/Pelinovsky (978-0-7637-3767-2) © 2008

Exploring Numerical Methods: An Introduction to Scientific
Computing Using MATLAB®
Linz (978-0-7637-1499-4) © 2003

Advanced Mathematics

A Transition to Mathematics with Proofs
Cullinane (978-1-4496-2778-2) © 2013

Mathematical Modeling with Excel®
Albright (978-0-7637-6566-8) © 2010

Clinical Statistics: Introducing Clinical Trials, Survival Analysis,
and Longitudinal Data Analysis
Korosteleva (978-0-7637-5850-9) © 2009

Harmonic Analysis: A Gentle Introduction
DeVito (978-0-7637-3893-8) © 2007

Beginning Number Theory, Second Edition
Robbins (978-0-7637-3768-9) © 2006

A Gateway to Higher Mathematics
Goodfriend (978-0-7637-2733-8) © 2006

For more information on this series and its titles, please visit
us online at http://www.jblearning.com. Qualified instructors,
contact your Publisher's Representative at 1-800-832-0034
or info@jblearning.com to request review copies for course
consideration.

The Jones & Bartlett Learning International Series in Mathematics

Essentials of Mathematical Statistics
Albright (978-1-4496-8534-8) © 2014

Basic Modern Algebra
Turner (978-1-4496-5232-6) © 2014

A Transition to Mathematics with Proofs
Cullinane (978-1-4496-2778-2) © 2013

Linear Algebra: Theory and Applications, Second Edition,
International Version
Cheney/Kincaid (978-1-4496-2731-7) © 2012

Multivariable Calculus
Damiano/Freije (978-0-7637-8247-4) © 2012

Complex Analysis for Mathematics and Engineering,
Sixth Edition, International Version
Mathews/Howell (978-1-4496-2870-3) © 2012

A Journey into Partial Differential Equations
Bray (978-0-7637-7256-7) © 2012

Association Schemes of Matrices
Wang/Huo/Ma (978-0-7637-8505-5) © 2011

Advanced Engineering Mathematics, Fifth Edition,
International Version
Zill/Wright (978-1-4496-8980-3) © 2014
Canada/UK (978-1-4496-9302-2)

Calculus: Early Transcendentals, Fourth Edition,
International Version
Zill/Wright (978-0-7637-8652-6) © 2011

Real Analysis
Denlinger (979-0-7637-7947-4) © 2011

Mathematical Modeling for the Scientific Method
Pravica/Spurr (978-0-7637-7946-7) © 2011

Mathematical Modeling with Excel®
Albright (978-0-7637-6566-8) © 2010

An Introduction to Analysis, Second Edition
Bilodeau/Thie/Keough (978-0-7637-7492-9) © 2010

Basic Real Analysis
Howland (978-0-7637-7318-2) © 2010

For more information on this series and its titles, please
visit us online at http://www.jblearning.com. Qualified
instructors, contact your Publisher's Representative
at 1-800-832-0034 or info@jblearning.com to request
review copies for course consideration.

exploring Mathematics

Mathematics

INVESTIGATIONS WITH FUNCTIONS

CRAIG M. JOHNSON
Department of Mathematics
Marywood University
Scranton, Pennsylvania

JONES & BARTLETT
LEARNING

World Headquarters
Jones & Bartlett Learning
5 Wall Street
Burlington, MA 01803
978-443-5000
info@jblearning.com
www.jblearning.com

Jones & Bartlett Learning books and products are available through most bookstores and online booksellers. To contact Jones & Bartlett Learning directly, call 800-832-0034, fax 978-443-8000, or visit our website, www.jblearning.com.

Substantial discounts on bulk quantities of Jones & Bartlett Learning publications are available to corporations, professional associations, and other qualified organizations. For details and specific discount information, contact the special sales department at Jones & Bartlett Learning via the above contact information or send an email to specialsales@jblearning.com.

Production Credits

Executive Publisher: William Brottmiller
Publisher: Cathy L. Esperti
Acquisitions Editor: Laura Pagluica
Associate Production Editor: Sara Kelly
Marketing Manager: Cassandra Peterson
Art Development Editor: Joanna Lundeen
Art Development Assistant: Shannon Sheehan

VP, Manufacturing and Inventory Control: Therese Connell
Composition: diacriTech
Cover Design: Timothy Dziewit
Rights and Photo Research Coordinator: Ashley Dos Santos
Cover Image: © Noel Powell, Schaumburg/ShutterStock, Inc.
Printing and Binding: Courier Companies
Cover Printing: Courier Companies

To order this product, use ISBN: 978-1-4496-8854-7

Library of Congress Cataloging-in-Publication Data
Johnson, Craig, 1953-
   Exploring mathematics : investigations with functions / Craig Johnson.
      p. cm.
   ISBN-13: 978-0-7637-8116-3 (paperback)
   ISBN-10: 0-7637-8116-9 (paperback)
   1. Functional analysis—Textbooks. 2. Mathematics—Textbooks.
   I. Title.
   QA320.J55 2014
   515'.7—dc23
                            2012015801
6048

Printed in the United States of America
18 17 16 15 14   10 9 8 7 6 5 4 3 2 1

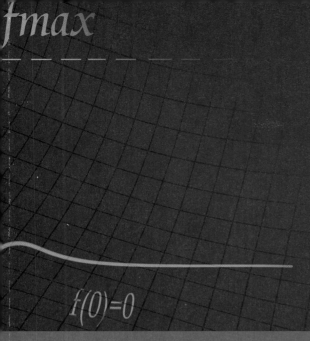

$f(0)=0$

Table of Contents

Foreword		*xi*
Preface		*xv*
Acknowledgments		*xix*
Chapter 1	**The Concept of Function**	**3**
	1.1 Functions Around Us	4
	Exercise Set 1.1	18
	1.2 Creation of Functional Expressions	23
	Exercise Set 1.2	32
	1.3 The Graph of a Function	38
	Exercise Set 1.3	45
	1.4 Interpretation of Graphs	50
	Exercise Set 1.4	59
	Chapter Review Test 1	65
Chapter 2	**Functions of Personal Finance**	**71**
	2.1 Interest and Effective Yield	72
	Exercise Set 2.1	81
	2.2 Annuities	85
	Exercise Set 2.2	91
	2.3 Amortization of Loans	95
	Exercise Set 2.3	103
	Chapter Review Test 2	106

Chapter 3	**Logic and Computer Science**	**109**
	3.1 Statements and Connectives	110
	Exercise Set 3.1	117
	3.2 Truth and Consequences	121
	Exercise Set 3.2	132
	3.3 Tautologies and Syllogisms	136
	Exercise Set 3.3	146
	3.4 Computer Programming Structures	151
	Exercise Set 3.4	163
	Chapter Review Test 3	170
Chapter 4	**Astronomy and the Methods of Science**	**175**
	4.1 Ancient Milestones	177
	Exercise Set 4.1	186
	4.2 The Two Great Systems	190
	Exercise Set 4.2	202
	4.3 The Defense of Copernicanism	208
	Exercise Set 4.3	218
	4.4 And All Was Light	221
	Exercise Set 4.4	235
	Chapter Review Test 4	240
Chapter 5	**Graph Theory**	**245**
	5.1 Graphs and Paths	246
	Exercise Set 5.1	253
	5.2 Euler Circuits	260
	Exercise Set 5.2	266
	5.3 Hamilton Circuits	271
	Exercise Set 5.3	280
	Chapter Review Test 5	286
Chapter 6	**Social Choice and Voting Methods**	**291**
	6.1 Plurality and the Borda Count	292
	Exercise Set 6.1	299
	6.2 The Hare System	303
	Exercise Set 6.2	308
	6.3 Approval Voting	314
	Exercise Set 6.3	320
	Chapter Review Test 6	324
Chapter 7	**Probability**	**329**
	7.1 Sample Spaces and Events	330
	Exercise Set 7.1	335
	7.2 Introduction to Sets	338
	Exercise Set 7.2	346

	7.3	Basic Probability	350
		Exercise Set 7.3	359
		Chapter Review Test 7	365
Chapter 8	**Statistics**		**369**
	8.1	Representation of Data	371
		Exercise Set 8.1	382
	8.2	Dispersion of Data	391
		Exercise Set 8.2	398
	8.3	The Normal Distribution	405
		Exercise Set 8.3	417
	8.4	Confidence Intervals	423
		Exercise Set 8.4	433
		Chapter Review Test 8	437
Chapter 9	**Mathematics in Music and Cryptology**		**441**
	9.1	Residue Classes and Notes	442
		Exercise Set 9.1	452
	9.2	Transformations	459
		Exercise Set 9.2	472
	9.3	The Circle of Fifths	478
		Exercise Set 9.3	485
	9.4	Modular Arithmetic and Cryptology	489
		Exercise Set 9.4	500
		Chapter Review Test 9	503
Chapter 10	**Mathematical Modeling**		**507**
	10.1	Linear Interpolation	509
		Exercise Set 10.1	518
	10.2	Linear Prediction with the Wald Function	522
		Exercise Set 10.2	530
	10.3	Nonlinear Modeling Functions	537
		Exercise Set 10.3	547
		Chapter Review Test 10	552
Appendices			*555*
Solutions to Odd-Numbered Exercises			*577*
Glossary			*629*
References			*647*
Index			*651*

Foreword

Exploring Mathematics: Investigations with Functions

This text is the accumulation of approximately twenty years of thinking, teaching, writing, and editing. When Professor Johnson first presented this idea to me, I was amazed at the forward thinking expressed by a survey of mathematics text universally centered on the function concept. Many textbooks at this level were, and still are, survey books. They tend to be books covering a multitude of topics, with no central theme.

Thus began the writing of *Exploring Mathematics: Investigations with Functions*. Dr. Johnson's idea was to give a sampling of mathematics beyond the intermediate algebra level, which presented explorations with different applications of mathematics, all based on the function concept as the unifying theme. Throughout the years I have taught most of these topics and truly love them. Dr. Johnson's informal style of writing is a breath of fresh air in the typical definition-example-theorem approach we are all familiar with in mathematics textbooks. This book reads more like a nonfiction novel than a textbook. It appeals to students.

Chapter 1, *The Concept of Function*, sets the stage by developing the concept of a function, not by definition, but by examples of naturally occurring functions as actions. What does domain really mean? How are domain and range related to each other? The chapter concludes with some basic graphing concepts from calculus, perhaps as a teaser to students who might want to study further.

Chapter 2, *Functions of Personal Finance*, is one which students respond to quite favorably. Their background on interest, annuities, and loans is nonexistent. This information will be useful to them the rest of their lives. While a teacher of this course often has choices on what to cover (except chapter 1), I would always include this chapter in my course for its clear application to the student's future life.

Chapter 3, *Logic and Computer Science,* is written in a style that is very accessible (and useful) to a student at this level. Students rarely have much experience with programming routines or any kind of logical structure, and this brief introduction to the programming idea would certainly be of great benefit to some students.

Chapter 4, *Astronomy and the Methods of Science,* is my favorite chapter in this book. Knowing nothing about astronomy, I feared teaching this chapter to a mathematics class. Written historically, this chapter takes the students through the major developments of astronomy, which has been such a significant driving force in mathematical development. The chapter also wonderfully describes the scientific method, the challenge of long-held beliefs, and the human nature of the great mathematician-astronomers Hipparchus, Ptolemy, Copernicus, Kepler, Galileo, and finally Newton. The chapter reads like a murder-mystery novel. I couldn't wait to get to the "end," which of course only inspired me to do more reading on the subject. I hope it has a similar effect on some of our students.

Chapter 5, *Graph Theory*, is a fun and easy chapter to teach. Most texts at this level do cover basics of graph theory, so in some ways this chapter varies a bit from the function theme. I really like the inclusion of complete graphs and the traveling salesman problem.

Similarly, Chapter 6, *Social Choice and Voting Methods,* is also an interesting chapter to teach, especially if you explore scenarios (as Dr. Johnson does) where the voting method affects the winner of the election. As students in a democracy, this chapter benefits both students and the instructor in understanding why certain voting methods are advantageous over others.

Chapters 7, *Probability,* and 8, *Statistics,* are natural inclusions in a mathematics text, with a continued emphasis on functions as a primary mechanism for defining probabilities. While many survey texts at this level focus on reading graphs and computational statistics, I love Dr. Johnson's approach of the end prize: confidence intervals for sample means and sample proportions. A traditional statistics book takes 5-6 chapters to get to inferential statistics, while Dr. Johnson gets to the same idea in just two chapters. In my opinion, this focused approach serves students at this level well.

Chapter 9, *Mathematics in Music and Cryptology,* (like Chapter 4), arises out of another passion of Dr. Johnson's: music and harmonics. I had an absolute blast teaching this chapter; again, this is a chapter in which I had little background knowledge. The blending of music notes and modular arithmetic is beautiful and natural. I suspect music theorists understand this, but to me this material was all new. I know our students will have little experience with these connections. Dr. Johnson's sections on transformations and the circle of fifths, with specific examples in music, is a great read. Concluding the chapter with a connection to cryptology is a stroke of genius for students at this level. Again, Dr. Johnson has taken a complex topic (cryptography), and removed the linear algebra to make it accessible to the introductory student.

Chapter 10, *Modeling*, is a brief but thorough introduction to mathematical modeling. Rather than getting bogged down into complicated regression formulas, Dr. Johnson introduces the Wald function, which is simply a line between the lower average point and the upper average point of a set of data. This remarkably simple approach to linear regression is surprisingly accurate, and allows the text to focus on the applications of modeling. I especially like the transformation of exponential data to linear data, which will serve the students well in other courses (chemistry and biology). And, as a fitting conclusion, this chapter again ties the concept of a function (linear or exponential) to real-world problem solving.

There is a lot of deep material in this book, and as such a semester has to be planned accordingly. For a typical one semester course, 3 to 4 chapters are sufficient. I always start with chapter 1 to set the theme. I've taught a "traditional" course of Chapter 1, Chapter 7 Chapter 8, and Chapter 10. Less traditional is Chapter 1, Chapter 4, Chapter 5, and Chapter 9. I could imagine a "discrete" version of the course as Chapter 2, 3, 5, and 8 or Chapters 2, 4, 6, and 7. The options are endless. If you have a two-semester option, you could include most of the text. At our school the course has a flexible course outline, which makes this an exciting course to teach out of Dr. Johnson's text.

Finally, I want to thank Dr. Johnson for having the conviction in this project to follow it through to publication. Twenty years of teaching and writing is a long time, and it is easy to give up trying to share this with the mathematics community. I believe, just as I did twenty years ago, that this is as fresh a mathematics text now as it was then. After all, when was the last time you picked up an undergraduate mathematics text and said, "Well, this is interesting. I wonder how I could teach out of this book?"

And better yet, when your students pick this text up at the bookstore and page through it, they just might ask: "I wonder if my professor will cover astronomy? Why is that in here? What does music have to do with number theory?"

We need more of this type of mathematics content. I hope you like it as much as I do.

Ross Rueger

College of the Sequoias

$f(0)=0$

Preface

This textbook is intended for a one- or two-semester course in mathematics for college students majoring in education, English, history, music, art, foreign language, philosophy, or any other liberal arts discipline. It showcases modern applications of mathematics in many areas as well as the aesthetic features of this very rich facet of the history and ongoing advancement of society. The prerequisite for taking a course that uses this book is intended to be competency in intermediate algebra, although most algebra skills needed, beyond rudimentary computation, are reviewed in the many examples that have been worked out explicitly in each section.

Philosophy

I have often heard prominent mathematicians express the sentiment that *functions*, more than any other single topic, is a central concept that threads through most areas of mathematics. Therefore, the concept of function has been chosen as a unifying theme for the entire book. The first chapter sets the stage for the rest of the text by exploring the true nature of a function presenting a wide variety of examples, demonstrating the pattern recognition used in the creation of a function, and studying how the graph of a function aids in examining its behavior. Early development of this concept helps strengthen the ability of the student to understand later topics at a deeper level.

I believe that both the use of the concept of function as a central theme and the inclusion of several chapters seldom seen in other texts of this genre, such as the chapters on astronomy (Chapter 4) and music (Chapter 9), considerably broaden the appeal of this book. It provides a sample of the next level of mathematics for which a college student who has passed intermediate algebra should be able to master. The book attempts to answer the questions, "How does mathematics help us to better our society and understand the world around us?" and "What are some of the unifying ideas of mathematics?" The central theme helps to impress upon the student the feeling that mathematics is more than a disconnected potpourri of rules and tricks. Although it would be inappropriate to force a functional connection in every single section, the theme is used whenever possible to provide conceptual bridges between chapters. Developing the concept of a function facilitates the presentation of many topics in every chapter, including:

- Chapter 1, *The Concept of Function*: The study of growth models and optimal values of functions.
- Chapter 2, *Functions of Personal Finance*: Computation of the future value of savings accounts and annuities. This chapter also covers the amortization of loans.
- Chapter 3, *Logic and Computer Science:* Analysis of statement forms according to the logical values of the primary statements. This chapter can serve as a clear introduction to the use of algorithms for computer programming.
- Chapter 4, *Astronomy and Methods of Science:* Many of the main ideas in astronomy such as the determination of distance, velocity, and acceleration, and the construction of planetary models.
- Chapter 5, *Graph Theory:* The search for optimal circuits in a graph by using vertex degree in finding Euler circuits.
- Chapter 6, *Social Choice and Voting Methods:* The arrival at different election results depending on the voting method. This chapter also features analysis of the Saari triangle to determine election results.
- Chapter 7, *Probability:* The study of probability distributions and the explanation of the rules of probability.
- Chapter 8, *Statistics:* Use of a density function to find probabilities and portions of distributions. In addition, the construction of confidence intervals in statistics.
- Chapter 9, *Mathematics in Music and Cryptology:* The notion of geometric transformations and number-theoretic functions in music theory and the use of the modulo function to create encryption and decryption algorithms.
- Chapter 10, *Modeling:* A discussion of how real mathematical models are created. The emphasis is on *finding* the model. This differs fundamentally from the approach of those texts that simply *present* a model to study with no notion of how the functional expression was formed.

Approach

This text was designed to implement NCTM (National Council of Teachers of Mathematics) curriculum standards. The topics throughout the text were chosen specifically to:
- strengthen estimation and computational skills
- utilize algebra concepts
- emphasize problem-solving and reasoning
- emphasize pattern and relationship recognition
- highlight importance of units in measurement
- highlight importance of the notion of a mathematical function
- display mathematical connections to other disciplines

However, it must be emphasized that the main objective of this book is to provide the student with an exposure to *both* the widespread application power of mathematics as well as a perspective on its aesthetically pleasing characteristics. My experience has been that a full diet of the strictly "practical approach" often appears to the student as a dreary unbroken series of formulas to be memorized and soon forgotten. The primary objectives for each individual chapter are listed at the beginning of the chapter in order to provide a guideline and set of goals and expectations for the student.

Key Features

- *Writing style*

The approach and writing style of this text is informal and colloquial. Much of the exposition is example-driven. A formal definition is given as a stand-alone statement only when it is essential in crystallizing the concept for the student. The goal is for the student to see where the mathematics "fits," how it encapsulates knowledge, and why its ability to predict is so powerful. Hopefully problems will be tackled not just because they are assigned, but because the answer seems intriguing and relevant. Situations have been presented to pique curiosity, inspire investigation without a sense of drudgery, or, at the very least, to occasionally raise an eyebrow and prompt the comment "No kidding?"

- *Abundance of examples*

In each chapter, examples of particular concepts often precede the presentation of the mathematical formalism in order to provide motivation for how situations are modeled mathematically. They have been carefully chosen to demonstrate precisely how mathematics is used to both find answers to practical problems and to provide a language that describes the behavior of natural phenomena. The units that accompany answers are stressed. The problem sets are arranged in parallel with the development in the text and ordered by increasing complexity. Many applications concern real-world problems with real data.

- *Nonstandard material*

Several chapters are included that are seldom seen in other books of this genre. Astronomy, in particular, serves well as a vehicle to showcase the historical development of science, its attendant methodologies, and the key role that mathematics plays as the language in which new science is created. Additionally, it offers an opportunity to discuss some key moments in the lives of some of the main contributors (Copernicus, Kepler, Galileo, Newton) and reveal some very human characteristics of these towering figures. The student therefore also learns that new mathematics and science do not appear suddenly in isolated bursts but are much affected by the humans who pioneer them and the culture in which they arise.

The chapter on connections of math to music also is a novel topic enjoyed by many of the students who populate this type of class. The various uses of number theory and geometry are explored in identifying and creating musical patterns and in musical composition. Since most people never associate applications of mathematics to disciplines other than the mainstream examples in engineering and science, many students are delighted with this chapter. I have witnessed many grins and heard many favorable comments from art and theater majors about how much they enjoyed this material. Many examples are presented from well-known musicians and composers (Bach, Brahms, Beethoven, Joplin) to provide a human element that cannot be so readily examined in the other more standard topics.

- *Key Terms*

Every discipline has its own vernacular that is essential for accurate explanations of the central ideas. Definitions of key terms are set out and provided in the margins of the text as well as in a complete glossary at the end of the text. Each glossary word is indicated in colored and bold font in the text the first time it is used and defined.

- *Chapter Review Tests*

Students appreciate knowing the types of problems they should be expected to know how to solve and the format in which they will be given. It has been shown that good pedagogy includes providing clear expectations. The chapter review tests give students a framework to help them properly prepare and eases the anxiety that can arise from "fear of the unknown."

- *Historical Figures*

Brief accounts of the lives of the key mathematical contributors are presented in most chapters, directly embedded in the text rather than boxed and set to the side in the style used in most books. The idea is to emphasize that mathematical development definitely occurs *within* the human experience rather than up in the attic apart from one's daily life. I strongly believe that the student is well served by a healthy dose of appreciation for the historical development of mathematics within the human experience and its associated impact on how we view the world.

- *Flexibility*

A semester course usually consists of three or four chapters selected according to student need or interest. Coverage of four chapters is usually only possible if at least two of the chapters consist of three rather than four sections. A good rule of thumb is to limit the total number of sections to less than or equal to fourteen. If *Logic* or *Mathematics in Music* is included in a four-chapter semester, the fourth section can be omitted if time does not allow. Over the past fifteen years I have used many different groupings including *Functions, Astronomy,* and *Music*; *Functions, Graph Theory, Voting Methods,* and *Music*; *Graph Theory, Voting Methods, Probability,* and *Statistics*; *Functions, Logic,* and *Personal Finance*; *Astronomy, Logic* (excluding the fourth section), *Voting Methods,* and *Music*; and many others. My habit has been to usually begin with *Functions,* but it is certainly not essential. The instructor can expand upon the function theme when the concept emerges in a particular chapter.

Calculators

Calculators are assumed to be available to the student for computation throughout the textbook. If the number of significant digits used in answers is not indicated, it is assumed to be three or four, depending on the given data. One exception is the chapter on *Personal Finance*, where all answers are rounded to the nearest cent. A few exercises in the *Functions* chapter explore graphing calculator capabilities.

Additional Resources

Access to a student companion website is available with every new copy of the text. This includes Interactive Flashcards, Interactive Glossary, and a Laboratory Manual. An electronic Student Solutions Manual with worked-out solutions to selected problems that appear in the text is also available on the student companion website.

Instructors may access a Complete Solutions Manual that contains solutions to all problems in the text, as well as Lecture Outlines in PowerPoint format and Review Tests.

f max

f(0)=0

© Dejan Lazarevic/ShutterStock, Inc.

Acknowledgments

I wish to thank the following people for their help and many valuable suggestions. I am deeply grateful. In particular, I would like to express my deep appreciation to Randy Moser of Horry Georgetown College for his hard work and attention to detail while producing the excellent ancillary materials for this book. Additionally, he and Jill Juka of Marywood University provided much valuable assistance in editing and correcting errors in the problems and solutions.

Sara Kelly, *Production Editor, Jones & Bartlett Learning*
Laura Pagluica, *Acquisitions Editor, Jones & Bartlett Learning*
Tiffany Sliter, *Senior Production Editor, Jones & Bartlett Learning*
David Bock, *Statistics textbook author*
Victor Cifarelli, *University of North Carolina at Charlotte*
Owen Gingerich, *Harvard University*
Leon Harkelroad, *Bowdoin College*
Thomas Kent, *Marywood University*
Martin Parise, *Marywood University*
Anthony Pusateri, *Marywood University*
Ross Rueger, *College of the Sequoias*
Beverly West, *Cornell University*
Chaogui Zhang, *Marywood University*

Dedication

To my wife, Vivian and my sons, Nick and Scott,
whose support and witty repartee made this book possible.
And to Anne and Casey, who have always been in my corner.

Chapter Objectives ☑◯◯ ← check off when you've completed an objective

☐ Identify a function and its domain and range

☐ Identify a linear or exponential relationship from a table of paired values and create the functional expression

☐ Use the natural exponential function (with base *e*)

☐ Graph basic linear and exponential functions

☐ Use the vertical line test

☐ Determine the domain, range, and zeroes from the graph of a function

☐ Determine maximum and minimum values of a function from its graph

☐ Determine intervals where a function increases and decreases from its graph

☐ Determine an optimal value in applied problems

Navigate Companion Website

go.jblearning.com/johnson

Visit go.jblearning.com/Johnson to access a Laboratory Manual, Student Solutions Manual, Interactive Glossary, and Interactive Flashcards.

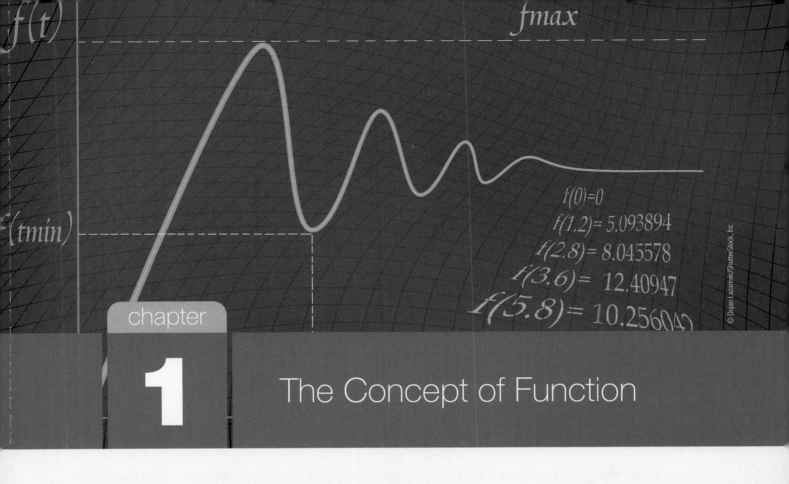

f(t)

fmax

f(tmin)

f(0)=0
f(1.2)= 5.093894
f(2.8)= 8.045578
f(3.6)= 12.40947
f(5.8)= 10.25604

chapter

1 The Concept of Function

1.1 Functions Around Us

1.2 Creation of Functional Expressions

1.3 The Graph of a Function

1.4 Interpretation of Graphs

Form and function should be one, joined in a spiritual union.

—Frank Lloyd Wright

It has been argued by some educators that the concept of *function* is the single most important mathematical idea from kindergarten to graduate school. To begin by presenting a dictionary-style definition, however, may not offer any true insight into the heart of the matter. A subtler approach is needed. When teaching the meaning of a new word, an English teacher often gives a definition immediately followed by the use of the word in a sentence. Most people learn new concepts not by memorization of a formal definition but rather by exposure to a set of clear examples coupled with personal experience that is connected to those examples. Understanding the meaning of a definition comes later, followed later still by perhaps more subtle ramifications of that definition. The learning of mathematical concepts usually follows a similar path. Therefore, we begin this chapter with a section that presents a collection of "experience" examples of functions—apparent relationships, those between sets of objects, that may or may not be expressible in algebraic form. These examples illustrate the notion of a strict correspondence between sets— an idea that leads in the next section to the examination of a function between sets of real numbers as a process or action. The last two sections explore the representation of real-valued functions as graphs and the powerful tool such graphs provide in analyzing the behaviors of functions.

1.1 Functions Around Us

In the study of elementary algebra, letters and other symbols are introduced as convenient representatives for numbers. The first demonstrated use of a letter in mathematics is typically as a stand-in or a holder for a value that is unknown until the completion of some solution process—e.g., $2x + 7 = 19$. There is a fundamental distinction between the use of the letter x in an equation as a holder and its use as a *variable* quantity whose value is associated with a second variable quantity at all times—e.g., $y = 2x + 7$. In the first case, by either observation or some formal technique, we quickly reach the conclusion that the value of x is 6. The second case is very different. Here we are free to assign to x whatever value we like, and this choice then dictates a corresponding value for y. When x is 2, y is 11; when x is 3, y is 13; and so on. It is this second case that helps us get to the definition of a function. Consider the following example.

? Example 1

Phillip has accepted a salesman's position at a computer store and is offered a choice of compensation: either $400 per week plus 12% commission on his sales, or $450 per week plus 10% commission on his sales. In order to reach a decision, Phillip asks himself these two questions:

1. For what dollar amount of sales will these options yield the same payment?
2. If Phillip feels he will consistently sell $3,500 of computer equipment per week, which option of compensation will pay him the most money?

⚙ Solution

The first question is definitely asking for a single numerical answer. It can be solved by the use of a single symbol. If s is the amount of sales that will yield identical payments according to each option, then s must satisfy the equation

$$0.12s + 400 = 0.10s + 450$$
$$0.12s = 0.10s + 50$$
$$0.02s = 50$$
$$s = 2500.$$

FIGURE 1.1.1   Should Phil sell these?

Phillip would be paid the same amount by both options if he sells $2500 worth of computer equipment per week.

The second question is *not* concerned with a single number response. A comparison needs to be made between the two amounts that Phillip would earn, assuming he made sales of $3500 per week. We already know these amounts *must* be different because they are the same only for sales of $2500. Each amount is to be computed according to a different option. In other words, a *relationship* needs to be written down in a formula-type format for each scheme so that Phillip can compute each payment based on a particular sales value. Let *y* represent the amount of payment Phillip receives based on *x* dollars' worth of sales.

Under the first option, $y = 0.12x + 400.$

Under the second option, $y = 0.10x + 450.$

Now we simply replace *x* with 3500 in each formula and calculate the corresponding amounts.

Under the first option, $y = 0.12(3500) + 400 = 820.$

Under the second option, $y = 0.10(3500) + 450 = 800.$

We see that Phillip will make more money under the first option if he sells $3,500 worth of computers per week.   ◆

---

variable A letter or symbol that represents any one of a variety of quantities or objects.

independent variable The variable in a two-variable relationship or a function that varies freely among the allowed values.

dependent variable The variable in a two-variable relationship whose value depends on the value assigned to the independent variable.

Answering the second question needed a significantly different approach, one requiring the identification of related *varying* quantities. A letter or symbol that may represent any one of a variety of different quantities is called a **variable**. Two variables were needed to form each relationship in the previous example, indicating that as one quantity changed, the other quantity also changed in a well-defined way. Usually, it is helpful to think of one variable freely assuming any one of a host of different values, and so we call it the **independent variable**. Once the independent variable is assigned a particular value, the other variable immediately obtains its own value as a direct consequence and hence is known as the **dependent variable**. In the relationship given by $y = 0.12x + 400$ in the previous example, *x* is the independent variable, and *y* is the dependent variable. If Phillip is paid according to this scheme, his payment *y* depends strictly on his sales *x* for the week, which would be changing from week to week. In essence, the construction of the two-variable relationship saves us the time and energy that would be needed to repeatedly solve for an unknown quantity.

Okay, so what exactly is a *function*? In common, everyday conversation, this word gets used in different ways according to the various forms of its definition. One might say, for example, that "the main function of the U.S. Supreme Court is to interpret the Constitution," meaning that interpretation is the characteristic action of the Supreme Court. A weather forecaster might say, "This drought is a function of the atmospheric pressure patterns," indicating

FIGURE 1.1.2 Do more people vote on a warm day?

domain The set of elements that are the inputs to a particular function.

range The set of elements that are the outputs of a particular function.

function A relationship between two sets that assigns to every element in the first set (domain) exactly one element in the second set (range).

set A collection of objects.

formula The precise expression for a function that describes its action in terms of the independent variable.

that a connection exists between atmospheric pressure and the lack of rain. The word *result* would be just as appropriate to use as *function* here, because the implied connection is a rather loose one. On the other hand, in a statement such as "in our county, voter turnout is always a function of temperature," a more direct tie could exist. It may even be possible that the connection is one that could be approximately modeled by an equation—e.g., $v = 100t + 1,500$, where v is the number of voters and t is any temperature between 10 and 80°F. If t is only 30°F, then 4,500 people vote, but if it is a more balmy 70°F, then 8500 show up at the polls. This is a relationship between an independent variable t and a dependent variable v. The set of temperatures (real numbers between 10 and 80) for which this equation holds is referred to as its **domain**, and the resulting collection of numbers of voters is called its **range**. Because mathematics is the language used to accurately record and describe many types of knowledge, mathematicians cannot afford to be ambiguous in their definitions. The function is a concept so central to the realm of mathematics that it will be one of the guiding themes of this book.

A **function** is a relationship between two groups of objects (usually numbers) such that every object in the first group, called the *domain*, is unambiguously matched up with exactly one object in the second group, called the *range*.

Domains and ranges are generally designated by the use of set or interval notation. (See Appendix V for a description of interval notation.) A **set** is a collection of objects, usually possessing some similar characteristic, such as a set of cities, a set of cars, or a set of numbers. A pair of brackets { } is placed around a listing of the members of the set, and the individual members of the set are separated by commas; this is the standard method of representation. For instance, the set of capital letters from A to F is {A, B, C, D, E, F}. Very large sets need some abbreviation. The set of capital letters from A to Z can be expressed {A, B, C, …, Z}. Infinite sets trail off with an ellipsis, …, after a pattern has been established. For instance, the even positive integers can be denoted by {2, 4, 6, 8, …}.

The world around us is brimming with functions, many of which are specific enough to be represented with independent and dependent variables as in the previous section. We shall call such an equation a **formula**, and the important skill of creating a formula for a function by identifying the *action* it performs on elements in its domain will be studied in the next section. For the present, however, we concern ourselves with examples of naturally occurring functions, many of which cannot be translated to a neat, concise formula but are functions nonetheless. In this section, we simply want to identify the function in each example along with its domain and range.

? Example 2

FIGURE 1.1.3 A pedometer.

Suppose you are taking a walk with a pedometer on your wrist, one that keeps track of your travel time and distance covered and computes your walking speed. Then the distance d you travel as measured from some starting point is a function of the elapsed time t. If your current walking speed is 3.4 miles per hour, then for $t = 1$ hour, $d = 3.4$ miles; for $t = 2$ hours, $d = 6.8$ miles; and so forth. If t is measured in hours and d is measured in miles, then we can write

$$d = 3.4t$$

to express this function. If your longest walk takes 3 hours, then a reasonable domain would be the interval **D** = [0, 3] = {t any real number $| \ 0 \le t \le 3$} and the associated range would be **R** = [0, 10.2]. (See Appendix V for a definition of intervals.)

? Example 3

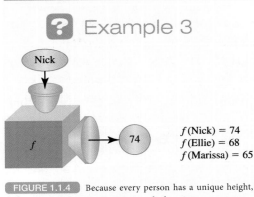

f(Nick) = 74
f(Ellie) = 68
f(Marissa) = 65

FIGURE 1.1.4   Because every person has a unique height, these assignments are representative of a function.

Suppose we form a set **D** consisting of the people in your mathematics class (all having different first names) and another set **R** containing all their heights (in inches). Because every person has one and only one height, we can define a function f by assigning to each person the number corresponding to his or her height. Then **D** would be the domain of f and **R** would be the range. **Figure 1.1.4** shows a schematic diagram displaying some sample assignments of this function. We think of the function as an assignment box. One domain element that goes into the box yields one range element that comes out of the box.

Suppose Nick is 74 inches tall. To display that specific assignment of a domain member to his corresponding range value, the conventional notation is to write f(Nick) = 74. Similarly, if Ellie and Marissa were 68 and 65 inches tall, respectively, we would write f(Ellie) = 68 and f(Marissa) = 65, and we would then continue in this manner. Note, incidentally, that it is impossible to define a function having a domain of heights and a range consisting of your classmates. Two or more people, say Claudia and Rafer, could have the same height, say 70 inches, in which case the assignment of a range member to 70 would require an arbitrary choice—either Claudia or Rafer. If g were going to represent a possible function, g(70) would not be unique, and so our definition of a function would not be satisfied.

? Example 4 (Chemistry)

Pythagoras (c. 580–500 BC) was an early Greek mathematician and philosopher who felt that numbers held the key to any deep understanding of the world. He would have been thrilled with the Periodic Table of the Elements—a categorical listing of the known elements that is essential to the study of chemistry. The nucleus of an atom of each element has a unique number of protons known as the *atomic number* of the element. The numbers listed above the element symbols in the Periodic Table in **Figure 1.1.5** are atomic numbers.

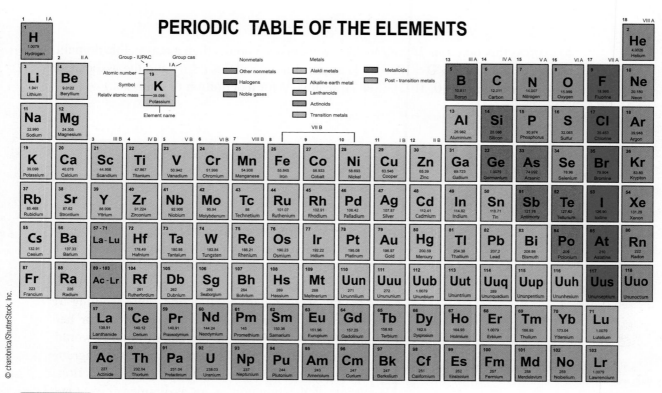

FIGURE 1.1.5   The Periodic Table of the Elements.

The assignment of an atomic number to each element is a functional relationship at its heart because the basic definition of what constitutes an element starts, in fact, with its atomic number, and so it is unique. The first three assignments would be:

$$H \rightarrow 1 \quad (\text{Hydrogen}),$$
$$He \rightarrow 2 \quad (\text{Helium}),$$
$$Li \rightarrow 3 \quad (\text{Lithium}).$$

Borrowing the practice from algebra of using letters such as f and h for representing functions, we use f for our function and write these pairings as $f(H) = 1$, $f(He) = 2$, $f(Li) = 3$, and so on. This notation is commonly used because it is less cumbersome. The domain is the set $\mathbf{D} = \{H, He, Li, Be, B, C, \dots, Uuo\}$, and the range is $\mathbf{R} = \{1, 2, 3, 4, 5, 6, \dots, 118\}$.

A helpful analogy to use when thinking about functions is that of a bow shooting arrows onto a target. The quiver of arrows is the domain, the bow is the function itself, and the points on the target where the arrows are shot represent the members of the range. The key to this picture is to realize that the bow (function) provides the connection from quiver (domain) to target points (range) as in **Figure 1.1.6**.

The word **image** is also used when speaking of a specific member of the range. In the prior example, 1 is the image of H with respect to this function, 2 is the image of He, 3 is the image of Li, and so on. For any function denoted by f, we shall always think of $f(x)$ as the image of x after application of that function. As the independent variable x varies in the domain, its image $f(x)$ varies throughout the range.

image The specific functional value in the range of a function associated with a particular value in the domain of the function.

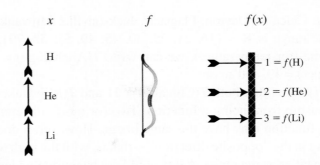

x   *f*   *f(x)*

H

He

Li

1 = *f*(H)

2 = *f*(He)

3 = *f*(Li)

FIGURE 1.1.6 The function represented as a bow shooting arrows (members of the domain) onto target points (members of the range).

❓ Example 5

In contrast to the previous example, the number of *neutrons* contained in a nucleus is *not* a function of the chemical element, and this is because of the presence of isotopes. The isotopes of an element are atoms whose nuclei contain the same number of protons but different numbers of neutrons. Although carbon (C), for instance, normally has six protons and six neutrons in its nucleus, approximately one in 10^{12} atoms contains six protons and eight neutrons and is referred to as carbon 14 (used in the dating of carbonaceous materials). A function does not exist if the domain is allowed to contain isotopes because it is not possible to unambiguously select the image of C to be 6 or 8. A similar situation exists for all the other elements, because every known element possesses at least two isotopes (see **Figure 1.1.7**).

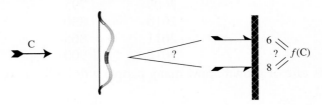

FIGURE 1.1.7 The assignment of an element to the number of neutrons in its nucleus is not a function.

❓ Example 6 (Meteorology)

The high temperatures in several American cities for a typical Saturday in January are given below.

Anchorage	25	Flagstaff	58
Birmingham	45	Jacksonville	53
Casper	49	Milwaukee	21
Chicago	21	Phoenix	79
Dayton	16	Sioux Falls	30

FIGURE 1.1.8 High temperatures in some American cities.

Because every city can only have one high temperature for the day, we have a function between cities and temperatures. The domain here is **D** = { Anchorage, Birmingham,

Casper, Chicago, Dayton, Flagstaff, Jacksonville, Milwaukee, Phoenix, Sioux Falls}, and the range is **R** = {16, 21, 25, 30, 45, 49, 53, 58, 79}. Using T for the function this time (for temperature), we can write T(Anchorage) = 25, T(Birmingham) = 45, T(Casper) = 49, and so on.

The fact that both T(Chicago) = 21 and T(Milwaukee) = 21 does not keep this relationship from being a function. Two (or more) different members of the domain of any function may have the same image. However, it does prevent a function from existing in the "opposite direction"—that is, with the high temperatures as the domain and the cities as the range. A city could not be uniquely identified by its high temperature. What, for instance, would be assigned to 21? The presence of an arbitrary choice for that assignment eliminates the possibility of a functional relationship.

? Example 7

The population of the village of Beushane has increased in recent years. In 2008, there were only 760 people. However, the construction of a new highway near the village increased the population P by a constant annual amount according to the following table.

Year	Population
2008	760
2009	795
2010	830
2011	865
2012	900

At this rate of growth, how many people will be living in Beushane in 2015? 2017? 2020?

⚙ Solution

If we let t (for time) be our independent variable here, it is convenient to let the initial year of 2008 correspond to $t = 0$. Then t stands for the number of years after 2008.

Year	t	Population
2008	0	760
2009	1	795
2010	2	830
2011	3	865
2012	4	900

Because the population grows by 35 people every year, we see that we have a two-variable relationship given by $P = 35t + 760$. The years 2015, 2017, and 2020 correspond to $t = 7$, 9, and 12 years after 2008, and so the population (assuming the growth rate remains the same) would be:

$$P = 35(7) + 760 = 1,005 \quad \text{for } t = 7,$$

$$P = 35(9) + 760 = 1,075 \quad \text{for } t = 9,$$

$$P = 35(12) + 760 = 1,180 \quad \text{for } t = 12.$$

The use of tables illustrates one of the oldest methods for representing a relationship between two varying quantities. Your familiarity with the creation of tables probably comes from experiments done in a science class or from the style with which information is often given in newspapers and magazines. Usually, it is tacitly assumed that such tables are only a small subset of a much larger set of paired numbers, and so we may extend the relationship implied in the table for predictive purposes. Whenever we have paired data of any type available, predictions can be made for other instances with the assumption that the relationship appearing in the listing or table holds steady. This practice is at the heart of one of the main concerns of this book and shall be explored in more detail. ◆

❓ Example 8 (Geometry)

The circumference C of any circle is related to its diameter d by $C = \pi d$. Because the diameter is equal to twice the radius r, this is also commonly written as $C = 2\pi r$. Clearly, the area A of any circle must also depend on the size of its radius and, in fact, the specific relationship is given by $A = \pi r^2$. For instance, a circle of radius 3.00 ft has a circumference and an area of

$$C = 2\pi(3.00) \approx 18.8\,\text{ft} \quad \text{and} \quad A = \pi(3.00^2) \approx 28.3\,\text{ft}^2.$$

It is clear that one radius leads to unique values for both C and A. So we see that the circumference and area of a circle are functions of its radius. The domains and ranges of both functions are all given by the set of positive real numbers. In **Figure 1.1.9**, we note further that both of these physical quantities must get larger as the radius gets larger. The mathematical way to describe this property is to say that circumference and area are *increasing* functions of radius.

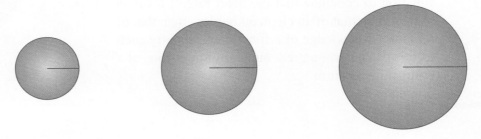

FIGURE 1.1.9   The area of a circle is an increasing function of radius.

increasing
A function f is said to be increasing over an interval if $f(a) < f(b)$ whenever $a < b$ for any pair of numbers a, b in the interval.

decreasing
A function f is said to be decreasing over an interval if $f(a) > f(b)$ whenever $a < b$ for any pair of numbers a, b in the interval.

It is possible for a function whose domain **D** and range **R** both consist entirely of numbers to have certain properties that help characterize the function. A function is called **increasing** if $f(x)$ in **R** always increases whenever x in **D** increases. One might compare this to the growth process of a child—the older she gets, the taller she gets. On the other hand, if $f(x)$ decreases whenever x increases, we call f a **decreasing** function. Any mountain climber will tell you that atmospheric pressure and air density are decreasing functions of altitude.

? Example 9 (Biology)

Biologists are concerned with the concentration of hydrogen ions (H^+) and hydroxide (OH^-) in substances because of their effect on many organic processes. In a solution, the concentration of H^+ times the concentration of OH^- is always equal to the constant value 10^{-14} mole per liter (mol/L). A typical carbonated soft drink has an H^+ concentration of 10^{-3} mol/L, and therefore its OH^- concentration must be 10^{-11} mol/L. Clearly, each of these concentrations is a function of the other. As one increases, the other decreases in a well-defined manner, which we could actually write as

$$H^+ = \frac{10^{-14}}{x}.$$

Here we see that the hydrogen ion concentration H^+ is a decreasing function of the hydroxide concentration x.

? Example 10 (Zoology)

The rate at which a mammal's heart beats is approximately a decreasing function of the size of the mammal. This means that the heart rate decreases as the size increases. Even if we lack a specific formula to make specific matches between domain and range numbers, we can still use this function to conclude that the heart rate of a fox is greater than that of an elephant but less than that of a mouse. Knowledge of a functional property such as this can answer a great number of questions of a comparative nature.

© Igor Zakowski cartooniz_com/ShutterStock, Inc.

? Example 11 (Astronomy)

The Greek astronomer **Hipparchus** (c. 2nd century BC) was the first person to catalog over a thousand stars according to size. Because all the stars at that time were thought to be the same distance from Earth, this amounted to a classification by brightness. He created six categories: The brightest stars were of class 1, the second brightest of class 2, and so on. Latin translators of the catalog used the word *magnitudo*, the word for *size*, for a class number, and so today we say, for example, that the North Star, Polaris, has magnitude 2. **Figure 1.1.10** displays some magnitudes (of greater accuracy because of modern instruments) for the stars in the constellation Pegasus.

FIGURE 1.1.10 Apparent magnitudes of some stars in Pegasus.

Alpheratz	2.06	Mu Pegasi	3.49
Markab	2.50	Upsilon Pegasi	4.41
Algenib	2.83	Psi Pegasi	4.64
Homam	3.39	Phi Pegasi	5.05

To an astronomer in the time of Hipparchus, the size of a star was a decreasing function of magnitude, and so the table here revealed that Psi Pegasi is a smaller star than Markab. Today, however, we know that the stars in the skies above us exist at many varying distances. Just as we perceive the beams differently from two similar flashlights at two different distances, so too is the apparent brightness of a star to our eyes here on Earth related not just to its size but also to its distance from us (as well as its temperature). Today, the values in Figure 1.1.10 are known as *apparent magnitudes*, and we may thus only state that apparent brightness, rather than size, is a decreasing function of apparent magnitude. In fact, Markab appears to be the brighter star in large part because its distance (73 light-years) is less than 7% the distance to Psi Pegasi (1080 light-years).

Example 12 (Music)

Every musical pitch has a vibrational frequency associated with it. For instance, the A string on most acoustic guitars vibrates at 440 hertz (cycles/sec). This is the case regardless of the system—a string, column of air, or bar—that is being used to produce the vibrations. The values in **Figure 1.1.11** correspond to one type of musical scale (chromatic) and are not exact.

Frequency Table

	Octave				
	0	1	2	3	4
C	65.4	130.8	261.6	523.2	1,046.4
D	73.4	146.8	293.6	587.2	1,174.4
E	82.4	164.8	329.6	659.2	1,318.4
F	87.3	174.6	349.2	698.4	1,396.8
G	98.0	196.0	392.0	784.0	1,568.0
A	110.0	220.0	440.0	880.0	1,760.0
B	123.5	246.9	493.9	987.8	1,975.6

FIGURE 1.1.11 Frequencies of the pitches in the chromatic scale for several octaves.

Several functions can be produced from this chart, depending on how we wish to display the relationships. For example, because each frequency produces one and only one pitch, we may write the first few pitch names as a function of various frequencies.

$$f(65.4) = C \quad f(130.8) = C \quad f(261.6) = C \quad \ldots$$
$$f(73.4) = D \quad f(146.8) = D \quad f(293.6) = D \quad \ldots$$
$$f(82.4) = E \quad f(164.8) = E \quad f(329.6) = E \quad \ldots$$
$$\vdots$$

So we have a relationship in which pitch is a function of frequency. Every musician who has ever played an instrument has utilized this function. The domain here consists of $\mathbf{D} = \{65.4, 73.4, 82.4, \ldots\}$ and the range $\mathbf{R} = \{C, D, E, F, G, A, B\}$. Observe also that this function takes more than one element in the domain to the same element in the range. *This is okay!* This does not violate the definition of a function, because every domain member does get sent to a *unique* range member. Using our bow and arrow analogy, we would say that more than one arrow would be shot to the same point on the target (**Figure 1.1.12**).

FIGURE 1.1.12   A function may send more than one domain element to the same range element.

❓ Example 13 (Cryptology)

The study of integers and their related properties is an area of mathematics known as *number theory*, and any function whose domain is the set of positive integers is called a *number-theoretic* function. One such function is denoted by the Greek symbol τ ("tau") and defines $\tau(n)$ to be the number of positive divisors of n, where n is any positive integer. Thus, $\tau(10) = 4$, because 10 has four positive divisors: 1, 2, 5, and 10. Similarly, $\tau(30) = 8$, because the divisors of 30 are 1, 2, 3, 5, 6, 10, 15, and 30. Other examples are given in **Figure 1.1.13**.

We see that this is another example of a function that assigns the same (range) element to more than one element in its domain. In fact, it is true that that for every $k > 1$, there are an infinite number of integers that have k positive divisors.

cryptology The study of the techniques used for creating secret codes.

Number theory is used extensively in **cryptology**—the study of the techniques used in the creation of secret codes. Long used primarily for military communications, applications of cryptology have recently increased

n	**Positive Divisors of n**	$\tau(n)$
4	1, 2, 4	3
7	1, 7	2
8	1, 2, 4, 8	4
9	1, 3, 9	3
12	1, 2, 3, 4, 6, 12	6
15	1, 3, 5, 15	4
19	1, 19	2
40	1, 2, 4, 5, 8, 10, 20, 40	8
100	1, 2, 4, 5, 10, 20, 25, 50, 100	9
200	1, 2, 4, 5, 8, 10, 20, 25, 40, 50, 100, 200	12

FIGURE 1.1.13 $\tau(n)$ for various values of n.

dramatically as the need has arisen for security in the ever more sophisticated computer systems that contain and control large databases.

FIGURE 1.1.14 Leonhard Euler.

> A function of a variable quantity is an analytic expression composed in whatever manner of this same quantity and numbers or constant quantities.
>
> –Leonhard Euler

algorithms
Sequence of specific steps to compute a value or solve a problem.

When both the domain and range of our function consist of real numbers, often the relationship can be conveniently expressed using an independent variable in an equation. This takes the form of a *formula* of the type you are accustomed to from algebra. Sometimes, a dependent variable is also used in the equation to represent values in the range, but more often, we continue to use the functional notation device $f(x)$ to stand for the value of the function corresponding to x. If $f(x) = 3x^2 + 2x - 5$, for instance, then

$$f(-4) = 3(-4)^2 + 2(-4) - 5 = 35,$$
$$f(1) = 3(1)^2 + 2(1) - 5 = 0,$$
$$f(2) = 3(2)^2 + 2(2) - 5 = 11,$$

and so forth. This ideal notation was first introduced by the prolific mathematician extraordinaire **Leonhard Euler** (1707–1783), a man who did mathematics "without apparent effort, as men breathe, or as eagles sustain themselves on the wind." Born and educated in Switzerland, Euler had received two degrees by the age of sixteen at the local university. His prodigious and restless intellect was given free reign in his career as royal mathematician at the courts of Russia and Berlin to produce an amount of new mathematics so enormous that people are still examining portions of his works today. His memory was legendary—in his old age, he could still recite every word of the *Aeneid* by Virgil, even knowing the first and last sentence of every page! He was one of the first to utilize the full power of the calculus developed by **Isaac Newton** (1642–1727), and he realized the importance of functions to concisely explain relationships and patterns among physical phenomena. Because one of his specialties was the creation of **algorithms**, or step-by-step procedures for the solution of tough problems, the clear definition of a function was an important tool to him.

? Example 14 (Physics)

Pendulums were an important early mechanism for timekeeping and therefore an object of study by the scientist and mathematician **Galileo Galilei** (1564–1642). (See Section 4.3 for more details on his life.) The time for a pendulum to complete one swing back and forth is called its period of oscillation T. Our common experience would indicate that the longer the length x of the pendulum, the greater is the value of T. In fact, for small oscillations, it was determined that

$$T(x) = 2\pi\sqrt{\frac{x}{g}},$$

where g is the acceleration due to gravity, x is measured in meters, and T is in seconds. On the surface of the Earth, $g \approx 9.8$ m/sec², giving $\dfrac{2\pi}{\sqrt{9.8}} \approx 2.0$. So the formula for the function reduces to $T(x) = 2.0\sqrt{x}$. The periods for pendulums of lengths 0.25 m, 0.8 m, and 1.5 m are

FIGURE 1.1.15   Galileo.

$$T(0.25) = 2.0\sqrt{0.25} = 1.0 \text{ sec},$$
$$T(0.8) = 2.0\sqrt{0.8} = 1.8 \text{ sec},$$
$$T(1.5) = 2.0\sqrt{1.5} = 2.4 \text{ sec}.$$

The domain and range both must consist of all positive real numbers. Note that T is used here to represent the function instead of f. It is quite common to use letters for functions that remind us of the quantity being symbolized.

> Legends of Galileo recount an early mystical experience in church that fostered his profound insights about the pendulum as timekeeper: [He was] mesmerized by the to-and-fro of an oil lamp suspended from the nave ceiling and pushed by drafts. Timing the motion of the lamp by his own pulse, Galileo saw that the length of a pendulum determines its rate.
>
> –Dava Sobel in *Longitude*

? Example 15 (Business)

If you invest a principal of P dollars in a savings account that earns *simple* interest at an annual rate of r, then the sum of money $S(t)$ you will have accrued in t years is a function of t given by

$$S(t) = P(1 + rt),$$

where r has first been converted from a percentage to its decimal equivalent. (Note that t is used for the independent variable because it represents time.) Suppose you invest a principal of \$700 in an account earning simple interest at an annual rate of 3.5%. Then the above formula becomes the increasing function

$$S(t) = 700(1 + 0.035t).$$

In 2 years, you will have accrued $S(2) = 700(1 + 0.035 \cdot 2) = 700(1.07) = \$749.00$.
In 3 years, $S(3) = 700(1 + 0.035 \cdot 3) = 700(1.105) = \$773.50$.
In 5 years, $S(5) = 700(1 + 0.035 \cdot 5) = 700(1.175) = \$822.50$.
In 10 years, $S(10) = 700(1 + 0.035 \cdot 10) = 700(1.35) = \$945.00$.

Name _____

— Exercise Set **1.1**   ⬭⬭⬭←————————————————————

1. The low temperatures in several cities on a Saturday in January are given below. Describe a function, and give the domain and range in set notation. Also give three examples of functional pairings using the "f(domain element) = range element" notation.

Amsterdam	39	Oslo	24
Calgary	14	Rome	43
Dublin	41	Singapore	75
Lisbon	43	Tokyo	36
Manila	66	Vienna	27

2. Could you define a function from a domain of the above temperatures to a range of the corresponding cities? Explain.

3. A typical program at a professional sports contest contains a list of the players with both their heights and weights. Does a function exist from a domain of weights to a range consisting of the players? Explain.

4. The rate at which a mammal's heart beats is approximately a *decreasing* function of the size of the mammal. A baboon's heart rate is about 100 beats per minute, a lion's about 50, and an elephant's about 25. This information does not permit us to construct specific matches of mammals with heart rates for the function, but we can still answer the following questions. In what interval do you think the heart rate of a full-grown rhinoceros would be: 0–40, 40–80, or over 80? A zebra? A rabbit? Which has the faster heart rate, a ten-year-old child or an adult? A ten-year-old child or an infant?

5. Suppose your teacher for a certain class posts the grades for each test next to a list of social security numbers. Use the notion of a function to explain why this is sufficient for each student to be informed of his or her grade.

6. Suppose that during this year, July 21 is a Friday. Do you think July 21 is a Friday every year? Is it possible to create a single function from date to day of the week that is good for all years?

7. The multiple of 4 that is closest to 51 is 52. The closest multiple of 4 to 81 is 80. If we define a method for assigning numbers to an arbitrary integer x by

$$x \rightarrow \text{closest multiple of } 4,$$

is this a function? Explain.

8. At the end of the season, the scoring average per game is computed for each player in the National Basketball Association. If you were to define a function between the set of players and the set of scoring averages, which set must be the domain and which must be the range?

9. The prices for a six-pack of soda at the local grocery store are given below. Could you define a function with a domain consisting of the prices and a range consisting of the soda brands? Why or why not?

$2.58   Cran-Cola
$2.45   Ginger Ale
$2.25   Grape Surprise
$2.62   Raspberry Sluice
$2.25   Al's Root Beer
$2.08   Orange Fizz

10. The frequency v (Greek "nu") and wavelength λ ("lambda") of any type of radiation in the electromagnetic spectrum are related by the equation $v = \dfrac{c}{\lambda}$, where c is the speed of light and therefore a constant value. Is v an increasing or a decreasing function of λ?

11. When you stand close to a roaring campfire, you feel a great deal of heat, but as you step back, you notice a rapid lessening of the warmth. Is the warmth you feel an increasing or a decreasing function of your distance from the fire?

12. In Example 13, the number-theoretic function τ is defined by:

$\tau(n)$ = number of positive divisors of n, where n is any positive integer.

Determine $\tau(6)$, $\tau(11)$, $\tau(28)$, $\tau(33)$, $\tau(49)$, $\tau(53)$, $\tau(64)$, and $\tau(87)$.

13. A number-theoretic function often associated with the τ function is denoted by the Greek letter σ ("sigma") and is defined by:

$\sigma(n)$ = sum of the positive divisors of n, where n is any positive integer.

For example, the divisors of 10 are 1, 2, 5, and 10. Therefore, $\sigma(10) = 1 + 2 + 5 + 10 = 18$. Use Figure 1.1.13 to find $\sigma(4)$, $\sigma(12)$, $\sigma(19)$, and $\sigma(100)$.

14. Use your work from Exercise #12 to help determine $\sigma(6)$, $\sigma(11)$, $\sigma(28)$, $\sigma(33)$, $\sigma(49)$, $\sigma(53)$, $\sigma(64)$, and $\sigma(87)$.

15. If p is an integer greater than 1 whose only positive divisors are itself and 1, then p is called a *prime* number. The first several primes are 2, 3, 5, 7, 11, 13, 17, 19, 23, 29, 31, and so on. Find expressions for $\tau(p)$ and $\sigma(p)$ if p is a prime.

Find the images of each of the following functions at the given values for the independent variable.

16. $f(x) = 9x + 2$, $x = -2, 0, 3, 10, 70$

17. $f(x) = x^2 + 4x - 3$, $x = -6, -1, 0, 2, 10$

18. $g(x) = 5x + 3\sqrt{x}$, $x = 4, 9, 25$

19. $h(t) = \dfrac{2t + 7}{10}$, $t = -1, 1, 8, 25, 100$

20. $A(r) = \pi r^2$, $r = 2, 7, 15, 20$

←

21. $D(y) = 6\sqrt{y^2 + 7}$, $y = -3, -2, 0, 5$

22. $F(x) = 5x^3 - 2x^2 - 0.4x + 3$, $x = 1.2, 3.7, 4.5$

23. $G(t) = \dfrac{t+7}{t^2+3}$, $t = 0, 1, 2, 3$

24. Recall from Example 14 that the period of oscillation T (in seconds) for a pendulum is a function of its length x (in meters) given by

$$T(x) = 2.0\sqrt{x}.$$

Determine $T(0.2)$, $T(0.8)$, and $T(1.3)$. Round your answers to the nearest tenth. Is T an increasing or a decreasing function of x?

25. In Example 15, we saw that the amount of money accrued in t years by investing P dollars in a savings account earning simple interest at an annual rate r is given by $S(t) = P(1 + rt)$. For an investment of \$7,300 at an annual rate of 4.5%, find:
(a) the specific function giving the amount of money accrued in t years.
(b) the amount of money accrued in 3 years.
(c) the amount of money accrued in 5 years.

26. Deion invests \$325,000 in a savings account earning simple interest at an annual rate of 5.25%. Use the formula in the exercise above to find:
(a) the specific function giving the amount of money accrued in t years.
(b) the amount of money accrued in 2 years.
(c) the amount of money accrued in 10 years.

27. Some people reduce the size of their taxable estate before they die by giving money to friends or relatives. However, in some instances, a unified transfer tax T is assessed on these types of gifts. If the tax on any monetary gift x between \$40,000 and \$60,000 is computed according to the function $T(x) = 8200 + 0.24(x - 40,000)$, find $T(45,000)$ and $T(52,500)$ to the nearest cent.

28. The annual number (in millions) of new homes built is a function of the average interest rate $r\%$ on new home loans and is given by $h(r) = \dfrac{49 - 2r}{15}$. Find the number of new homes built when $r = 5, 8, 10$. Is this an increasing or a decreasing function?

29. The cost, in dollars, of renting a car for a day is given by $C(m) = 0.3m + 10$, where m is the number of miles driven. What would it cost to drive the rental car 70 miles? 100 miles? 450 miles? Is the cost an increasing or a decreasing function of the miles driven?

30. The resistance R, measured in ohms, of any aluminum wire to the flow of electric current is proportional to the wire's length x and is inversely proportional to its cross-sectional area a. Symbolically, this is written

$$R(x) = kx/a.$$

For an aluminum wire ($k = 3.2 \times 10^{-8}$) that has a cross-sectional area of 10^{-6} square meter, what is the resistance of the wire (to the nearest tenth) if it has a length of 0.5 m? 50 m? 100 m? Is the resistance of a wire an increasing or a decreasing function of its length? (A review of scientific notation is in Appendix III.)

31. The velocity v (in meters/sec) required for a satellite to remain in a circular orbit around the Earth is a function only of the distance r (in meters) from the center of the Earth given by

$$v(r) = \frac{2.0 \times 10^7}{\sqrt{r}}.$$

Is this an increasing or a decreasing function? Find the velocities (in m/sec) for $r = 9.0 \times 10^6$ m, 16.0×10^6 m, and 20.0×10^6 m.

32. The temperature B in degrees centigrade at which water boils is approximated by the function

$$B(h) = 100.86 - 0.04\sqrt{h + 431.03},$$

where h is the altitude (meters) above sea level. Find $B(1000)$ and $B(8500)$ to the nearest hundredth. Is this an increasing or a decreasing function?

33. Several diseases are associated with resistance to the flow of blood to the fetal spleen in humans. Researchers have discovered that the index of splenic artery resistance can be modeled by

$$f(x) = 0.057x - 0.001x^2,$$

where x is the number of weeks of gestation. What is the index of resistance after 8 weeks? After 30 weeks? [**Source:** Abuhamad, A. Z. et al., "Doppler Flow Velocimetry of the Splenic Artery in the Human Fetus: Is It a Marker of Chronic Hypoxia?" *American Journal of Obstetrics and Gynecology* 172, no. 3 (March 1995): 820–825]

34. According to the National Highway Traffic Safety Administration (U.S. Department of Transportation), if the brakes are applied to a car traveling at a velocity v, the distance d required to stop is related to the constant deceleration a on the car by

$$d = -v^2/2a.$$

Is d an increasing or a decreasing function of a? Low tire pressure can lower the deceleration achieved by the car upon application of the brakes. If your tire pressure is lower than the advised standard, will the car travel a longer or shorter distance than it would with your tires at the correct pressure?

35. The Greek astronomers in the time of Hipparchus assumed all stars to be the same distance from us (see Example 11), and so the magnitude numbers assigned to the stars were meant to be indicators of their sizes (see Figure 1.1.10). Dimmer (and therefore smaller) stars are classified by larger numbers. Suppose another ancient civilization instead adopted the hypothesis that all stars were the same size and therefore their differences in brightness in the night sky were the result of varying distances. If these people also created a function from the stars to numbers by assigning larger magnitudes to dimmer stars, would a magnitude 2 star be closer or farther away than a magnitude 4 star? As magnitude increases under this scheme, does distance increase or decrease?

36. The elements in the Periodic Table of the Elements are grouped in rows and columns according to certain shared properties. One of these properties is *electronegativity*, or the tendency of an element to acquire electrons in a chemical interaction. Electronegativity, a function of the atomic number, increases as atomic number advances along any row in the table (left to right) and decreases as atomic number advances along any column (top to bottom). The following list is a subset of the Periodic Table.

Li	Be	B	C	N	O	F
3	4	5	6	7	8	9

Na	Mg	Al	Si	P	S	Cl
11	12	13	14	15	16	17

K	Ca	Ga	Ge	As	Se	Br
19	20	31	32	33	34	35

According to the description of the function, does C (carbon) have more or less electronegativity than O (oxygen)? B (boron) have more or less than Ga (gallium)? Mg (magnesium) more or less than Cl (chlorine)? Rank the following elements from lowest to highest electronegativity.

Ca, As, F, N, P, K

37. *The Energy Book* (U.S. Government Printing Office) tells us that the average fraction of the wind's energy that can be converted by a small wind turbine into electricity is given by $f(x) = \frac{1}{2}x(2-x)^2$, where x is the fraction by which the wind's speed is decreased as it strikes the turbine blades. Note that $0 \leq x \leq 1$. Find the fraction of the wind's energy converted for $x = \frac{1}{4}$; $x = \frac{1}{2}$; $x = \frac{3}{4}$.

38. In his classic *Moby Dick*, Herman Melville wrote about the use of the quadrant.

It was hard upon high noon; and Ahab, seated in the bows of his high-hoisted boat, was about taking his wonted daily observation of the sun to determine his latitude.

What other inputs might Ahab need in order to determine his latitude at sea as a function of the elevation of the noon sun?

However, Ahab grew weary of this equipment and settled for a much cruder formula.

"Curse thee, thou quadrant!" dashing it to the deck, "no longer will I guide my earthly way by thee; the level ship's compass, and the level dead-reckoning, by log and by line; these shall conduct me, and show me my place upon the sea."

What do you think is meant by "by log and by line"?

1.2 Creation of Functional Expressions

action The steps taken by a function to process an element *x* in its domain.

We referred previously to the **action** of a function. By this we mean the steps taken for the function to process an element *x* in its domain in order to obtain the corresponding functional value $f(x)$. Suppose, for instance, we start with a number, say 2, and we are asked to perform some arithmetic operation or sequence of operations in order to transform it into 11. What might we do? A variety of options exist. For instance, we could simply add 9. Or we could square 2 and then add 7. Or we could multiply by 4 and add 3. All of these are acceptable procedures and, in fact, there exist a great many more.

Now, however, suppose we are asked to pick one of the above procedures that not only turns 2 into 11, but also 1 into 6, 3 into 16, 4 into 21, and 5 into 26. In other words, we need to identify the action of the correct function that assigns images to domain elements according to **Figure 1.2.1**.

This type of problem is often faced by scientists, businessmen, engineers, and people from a variety of professions. Upon observing a list or table of paired values that are representative of an existing function, they create an algebraic expression for it. We have already considered this intuitively in the first section, and now we look at it again a bit more formally.

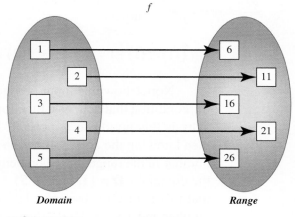

FIGURE 1.2.1   The assignments of a function *f*.

Linear Functions

In the present case, we must try to create a *formula* that represents the action of the function not just for an input of 2, but also for 3, 4, 5, and perhaps many other numbers. So this formula must encode the steps involved that will successfully produce the desired *outputs* of 11, 16, 21, 26, etc. The word *encode* implies use of a language—in this case, the language of *algebra*. What must our formula do? It must pair each element in the domain of the function with the proper image. We need an independent *vari*able to stand in for a *vari*ety of inputs in contrast to a single unknown quantity. Using the symbol *x*, we let $f(x)$ stand for the corresponding image and construct a table (**Figure 1.2.2**) displaying the correspondences $f(1) = 6, f(2) = 11, f(3) = 16, f(4) = 21,$ and $f(5) = 26$.

x	*f(x)*
+1 ⌈ 1	6 ⌉ +5
+1 ⌈ 2	11 ⌉ +5
+1 ⌈ 3	16 ⌉ +5
+1 ⌈ 4	21 ⌉ +5
5	26

FIGURE 1.2.2   A table of paired values for a linear function *f*.

Now what do we do to *x*? Adding 9 to *x* works just fine if *x* will only be replaced by 2, but the fact that *x* will also be replaced by 3, 4, 5, and so on spoils that process. Squaring and adding 7 doesn't work for all the inputs either. It is far more fruitful to examine how the

rate The change in the functional value of a linear function per unit change in the independent variable.

slope Characteristic of a linear graph computed by $\dfrac{y_1 - y_0}{x_1 - x_0}$ for any two points (x_0, y_0) and (x_1, y_1) on the line. It is numerically equal to the rate.

graph of a function f The set of all ordered pairs (x, y) plotted on a Cartesian coordinate system in which the y-coordinate is equal to $f(x)$ and x is a member of the domain of f.

functional value $f(x)$ changes as x changes. Each time x increases by 1, $f(x)$ increases by 5. The change in the functional value per unit change in the independent variable is called the **rate** (equal to the **slope** of the associated **graph of a function f** to be considered in the next section). In order to ensure a constant rate of 5, we write

$$f(x) = 5x + k.$$

Now all we need is to determine the value of the constant term k by substituting any of the pairs from the table. For instance, $16 = f(3) = 5(3) + k$ means that $k = 1$. Therefore, the desired function is

$$f(x) = 5x + 1.$$

We can then check to see that this formula also correctly produces the other outputs:

$$f(1) = 5(1) + 1 = 6, \, f(2) = 5(2) + 1 = 11, \, f(4) = 5(4) + 1 = 21, \text{ and } f(5) = 5(5) + 1 = 26.$$

Notice how one input produces one output and *no more*. Recall how this is precisely the condition that defines a function. Functions provide definitive connections between sets. Why is this notion so important? If it is possible to construct a formula for the function, then knowing the correct function gives you the power to make an accurate *prediction* of a number in the range, based on being given a number in the domain. In our current example, the domain is $\mathbf{D} = \{1, 2, 3, 4, 5\}$, and the range is $\mathbf{R} = \{6, 11, 16, 21, 26\}$. Suppose instead that this is just a subset of a larger domain, say the interval $\mathbf{D} = [1, 100] = \{x \mid x$ is a real number and $1 \leq x \leq 100\}$? (See Appendix V for a complete description of intervals.) If we *assume that our rate remains constant* regardless of the input, then our new range is also an interval obtained by simply substituting these values for x in our function $f(x) = 5x + 1$ in order to get $\mathbf{R} = [6, 501]$.

linear functions A function of the form $f(x) = mx + k$.

Any function with a constant rate is a member of an important class of functions known as **linear functions**. A linear function has the general form

$$f(x) = mx + k,$$

where m is the rate and k is a fixed constant. If the rate is a positive number as in the current case, then $f(x)$ gets larger with increasing x, and f is therefore an increasing function. On the other hand, if the rate is negative, then f is a decreasing function. A good example of a decreasing linear function concerns *depreciation* of machinery or other materials that are involved in running a business. If the value of a tractor on a farm decreases by $700 each year, the rate is -700 dollars/year, and so we could write $V(t) = -700t + k$, using t (for time) as the independent variable. If the tractor were originally worth $42,000$, $V(0) = 42,000 = k$, and so the function would be $V(t) = -700t + 42,000$. Linear functions occur in a large variety of applications.

❓ Example 1

Examine the following tables of paired values, and create a formula for each to express the action of the functions that they represent. Assuming an expanded domain and constant rates, find $f(1.3)$, $g(20)$, and $h(45.6)$.

x	f(x)	x	g(x)	x	h(x)
−2	−2	−10	−67	1	4.4
−1	1	−5	−37	2	1.3
0	4	0	−7	3	−1.8
1	7	5	23	4	−4.9
2	10	10	53	5	−8.0

⚙ Solution

Recognition of the pattern is the key. As x increases by 1 in the first list, $f(x)$ increases by 3. This implies that the rate of change of $f(x)$ is 3, and we can initially write $f(x) = 3x + k$. Then substituting, say 0, for x, we get that $k = f(0) = 4$, and so $f(x) = 3x + 4$.

In the second list, $g(x)$ is increasing by a constant increment of 30 for each increase of 5 in x. Because the rate is defined as the change in the function per *unit* change in the independent variable, $m = \dfrac{30}{5} = 6$. We write $g(x) = 6x + k$ and then obtain k by substituting any of the given pairs. Using $(0, -7)$, we get

$$g(0) = 6(0) + k = -7$$
$$k = -7,$$

and so $g(x) = 6x - 7$.

In the last table, $h(x)$ is *decreasing* by 3.1 for each unit increase in x, and so $m = -3.1$. Proceeding as before, we get $h(1) = -3.1 + k = 4.4$, and so $k = 7.5$, giving us $h(x) = -3.1x + 7.5$.

Finally, assuming the rates remain constant in expanded domains,

$$f(1.3) = 3(1.3) + 4 = 7.9.$$
$$g(20) = 6(20) - 7 = 113.$$
$$h(45.6) = -3.1(45.6) + 7.5 = -133.86.$$

❓ Example 2

In a psychological study, one researcher recorded the following times, in which a sample of five adults needed to memorize a 9-digit sequence of numbers and letters according to age. The results are given as follows:

Age (yr)	Time (sec)
38	29
42	31
46	33
50	35
54	37

The researcher would like to predict the memorization times for another, older set of people before actually testing them. Based on the above sample, what would be the predictions for people of ages 56, 65, and 70?

⚙ Solution

An examination of these particular data reveals a *linear* connection between age a and memorization time T. The values of our independent variable here are not increasing by 1 at a time, but rather by increments of 4. The rate of increase of time per year is

$$m = \frac{2 \text{ seconds}}{4 \text{ years}} = 0.5 \text{ sec/yr.}$$

We write $T(a) = 0.5a + k$ and find k by seeing that $T(38) = 0.5(38) + k = 19 + k = 29$, which implies that $k = 10$. We use our linear function $T(a) = 0.5a + 10$ to make the predictions.

$$T(56) = 0.5(56) + 10 = 38 \text{ sec.}$$
$$T(65) = 0.5(65) + 10 = 42.5 \text{ sec.}$$
$$T(70) = 0.5(70) + 10 = 45 \text{ sec.}$$

Of course, these values may not match the empirical results of the new survey very well, but they still provide an interesting source of comparative numbers for the researcher.

The creation of a linear function does not have to arise from a table of values if the rate is available in some other way. In particular, if we *already* know that some function is linear, then we only need to know two data points in our table—that is, two sets of paired values.

❓ Example 3

Acme Hardware manufacturing company has purchased a new drill press for $17,000. If it depreciates linearly and is worth $15,000 four years later, find a function that gives the value of the press in t years after it was purchased.

⚙ Solution

We only have two data points for this function, namely (0, 17,000) and (4, 15,000), but we have the advantage of *knowing* that the function is linear rather than having to determine that fact from some larger table. Hence, we find the constant rate by computing

$$m = \frac{15,000 - 17,000}{4 - 0} = \frac{-2000}{4} = -500.$$

Now since the value of the press is $V(t) = -500t + k$ and $V(0) = 17,000$, we must have $k = 17,000$, and so $V(t) = -500t + 17,000$.

nonlinear functions Any function that is not linear.

Many types of functions do not change at a constant rate and are therefore known as **nonlinear functions**. However, they can still behave in a manner that allows us to create a predictive expression to represent them. There are numerous varieties of nonlinear functions, but we shall limit ourselves to looking at just one major category of important examples—namely, those functions in which the independent variable is an exponent.

Exponential Functions

We now return to the frequency table for musical pitches from the last section, but we list the values differently in order to construct a common type of nonlinear function. (See **Figure 1.2.3**.)

Frequency Table

Octave	C	D	E	F	G	A	B
0	65.4	73.4	82.4	87.3	98.0	110.0	123.5
1	130.8	146.8	164.8	174.6	196.0	220.0	246.9
2	261.6	293.6	329.6	349.2	392.0	440.0	493.9
3	523.2	587.2	659.2	698.4	784.0	880.0	987.8
4	1,046.4	1,174.4	1,318.4	1,396.8	1,568.0	1,760.0	1,975.6

FIGURE 1.2.3 Frequencies (Hz) of pitches in the chromatic scale for several octaves.

? Example 4

Recall our table from the first section that identified each frequency by a pitch name. A second function can be produced from this chart by observing the well-known musical fact that, for any particular pitch, the next higher octave is obtained by doubling the frequency. For the pitch A, if we arbitrarily identify octave 0 with 110 Hz, we have a function f such that

$$f(0) = 110, \quad f(1) = 220, \quad f(2) = 440, \quad f(3) = 880, \quad \ldots \text{ etc.}$$

We can see a clear pattern here—namely, that each frequency has 110 as a factor and a power of 2 as a factor.

$$f(0) = 110 \cdot 2^0 = 110$$
$$f(1) = 110 \cdot 2^1 = 220$$
$$f(2) = 110 \cdot 2^2 = 440$$
$$f(3) = 110 \cdot 2^3 = 880$$

As we shall see repeatedly, the recognition of patterns plays a major role in the observation and recording of the physical world around us. It is basically a three-part process. We use our five senses to observe a phenomenon, our intellect to sort and categorize the information, and mathematics to create a written record. In a very real sense, mathematics allows us to both extend and perfect our observations. In the current case, the domain of this function consists of the nonnegative integers, and the range consists of values that start at 110 and then increase by a *factor* of 2 each time the domain value increases by an additional increment of 1. Such a function can be written down in a general form,

$$f(x) = 110 \cdot 2^x.$$

We can appreciate that this function "extends" our senses. Our ears have an upper limit for the frequencies they can hear, but our function tells us of the frequency associated with any octave of A, no matter how high.

---

exponential function A function of the form $f(x) = kb^x$.

The function demonstrated in the previous example is known as an **exponential function** and is an accurate model of a wide variety of scientific and social phenomena. The general form is

$$f(x) = kb^x,$$

where the coefficient k is any constant value and b is any positive number not equal to 1, referred to as the **base**. An exponential function can be identified by the unique way its values increase. Consider the assignments in the table in **Figure 1.2.4**.

base The factor by which an exponential function is consistently multiplied per unit change in the independent variable. It is denoted by b in the general formula for the function. It is necessary for $b > 0$ and $b \neq 1$.

x	$f(x)$
-2	$1/9$
-1	$1/3$
0	1
1	3
2	9

(+1 each in x; ×3 each in $f(x)$)

FIGURE 1.2.4   A table of paired values for an exponential function.

For every unit increase in the independent variable here, the functional value increases *by a factor of* 3 rather than just an additional increment of 3, which is the case for a linear function. This implies that the base is 3 in this case, and this characteristic is the hallmark of an exponential function. So we know initially that $f(x) = k \cdot 3^x$. As in the case of linear functions, we can then proceed to determine the value of k by substituting any of the given pairs of values. The easiest choice is to use $f(0)$ when it is available, since $f(0) = kb^0 = k$. In the present case, $k = f(0) = 1$, and so our function must be $f(x) = 3^x$, which is consistent with the other pairings:

$$f(-2) = 3^{-2} = \frac{1}{9},$$
$$f(-1) = 3^{-1} = \frac{1}{3},$$
$$f(1) = 3^1 = 3,$$
$$f(2) = 3^2 = 9.$$

Other values would be $f(-3) = 3^{-3} = \frac{1}{27}, f(3) = 3^3 = 27$, and so forth.

? Example 5

Suppose our range values are slightly different from those above.

x	$g(x)$
-2	$5/9$
-1	$5/3$
0	5
1	15
2	45

$+1$ on the left of each x-step; $\times 3$ on the right of each $g(x)$-step.

What has happened? As before, we observe that the function here increases by a factor of 3 per unit increase in x. Therefore, it is exponential, with a base of 3, the same as above. Because each image under the previous function has been multiplied by 5, our new function must be $g(x) = 5 \cdot 3^x$. However, we don't need the first function handy in order to deduce the second one. Because we know $g(x) = k \cdot 3^x$, we can determine k from $5 = g(0) = k \cdot 3^0 = k \cdot 1 = k$. Alternatively, we could choose any other ordered pair to substitute into the function. For instance, because $g(2) = 45$, we get

$$45 = g(2)$$
$$= k \cdot 3^2$$
$$= k \cdot 9,$$

which again gives us

$$k = 5.$$

Exponential functions appear in a wealth of diverse applications, ranging from describing the decay of radioactive elements to the growth of money saved in a bank account. Originally, most banks only added interest to a depositor's savings account once a year, known as annual compounding. (See the chapter on personal finance for more details.) Suppose you deposit \$200 in a bank offering an annual interest rate of 5%, and the interest is computed annually. At the end of 1 year, you would earn $(0.05)(\$200) = \$10.00$, which would increase your account to \$210. Note that we could also obtain this figure by factoring.

$$200 + (0.05)(200) = 200(1 + 0.05) = 200(1.05) = 210$$

If you then leave that amount in the bank for another year, this time you will earn $(0.05)(\$210) = \$10.50$, the extra \$0.50 coming from the fact that now you are earning 5% on your first year's interest. This brings your total savings to $210 + 10.50 = \$220.50$. Again note that

$$210 + (0.05)(210) = 210(1 + 0.05) = 210(1.05) = \$220.50,$$

which can be written

$$210(1.05) = 200(1.05)(1.05) = 200(1.05)^2,$$

where we have shown the replacement in boldface. Observe the pattern that develops from continuing to leave your money in the same account for 6 years. Let $A(t)$ stand for the amount present at time t (in years).

　　At $t = 0$ years, $A(0) = \$200.00$.
　　After $t = 1$ year, $A(1) = 200(1.05) = 210$.
　　After $t = 2$ years, $A(2) = \mathbf{210}(1.05) = \mathbf{200(1.05)}(1.05) = 200(1.05)^2 = 220.50$.
　　After $t = 3$ years, $A(3) = \mathbf{220.50}(1.05) = \mathbf{200(1.05)^2}(1.05) = 200(1.05)^3 = 231.53$.
　　After $t = 4$ years, $A(4) = \mathbf{231.53}(1.05) = \mathbf{200(1.05)^3}(1.05) = 200(1.05)^4 = 243.10$.

We see that we can easily develop our own formula here because the exponent for 1.05 in each case matches the number of years that have passed.

$$A(t) = 200(1.05)^t$$

In general, the amount $A(t)$ accumulated after t years from an initial principal P by compounding annually in an account earning interest at an annual rate r (as a decimal) is given by the following *exponential* function.

$$A(t) = P(1 + r)^t$$

This function needs only minor modifications to conform to the modern practice of compounding more than once a year—e.g., quarterly, monthly, daily, and so on. For a fixed rate r, increasing the number of times per year that compounding is done causes the amount of money to grow more rapidly. As the number of compounding times per year gets infinitely large, the expression for $A(t)$ becomes

$$A(t) = Pe^{rt},$$

natural exponential function An exponential function with base given by a power of e.

where the number $e \approx 2.71828$ is a special number that occurs frequently in a wide variety of applications. Because e (named for the mathematician Leonhard Euler, mentioned in the last section) occurs often in nature, e^{rt} is known as the **natural exponential function** with rate r. (Note that e^r serves as the base.) The function e^x is found on all modern calculators.

　　The growth of most organic populations from bacteria to rabbits to humans living in an environment of unlimited resources can also be modeled by the natural exponential function. If we think of new members of a population being produced almost continuously, then this function provides a reasonable first estimation of how an initial ($t = 0$) population P increases if it has an annual growth rate r.

© Ints Vikmanis/ShutterStock, Inc.

? Example 6

In 1900, the population of the United States was 76 million people. Assuming natural exponential growth and a constant annual growth rate of 1.3%, estimate the population in the years 2010 and 2020.

⚙ Solution

The value of P always corresponds to an initial ($t = 0$) amount. If we let $t = 0$ correspond to the year 1900, then $P = 76$ (million) and $r = 0.013$ in the earlier function $A(t) = Pe^{rt}$. This gives us

$$A(t) = 76e^{0.013t}$$

in units of millions of people t years after 1900. Hence, our estimate for the year 2010 is

$$A(110) = 76e^{0.013(110)}$$
$$= 76e^{1.43}$$
$$= 76(4.18)$$
$$= 318 \quad (\textit{Rounding to three significant digits.})$$

or a population of 318 million people. Similarly, we estimate a population in the year 2020 of $A(120) = 76e^{0.013(120)} = 76e^{1.56} = 362$ million people.

Our travels through this text will take us many places, but we will have the same traveling companion throughout—that champion tool for expressing in a definitive manner how one variable entity depends expressly on another: the *function*. The ability to create a function connecting two sets of data is a valuable skill—one that allows us to describe, analyze, make predictions, and draw conclusions about phenomena that affect us and the world around us.

Name _____

Exercise Set **1.2** ◯◯◯←

Each of the following tables displays a sample of pairings under a function f. Identify each function as linear or exponential. In each case, construct a formula for f(x) under the action of the function.

1.

x	f(x)
1	11
2	12
3	13
4	14
5	15

2.

x	f(x)
1	2
2	4
3	6
4	8
5	10

3.

x	f(x)
1	9
2	11
3	13
4	15
5	17

4.

x	f(x)
1	6
2	12
3	18
4	24
5	30

5.

x	f(x)
1	5
2	11
3	17
4	23
5	29

6.

x	f(x)
1	−4
2	−8
3	−12
4	−16
5	−20

7.

x	f(x)
1	1
2	−3
3	−7
4	−11
5	−15

8.

x	f(x)
1	−3.5
2	2.5
3	8.5
4	14.5
5	20.5

9.

x	f(x)
−2	1.0
−1	4.5
0	8.0
1	11.5
2	15.0

10.

x	f(x)
−2	13.8
−1	12.9
0	12.0
1	11.1
2	10.2

11.

x	f(x)
−2	0.6
−1	−1.2
0	−3.0
1	−4.8
2	−6.6

12.

x	f(x)
0	2.8
2	9.2
4	15.6
6	22.0
8	28.4

13.

x	f(x)
0	−11.5
2	−1.3
4	8.9
6	19.1
8	29.3

14.

x	f(x)
−10	120.4
−5	60.4
0	0.4
5	−59.6
10	−119.4

15.

x	f(x)
−10	49.0
−5	25.5
0	2.0
5	−21.5
10	−45.0

16.

x	f(x)
−2	1/4
−1	1/2
0	1
1	2
2	4

17.

x	f(x)
−2	3/4
−1	3/2
0	3
1	6
2	12

18.

x	f(x)
−2	0.01
−1	0.1
0	1
1	10
2	100

19.

x	f(x)
−2	0.04
−1	0.4
0	4
1	40
2	400

20.

x	f(x)
−2	7/9
−1	7/3
0	7
1	21
2	63

21.

x	$f(x)$
−2	0.4
−1	2
0	10
1	50
2	250

22.

x	$f(x)$
−2	4
−1	2
0	1
1	0.5
2	0.25

23.

x	$f(x)$
−2	12
−1	6
0	3
1	1.5
2	0.75

24.

x	$f(x)$
−2	63
−1	21
0	7
1	7/3
2	7/9

25.

x	$f(x)$
−2	25,000
−1	5000
0	1000
1	200
2	40

26.

x	$f(x)$
−2	50,000
−1	5000
0	500
1	50
2	5

27. The value of a copying machine is $5,200 when it is purchased. After 2 years, its value is $4,750. Create a formula giving the value of the machine as a linear function of time, and use it to find the value after 5 years and 8 years.

28. You hail a taxi and inquire about the fare. The taxi driver informs you that your ride will cost a flat fee of $2.00 plus $1.25 per minute. Create a formula that gives you the cost of the ride as a function of the length of time spent in the car. What is the cost of a 28-minute ride?

29. The pipes in your laundry room are behaving mysteriously, and you need to hire someone to fix them. The local plumber charges a minimum of $25 for any visit, plus $32.50 per hour. Create a formula that gives you the cost as a function of the number of hours spent fixing your plumbing. How much will it cost if the plumber needs 4.5 hours to fix your pipes?

30. The numbers of pairs of birds (all species) that nested in the Franklin Bay rookery over 4 years are given according to the following table. Create a formula that gives the number of nesting pairs as a function of years after 2006. At the current rate, how many pairs will be nesting there in 2012? 2015? By what year will no birds be nesting in this rookery?

Year	Nesting Pairs
2006	830
2007	752
2008	674
2009	596

31. In one South American country, the federal income taxes for working residents are given, as shown here, for some income brackets.

Annual Income ($)	Tax ($)
25,000 up to 29,999	2500 + 25% of excess over 25,000
30,000 up to 34,999	3750 + 28% of excess over 30,000
35,000 up to 39,999	5150 + 32% of excess over 35,000
40,000 up to 44,999	6750 + 36% of excess over 40,000

Find the expression that gives the tax as a function of annual income for the 30,000 to 34,999 bracket, and use it to compute the tax owed by someone who earned $32,450.

32. Use the table in the previous exercise to find the expression that gives tax as a function of annual income for the 35,000 to 39,999 bracket, and use it to compute the tax owed by someone who earned $37,250. What is the tax owed by someone who earned $41,900? $44,730?

33. The table here shows approximations to the life expectancy of a 25-year-old male smoker based on his daily cigarette consumption. Find a linear expression that estimates the life expectancy $f(x)$ as a function of the number of cigarettes smoked per day. Interpret the meaning of the rate.

Daily Cigarettes Smoked	Life Expectancy (years)
0	73.7
10	71.6
20	69.5
30	67.4
40	65.3

34. Use your function from the previous exercise to estimate the life expectancy of a 25-year-old male who smokes 25 cigarettes per day; 32 cigarettes per day.

35. If x is the temperature in degrees Celsius and $F(x)$ is the corresponding temperature in degrees Farenheit, then we know that $F(0) = 32$ and $F(100) = 212$. Find the formula that gives the action of this linear function, and use it to find $F(45)$.

36. Angelita owns and manages a print shop. She specializes in printing brochures and charges customers a flat set-up fee of $45 plus a cost per brochure according to the quality of paper used. The table here lists total cost per type for quantities of 50 and 100. A customer requests 2,000 brochures printed on the best possible paper without exceeding $450. Find the three linear functions that represent the cost of using each type of paper. Which type of paper should Angelita use?

Quantity	Type A	Type B	Type C
50	$52.50	$55.00	$57.50
100	$60.00	$65.00	$70.00

37. Suppose you deposit $750 in a savings account at a bank offering an annual interest rate of 4.2%, compounded annually. Form the expression that gives the amount of money accrued as a function of time t (in years). What amount of money will you have in that account in 3 years? 5 years?

38. George deposits $2,800 with an investment firm that guarantees an annual return of 6.5%. If we assume annual compounding, form the exponential function showing the amount of money George will have in t years. How much will he have accumulated in 4 years? 7 years?

39. To attract business, Joel's Bank is advertising an annual interest rate of 10% for the first year. After that, the rate drops to 3.5%. If the compounding is done annually, compute the amount accrued in 6 years from an initial deposit of $2,000.

40. Maura's Bank (Joel's competitor in the previous exercise) does not have a higher 1-year interest rate, but guarantees a 4.5% annual rate every year. How much money would you accumulate from a deposit of $2,000 in 6 years at this bank? In which bank would you deposit your money? What if you were going to leave it there for 7 years?

41. In 2012, the country of Herzon had a population of 1.85 million and was growing according to a natural exponential function, with an annual growth rate of 2.3%. If that growth rate remains constant, what will be the population of Herzon in the year 2014? 2016? 2018?

42. In 2007, the county seat of Abington Township had a population of 1,035 and was growing according to a natural exponential function, with an annual growth rate of 0.4%. If that growth rate remains constant, what will be the population in the year 2010? 2017? 2030?

43. Statistics produced by the United Nations forecast an annual approximate world population growth rate. Assuming an annual rate of 1.2% and a world population of 7.0 billion in 2011, estimate the world population in 2015 and 2020. If this annual rate then drops to 1.1%, predict the world population in 2025.

44. The population of the country of Uganda in 2011 was 34.5 million people, and its growth rate was 3.4%. [**Source**: World Population Data Sheet. (2011). Washington, DC: Population Reference Bureau, Inc.] Estimate the population in 2015 and 2020.

45. The population of the country of Mexico in 2011 was 114.8 million people, and its growth rate was 1.4%. [**Source**: World Population Data Sheet. (2011). Washington, DC: Population Reference Bureau, Inc.] Estimate the population in 2015 and 2020.

46. The population of the country of Finland in 2011 was 5.4 million people, and its growth rate was 0.2%. [**Source**: World Population Data Sheet. (2011). Washington, DC: Population Reference Bureau, Inc.] Estimate the population in 2015 and 2020.

47. The *doubling time* a population is the amount of time it requires for that population to double in size. The *Rule of 70* states that if the annual growth rate is r percent, then the doubling time is approximately equal to $\dfrac{70}{r}$ (where r is left as a percentage and not converted to a decimal). Estimate the doubling times for Uganda, Mexico, and Finland using the rates from the previous exercises.

48. If the doubling time for a population is 35 years, what is its annual growth rate? Use the rate you compute in the natural exponential function to verify that a population P grows to $2P$ in 35 years.

49. Certain *decreasing* populations can be modeled by exponential functions having a base that is less than 1. Suppose 512 cupcakes were placed on a table in the cafeteria of Marywood University at 8:00 a.m. If only half of the cupcakes present at the beginning of each hour are still remaining at the end of the hour, how many were left by 5:00 p.m.? Find an expression that gives the number of cupcakes as an exponential function of the number of hours past 8:00 a.m.

50. Here are the population figures for three cities over the same time period. Two of these cities grew linearly, and one grew exponentially. Create the expression giving the population for each city as a function of time, letting $t = 0$ correspond to 2005.

Year	Sometown	Anytown	Boomtown
2005	10,500	37,000	40,000
2006	11,200	34,700	60,000
2007	11,900	32,400	90,000
2008	12,600	30,100	135,000
2009	13,300	27,800	202,500

51. Suppose Smooth as Silk paint manufacturer buys a new large mixing vat for $22,000. This vat depreciates nonlinearly because its value at the end of each year is 90% of what it was worth at the start of the year. Find the decreasing exponential function that gives the value of the vat V in terms of time t (in years) after the initial purchase. How much will the vat be worth in 12 years?

52. If the temperature is constant, then the atmospheric pressure p (lb/in^2) depends on the altitude x (ft) above sea level according to $p = 15(0.8)^{(x/5000)}$. Is this a decreasing function? What is the atmospheric pressure at sea level ($x = 0$)? What is the pressure at 10,000 ft?

53. Radioactive substances decay over time, and so the amount A at any time t is a decreasing function of t. Given an initial amount P, the amount A remaining at time t can be given by the natural exponential function that has a negative rate r. For Plutonium 241, the function is given by

$$A(t) = Pe^{-0.053t}.$$

Starting with an initial amount of 20 gm of Plutonium 241, how much will remain after 5 years? After 8 years?

54. In a city with population P, the number of people N who have heard about a news bulletin broadcast over radio and television after t hours is $N(t) = P(1 - e^{-0.3t})$. In a city of 50,000 residents, how many people have heard of a major earthquake 5 hours after it was broadcast? 8 hours?

55. Studies have shown that income is affected by the number of mathematics courses an individual has taken. For example, if x represents the number of years of calculus classes taken, then the salary S earned by that person might be given by $S(x) = 45,000e^{0.195x}$ dollars. What is the salary of someone who has had 1.5 years of calculus? [**Source**: *Review of Economics and Statistics* 86. Cambridge, MA: MIT Press, 2004]

It is useful to be able to formulate the action of a function from a verbal description. For instance, $f(x) = (2x + 7)^2$ is the function described by the following procedure.

1. *Pick any number.*

2. *Double it and then add 7.*

3. *Square the result.*

Formulate the action of the function described by each of the following procedures.

56. 1. Pick any number.

   2. Square it and then multiply by 13.

   3. Add 20 and then divide the result by 9.

57. 1. Pick any number.

   2. Subtract 4 and then square it.

   3. Multiply the result by 5.

58. 1. Pick any number.

   2. Add 45 and then raise it to the fourth power.

   3. Multiply the result by 28.

59. 1. Pick any number (greater than or equal to −3).

   2. Add 3 and then divide by 8.

   3. Take the square root of the result.

60. 1. Pick any number and square it.

   2. Multiply the result by 5 and then add 3 times the original number you picked.

   3. Add 12 to the result.

61. 1. Pick any number (except 1) and add 6.

   2. Raise the result to the fourth power and add 7.

   3. Take the square root of the result.

   4. Divide that result by the difference of the original number and 1.

1.3 The Graph of a Function

Now that we have had some experience forming expressions for functions from raw data, our next step is to acquire skill at analyzing the formula we have created. The whole point in knowing a formula is that it assists in knowing how the function behaves. We want to be able to answer questions like these: "Where does it increase?" "Where does it decrease?" "Does it achieve a maximum value and, if so, where?" "Does it achieve a minimum value and, if so, where?" "Does it ever become zero?" Your symbolic expression is a description of a real physical process, and answers to questions like these form the basis for a thorough understanding of that process. In fact, finding these answers is one of the main reasons a person formulates a function in the first place.

That brings us to the role of a *graph of a function f*. As you learned in algebra, the graph of an equation with two variables consists of ordered pairs of numbers satisfying the equation, and it is usually pictured as a set of points existing in a two-dimensional coordinate system. The use of coordinate systems is credited in part to the great French mathematician and philosopher **Rene Descartes** (1596–1650) because of his classic work *La Geometrie*, which brought the force of algebra to bear on problems in geometry.

Descartes had the good fortune, in an era filled with poverty and suffering, to be born to an enlightened, caring, and wealthy father. The combination led to a catering to Descartes, who was in frail health during his early years, and so this bright, curious boy was allowed to spend the majority of the mornings of his youth at home. These times of solitude fostered a view of the world rooted in rational skepticism. Hence, along with such famous contemporaries as Galileo and Pascal, he was a pioneer in accruing knowledge strictly through experimentation and the application of mathematical reasoning. As a young man with his health restored, he split his time, amazingly, between contemplating science and mathematics and being a mercenary soldier for various kings and despots across the breadth of Europe. He doubted most of the standard axioms of knowledge of his day and assembled his own view of the universe, brick by logical brick.

FIGURE 1.3.1   Vintage French stamp of Descartes.

Although Descartes never fully employed the coordinate system that bears his name, it is said that the wondrous idea of representing geometric figures by equations came to him on a Bavarian battlefield in a dream of almost mystic potency on November 10, 1619. This date is now often referred to as the birthday of analytic geometry. By associating equations with curves, Descartes and others, notably the great **Pierre de Fermat** (1601–1665), used the new coordinate geometry as a tool to solve geometric construction problems that were considered essential to understanding some basic scientific questions. As it turned out, this tool had a tremendous impact on how mathematics came to be used to describe nature, and that impact is still being felt today. For that reason, the standard rectangular coordinate system of algebra is often referred to as the **Cartesian coordinate system** or plane.

> **Cartesian coordinate system** A system of perpendicular number lines called axes used to graph functions. Also referred to as rectangular coordinate system.

The graph of a function gives a visual representation of how the functional values in the range respond to changes in values in the domain. If x is the independent variable for a function f and y is the dependent variable, then we say that "y is a function of x," and we define the graph of f to be the set $\{(x, y) \mid x \in$ domain of f and $y = f(x)\}$. In other words, the graph is the set of ordered pairs of values where the pairing is done according to the functional assignments.

Let's pause and make clear the meaning of this definition. If $(2, 9)$ is a point on the graph of the function f, then that means precisely that $f(2) = 9$. If $(3.2, 51.6)$ is on its graph, then $f(3.2) = 51.6$. If $f(-5.8) = 101$, then $(-5.8, 101)$ must be on the graph. This one concept provides the framework for representing a functional relationship between two sets of numbers as a picture in the Cartesian plane.

? Example 1

Graph the linear function $f(x) = 3x - 4$.

⚙ Solution

From our definition, we see that our graph consists of ordered pairs (x, y) such that $y = 3x - 4$. The easiest approach is to set up a table with a few arbitrary values of x. Suppose we let x equal each of the integers running from, say, -2 to 2.

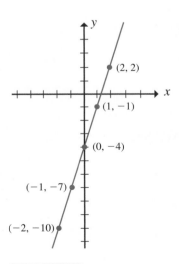

x	$y = f(x) = 3x - 4$	Point (x, y)
-2	$3(-2) - 4 = -10$	$(-2, -10)$
-1	$3(-1) - 4 = -7$	$(-1, -7)$
0	$3(0) - 4 = -4$	$(0, -4)$
1	$3(1) - 4 = -1$	$(1, -1)$
2	$3(2) - 4 = 2$	$(2, 2)$

Now we plot these points in the Cartesian plane. After plotting the five points from our table, we extend the domain to all real numbers by connecting those points with a smooth curve. (See **Figure 1.3.2**.) In this case, a line results because our function is linear. ◆

FIGURE 1.3.2 Graph of $f(x) = 3x - 4$.

An analysis of the table reveals that the corresponding constant rate of change is 3. It is no coincidence that 3 is also the coefficient of the x-term, which you may recall from algebra is known as the *slope* of the graph of the function. So the rate of a linear function and the slope of its graph are always the same value. Because this implies that the slope m is the rate of increase of the functional value per unit increase of the independent variable, it can always be computed from *any* two points (x_0, y_0) and (x_1, y_1) on the graph.

$$m = \frac{\text{vertical change}}{\text{horizontal change}} = \frac{\text{change in } y\text{-coordinates}}{\text{change in } x\text{-coordinates}} = \frac{y_1 - y_0}{x_1 - x_0}$$

In the present case, if we choose, say, $(2, 2)$ and $(-1, -7)$, we compute the slope

$$m = \frac{-7 - 2}{-1 - 2} = \frac{-9}{-3} = 3.$$

We recognize this function as an increasing function, which in turn corresponds to the fact that it has a positive slope. If a linear function possesses a negative slope, it must necessarily be a decreasing function, since a negative rate implies that the functional values are dropping as the independent values increase.

The graph of an exponential function looks very different.

❓ Example 2

Graph the exponential function $g(x) = 3 \cdot 2^x$.

⚙ Solution

Again we set up a table with integer values of x running from -2 to 2.

x	$y = g(x) = 3 \cdot 2^x$	Point (x, y)
-2	$3 \cdot 2^{-2} = 3 \cdot \dfrac{1}{4} = \dfrac{3}{4}$	$\left(-2, \dfrac{3}{4}\right)$
-1	$3 \cdot 2^{-1} = 3 \cdot \dfrac{1}{2} = \dfrac{3}{2}$	$\left(-1, \dfrac{3}{2}\right)$
0	$3 \cdot 2^0 = 3 \cdot 1 = 3$	$(0, 3)$
1	$3 \cdot 2^1 = 3 \cdot 2 = 6$	$(1, 6)$
2	$3 \cdot 2^2 = 3 \cdot 4 = 12$	$(2, 12)$

We plot these points on the xy-plane and draw a smooth curve. (See **Figure 1.3.3**.)

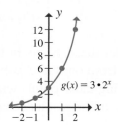

FIGURE 1.3.3   Graph of $g(x) = 3 \cdot 2^x$.

As x increases by an increment of 1, the linear function in Example 1 increases by an *increment* of 3 (the slope), while the exponential function in Example 2 increases by a *factor* of 2 (the base). As explained previously, this property distinguishes linear from exponential functions, and Figures 1.3.2 and 1.3.3 illustrate how it affects the shapes of the corresponding graphs. Notice, in particular, that the graph lies entirely above the x-axis, because an exponential function $f(x) = kb^x$ ($k > 0$, $b > 0$) is always positive. More specific features that distinguish the graph of every exponential function will be discussed in more detail in the next section.

 Example 3

quadratic function
Any function of the form
$f(x) = ax^2 + bx + c.$

Every function of the form $f(x) = ax^2 + bx + c$ (where $a \neq 0$) is called a **quadratic function**. Graph the quadratic function $f(x) = x^2 - 4$. (In this case, $a = 1$, $b = 0$, and $c = -4$.)

 Solution

We form a table of values with x running through the integers from -3 to 3. (See **Figure 1.3.4**.)

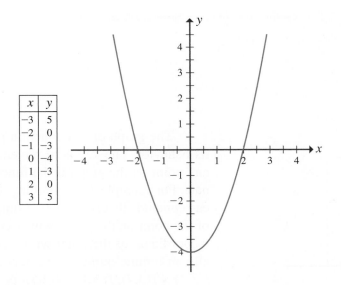

x	y
-3	5
-2	0
-1	-3
0	-4
1	-3
2	0
3	5

FIGURE 1.3.4   Graph of $f(x) = x^2 - 4$.

parabola The graph of a quadratic function.

intercepts Points on the coordinate axes where a graph intersects them.

The graph of every quadratic function is a curve known as a **parabola**. Note that this parabola crosses the x-axis at $x = -2$, and 2 and crosses the y-axis at $y = -4$. These are known as the **intercepts** of the graph. The graphs of quadratic functions will be studied in greater detail in the next section. ♦

Functions can also be graphed with a domain that has been restricted. In this case, we simply do not include any points on the graph whose x-coordinate lies outside the given domain.

? Example 4

Graph the function $g(x) = -\dfrac{1}{2}x + 5$ over the closed interval domain $[-2, 6]$.

⚙ Solution

This is a linear function with slope $-\dfrac{1}{2}$ and crossing the y-axis at 5. Note that no points are included whose x-coordinate is less than -2 or greater than 6. (See **Figure 1.3.5**.)

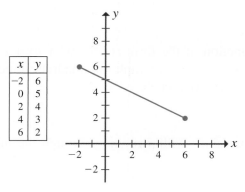

x	y
-2	6
0	5
2	4
4	3
6	2

FIGURE 1.3.5   Graph of $g(x) = -\dfrac{1}{2}x + 5$ over the restricted domain $[-2, 6]$.

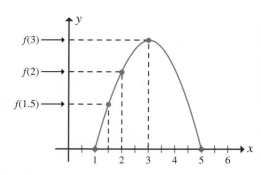

FIGURE 1.3.6   From this graph, we see that $f(1.5) < f(2)$ and that $f(3)$ is the maximum value of the function.

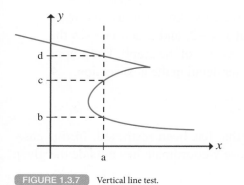

FIGURE 1.3.7   Vertical line test.

The graph of any function presents us with a picture of its behavior. The key is to remember that the y-coordinate of any point on the graph is the functional value of the x-coordinate. For example, by examining the graph in **Figure 1.3.6**, we can conclude that $f(1.5) < f(2)$ simply because the y-coordinate of the point on the graph with x-coordinate 1.5 is less than the y-coordinate of the point with x-coordinate 2. There are literally an infinite number of observations we could make—e.g., $f(3.2) > f(3.7)$, $f(2.5) > f(4.8)$, or even that $f(3)$ appears to be the largest value the function attains. Such a value is called the *maximum value of the function* and will be discussed further in the next section.

Be aware that not every curve in the Cartesian plane is the graph of a function. Consider **Figure 1.3.7**. Is this the graph of a function?

If the curve in Figure 1.3.7 did represent a function, say $g(x)$, would $g(a)$ be equal to b, c, or d? There is no way to choose. The requirement for a relation to be a function is that the functional pairing with each value of the independent variable be unambiguous. In this case, there is not a unique image that we can assign to a. Pictorially, we see that the vertical line above a intersects the curve more than once,

vertical line test
A graph represents a function of the variable associated with the horizontal axis if no vertical line intersects the graph more than once.

and so we have the condition that is the basis for a handy check known as the **vertical line test.**

A curve in the Cartesian plane is the graph of a function if and only if no vertical line intersects the curve more than once.

We make one final observation about finding the domain and range of any function. Consider the graph of the linear function $f(x)$ in **Figure 1.3.8**. To determine the domain, we must ask, "What are the x-coordinates of the points on the graph?" Imagine a flashlight being aimed at the graph from the top (and the bottom if it crosses the x-axis). The shadow projected on the x-axis is precisely that set of x-coordinates and therefore constitutes the domain. From the figure, convince yourself that the domain of $f(x)$ is the interval $[1, 6]$. Similarly, we determine the range by asking, "What are the y-coordinates of the points on the graph?" By shining our flashlight from the right (and the left if the graph crosses the y-axis) this time, we can determine the range by looking at the curve's shadow on the y-axis. In the figure, we see that the range $f(x)$ is the interval $[2, 4]$. In the next section, we will learn more about how the graph of a function reveals a great deal of information about its behavior.

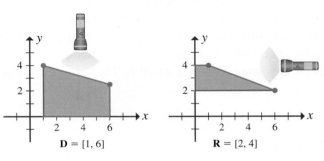

FIGURE 1.3.8   Determining the domain and range of a function.

Example 5

Find the domain and range of the function whose graph is given in **Figure 1.3.9**.

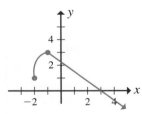

FIGURE 1.3.9

Solution

The arrow in the above graph indicates that the line continues forever in that direction. Consequently, the imaginary shadows produced by our flashlight would yield the interval $[-2, \infty)$ as the domain and $(-\infty, 3]$ as the range of this particular function.

Graphing Calculators

Calculator technology has greatly expanded our abilities to draw and use graphs in order to help us determine the behavior of functions. Graphs that are difficult and time-consuming to create by hand can be done in minutes on a calculator with a graphing capability. For instance, the five buttons beneath the window of a typical Texas Instruments calculator are shown in **Figure 1.3.10**.

| Y = | WINDOW | ZOOM | TRACE | GRAPH |

FIGURE 1.3.10

The first button allows you to input any function that can be composed from the functions available on the calculator. The second button allows scaling of the axes that will appear in the window. When you have assigned the necessary commands with these two features, pressing the **GRAPH** button produces the graph of your function. After the curve appears, you can check the coordinates of any of the points on it by pressing **TRACE** and moving the flashing point with the cursor arrows. Finally, you may zoom in and out to view the graph with different levels of magnification with **ZOOM**.

The speed of graphing with a calculator allows us to more easily learn about the effects on any graph brought about by changing the value of a coefficient, power, or other parameter.

? Example 6

Use a calculator to graph $y_1 = 0.25x^2$, $y_2 = 0.5x^2$, $y_3 = x^2$, and $y_4 = 3x^2$ all in the same window. (See **Figure 1.3.11**.) Adjust your axes so that **Xmin** $=-5$, **Xmin** $= 5$, **Xsc1** $= 1$, **Ymin** $= -1$, **Ymax** $= 8$, and **Ysc1** $= 1$. (**Xsc1** and **Ysc1** set the distance between tick-marks on the axes.) What do you conclude about the effect on the graph of $y = ax^2$ as a increases?

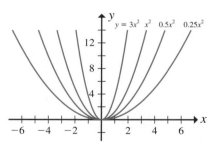

FIGURE 1.3.11   Graphing on a calculator.

⚙ Solution

The width of the parabola decreases as the coefficient a increases.

Name _____

Exercise Set **1.3** ◯◯◯ ←─────────────────

Graph each of the following functions.

1. $f(x) = -2x + 3$

2. $g(x) = \frac{2}{3}x - 4$

3. $h(x) = -\frac{2}{3}x + 5$

4. $y(x) = x^2 + 2$

5. $A(x) = 2x^2 - 5$

6. $P(x) = \left(\frac{1}{2}\right)x^2 - 3$

7. $f(x) = -x^2 + 9$

8. $g(x) = \left(\frac{1}{2}\right)^x$

9. $f(x) = 3^x$

10. $w(t) = 6\left(\frac{3}{4}\right)^t$

11. $r(t) = 2(0.8)^t$

12. $F(x) = 3 \cdot 2^x$

Graph each of the following functions over the given restricted domain.

13. $k(x) = -3x + 1;\ \mathbf{D} = [-3, 2]$

14. $g(x) = \left(\frac{3}{4}\right)x - 5;\ \mathbf{D} = (-\infty, 8]$

15. $r(t) = -t^2 + 9;\ \mathbf{D} = [-4, \infty)$

16. $f(x) = 2x^2 - 6;\ \mathbf{D} = [-1, 2]$

17. $h(t) = 5\left(\frac{1}{2}\right)^t;\ \mathbf{D} = [-2,3]$

18. $y(t) = 4(0.3)^t;\ \mathbf{D} = [-3, 6]$

In the next four exercises, determine whether the accompanying statements concerning the function of the corresponding graph are true or false. (See accompanying figures.)

19. (a) $f(-3) < f(-2)$

(b) $f(1) < f(3)$

(c) $2 < f(0) < 3$

(d) $f(3) > 0$

(e) $f(-1)$ is the maximum value of the function.

(f) $f(2) = 0$

(g) $f(0.5) < f(2.5)$

(h) $f(4) < 0$

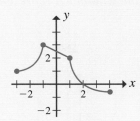

20. (a) $f(-1) < f(1)$

(b) $f(1) < f(2)$

(c) $f(2) < f(3)$

(d) $f(0) > 0$

(e) $f(-2) < 0 < f(-1)$

(f) $f(-1.9) > f(1.9)$

(g) $f(2)$ is the maximum value of f.

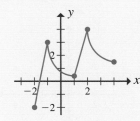

21. (a) $f(-2) < 0$ (b) $f(0) < f(1)$ (c) $f(0) > 0$ (d) $f(2) < f(3)$

(e) $f(0)$ is the minimum value of f.   (f) $f(-2)$ is the maximum value of f.

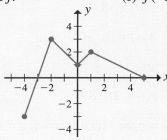

22. (a) $f(x) = f(-3)$ for $x < -3$ (b) $f(-3) < f(-2)$ (c) $f(2) > f(3)$

(d) $f(3) < f(4)$ (e) $f(x)$ has no maximum value.   (f) $f(x) < 0$ for $x > 0$

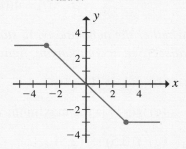

For each of the following graphs, determine whether it represents a function and, if so, find the domain and range.

23.

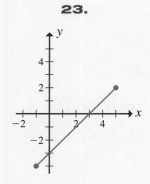

24.

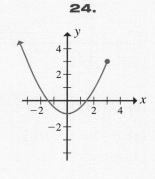

25.

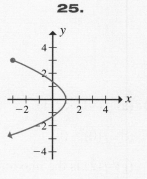

26.

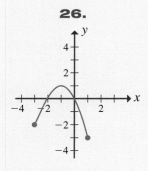

27.

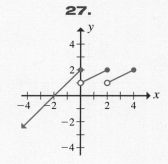

28.

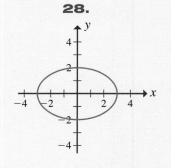

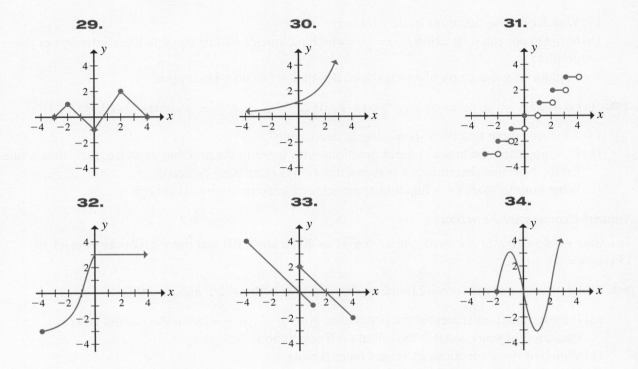

Sketch the graph of each of the following functions by reproducing the graph obtained using a calculator. Graph each of the three functions for each exercise in the same window and sized according to the directions. The accompanying questions are concerned with associating features of the graph with values of parameters in the function.

Linear Functions

Size your window so that both axes run from −4 to 4 and the scale is set to 1.

35. $f(x) = 2x + 1$ $g(x) = 2x + 3$ $h(x) = -0.5x + 2$

    (a) Which two of these graphs are parallel? What do those functions have in common?
    (b) Which graph has a y-intercept of 2?
    (c) Which of the functions is decreasing?

36. $f(x) = 3x + 3.5$ $g(x) = -2x - 4$ $h(x) = 3x + 1$

    (a) Which two of these graphs are parallel? What do those functions have in common?
    (b) Which graph has an x-intercept of −2?
    (c) Which of the functions is decreasing? What parameter indicates that?

Exponential Functions

Size your window so that the x-axis runs from −4 to 20 and the y-axis runs from −1 to 10. Set both scales to 4.

37. $f(x) = 2 \cdot (1.1)^x$ $g(x) = 2 \cdot (1.5)^x$ $h(x) = 1.5 \cdot 2^x$

(a) Which of these functions has a y-intercept of 2?

(b) In the exponential function $f(x) = kb^x$, which parameter will be equal to the y-intercept of the graph?

(c) Explain why the graph of an exponential function has no x-intercepts.

38. $f(x) = (0.25)^x$ $\qquad\qquad$ $g(x) = 2 \cdot (0.25)^x$ $\qquad\qquad$ $h(x) = 3 \cdot (0.5)^x$

(a) Are these three functions increasing or decreasing?

(b) By comparing the bases of these functions with those in the previous exercise, formulate a rule for the base that determines whether a function is increasing or decreasing.

(c) What number does each function approach as x becomes infinitely large?

Natural Exponential Functions

Size your window so that the x-axis runs from −1 to 40 (scale = 10) and the y-axis runs from −1 to 15 (scale = 5).

39. Graph the natural exponential function $f(t) = 5e^{rt}$ for both $r = 2\%$ and $r = 3\%$.

(a) If these functions represent the population growth of two towns (in thousands) and t is measured in years, what is the initial $t = 0$ population?

(b) Which of these functions increases more rapidly?

(c) How many years would it take for $f(t)$ to reach 10 for each of the given values for r?

40. Graph the natural exponential function $f(t) = 7.5e^{.045t}$. Use the graph to estimate the values of t for which $f(t) = 5$, 10, and 15.

Quadratic Functions

Size your window so that both axes run from −4 to 4 and the scale is set to 1.

41. $f(x) = x^2 - 3$ $\qquad\qquad$ $g(x) = x^2 + 2$ $\qquad\qquad$ $h(x) = -x^2 + 4$

(a) What main feature of the third parabola differs from the first two?

(b) What is different in the expression for the third function that causes this difference?

(c) What is the y-intercept of each graph?

42. $f(x) = x^2 + 2x - 3$ $\qquad\qquad$ $g(x) = -0.75x^2 + x + 4$ $\qquad\qquad$ $h(x) = 2x^2 - 3x - 1.5$

(a) Estimate the vertex of each parabola using the TRACE button on your calculator.

(b) If the coefficient of the x^2 term is positive, does the graph open up or down? Is it negative?

(c) Is the y-intercept of the graph of a quadratic function $f(x) = ax^2 + bx + c$ equal to a, b, or c?

43. Most graphing calculators can draw more than one graph on the same screen. Graph the following linear functions on the same set of axes:

$$y_1 = -2.6x \quad y_2 = -0.5x \quad y_3 = 0.3x \quad y_4 = x \quad y_5 = 2x$$

What happens to the graph as the slope of the function increases?

44. Graph the following exponential functions on the same set of axes. Size your window so that the x-axis runs from -5 to 5 and the y-axis runs from -1 to 8.

$$y_1 = 1.2^x \quad y_2 = 1.4^x \quad y_3 = 1.7^x \quad y_4 = 2^x$$

What happens to the shape of the graph as the value of the base increases? How does this correspond to the behavior of the function?

Find a formula (by experimentation) for each function whose graph is given below.

45.

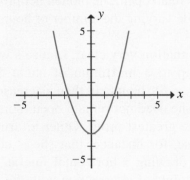

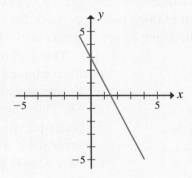

(a) $f(x) =$ _____   (b) $f(x) =$ _____

46.

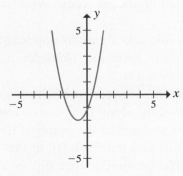

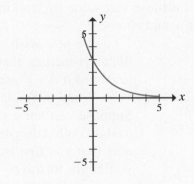

(a) $f(x) =$ _____   (b) $f(x) =$ _____

1.4 Interpretation of Graphs

Have you ever tried to describe the physical appearance of another person to someone? If you have a picture of that person, you don't even bother with the verbal description. You just show her the picture. Much the same thing can be said for attempting to describe the behavior of a function between two sets. Because the graph of a function visually displays its key characteristics, the phrase *a picture is worth a thousand words* has never been more appropriate. The rises, falls, peaks, valleys, and corners of the curve provide the observer important information at a glance that may not otherwise be so apparent by just studying the functional pairing of numbers in a table.

Consider the situation of Louise, owner and manager of a local pet store. When she first opened her store, Louise experimented with the number of hours per week that she kept the store open. Initially, it seemed to her that the more hours her store was open, the more profit she would realize. But later, it became apparent that if the store was open at times with very few customers, the operation costs—for heating, cooling, lights, employee salaries, and so on.—were greater than her revenue, and so her profits actually decreased. In order to determine the optimal number of hours to remain open, she plotted her profits as a function of open hours per week for six months while varying the number of hours each week. The graph of her function is given in **Figure 1.4.1**.

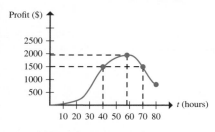

FIGURE 1.4.1 Profit per week as a function of open business hours.

This picture makes the situation very clear. Louise's weekly profit continues to increase up to a maximum of about $1900 when she remains open for $t = 58$ hours. It then decreases for $t > 58$. Thus, Louise sees that she need not remain open more than 58 hours per week to attain the greatest profit. Other information presents itself as well. Suppose, for instance, that she is content to earn $1,500 per week. By drawing a horizontal line at 1,500 on the y-axis, we see that the points of intersection with the graph are (40, 1,500) and (70, 1,500), meaning that $f(40) = 1,500$ and $f(70) = 1,500$ if f is the profit function. Thus, the curve tells us that she could achieve this value by working either 40 or 70 hours per week. Which do you think she would choose?!

We now wish to get more specific about certain important functional properties that characterize their behavior. Consider next the graphs of $f(x) = x^2$ and $g(x) = x^3$ in **Figure 1.4.2**.

What is fundamentally different between these two graphs? Suppose you are a small insect traveling along each curve from left to right. What do you experience? If you are on the graph of $f(x) = x^2$, note that you first travel "down" until you reach (0, 0); at that point, you begin to travel "up." This would be similar to a dip in a roller coaster. On the graph of $g(x) = x^3$, however, you first travel up, seem to level off at (0, 0), then continue traveling up again. Hikers often experience this when they realize their path continues upward after reaching a pass. These are the graphical expressions of the terms *increasing* and *decreasing*, which we have used previously to describe this behavior, except that now we need to associate each region of increase or decrease with the appropriate interval. We would say that the function f is decreasing over $(-\infty, 0]$ and increasing over $[0, \infty)$. Because g is "going up" everywhere, we say that it is increasing over $(-\infty, \infty)$—that is, for all real numbers. Although a calculus textbook

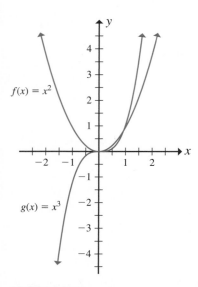

FIGURE 1.4.2 Graphs of $f(x) = x^2$ and $g(x) = x^3$.

may have a more specialized definition, we will say that a function f is ***increasing*** over an interval if, for every pair of elements $a < b$ in the interval, it is true that $f(a) < f(b)$. Similarly, f is said to be ***decreasing*** over an interval if, for every $a < b$ in the interval, it is true that $f(a) > f(b)$. The essence of this definition is captured in **Figure 1.4.3**.

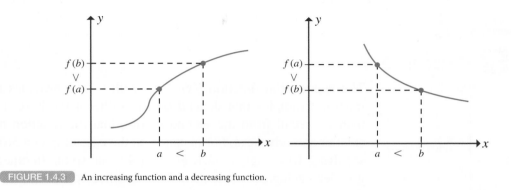

FIGURE 1.4.3   An increasing function and a decreasing function.

? Example 1

The average temperature at any particular geographic location is determined primarily by the latitude and altitude of the location as well as its proximity to large bodies of water. The graph in **Figure 1.4.4** depicts mean temperature at sea level as a decreasing function of latitude, in the northern hemisphere.

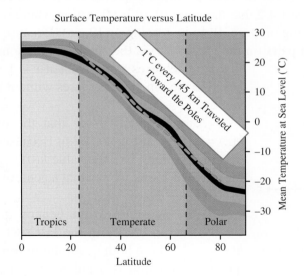

FIGURE 1.4.4   Average temperature as a function of latitude.

Approximately what is the temperature for most locations between 0 and 20 degrees latitude? By about how much does the temperature decrease as you move from a latitude of 20° to one of 80°? If you wish to live in a region where the mean temperature is above freezing (0°C), below what latitude must you live?

⚙ Solution

We can see from the graph that the mean temperature in the tropics remains roughly around 20°C. The temperature decreases from about 20° to about −22° over the interval

20° to 80° latitude, for a net change of about 42°. And by drawing a horizontal line through 0°C, we see that locations between the equator and about 55° latitude have a mean temperature above freezing.

? Example 2

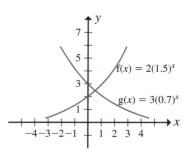

The graphs of $f(x) = 2(1.5)^x$ and $g(x) = 3(0.7)^x$.

The exponential function $f(x) = kb^x$ with base b provides for contrasting examples that depend on the value of the base. If $b < 1$, then we recall from the second section that the function must get smaller as x advances. For instance, in the case of $g(x) = 3(0.7)^x$, we see that $g(0) = 3$, $g(1) = 2.1$, $g(2) = 1.47$, and so on. In other words, g is decreasing over $(-\infty, \infty)$. This is reflected in **Figure 1.4.5** by the fact that the graph of g is falling as we scan the picture from left to right. On the other hand, if $b > 1$, then the function will continue to get larger as x gets bigger. For $f(x) = 2(1.5)^x$, we note that $f(0) = 2$, $f(1) = 3$, $f(2) = 4.5$; and correspondingly, in Figure 1.4.5, the graph of f rises from left to right. It is an increasing function over $(-\infty, \infty)$.

We now turn our attention to points where a function *reverses* its direction. We saw in Figure 1.4.2 that the point $(0, 0)$ is a location where the curve of $f(x) = x^2$ switches from decreasing on $(-\infty, 0]$ to increasing on $[0, \infty)$. Because this is the lowest point for all the ordered pairs on the graph, $f(0)$ is the smallest functional value in the range of f, and we say that f achieves a *minimum* at $x = 0$, whose value there is $f(0) = 0^2 = 0$. Because the function increases without bound forever, it has no maximum value.

Alternatively, we see that the opposite is true for the graph of $h(x) = -x^2 + 4$ in **Figure 1.4.6**. Because $h(x)$ is increasing on $(-\infty, 0]$ and decreasing on $[0, \infty)$, the value $h(0)$ is greater than all the other values in the range of h, and we say that h achieves a *maximum* at $x = 0$, whose value is $h(0) = 4 - 0^2 = 4$.

We now return to the general quadratic $f(x) = ax^2 + bx + c$ that we considered briefly in the last section. (The functions $f(x) = x^2$ and $h(x) = -x^2 + 4$ are both quadratic in form with $b = 0$.) The graph of every quadratic function is a bullet-shaped curve known as a *parabola*, and every parabola has a turning point called the **vertex**. The vertex of $f(x) = x^2$ was $(0, 0)$, and that of $h(x) = -x^2 + 4$ was $(0, 4)$. In general, the x-coordinate of the vertex of the graph of $f(x) = ax^2 + bx + c$ is given by $-\dfrac{b}{2a}$, and the corresponding y-coordinate is obtained by substituting this number into the function. The resulting value $f\left(-\dfrac{b}{2a}\right)$ must be its minimum or maximum value, depending on whether the parabola opens up or down. If $a > 0$, the parabola opens upward, and the value is a minimum. If $a < 0$, it opens downward, and so a maximum must result. For instance, the vertex of the graph of $f(x) = x^2 - 2x - 3$ must have the following coordinates:

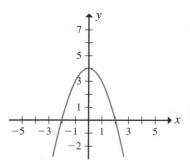

Graph of $h(x) = -x^2 + 4$.

vertex The point on a parabola at which the curve turns around. The x-coordinate of the vertex is given by $x = -\dfrac{b}{2a}$, where $f(x) = ax^2 + bx + c$ is the associated quadratic function.

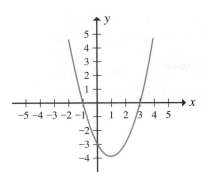

$$x = -\frac{b}{2a} = -\frac{-2}{2(1)} = 1 \quad \text{and} \quad y = f(1) = 1^2 - 2(1) - 3 = -4.$$

Therefore, it is located at the point $(1, -4)$. Plotting a few more points quickly leads us to the picture in **Figure 1.4.7** where the vertex appears as the bottom of the valley of the parabola.

We readily see that $f(x)$ is equal to 0 when $x = -1$ and $x = 3$ by factoring the expression $x^2 - 2x - 3 = (x + 1)(x - 3)$, equating to 0, and solving for x. Values in the domain of a function for which the function equals 0 are known as the **zeroes of the function**. So in this case, the zeroes of f are $x = -1$ and 3. We also observe that f is decreasing over the interval $(-\infty, 1]$, has a minimum value of -4 for $x = 1$, and is increasing over $[1, \infty)$. The domain for this function is the set of all real numbers $\mathbf{D} = (-\infty, \infty)$, while the range includes all numbers greater than or equal to -4 and so consists of the set $\mathbf{R} = [-4, \infty)$.

FIGURE 1.4.7 Graph of $f(x) = x^2 - 2x - 3$.

zeroes of the function The value or values r in the domain of a function f such that $f(r) = 0$.

? Example 3

A rock is thrown from the edge of a cliff h_0 ft high above ground level with a velocity having a vertical component given initially by v ft/sec. (The vertical component of velocity is the rate at which the rock is gaining or losing height.) The height $h(t)$ of the rock above the ground at the base of the cliff must be a function of the time t after the rock is thrown. (Note that $t = 0$ marks the instant the rock leaves the thrower's hand.) It is a known fact that if we ignore the effects of air resistance, this relationship is given by

$$h(t) = -16t^2 + vt + h_0,$$

where h is measured in feet and t in seconds. Note that in the absence of gravity, the height of the object would be $vt + h_0$. The $16t^2$ term refers to the distance an object drops in a free fall after t seconds. The negative sign is needed because the positive direction for height is up in this case, and gravity is working opposite to the direction of motion of the object. If $v = 48$ ft/sec and $h_0 = 64$ ft, what is the maximum height attained by the rock? How much time does it take for the rock to hit the ground?

⚙ Solution

The height function here is $h(t) = -16t^2 + 48t + 64$.

Because the coefficient -16 of the t^2 term is negative, the graph of this function must be a parabola opening downward. The first coordinate of the vertex is $t = -\frac{b}{2a} = -\frac{48}{-32} = 1.5$, and the second coordinate must be $h(1.5) = -16(1.5)^2 + 48(1.5) + 64 = 100$. The maximum height attained by the rock must be 100 ft. To answer the second question, we see that a height of 0 corresponds to the rock hitting the ground, and so we factor the quadratic and equate to 0.

$$-16(t^2 - 3t - 4) = 0$$

$$-16(t + 1)(t - 4) = 0$$

$$(t + 1)(t - 4) = 0$$

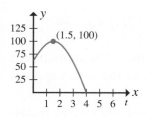

FIGURE 1.4.8 Graph of
$h(t) = -16t^2 + 48t + 64$.

So the zeroes of h are $t = -1$ and $t = 4$. However, the value of -1 has no meaning here because it would represent a negative time! Thus, the domain for h is $\mathbf{D} = [0, 4]$, and the range is $[0, 100]$. The rock hits the ground 4 seconds after it was thrown. Noting that this parabola must cross the vertical axis at $h(0) = 64$ gives us enough information to complete a sketch of the graph, as shown in **Figure 1.4.8**.

Certainly, anybody needing to use a function to draw conclusions about how changes in the independent variable affect the values of the dependent variable would want to have a copy of its graph close at hand. A graph succinctly displays what is important about the relationship between the two variables, no matter what types of physical quantities are involved.

? Example 4

Find the maximum and minimum values of $f(x) = 0.75x + 3.5$ for the domain $\mathbf{D} = [1, 10]$. Is this an increasing or a decreasing function?

⚙ Solution

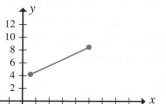

FIGURE 1.4.9   Graph of
$f(x) = 0.75x + 3.5$.

The graph is given in **Figure 1.4.9**.

We can see that the minimum is given by the y-coordinate of the left endpoint on the line segment, which would be $f(1) = 4.25$. The maximum value of the function over $[1, 10]$ is the y-coordinate of the right endpoint, or $f(10) = 11$. Clearly, f is increasing and is never 0 for any x-value in the given interval. Note additionally how the range must include every real number between 4.25 and 11, and so $\mathbf{R} = [4.25, 11]$.

absolute maximum   The largest value a function attains in its range.

strictly local maximum   The largest value of a function for an interval of "nearby" domain values. It is not an absolute maximum. It appears on the graph as the y-coordinate of a point at the top of a minor hill.

absolute minimum   The smallest value a function attains in its range.

Sometimes, there is more than one so-called hill in the graph of a function, creating a situation where the function may have one or more values that are the largest in the "neighborhood" but not necessarily the biggest value in the entire range of f. When this is the case, we distinguish the functional value of the point at the top of the highest hill by calling it the **absolute maximum** and each of those at the top of a minor hill a **strictly local maximum**. Likewise, the **absolute minimum** is the smallest value the function achieves and is positioned at the bottom of the lowest valley, while a **strictly local minimum** is the bottom of any valley not quite as deep. All of these concepts are on display in the graph of the function f in **Figure 1.4.10**. Without even knowing the formula for f, we may still glean an abundance of information about its behavior from this picture.

1. f is increasing over $(-\infty, x_2]$ and over $[x_3, x_4]$. Notice that the curve is rising from left to right in each of these intervals.

2. f is decreasing over $[x_2, x_3]$ and over $[x_4, \infty)$. Notice that the curve is falling from left to right in each of these intervals.

3. *f* achieves a strictly local maximum at $x = x_2$. The local maximum value is $f(x_2) = L$, since L is the y-coordinate of the point at the top of the hill.

4. *f* achieves a strictly local minimum at $x = x_3$. The local minimum value is $f(x_3) = K$, since K is the y-coordinate of the point at the bottom of the valley.

5. *f* achieves an absolute maximum at $x = x_4$. The absolute maximum value $f(x_4) = M$, since the top of the highest hill has M as its y-coordinate.

6. *f* never achieves an absolute minimum. The curve plunges down forever as *x* becomes smaller (more and more negative) and as *x* becomes larger. If *f* did have an absolute minimum, it would appear on the graph as the y-coordinate of the deepest valley.

7. The zeroes of *f* are $x = x_1$ and x_5.

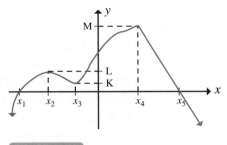

FIGURE 1.4.10

The horizontal axis of many graphs used in the literature of science and business is meant to only contain the domain of the graphed function *as a subset*. It often happens that there is a finite number of values for which the function can be evaluated, and once the pairings have been plotted, a smooth curve is extended from point to point to "fill in" the graph. The purpose is only a cosmetic one, with no intention to imply that functional values exist at those filled-in areas. These notions have to do with the mathematically technical concept of *continuity*, with which we will not concern ourselves. However, to distinguish such graphs from ones for which the horizontal axis does constitute the whole domain, we shall refer to such graphs as **extended graphs**. Extended graphs are widely used to convey information.

? Example 5

The distribution of precipitation across the surface of the Earth depends, in large part, on the general circulation of the atmosphere. The circulation at any particular location, in turn, depends on the latitude of the location, and so we can plot annual precipitation as a function of latitude for a few select latitudes and then create an extended graph, as seen in **Figure 1.4.11**. At what latitudes does the annual precipitation acquire strictly local maximum and minimum values? At what latitude does the precipitation achieve an absolute maximum?

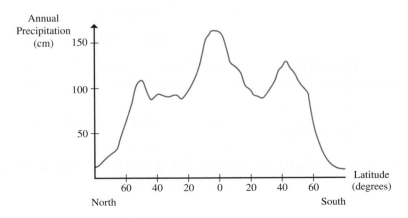

FIGURE 1.4.11   Annual precipitation as a function of latitude.

⚙ Solution

We see from the graph in Figure 1.4.11 that there are several strictly local maxima. A prominent one of about 110 cm of precipitation occurs around 50°N latitude and another one of about 120 cm at around 42°S latitude. A strictly local minimum value of about 80 cm occurs for three latitudes between 45°N and 20°N and again at 30°S latitude. An absolute maximum precipitation of more 150 cm occurs just a few degrees north of the equator.

❓ Example 6

Elof is a wood-carver who specializes in making whistles. Elof knows that, up to a point, the more whistles he sells each month, the cheaper price he can charge. This, in turn, is good for business. In fact, the price p is a decreasing function of the number x of whistles he sells per month, according to $p = -0.08x + 12$. What is the maximum revenue Elof can make per month from the sale of his whistles, and how many must he make to achieve this maximum?

© Maroš Markovic/ShutterStock, Inc.

⚙ Solution

The revenue R realized by Elof is given by the product of the number x of whistles that he sells each month and the price p per whistle. We make R a function of x.

$$R(x) = xp$$
$$= x(-0.08x + 12)$$
$$= -0.08x^2 + 12x$$

This is a quadratic function. Because x can only have integer values, the parabola we draw will be an extended graph. The vertex occurs at

$$x = \frac{-b}{2a} = \frac{-12}{-0.16} = 75 \quad \text{and} \quad y = R(75) = -0.08(75)^2 + 12(75) = 450.$$

We equate the factored version of $R(x)$ to 0 to find the zeroes of the function.

$$x(-0.08x + 12) = 0 \quad \text{for } x = 0 \text{ and } x = \frac{12}{0.08} = 150$$

The extended graph appears in **Figure 1.4.12**. If Elof carves and sells 75 whistles each month, he will make a maximum revenue of \$450. If he makes more than 75 whistles per month, his revenue will decrease. In fact, if he makes 150 whistles, his revenue will drop to nothing!

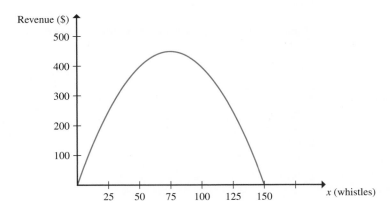

FIGURE 1.4.12   Elof's revenue as a function of the number of whistles.

The previous example illustrates one the basic concepts of conventional economic theory known as the *Law of Demand*. The price of a product decreases as the quantity that consumers demand (or may purchase at that price) increases. Note that this matches our intuitive observation that as price increases, the demand decreases. So this law may be stated thus:

Demanded quantity is a decreasing function of the unit price.

A related relationship is the *Law of Supply*. This embodies the principle that the manufacturers of a particular item will tend to devote more resources to its production as the price increases. This is concisely stated as:

Supplied quantity is an increasing function of the unit price.

By graphing both the demand curve and the supply curve for the same product on the same set of axes, we can locate the point of intersection (p_0, q_0) of the two graphs known as the *equilibrium point*. Because this identifies the price at which the demand is equal to the supply, the price p_0 is called the **equilibrium price**, and the quantity q_0 is called the **equilibrium quantity**. These concepts are illustrated in **Figure 1.4.13**. Note that for any unit price greater than p_0, the supply exceeds the demand resulting in a *surplus* of the product. On the other hand, if the unit price is less than p_0, the supply is less than the demand creating a situation called a product *shortage*.

equilibrium price
The price for which the supply quantity is equal to the demand quantity.

equilibrium quantity The quantity attained by both the supply and demand functions at the equilibrium price.

We see that graphs are capable of conveying large amounts of information quickly and succinctly. They are indispensable tools in modern society for efficiently charting relationships between varying quantities. Mastery of the ability to make an accurate analysis of a variety of different types of graphs improves your ability to make decisions concerning your job, your home, and your life.

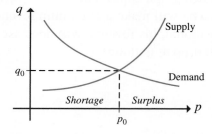

FIGURE 1.4.13   Supply and demand curves graphing quantity q as a function of price p.

Name _____

Exercise Set **1.4** ⬜⬜⬜←—————————

Use each of the following graphs to determine:

(a) *the domain and range*
(b) *over what intervals the function increases or decreases*
(c) *maximum or minimum values and where they occur*
(d) *the zeroes of the function*

1.
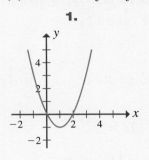

2.
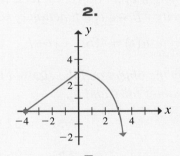

3.
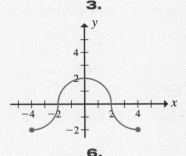

4.
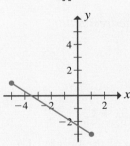

5.
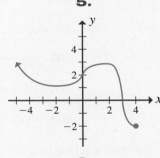

6.
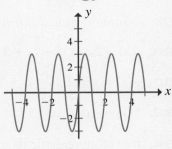

7.
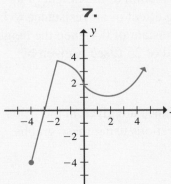

8.
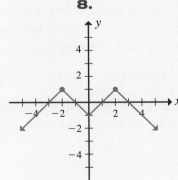

9.
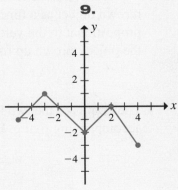

10. Figure 1.4.11 gives the annual precipitation as a function of latitude. Is the annual precipitation increasing or decreasing between 0° and 30°S? Above what northern latitude is the precipitation less than 50 cm?

Graph the following functions. In each case, use the graph to locate any maxima, minima, or zeroes, and determine where the function increases or decreases.

11. $f(x) = 2x + 3$ over $[-2, 1.5]$

12. $f(x) = 8x - 5$ over $[-3, 6.5]$

13. $g(t) = -1.6t + 7$ over $[2, 3]$

14. $h(x) = -2.5x + 4$ over $[1.3, 5]$

15. $f(x) = -x^2 + 4$

16. $g(x) = x^2 + 1$

17. $f(x) = -x^2 - 2x$

18. $f(t) = t^2 + 4t$

19. $h(x) = x^2 - 5x + 7$

20. $h(x) = 3x^2 - 4x + 2$

21. $g(t) = -t^2 - 2t + 8$

22. $f(x) = 2x^2 - 9x - 5$

23. $h(t) = 3t^2 - 5t - 2$

In the absence of air resistance, the height of an object (in feet) thrown from a cliff h_0 feet high having an initial vertical component of velocity v ft/sec is a function of time t (in sec) given by

$$h(t) = -16t^2 + vt + h_0.$$

Find the maximum height attained by the object and the elapsed time it takes for the object to strike the ground given each of the following values for v and h_0.

24. $v = 16$ ft/sec; $h_0 = 96$ ft

25. $v = 32$ ft/sec; $h_0 = 128$ ft

26. $v = 48$ ft/sec; $h_0 = 160$ ft

27. $v = 96$ ft/sec; $h_0 = 112$ ft

Use a graphing calculator to estimate any maxima, minima, or zeroes to the nearest tenth.

28. $f(x) = 2x^3 - x^2 - 2x + 3$

29. $f(x) = x^4 - 2x^2 + 3$

30. $g(x) = x^4 + 3x^2 - 4x + 1$

31. $g(x) = 2x^3 - 5x^2 + 2$

32. $h(x) = -x^3 + 2x - 1$

33. $h(x) = -2x^5 + 5x^3 - 3x + 1$

34. It is more realistic to consider air resistance when obtaining the equation of the height of a thrown object as a function of time. If we assume the decelerating effect of air resistance to be proportional to the velocity of the object with a proportionality constant of 0.4, then the height of the object thrown up from the ground ($h_0 = 0$) with an initial speed of 25 ft/sec is given by

$$h(t) = -80t - 262.5e^{-0.4t} + 262.5.$$

The graphs of the two height functions with and without air resistance are given in the accompanying figure. Estimate the responses to the following questions using these graphs.

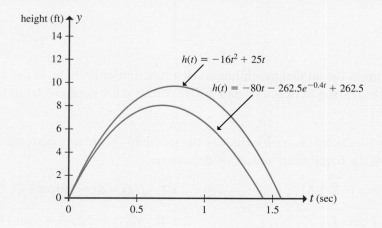

(a) What are the maximum heights attained with and without the air resistance assumption? Which assumption yields the bigger maximum?

(b) At what times are these maximum heights achieved?

(c) Under which assumption is the object in the air for the longest amount of time? What amount is that?

(d) For what interval of time is $h(t) = -16t^2 + 25t$ still increasing while $h(t) = -80t - 262.5e^{-0.4t} + 262.5$ is decreasing?

35. Assume the pressure under water exerted by the ocean changes linearly with the depth. If the pressure at sea level is 1 atm (atmospheres) and the pressure at 100 ft under water is 4 atm, find the water pressure as a function of depth. Use it to compute the maximum pressure that a submarine designed to dive down to 1,500 ft would have to endure.

36. Fahrad has learned that the monthly revenue R that he makes from his donut shop is a function of the number x of donuts he bakes each day according to

$$R(x) = -0.02x^2 + 9x,$$

where R is measured in dollars. Graph the function.

37. Use the graph of the function in the previous exercise to find the maximum revenue and the number of donuts that Fahrad must bake each day in order to achieve that maximum.

38. The *photosphere* is the name given to that layer of the sun that is visible to the human eye. At the photosphere, the temperature is about 6,000°K. Up to about 400 km above the photosphere, the temperature (in degrees Kelvin) decreases according to $T(x) = 6,000 - 5x$, where $x =$ height above the photosphere in kilometers. What is the minimum value of the temperature over the domain [0, 400]?

39. A land developer wishes to form a rectangular lot having the greatest possible area and such that the sum of the boundaries is equal to 600 ft. What are the length, width, and area of this rectangle? (*Hint:* If x is the length, then $300 - x$ must be the width.)

40. Whoopee Toys is marketing an action figure called the Rodent Warlock. The unit price has been determined to be $p = 6 - 0.25x$ dollars, where x is the number of warlocks to be sold. Find the maximum possible revenue and the number of warlocks that Whoopee Toys should produce to achieve that maximum. (*Hint:* See Example 6 in this section.)

41. The accompanying figure gives the supply and demand curves for (hypothetical) bushels of soyoats. Each curve is a graph of quantity (bushels per week) as a function of price (dollars).

(a) Which function is increasing? Decreasing?

(b) What are the equilibrium price and quantity?

(c) Over what interval of prices will there be a surplus of soyoats? Over what interval of prices will there be a shortage of soyoats?

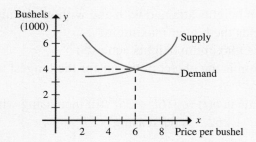

42. Steven G. Monk is a mathematician at the University of Washington who researches how students learn mathematical concepts. In an experiment, he presented the accompanying graph to a student giving the *velocities* of two cars (Red and Blue) as functions of time. He asked the question, "Between 5 and 15 seconds, is car Blue catching up to car Red, or are they getting farther apart?" How would you answer that question? Justify your answer.

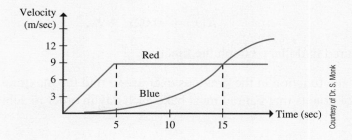

43. The intensity of radiation from a heated object depends on the temperature of the object and the wavelength of light being emitted. The accompanying graph shows the intensity as a function of wavelength (μm) for several temperatures in degrees Kelvin (1 μm = 10^{-6} meter).

(a) For each temperature, estimate the wavelength for which the object radiates at maximum intensity, usually denoted by λ_{max}.

(b) Is λ_{max} an increasing or a decreasing function of temperature? (This fact is known as Wein's Law.)

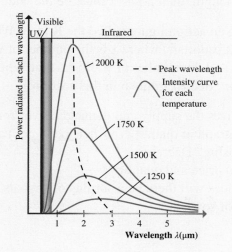

44. The accompanying graph represents the profitability of a product at different stages of its life cycle. Placing a product on this timeline suggests certain strategies for keeping sales high. For example, if sales sag even though the product is thought to be in a growth period, then action may be taken such as wider distribution or discounts for a large volume of sales. What do you think might be the units of time for a product such as a car? A type of sneaker? Seasonal clothes?

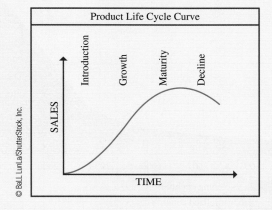

45. An *oscillator* is any system that exhibits periodic behavior. Oscillators are sudied by biologists since they are found extensively in living organisms: pacemaker cells in the heart, insulin-secreting cells in the pancreas, and neural networks in the brain. The pendulum mentioned in the first section is another example of an oscillator. The two graphs labeled 1 and 2 on the following axes represent the displacement from rest and the velocity of the associated moving pendulum as functions of time. Which graph corresponds to displacement? to velocity? (Consider the vertical axis to represent either distance or velocity units.)

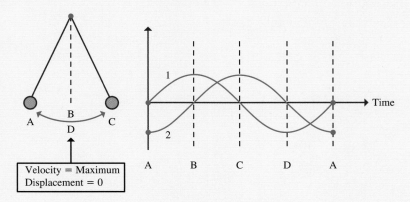

46. According to the accompanying graph, about what year will the world population reach 8 billion? In what year will it reach 10 billion? In 1950, the developing countries constituted about what percentage of the total? About what percentage is predicted for 2050?

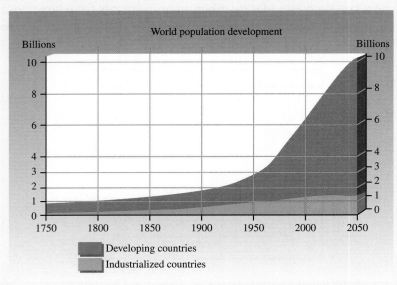

[**Source:** Reproduced from: Philippe Rekacewicz, UNEP/GRID-Arendal.]

Name _____

use the extra space
to show your work

Chapter Review Test **1** ⬜⬜⬜ ⟵

1. Suppose we wish to assign to each consonant in the alphabet the closest vowel. For example, h → i and q → o. Would this be a function? Why or why not?

2. Assume that everyone in a room has one of the distinct eye-colors blue, green, brown, or hazel. If you wish to define a function between the set of these eye-colors and the set of people in this room, which set must be the domain and which set must be the range?

3. The probability of heart disease of a person is an increasing function of the average amount of fat consumed daily by that person. Explain what this means.

4. The French mathematician-soldier credited with introducing the concept of a graph by using a rectangular coordinate system was _____.

5. The gravitational acceleration a imparted by a body of mass M to any object is inversely proportional to the square of the distance r of the object from the center of the mass. In symbols,

$$a = \frac{GM}{r^2},$$

where G is the gravitational constant. Is a an increasing or a decreasing function of r?

←

6. Given the function $f(x) = 350 + 6.2\sqrt{77 - x}$, compute $f(28), f(52)$, and $f(70)$ to the nearest tenth. Is f an increasing or a decreasing function (for $x \leq 77$)?

7. A milking matching at a local dairy was originally purchased for $7,200. If it depreciates linearly in value according to the following table, find the expression for the value $V(t)$ as a function of time t, and use it to forecast the value of the machine in 10 years.

Time (years)	Value (dollars)
0	7200
1	6900
2	6600
3	6300
4	6000
5	5700

8. Construct a formula for each of the functions that make the assignments indicated in the following tables of values.

(a)

x	$f(x)$
5	11
10	26
15	41
20	56
25	71

(b)

x	$g(x)$
−2	5/16
−1	5/4
0	5
1	20
2	80

9. Which of the following graphs do not represent functions?

(a)   (b)   (c)

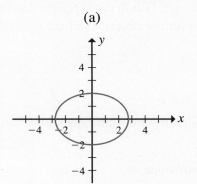 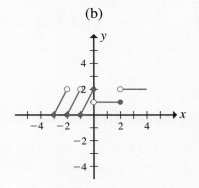 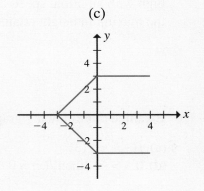

10. Graph each of the following functions.

(a) $f(x) = 2x + 3$ (b) $g(x) = \left(\dfrac{3}{4}\right)2^x$

11. Give the domain, range, absolute maximum value, and absolute minimum value for each of the following functions.

(a)   (b)

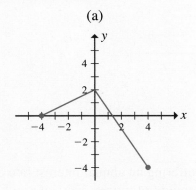 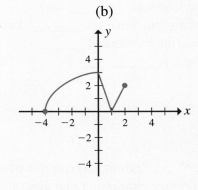

12. The function $h(t) = -16t^2 + 64t + 80$ gives the height (in ft) of an object thrown from a cliff 80 feet high with an initial speed of 64 ft/sec as a function of time t (in sec). Graph the function and find the maximum height attained by the object and the time it takes for the object to hit the ground.

13. Examine the following graph of the function f to answer the following true/false questions.

(a) $f(-3) < f(3)$ (b) $f(0) = 4$ (c) $f(-7) < 0$
(d) If $x > -1$, then $f(x) > 0$ (e) $f(-2)$ is the absolute minimum value of f.

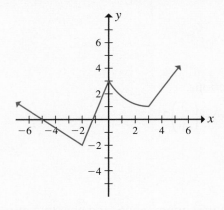

14. Fill in the blanks based on the above graph of the function f.

(a) f is increasing over the intervals _____ and _____.
(b) f has a strictly local minimum value of _____ at $x =$ _____.
(c) f has a strictly local maximum value of _____ at $x =$ _____.
(d) The zeroes of f occur at $x =$ _____.

15. Suppose you deposit $3200 in a savings account at a bank offering an annual interest rate of 4.7% compounded annually. Form the expression that gives the amount of money accrued as a function of time t (in years). What amount of money will you have in that account in 2 years? 5 years?

16. The population of Gotham City is currently 217,000 and is growing at an annual rate of 1.6%. If the growth is modeled by the natural exponential function, what will be the population of the city in 4 years? 8 years?

17. Create a formula for each function described by the following procedures.

(a) 1. Pick any number.
    2. Add 3 to the number and then square it.
    3. Multiply the result by 6.

(b) 1. Pick any number.
    2. Multiply the number by 7 and then subtract 4.
    3. Take the cube root of the result.

Chapter Objectives  ← check off when you've completed an objective

- ☐ Compute the future value, using a simple interest scheme.
- ☐ Compute the future value, using a compound interest scheme.
- ☐ Determine the effective yield of an investment.
- ☐ Compute the future value of a simple ordinary annuity.
- ☐ Determine the rent payment needed for a sinking fund.
- ☐ Compare annuities having different parameters.
- ☐ Compute the present value of an annuity.
- ☐ Determine an amortization schedule for a loan.
- ☐ Compute the unpaid balance remaining for a loan.

2

Functions of Personal Finance

2.1 Interest and Effective Yield

2.2 Annuities

2.3 Amortization of Loans

As a novelist, I tell stories and people give me money. Then financial planners tell me stories and I give them money.

—Martin Cruz Smith

As much fun as it is to learn of the many ways that mathematical functions appear in the world outside our daily lives, it is no less important to know about those functions that can assist us in our matter-of-fact financial decisions. Loans, savings accounts, certificates of deposit, annuities, amortizations, and mortgages are the topics of this chapter. Obviously, in addition to setting off on lofty explorations for knowledge and connections, we must also pay the bills. In fact, we might reasonably argue that the toils of daily life carry higher priority than more aesthetic quests. Educated decisions about our personal finances can greatly affect the quality of our lives. After all, we cannot search for truth if an unpaid electric bill leaves us in the dark!

People and business enterprises both invest and borrow sums of money. Each of the terms *investment* and *loan* simply depends on your end of the money transaction. The receiver of the money views it as a loan, whereas the lender views it as an investment. The key characteristic important to both parties is the *time value of the money*. In either case, the original amount is called the **principal** or **present value**, and it is eventually returned to the lender along with a profit charged for the use of the money. This profit, called the **interest**, is a fee for the use of the principal. The amount of interest, of course, depends on the **interest rate** and how much time has elapsed. All interest rates in this chapter will be given as annual percentage rates. The accumulation of the present value plus the interest at some point in the future is called the **future value**.

principal The original amount of money deposited in an investment or borrowed from a lender.

present value The original or current amount of money in an interest-bearing account or annuity.

interest An amount of money earned in an investment or a fee charged for the lending of money.

2.1 Interest and Effective Yield

interest rate The percentage to be applied to the accrued amount in an investment, or the remaining balance in a loan when determining the interest.

future value The amount of an annuity or some current sum of money that will accumulate with interest over some future period of time.

simple interest The interest earned on the principal P at an annual rate r over t years and given by Prt.

term Time interval over which the parameters of a loan or an annuity are in effect.

The first item on our agenda for the study of finance is the notion of **simple interest**. The simple interest earned (or owed) on a principal of P dollars invested (or borrowed) at an annual interest rate r for a time of t years is defined to be

$$I = Prt.$$

The total sum of money, S, accumulated (or owed) is therefore just

$$S = P + I$$
$$= P + Prt$$
$$= P(1 + rt).$$

Note that for a fixed P and rate r, the sum S is a *linear* function of time t. If we think of P as the present value, then S is the future value of P in t years. If P happens to be a loan amount, then the future value is the total amount to be repaid over the **term** t of the loan. Also, we see from this equation that if we are given the future value S at some time t in the future, then we can find the present value of the sum S by

$$P = \frac{S}{1 + rt}.$$

? Example 1

Thomas borrows $3,800 at a 5% interest rate under a simple interest scheme. Find the simple interest and the future value of the loan after (a) 3 years, (b) 9 months, and (c) 140 days.

⚙ Solution

(a) Here $P = 3,800$, $r = 0.05$, and $t = 3$. So the simple interest is $I = 3,800(0.05)(3) = \$570$. To find the future value of the loan, either we can add 570 to the principal to get

$$S = 3,800 + 570$$
$$= \$4,370,$$

or we can compute

$$S = 3,800(1 + 0.05 \cdot 3)$$
$$= 3,800(1.15)$$
$$= \$4,370.$$

(b) Although different months can be different lengths of time, we will assume that every month is equal to $\frac{1}{12}$ of a year. Therefore, $t = \frac{9}{12} = 0.75$, and so the simple interest and future value are, respectively,

$$I = 3,800(0.05)(0.75) = \$142.50$$
and
$$S = 3,800(1 + 0.05 \cdot 0.75)$$
$$= 3,800(1.0375) = \$3,942.50.$$

(c) Here $t = \frac{140}{365}$ of a year. So,

$$I = 3,800(0.05)\left(\frac{140}{365}\right) = \$72.88$$
and
$$S = 3,800\left(1 + 0.05 \cdot \frac{140}{365}\right)$$
$$= 3,800(1.0191781) = \$3,872.88. \quad ◆$$

? Example 2

How much money must be invested now under a simple interest scheme at 6.5% to achieve a future value of $19,200 after 18 months? What if the rate were 7.5%?

Solution

We must find the present value of $19,200 in 1.5 years. For $r = 0.065$,

$$P = \frac{19,200}{1 + (0.065)(1.5)}$$

$$= \$17,494.31.$$

For $r = 0.075$,

$$P = \frac{19,200}{1 + (0.075)(1.5)}$$

$$= \$17,258.43. \quad \blacklozenge$$

One common example of simple interest involves the use of credit cards. The monthly fee that most credit card companies assess an individual for their services is computed as simple interest on the average daily balance for the previous month.

Example 3

Isabella's balance on her Fast Credit card from last month is $200. In the current 30-day period, she charged a $27 dinner on the 10th day, a $43 pair of slacks on the 16th day, and $86 worth of groceries on the 21st day. If the annual interest rate is 9%, what is her finance charge for the month?

Solution

The average daily balance is computed by first multiplying each balance by the number of days it remained at that amount. It was $200 for 10 days; it increased to $227, at which it remained for 6 days; and so on. By summing these products and dividing by 30, we get the average daily balance. This is the principal used to compute the simple interest, which is the finance charge.

$$P = \frac{10 \cdot 200 + 6 \cdot 227 + 5 \cdot 270 + 9 \cdot 356}{30} = \$263.87$$

So,

$$I = (263.87)(0.09)\left(\frac{30}{365}\right) = \$1.95. \quad \blacklozenge$$

compound interest Interest that is repeatedly computed at the end of regular time intervals and added to the accrued amount.

interest period Period of time over which the interest rate is applied.

Far more common in the world of financial transactions is the concept of **compound interest** rather than simple interest. In an investment using simple interest, interest is never added to the account. In contrast, the *compounding* of interest means that the interest is repeatedly computed at the end of an interval of time called the **interest period**, and the interest is added to the total amount. As each period passes, the growth of the account benefits from new interest based on not only the original principal, but also the previously added interest.

For instance, let $A(t)$ be the functional notation for the future value of P after t years of annual compounding. After 1 year, the amount of interest rP is computed and added to the original principal P. So,

$$A(1) = P + rP = P(1+r).$$

Because $A(1)$ is the amount in the account at the beginning of the second year, at the end of the second year we have

$$A(2) = A(1)(1+r)$$
$$= P(1+r)(1+r)$$
$$= P(1+r)^2.$$

Similarly, after 3 years,

$$A(3) = A(2)(1+r)$$
$$= P(1+r)^2(1+r)$$
$$= P(1+r)^3.$$

We can generalize this result to obtain the future value $A(t)$ of an initial principal P compounded annually at an annual rate r after a time t (in years). For a fixed P and r, this is often expressed with functional notation:

$$A(t) = P(1+r)^t.$$

We note that this is an exponential function of t rather than the linear function that defines growth resulting from simple interest. The dramatic difference this makes in the rate of growth is shown by the graph in **Figure 2.1.1**. This displays the future values of $1,000 at a rate of 8% according to the two different interest schemes.

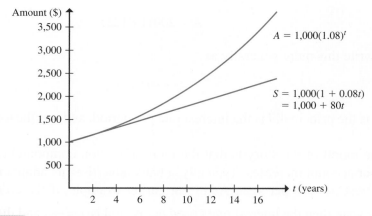

FIGURE 2.1.1   The graph of the future value of $1,000 at 8% interest using simple and compound interest.

? Example 4

Determine the future value of $1,000 in a savings account that compounds annually at a rate of 8% after 5 years. Compare this to the amount acquired if only simple interest is computed.

⚙ Solution

Here $P = 1,000$ and $r = 0.08$. Therefore, the appropriate function to use for annual compounding is $A(t) = 1,000(1.08)^t$. For $t = 5$, we get

$$A(5) = 1,000(1.08)^5 = \$1,469.33.$$

This is significantly greater than an amount based on simple interest. In that case, the amount in 5 years would be only

$$S = 1,000(1 + 0.08 \cdot 5) = \$1,400. \quad \blacklozenge$$

Our expression for annual compounding needs only minor modification to conform to the modern practice of compounding more than once a year. As an example, let $P = \$200$ and $r = 5\%$, and suppose the compounding is done quarterly (every 3 months). Then the applicable interest rate i to use for each quarter is just $i = \dfrac{0.05}{4} = 0.0125$. Although this is a lesser rate, after 1 year, interest will have been computed and added to the account four times. This means that the amount will increase to

$$200(1.0125)^4 = \$210.19.$$

At the end of 2 years, eight quarters will have passed, and so the amount is up to

$$200(1.0125)^8 = \$220.90.$$

Interest will be added 12 times during an interval of 3 years, 16 times over 4 years, 18 times over 4.5 years, and so on. We see that the future value A of this deposit after n quarters is given by

$$A = 200(1.0125)^n.$$

We can write this more generally as

$$A = P(1 + i)^n,$$

where P is the principal, i is the interest rate per period, and n is the total number of interest periods.

The moral of the story is that the more often you compound (at the same rate), the faster your account increases. Typically, a bank advertises its annual rate r and the number m of interest periods per year. As in our earlier example, if the compounding is done m times per year, then the interest rate i used per period is $i = \dfrac{r}{m}$, and the number n of interest periods after t years is $n = mt$. Therefore, as a practical matter, it is useful to think of the future value $A(t)$ of the principal P to be a function of time, t, in years. We can then express the above relationship by the following exponential function:

$$A(t) = P\left(1 + \frac{r}{m}\right)^{mt}.$$

? Example 5

A bank offers a $6\frac{1}{8}\%$ interest rate with monthly compounding.

(a) Find the future value of $2,500 after 3 years; 5 years, 6 months; and 8 years, 9 months.

(b) If you wish to accrue $7,000 in 4 years, how much money must you deposit now?

⚙ Solution

(a) Here $r = 0.06125$, $m = 12$, and the interest rate per period is $i = \dfrac{0.06125}{12} = 0.0051041667$. The required function is therefore

$$A(t) = 2,500\left(1 + \frac{0.06125}{12}\right)^{12t}$$

$$= 2,500(1.0051042)^{12t}.$$

So,

$$A(3) = 2,500(1.0051042)^{36} = \$3,002.89,$$

$$A(5.5) = 2,500(1.0051042)^{66} = \$3,498.42,$$

$$A(8.75) = 2,500(1.0051042)^{105} = \$4,266.82.$$

(b) Here we want the present value of the future value of $7,000 for $t = 4$. So we need to find the value of P such that $A(4) = 7,000$.

$$7,000 = P(1.0051041667)^{48}$$

$$= P(1.2768254)$$

$$P = \frac{7,000}{1.2768254}$$

$$= \$5,482.35. \quad \blacklozenge$$

? Example 6

You deposit $830 in a savings account at a bank offering an annual interest rate of 5.25%. What amount of money will have accumulated in that account in 6 years if the compounding is done annually? Quarterly? Monthly? Daily?

 Solution

For compounding annually, $m = 1$. Then

$$A = 830(1 + 0.0525)^6 = \$1{,}128.26.$$

For compounding quarterly, $m = 4$. Then

$$A = 830\left(1 + \frac{0.0525}{4}\right)^{24} = \$1{,}134.99.$$

For compounding monthly, $m = 12$. Then

$$A = 830\left(1 + \frac{0.0525}{12}\right)^{72} = \$1{,}136.53.$$

For compounding daily, $m = 365$. Then

$$A = 830\left(1 + \frac{0.0525}{365}\right)^{2{,}190} = \$1{,}137.29. \quad \blacklozenge$$

In Example 6, it is clear that as the number of interest periods per year increases, so does the amount of money earned. The function $A(t) = P\left(1 + \dfrac{r}{m}\right)^{mt}$ tells us that any principal P increases by the factor $\left(1 + \dfrac{r}{m}\right)^{m}$ in 1 year. For $r = 0.0525$ compounded monthly, for

> **effective rate or yield** The percentage increase of an investment attained in 1 year.

instance, P increases by $\left(1 + \dfrac{0.0525}{12}\right)^{12} = 1.0538$. Because this means that P has increased by 5.38%, we refer to 5.38% as the **effective rate or yield**. We see that the effective yield is given by

$$\left(1 + \frac{r}{m}\right)^{m} - 1$$

and is an increasing function of m. Banks often advertise both the annual interest rate and the effective yield, which we now see is a reflection of the number of interest periods per year.

It is interesting to also note in Example 6 that the amount of increase in A became successively smaller as m got bigger. This is so because the factor $\left(1 + \dfrac{0.0525}{m}\right)^{m}$ does not become infinitely large as m increases, but instead levels off and converges to the limiting value of $e^{0.0525} \approx 1.05390$, where $e \approx 2.71828$. (See **Figure 2.1.2**.) The number e is a very special number that appears in a wide variety of applications, and the exponential function e^x is an important function on your calculator. (See Exercise 27 for an example using this function.)

Compounding	m	$\left(1+\dfrac{0.0525}{m}\right)^m$	Effective Yield
Annually	1	1.0525	5.25%
Semi-annually	2	1.053189063	5.32
Quarterly	4	1.053542667	5.35
Monthly	12	1.053781887	5.38
Weekly	52	1.05387465	5.39
Daily	365	1.053898583	5.39

FIGURE 2.1.2   The effective yield as a function of m.

Example 7

Are the effective yields given in the advertisement (as annual yields) in **Figure 2.1.3** accurate?

**These Certificate of Deposit Rates
Available Through December 31!**

Rate	Yield	Term
8.25%	8.59%	12 months
9.00%	9.41%	30 months
10.00%	10.51%	60 months

- Fixed rate
- Terms from 3 months to 10 years
- $1,000 minimum investment
- Interest compounded daily

FIGURE 2.1.3   Typical bank advertisement showing annual rate and effective yield.

Solution

In the bulleted list, it is stated that compounding is done daily. For the 12-month term account, the rate $r = 0.0825$. Thus,

$$\text{Effective yield} = \left(1+\frac{0.0825}{365}\right)^{365} - 1$$
$$= 1.0859885 - 1$$
$$= 0.0859885$$
$$= 8.60\%.$$

(Note that this bank decided not to round up.)

For $r = 0.09$,

$$\text{Effective yield} = \left(1 + \frac{0.09}{365}\right)^{365} - 1$$
$$= 9.42\%.$$

For $r = 0.10$,

$$\text{Effective yield} = \left(1 + \frac{0.10}{365}\right)^{365} - 1$$
$$= 10.52\%. \quad \blacklozenge$$

Name _____

Exercise Set **2.1** ◯◯◯←────────────────────

All given interest rates are annual rates. Assume that every month is equal to $\frac{1}{12}$ of a year and that 1 year is equal to 365 days.

1. Find the simple interest owed for each of the following loans:
(a) $4,300 borrowed at 5.7% for 3 years.
(b) $65,000 borrowed at 7.25% for 15 months.
(c) $22,500 borrowed at 9% for 100 days.

2. Find the simple interest earned in each of the following investments:
(a) A principal of $17,500 earning 4.5% for 2 years.
(b) A principal of $7,800 earning 6% for 18 months.
(c) A principal of $133,000 earning $9\frac{5}{8}$% for 120 days.

3. Assuming a simple interest scheme, find the future value of each of the loans in Exercise 1.

4. Assuming a simple interest scheme, find the future value of each of the investments given in Exercise 2.

5. Assuming a simple interest scheme, what is the present value of each of the following future values?
(a) $8,900 after 2.5 years at 7.3% interest.
(b) $54,300 after 1 year 4 months at $8\frac{3}{8}$% interest.
(c) $125,000 after 195 days at 4.9% interest.

6. How much money must be invested now under a simple interest scheme at 4.25% to accumulate $9,500 in 3.5 years?

7. Your credit card balance from last month is $340. In the current 30-day month, you charged $45 for shoes on the 6th day, $32 for a round of golf on the 19th day, and $172 worth of camping equipment on the 25th day. If the annual interest rate is 12%, what is your finance charge for the month?

8. The balance on Albert's Wonder Gas credit card from last month is $122. In the current 31-day month, he charged $20 worth of gas every 7th day. If the annual interest rate is 18%, what is his finance charge for the month?

9. Find the future value of $8,700 earning 5½% compounded quarterly after
(a) 9 months.
(b) 3 years.
(c) 6.5 years.
(d) 11 years 3 months.

10. Find the future value of $15,000 after 7 years for each of the following compounding schemes:
(a) Compounded annually at 6.3%.
(b) Compounded weekly at 7.8%.
(c) Compounded bimonthly (every 2 months) at 8.1%.
(d) Compounded semi-annually (twice a year) at 11.2%.

11. Suppose you deposit $1,750 in a savings account at a bank offering an annual interest rate of 3.75%. What amount of money will you have in that account in 10 years if the compounding is done
(a) annually?
(b) monthly?
(c) daily?
(d) How much more money do you earn under daily compounding than under annual compounding?

12. Agatha needs to have saved $35,000 at the end of 6 years. How much would she need to deposit now in an account that compounds quarterly if the interest rate is one of the following?
(a) 8.2%
(b) 7.1%
(c) 6.3%
(d) 3.9%

13. What is the present value of each of the following future values?
(a) $8,500 accrued after 3 years, compounded weekly at 4.9%.
(b) $35,700 accrued after 5 years, compounded quarterly at 5.7%.
(c) $118,900 accrued after 10 years, compounded daily at 4⅜%.

14. Compute the effective yield for savings options with the following parameters.
(a) 8½% compounded bimonthly (every 2 months).
(b) 5% compounded monthly.
(c) 7.8% compounded quarterly.
(d) 4.6% compounded daily.

15. Compute the effective yield of an investment earning 6.3% and compounded according to each of the following schedules:
(a) Annually
(b) Quarterly
(c) Monthly
(d) Daily

16. Billy Joe is going to sign a contract with a baseball team and is offered a choice of either a bonus of $250,000 to be paid now or a deferred bonus of $400,000 to be paid in 15 years. Assuming he could deposit his bonus in a fund that compounds daily at an interest rate of 4%, which should he choose?

17. You are shopping for a bank that offers the most favorable yields on certificates of deposit. The Bank of Oz offers a 6% rate, compounded daily, while Auntie Em's bank offers a 6.1% rate, compounded quarterly. What is the effective yield of each bank? Which bank would you choose?

18. Ed's bank offers certificates of deposit at a rate of 7.5%, compounded daily, while Denise's bank offers them at a rate of 7.75%, but only compounded quarterly. What is the effective yield offered at each bank?

19. You know that you will need $12,500 in 10 years to help your child attend college. What principal must you invest in a savings account that compounds quarterly at an annual interest rate of 5.5%? How much do you need to invest if the compounding is done monthly? Daily?

20. Leif wants to save $150,000 in 8 years to buy a yacht and sail around the world. What principal must he invest in a savings account that compounds quarterly at 4.3%? How much does he need to invest if the rate is 4.5% and the compounding is daily?

21. The Jensens are saving for a car as a graduation present for their daughter Tiffany. They feel they will need $15,000 in 4 years. They have found a money market account advertising a 7.25% interest rate, compounded monthly. How much money will they still need if they deposit $8,000 now?

22. Hector has inherited a sum of money, part of which he wishes to set aside to buy a new boat in 5 years for $23,000. If he can earn 8.5% per year in his investment strategies, how much money does he need to invest now?

23. Emilio deposits $5,000 in an IRA with an effective yield of 6.3%. If he adds an additional $5,000 after 3 years, how much money will he have 6 years after his original deposit?

24. Gabriella deposits $2,500 in an IRA with an effective yield of 5.9%. If she adds $2,500 more after 5 years, how much money will she have 10 years after her original deposit?

25. Jochula deposits $18,500 in an IRA with an effective yield of 7.25%. If he adds $12,000 more after 4 years, how much money will he have 10 years after his original deposit?

26. Samantha has an option to receive either $35,000 5 years from now or $27,000 right now. What interest rate, compounded monthly, would make the second option preferable?

27. We saw that for any interest rate r (as a decimal), the factor $\left(1+\dfrac{r}{m}\right)^m$ gets closer to e^r as m gets larger. (See Figure 2.1.2.) As m gets larger, the length of the interest period becomes infinitesimally small and the compounding is said to occur *continuously*. Using a result obtained in calculus, we may then replace $\left(1+\dfrac{r}{m}\right)^m$ with e^r in the future value function to get

$$A(t) = Pe^{rt}$$

continuous compounding
Compounding of money according to the function $A(t) = Pe^{rt}$.

for a principal P after t years of **continuous compounding**. For example, the future value of $5,600 compounded continuously at 7.8% for 3.5 years is

$$A(3.5) = 5,600e^{0.078(3.5)}$$
$$= 5,600e^{0.273}$$
$$= \$7,357.84.$$

Find the future values for each of the following investments, using continuous compounding:

(a) $3,500 at 4.5% for 4 years.
(b) $128,000 at 6.1% for 10 years.
(c) $56,500 at $3\frac{1}{4}$% for 23 years.

28. In this advertisement, the annual percentage yield (which is the same as the effective yield), but *not* the annual interest rate, is given. Describe a procedure by which you could find the annual rate.

First-Rate CDs!

Term of CD	APY
1 Year	2.00%
2 Years	2.25%
3 Years	2.50%
4 Years	3.00%
5 Years	3.25%

Open a CD account

No penalty for early withdrawl

2.2 Annuities

An **annuity** is a sequence of equal payments paid or received at regular intervals of time. It is a popular financial strategy used by people from a wide spectrum of income brackets because it utilizes the power of compound interest applied over a long-term period. Knowing the functions that govern the operation of annuities allows you to make judicious choices about planning for your future financial needs. Two typical examples follow.

annuity A sequence of equal payments paid or received at regular intervals of time.

Case 1 You want to accrue some desired sum of money in the future by making small deposits on a regular basis. The sum F is called the future value of the annuity. For instance, upon the birth of your daughter, you might want to deposit some fixed amount of money R each month for the next 18 years at 4.75% annual interest in order to accumulate $40,000 for her college education. You want to determine the value of R.

Case 2 You want to make one large initial deposit right now for the purpose of receiving equal regular payments over an extended future period. For instance, you might want to receive $3,500 at the end of every quarter for the next 6 years, and so you wish to know the present value P that you need to invest now at 8% annual interest, compounded quarterly.

payment period or interval Period of time between rent or annuity payments.

rent Payment made or received in an annuity.

simple annuity Annuity in which the interest period is equal to the payment period.

ordinary annuity Annuity in which the rent is paid at the end of the period.

annuity due Annuity in which the rent is paid at the beginning of the period.

The time period between payments made or received in any annuity is known as the **payment period or interval**. Each payment is called the **rent**. The term is the length of time from the beginning of the annuity until it has expired. In Case 1, the payment period is 1 month, the rent is the unknown R, and the term is 18 years. In Case 2, the period is 1 quarter, the rent is $3,500, and the term is 6 years. In this section, we learn how to find the future value of an annuity and how to solve for R in Case 1. (See Exercise 15.) We defer the solution for P in Case 2 until the next section.

The growth of the money in any annuity depends on the compound interest scheme. A **simple annuity** has an interest period equal in length to the payment period. We shall work exclusively with this type. So, for example, if payments are made quarterly, then compounding is also done quarterly. Additionally, accountants make a distinction between an annuity in which the payment (rent) is made at the end of the period and an annuity in which payment is made at the beginning. The former is known as an **ordinary annuity** and the latter as **annuity due**. We restrict our study to the ordinary annuity, which, as the name suggests, is more common.

In the last section, we learned that if r is the annual interest rate and m is the number of interest periods per year, then the interest rate i per period must be $i = \dfrac{r}{m}$, and the number of periods n after t years must be $n = mt$. In the case of an annuity, the rent R is added to the account every period. The future value of the annuity after n periods can therefore be thought of as a sum of n separate amounts, each of which is the result of a payment that has been compounding for a different amount of time and therefore has increased by a different power of $(1 + i)$. The vertical time line in **Figure 2.2.1** shows how that sum (shown horizontally) is computed for several payment periods. The rent R that is added to the account

$$
\begin{array}{lllllll}
0 & \downarrow \\
1 & R & \downarrow \\
2 & R(1+i) & +R & \downarrow \\
3 & R(1+i)^2 & +R(1+i) & +R & \downarrow \\
4 & R(1+i)^3 & +R(1+i)^2 & +R(1+i) & +R & \downarrow \\
5 & R(1+i)^4 & +R(1+i)^3 & +R(1+i)^2 & +R(1+i) & +R \\
\vdots & \vdots & \vdots & \vdots & \vdots & \vdots & \downarrow \\
n & R(1+i)^{n-1} & +R(1+i)^{n-2}+ & \cdots & + & \cdots & +R(1+i)+R
\end{array}
$$

FIGURE 2.2.1   Computing the future value of an annuity.

each period is indicated by the arrow, and the column under each R shows how the value of that particular payment increases with the passage of each period.

We see from Figure 2.2.1 that we can write the expression for the future value F of an ordinary annuity after n periods as

$$
\begin{aligned}
F &= R + R(1+i) + R(1+i)^2 + R(1+i)^3 \ldots + R(1+i)^{n-2} + R(1+i)^{n-1} \\
&= R\left[1+(1+i)+(1+i)^2 +(1+i)^3 +\ldots+(1+i)^{n-2} +(1+i)^{n-1}\right].
\end{aligned}
$$

The long expression in square brackets on the right-hand side of the second equation is what mathematicians call a *geometric sum*, and it is characterized by the fact that each term in the sum is equal to the preceding term multiplied by some fixed number. The fixed number is called the *common ratio*, and if we represent this ratio by a, then a geometric sum is readily computed by use of this formula:

$$
1+a+a^2+a^3+\ldots+a^{n-2}+a^{n-1} = \frac{a^n-1}{a-1}.
$$

(See Exercise 30 for a derivation of this formula.) For instance, let $a = 3$ and $n = 4$. The formula gives us

$$
\begin{aligned}
1+3+3^2+3^3 &= \frac{3^4-1}{3-1} \\
&= 40.
\end{aligned}
$$

Letting $a = 1 + i$ in the above formula for geometric sum, the long bracketed sum in the expression for future value F can be replaced by $\dfrac{a^n-1}{a-1} = \dfrac{(1+i)^n-1}{1+i-1} = \dfrac{(1+i)^n-1}{i}$, and so we get something more manageable, that is,

$$
F = R\left(\frac{(1+i)^n-1}{i}\right),
$$

where R is the rent paid for n periods at an interest rate i per period.

Recall that when devising a financial plan, we often know the annual interest rate r and the number m of periods per year. As in the last section, we may think of R, r, and m

(and therefore i) as fixed quantities established by the bank (often referred to as *parameters*). By substituting $n = mt$ into the earlier expression, we write the future value as a function of the term t in years:

$$F(t) = R\left(\frac{(1+i)^{mt} - 1}{i}\right).$$

? Example 1

In planning for their retirement, Lucy and Antonio consider setting up a simple ordinary annuity of $50 per month at an annual rate of 3%. Calculate the future value in 15, 20, and 25 years.

⚙ Solution

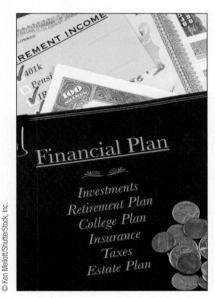

Here $R = 50$, $r = 0.03$, and $m = 12$. Hence, $i = \dfrac{0.03}{12} = 0.0025$, and so we have

$$F(t) = 50\left(\frac{(1.0025)^{12t} - 1}{0.0025}\right).$$

For the desired times, this function yields these future values:

$$F(15) = 50\left(\frac{(1.0025)^{180} - 1}{0.0025}\right) = 50\left(\frac{0.56743172}{0.0025}\right) = \$11{,}348.63,$$

$$F(20) = 50\left(\frac{(1.0025)^{240} - 1}{0.0025}\right) = 50\left(\frac{0.82075500}{0.0025}\right) = \$16{,}415.10,$$

$$F(25) = 50\left(\frac{(1.0025)^{300} - 1}{0.0025}\right) = 50\left(\frac{1.1150196}{0.0025}\right) = \$22{,}300.39. \quad ◆$$

The graph of the function in Example 1 is shown in **Figure 2.2.2** and helps us gain a perspective on the rate of increase of the future value of this particular annuity. The shape of the graph is characteristic of an exponential function.

Of course, our annuity function changes in response to replacing any of the three parameters R, r, or m. Suppose we change the annual interest rate to 6%. Then $i = \dfrac{0.06}{12} = 0.005$, and the new function is

$$F(t) = 50\left(\frac{(1.005)^{12t} - 1}{0.005}\right).$$

Figure 2.2.3 shows the comparison of the graph of this function to the previous one. The graphs make it easy to visualize the impact of the higher interest rate on the accumulation process. Similar comparisons can be made by varying the rent or the payment period or both.

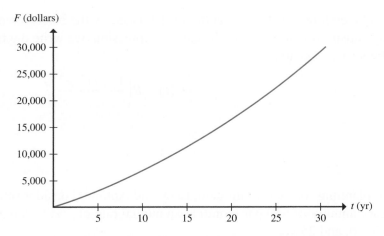

FIGURE 2.2.2 The future value of a simple ordinary annuity with monthly rent of $50 at an annual rate of 3%.

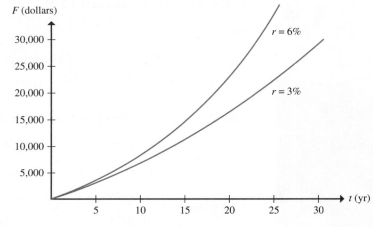

FIGURE 2.2.3 Comparison of the growth of two annuities with two different interest rates.

We are now in a position to solve for the unknown R as was desired in Case 1 at the beginning of the section. In that situation, we know the amount of the future value F that we wish to accrue at some specific time t in the future; given the interest rate and the period of the annuity, we want to determine the necessary rent R to make it happen. This financial structure is typically also known as **sinking fund**—that is, an annuity set up to obtain a specific future value. It is used to pay for some well-known future expense, such as a new car or a child's education, and it uses the same mathematics that we have already derived. We simply solve the relationship for R instead of for F.

sinking fund Annuity established for the purpose of accumulating a specific amount of money in the future.

 ## Example 2

Mr. Kolewski wants to set aside money in a sinking fund at 8.5% annual interest, with quarterly payments, to buy a $35,000 yacht in 10 years. How much rent must he deposit each quarter?

Solution

Here $m = 4$, and the interest per period is $i = \dfrac{0.085}{4} = 0.02125$. Therefore,

$$F(t) = R\left(\frac{(1.02125)^{4t} - 1}{0.02125}\right).$$

Because it is desired that $F(10) = 35,000$,

$$F(10) = R\left(\frac{(1.02125)^{40} - 1}{0.02125}\right)$$

$$35,000 = R(62.066074)$$

$$R = \frac{35,000}{62.066074}$$

$$= \$563.92. \quad \blacklozenge$$

? Example 3

The Jacoby County school board wants to have $750,000 available in 8 years to build a new elementary school. The board considers two possible sinking funds. One can be arranged with semi-annual payments at 6¾% annual interest and the other with quarterly payments at 6⅝%. Which of these funds would cost the board less money annually?

Solution

For the first fund, $m = 2$ and $i = \dfrac{0.0675}{2} = 0.03375$. Because the future value of the annuity must be 750,000 in 8 years, the exponent of 1.03375 must be $n = 2(8) = 16$. We therefore solve the following equation for the rent R.

$$750,000 = R\left(\frac{(1.03375)^{16} - 1}{0.03375}\right)$$

$$750,000 = R(20.763877)$$

$$R = \frac{750,000}{20.763877}$$

$$= \$36,120.42.$$

Because this rent would be paid twice a year, the annual cost is $72,240.84.

In the second option, $m = 4$ and $i = \dfrac{0.06625}{4} = 0.0165625$. This time, the exponent of 1.0165625 must be $n = 4(8) = 32$, and so we solve:

$$750,000 = R\left(\frac{(1.0165625)^{32} - 1}{0.0165625}\right)$$

$$750,000 = R(41.755345)$$

$$R = \frac{750,000}{41.755345}$$

$$= \$17,961.77.$$

Because this rent would be paid four times a year, the annual cost for this option is $71,847.08. The second option is less expensive and therefore preferable. ♦

Name _____

Exercise Set **2.2** ☐☐☐←

All annuities considered in these exercises are simple and ordinary, so the payment period is equal to the interest period and every payment is made at the end of a period. Additionally, every interest rate is an annual rate considered to be constant for the appropriate term.

1. Find the future value of each of these annuities:
    (a) $100 per month at 6% interest with a term of 30 years.
    (b) $125 per quarter at 7% interest with a term of 18 years.
    (c) $350 semi-annually at 9.5% interest with a term of 7 years.

2. Find the future value of each of these annuities:
    (a) $80 per month at 6.5% interest with a term of 20 years.
    (b) $175 every other month at 5% interest with a term of 15 years.
    (c) $110 every 2 weeks at 7.25% interest with a term of 25 years.

3. To help pay for sending their newborn twins to college, Mr. and Mrs. Williams set up an annuity of $200 per month that earns 4.8% interest. What is the future value of this annuity in 18 years?

4. Rita makes monthly deposits of $25 to a Christmas club account that operates as an annuity, earning 4% interest. How much money will Rita have at the end of 1 year?

5. Mondar starts his new job at age 25 and sets up a retirement account that operates as an annuity. If he deposits $1,500 per year in an account paying 7.5% interest, how much will be in the account when he retires at age 65?

6. Lopsang Textiles deposits $12,500 every quarter into an annuity paying 5.5% interest to finance the employee pension fund. What is the future value of the fund in 20 years?

7. An annuity pays 6% interest and requires a rent payment of $75 per month. Determine the future value F of the annuity as a function of time t in years.

8. An annuity pays 8% interest and requires a rent payment of $100 per month. Determine the future value F of the annuity as a function of time t in years.

9. Use the function you computed in Exercise 7 to compute the future value of the annuity in 5, 10, and 20 years. If you have access to a graphing calculator, use it to graph the function on appropriate axes.

10. Use the function you computed in Exercise 8 to compute the future value of the annuity in 5, 10, and 20 years. If you have access to a graphing calculator, use it to graph the function on appropriate axes.

11. Harold can afford to deposit a total of $1,560 per year in an annuity for his retirement. The first annuity he considers is $60 deposited biweekly at 7.25%. The second is $30 deposited weekly at 7%. By evaluating each future value in 1 year, determine which annuity offers the more favorable parameters. If you have access to a graphing calculator, use it to graph both functions on the same coordinate axes.

12. Karen can afford to deposit a total of $1,300 per year in an annuity for her retirement. The first annuity she considers is $50 deposited biweekly at 6¼%. The second is $25 deposited weekly at 6⅛%. By evaluating each future value in 1 year, determine which annuity offers the more favorable parameters. If you have access to a graphing calculator, use it to graph both functions on the same coordinate axes.

13. Find the rent payment required for each of the following sinking funds to yield the given future value:
(a) $100,000 after 20 years of monthly payments at 6%.
(b) $62,000 after 15 years of quarterly payments at 4.5%.
(c) $3,500,000 after 25 years of annual payments at 5%.

14. Find the rent payment required for each of the following sinking funds to yield the given future value:
(a) $650,000 after 17 years of semi-annual payments at 8%.
(b) $1,500,000 after 25 years of quarterly payments at 4.75%.
(c) $83,000 after 5 years of monthly payments at 7.75%.

15. We solve Case 1 from the beginning of the section: Upon the birth of your daughter, how much rent should you deposit each month in a sinking fund earning 4.75% interest to accumulate $40,000 in 18 years for her college education?

16. The city of Belford has a $2 million debt that comes due in 7 years. How much rent must it deposit quarterly in an annuity paying 6.5% interest to accrue this amount?

17. The chief financial officer of Brenner Manufacturing anticipates that $1.6 million will be needed in 20 years to meet the obligations of the employee pension fund. What quarterly rent will be necessary if the fund is an annuity paying 5.5%?

18. Angel has arranged a loan for a new house in which she will be paying a balloon payment of $13,500 in 5 years. To make her payment, what monthly rent must she deposit in a sinking fund paying 4.75%?

19. Eric wants his employer to deduct enough money from his paycheck every 2 weeks to provide him with $150,000 in 20 years. If the funds are deposited into an annuity at 7⅜% interest, what must be the amount of the deductions?

20. To pay off the $4.45 million bond issue in 15 years, the Cramen County government has arranged to make semi-annual rent payments of $85,000 into a sinking fund earning 7.5% interest. Will this fund yield enough money to retire the bonds?

21. The Cardinals wish to save $38,000 for an extensive trip through Europe 10 years from now. They have collected information about several annuities and wish to compare them. Annuity 1 has an annual interest rate of 5.75% and monthly payments. Annuity 2 has a smaller annual interest rate of 5.5%, but the payments are made every week. Which annuity would cost the Cardinals less money annually?

22. Dexter County wishes to accumulate $625,000 in 5 years to finance a new addition to the senior citizen center. Competing lending institutions *X* and *Y* have each offered an annuity with different options. The annuity from *X* earns 9.5% with semi-annual payments. The one from *Y* earns 9.25% with monthly payments. Which annuity should the county choose?

23. In 5 years, Sheryl wants to have a down payment of $27,000 to purchase a small bookstore. She asks two lending institutions about the parameters of a sinking fund to meet her goal. One earns $5\frac{7}{8}\%$, and payments are made every 2 weeks. The other earns 6%, and payments are made every month. Which fund should she choose?

24. Wawchester Coal Corporation needs to invest $437,000 in new mining equipment in 6 years. It compares two sinking funds with the following parameters:
    (a) $r = 7\frac{1}{4}\%$, $m = 12$.
    (b) $r = 7\frac{1}{8}\%$, $m = 26$.

    Which fund is less expensive?

aggregate contribution The total amount paid by the investor in an annuity. It is equal to the sum of the rent payments.

*In an annuity with rent R and m payments per year, the **aggregate contribution** of the investor after t years is Rmt. The total interest earned is the difference between the future value at time t and the aggregate contribution and is therefore equal to F(t) – Rmt.*

25. Find the aggregate contribution and the total interest earned in the annuities described in Exercise 1.

26. Find the aggregate contribution and the total interest earned in the annuities described in Exercise 2.

27. Find the aggregate contribution and the total interest earned in the annuities described in Exercise 13.

28. Find the aggregate contribution and the total interest earned in the annuities described in Exercise 14.

29. A *geometric sum* is the total of *n* terms, $1 + a + a^2 + a^3 + \cdots + a^{n-2} + a^{n-1}$, where each term is *a* times the previous one. The factor *a* is called the *common ratio*. It is true that this sum can be quickly computed by the formula

$$1 + a + a^2 + a^3 + \cdots + a^{n-2} + a^{n-1} = \frac{a^n - 1}{a - 1}.$$

We verify this for $a = 2$ and $n = 6$:

$$1 + 2 + 2^2 + 2^3 + 2^4 + 2^5 = 1 + 2 + 4 + 8 + 16 + 32 = 63 = \frac{2^6 - 1}{2 - 1}.$$

Use the formula to find the following sums. Note that *n* can be found by counting the number of terms.

(a) $1 + 5 + 25 + 125 + 625 + 3{,}125$

(b) $1 + 3 + 9 + 27 + 81 + 243 + 729 + 2{,}187$

(c) $1 + 7 + 7^2 + 7^3 + \cdots + 7^9 + 7^{10}$

(d) $1 + \dfrac{1}{2} + \dfrac{1}{4} + \dfrac{1}{8} + \dfrac{1}{16} + \dfrac{1}{32} + \dfrac{1}{64}$

30. We can derive the previous formula in the following way:

(a) Set $S = 1 + a + a^2 + a^3 + \cdots + a^{n-2} + a^{n-1}$.

(b) Multiply both sides by a and subtract S from aS.

$$aS = a + a^2 + a^3 + \cdots + a^{n-2} + a^{n-1} + a^n,$$

$$S = 1 + a + a^2 + a^3 + \cdots + a^{n-2} + a^{n-1},$$

$$aS - S = a^n - 1.$$

(c) Finish this argument by solving algebraically for S.

2.3 Amortization of Loans

In the last section, we learned how to calculate the annuity payments necessary to accumulate some desired amount of money in the future. We initiate our study of loans by considering the reverse situation. This was exemplified by Case 2, presented in the introduction to the last section:

> **Case 2** You want to make one large initial deposit right now for the purpose of receiving equal regular payments over an extended future period. For instance, you might want to receive $3,500 at the end of every quarter for the next 6 years, and so you wish to know the present value P that you need to invest now at 8% annual interest, compounded quarterly.

This concerns the determination of a lump sum of money P needed to generate a regular sequence of equal payments R back to the investor over some fixed period of time. This sequence is, in effect, an annuity in which the rents are paid by the bank to the investor. As in the last section, we will always assume that the payment period coincides with the interest period and that each payment is made at the end of the period. The amount P is the amount in the present time needed to finance these payments and is therefore called the present value of the annuity. The present value can be thought of as the amount deposited, which, if left untouched for n interest periods at a rate of i per period, would accumulate the same amount as a simple ordinary annuity with rent R based on the same parameters. Therefore, it must be true that

$$P(1+i)^n = R\left(\frac{(1+i)^n - 1}{i}\right).$$

Dividing both sides by $(1 + i)^n$, we arrive at an expression for the present value:

$$P = R\left(\frac{1-(1+i)^{-n}}{i}\right).$$

Note that the exponent used in this calculation is a negative number.

? Example 1

Solve for the present value P in Case 2.

⚙ Solution

Here we have $R = 3,500$, $m = 4$, the annual rate $r = 0.08$, and the term $t = 6$ years. Thus, $i = \dfrac{0.08}{4} = 0.02$, and $n = 4(6) = 24$. We want the present value of the corresponding annuity.

$$P = 3,500 \left(\frac{1 - (1.02)^{-24}}{0.02} \right)$$

$$= 3,500 \left(\frac{0.37827851}{0.02} \right)$$

$$= 3,500 (18.913926)$$

$$= \$66,198.74. \quad \blacklozenge$$

Given the present value P, the same equation can also be used to find the rent payment R.

? Example 2

Rhonda has just won $147,000 in a lottery and wants to use it as a deposit to provide her with a steady monthly income for the next 7 years. How much would she receive each month in an appropriate annuity at 6% interest? At 6½% interest?

⚙ Solution

In this case, we know the present value $P = \$147,000$, $m = 12$, $r = 0.06$, and $t = 7$. Thus, $i = \dfrac{0.06}{12} = 0.005$ and $n = 12(7) = 84$. We wish to compute the monthly payment R.

$$147,000 = R \left(\frac{1 - (1.005)^{-84}}{0.005} \right)$$

$$= R \left(\frac{0.34226521}{0.005} \right)$$

$$= R(68.453042).$$

Hence,

$$R = \frac{147,000}{68.453042}$$

$$= \$2,147.46.$$

When $r = 0.065$ and $i = \dfrac{0.065}{12} = 0.0054166667$, we get

$$147,000 = R \left(\frac{1 - (1.0054166667)^{-84}}{0.0054166667} \right)$$

$$147{,}000 = R\left(\frac{0.36477254}{0.0054166667}\right)$$

$$= R(67.342623).$$

Hence,

$$R = \frac{147{,}000}{67.342623}$$

$$= \$2{,}182.87. \quad \blacklozenge$$

In modern America, few people escape the necessity of borrowing money from a bank or similar institution to purchase a car, a home, or other large expensive item. It is helpful to know that the mathematics we have been exploring is also used in the creation of a schedule of payments for the purpose of repaying a bank loan, called an **amortization**. This schedule consists of a regular series of equal payments by the borrower, each consisting of an amount of interest plus an amount used to reduce the principal. The key fact to know is that this series is basically an annuity, one in which the loan amount (principal) is the present value P and each of the payments is equal to the rent R. Note that the money flow is reversed from the situation in Example 2. The individual is receiving a loan amount instead of making a deposit and is making regular payments instead of receiving them. However, the mathematics is the same.

amortization
A schedule of regular payments for the purpose of repaying a loan.

? Example 3

Francella wants to buy a new car and is going to borrow \$15,000 from a bank at 12% interest with monthly payments over a term of 2 years. What will be her monthly payment, and what will be the total interest paid after 2 years?

⚙ Solution

As before,

$$P = R\left(\frac{1-(1+i)^{-n}}{i}\right).$$

Here $P = \$15{,}000$, $i = \dfrac{0.12}{12} = 0.01$, and $n = 12(2) = 24$. Therefore,

$$15{,}000 = R\left(\frac{1-1.01^{-24}}{0.01}\right)$$

$$= R(21.243387)$$

$$R = \frac{15{,}000}{21.243387}$$

$$= \$706.10.$$

Because the loan will be repaid after 24 payments of $706.10 each, the total amount paid to the bank by Francella is 24(706.10) = $16,946.40. Therefore, the total amount of interest paid is $16,946.40 − $15,000 = $1,946.40. ◆

We now create an amortization schedule of the payments that Francella (from Example 3) will make over the 2-year term of the loan including the breakdown of each payment into interest plus the amount applied to the principal. (See **Figure 2.3.1**.)

Number	Payment	Interest	Applied to Principal	Balance
1	706.10	150.00	556.10	14,443.90
2	706.10	144.44	561.66	13,882.24
3	706.10	138.82	567.28	13,314.96
4	706.10	133.15	572.95	12,742.01
⋮	⋮	⋮	⋮	⋮
22	706.10	20.77	685.33	1,391.35
23	706.10	13.91	692.19	699.16
24	706.10	6.99	699.11	0.05

FIGURE 2.3.1   Amortization schedule: $P = 15,000$, $i = 0.01$, $n = 24$.

We make the following observations of this schedule, which apply to every loan amortization.

1. The periodic payment or rent R is computed first from the given parameters of P, i, and n. This payment remains constant for the duration of the term of the loan.

2. The interest portion of each payment is equal to i times the previous unpaid balance of the loan. So, for example, the interest portion of the second payment in Figure 2.3.1 is $0.01(14,443.90) = \$144.44$; the interest portion of the third payment is $0.01(13,882.24) = \$138.82$, and so on.

3. Each period, the amount applied to the principal is equal to the payment minus the interest.

4. With each successive payment, the interest portion necessarily decreases, and the amount applied to the principal increases.

5. The balance after the last payment may not be exactly zero, only because of the rounding done in previous steps.

mortgage Loan used to purchase real estate.

One of the most important types of loans is used to purchase a home or other real estate. Such a loan is called a **mortgage**, and the amortization of a mortgage can extend over a term of up to 30 years. When you go shopping for a mortgage for a new house, lending institutions will offer you options with different sets of parameters. The ability to analyze these options will assist you in making a judicious choice, one that best meets your financial needs.

? Example 4

The Conrads are considering buying a new house for $143,000. They can afford a $20,000 down payment and have qualified for a 30-year mortgage at 7¾% interest, with monthly payments. Determine the first three lines of the amortization schedule.

⚙ Solution

We first must find the monthly payment R for $P = 123,000$, $i = \dfrac{0.0775}{12} = 0.00645833$, and $n = 12(30) = 360$.

$$123,000 = R\left(\frac{1 - 1.00645833^{-360}}{0.00645833}\right)$$

$$= R(139.584)$$

$$R = \frac{123,000}{139.584}$$

$$= \$881.19.$$

The interest portion of the first payment must be equal to (0.00645833) $(123,000) = \$794.37$. Upon subtracting this from $\$881.19$, we get the amount applied to the principal, $\$86.82$. Each succeeding line is generated by first computing the interest as i times the previous balance and then subtracting this from the payment R to determine the amount applied to the principal.

Number	Payment	Interest	Applied to Principal	Balance
1	881.19	794.37	86.82	122,913.18
2	881.19	793.81	87.38	122,825.80
3	881.19	793.25	87.94	122,737.86 ◆

Although we can find the unpaid balance of a loan after any payment by creating an amortization schedule, it is convenient to be able to compute it as a *function* of the payment number. To derive the correct expression, let B_k represent the unpaid balance (also called the *outstanding principal*) after the kth payment. After the first payment, the balance would be

$$B_1 = P - (R - iP)$$

$$= P + iP - R$$

$$= P(1+i) - R.$$

Similarly, after the second payment, the balance would be

$$B_2 = B_1(1+i) - R.$$

After we substitute for B_1, this becomes

$$B_2 = [P(1+i) - R](1+i) - R$$

$$= P(1+i)^2 - R(1+i) - R.$$

Doing this one more time, we get

$$B_3 = B_2(1+i) - R$$
$$= \left[P(1+i)^2 - R(1+i) - R \right](1+i) - R$$
$$= P(1+i)^3 - R(1+i)^2 - R(1+i) - R$$
$$= P(1+i)^3 - R\left[(1+i)^2 + (1+i) + 1 \right]$$
$$= P(1+i)^3 - R\left[\frac{(1+i)^3 - 1}{i} \right],$$

where the last step comes from using the formula for a geometric sum. By a similar argument, we can conclude that the balance of a loan after the kth payment is

$$B_k = P(1+i)^k - R\left(\frac{(1+i)^k - 1}{i} \right).$$

We confirm this by finding the unpaid balance of the loan in Figure 2.3.1 after the fourth payment:

$$B_4 = 15{,}000(1.01)^4 - 706.10\left(\frac{1.01^4 - 1}{0.01} \right)$$
$$= 15{,}609.06 - 706.10(4.060401)$$
$$= 15{,}609.06 - 2{,}867.05$$
$$= \$12{,}742.01.$$

Note that this matches the balance given in Figure 2.3.1. We also cannot help but notice that the expression for B_k is the difference between two functions with which we are familiar. Here $P(1+i)^k$ is the future value of the loan amount P after k interest periods, and $R\left(\frac{(1+i)^k - 1}{i} \right)$ is the future value of an annuity with rent R after k periods. By graphing $y = P(1+i)^k$ and $y = R\left(\frac{(1+i)^k - 1}{i} \right)$ on the same axes with k as the independent variable,

we can visually identify the balance after k payments as the difference between the corresponding y-coordinate values on the two graphs. **Figure 2.3.2** displays these graphs for the loan in Example 3. Note, in particular, that the two graphs must necessarily intersect at $k = 24$, because $B_{24} = 0$ (neglecting round-off error).

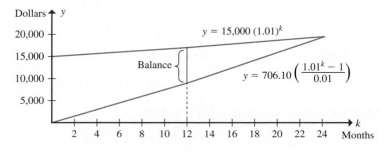

FIGURE 2.3.2 The unpaid balance on a loan is displayed as the difference between two graphs. This loan is for $15,000 at 12% interest, with monthly payments for 2 years.

? Example 5

What is the balance of the loan from Example 3 after 1 year? After 1.5 years? By what amount is the principal reduced between 1 and 1.5 years? What is the total interest paid during that same interval?

⚙ Solution

We need the balance after 12 payments and also after 18 payments:

$$B_{12} = 15,000(1.01)^{12} - 706.10\left(\frac{1.01^{12} - 1}{0.01}\right) = \$7,947.26,$$

$$B_{18} = 15,000(1.01)^{18} - 706.10\left(\frac{1.01^{18} - 1}{0.01}\right) = \$4,092.24.$$

So during that 6-month period, the principal is reduced by $7,947.26 - 4,092.24 = \$3,855.02$. The interest paid in that same time must then be

$$6(706.10) - 3,855.02 = 4,236.60 - 3,855.02$$
$$= \$381.58. \quad \blacklozenge$$

? Example 6

The interest paid on your home mortgage each year can be deducted from your gross income when you compute the income tax you owe the federal government. As Example 4 indicates, this can be a sizable amount of money, especially in the first few years of a mortgage. Determine the interest paid in the 1st year and the 10th year of the loan in that example.

⚙ Solution

The balance of the loan after the first year is

$$B_{12} = 123,000(1.00645833)^{12} - 881.19\left(\frac{1.00645833^{12} - 1}{0.00645833}\right)$$
$$= 132,878.49 - 10,958.09$$
$$= \$121,920.40.$$

So the principal has been reduced by $123,000 - 121,920.40 = \$1,079.60$. The amount of interest paid during the first year is $12(881.19) - 1,079.60 = \$9,494.68$.

To determine the total reduction in principal during the 10th year of the mortgage, we need to subtract the balance after the 10th year from the balance after the

9th year. The balance after the 9th year occurs after 108 payments, and the balance after the 10th year occurs after 120 payments. Using the balance function, we get

$$B_{108} = \$109,500.83$$

and

$$B_{120} = \$107,337.07.$$

Therefore, the amount paid toward the principal during the 10th year is $109,500.83 - 107,337.07 = \$2,163.76$, and the amount of interest paid that year is $12(881.19) - 2,163.76 = \$8,410.52.$ ◆

Name _____

Exercise Set **2.3** ☐☐☐←⎯⎯⎯⎯⎯⎯⎯⎯⎯⎯⎯⎯⎯⎯⎯⎯⎯⎯

In each exercise, the payment period is equal to the interest period, and every payment is made at the end of a period. Additionally, every interest rate is an annual rate and is considered to be constant for the appropriate term.

1. Find the present value of each of the following annuities:

(a) $100 per month at 6% interest with a term of 30 years.
(b) $125 per quarter at 7% interest with a term of 18 years.
(c) $350 semi-annually at 9.5% interest with a term of 7 years.

2. Find the present value of each of the following annuities:

(a) $80 per month at 6.5% interest with a term of 20 years.
(b) $175 every other month at 5% interest with a term of 15 years.
(c) $110 every 2 weeks at 7.25% interest with a term of 25 years.

3. What amount of money should Tucker deposit now in an account earning 3% interest in order to receive quarterly payments of $2,500 for the next 10 years?

4. You might want to establish an annuity in which your elderly aunt receives $7,500 per year for the next 10 years. How much money must you place in the fund now at 6.25% interest?

5. Mittola just won $383,000 in a state lottery. He invests it in an annuity earning $4\frac{7}{8}$% interest in order to receive monthly payments for 20 years. What is the amount of each payment?

6. The will of a deceased relative bequeaths $75,000 to a trust fund, enabling you to receive equal payments every 4 months for the next 10 years. If the fund earns 5.1% interest, what is the amount of each payment?

7. Compute the rent payment and total interest paid for each of the following loans:

(a) Loan amount of $35,000 at 4.8% paid monthly for a term of 6 years.
(b) Loan amount of $217,000 at 11.5% paid quarterly for a term of 25 years.
(c) Loan amount of $160,000 at $7\frac{5}{8}$% paid annually for a term of 18 years.

8. The city council of Green Valley needed some quick capital to repair a collapsed bridge. It borrowed $1 million at 5.7% interest to be paid back with 4 annual payments. How much is each payment, and how much total interest will be paid?

9. Francella (from Example 3) is considering a loan of $15,000 at 12% interest paid monthly. She is undecided about whether to make the term for 2 or 3 years. Find the monthly payment and the total interest paid if the term is 3 years. Compare these values to those computed in Example 3. If she desires a smaller monthly payment, which should she choose? If she desires to pay less total interest, which should she choose?

10. If Francella (from Example 3) chooses to make the term of her loan (with parameters given in the previous exercise) equal to 4 years, will the monthly payment be less or greater than that computed for a 2-year term? What about the total interest paid?

11. Gary is considering a loan of $25,000 at 9% interest paid quarterly. Compute the quarterly payment and the total interest paid if the term of the loan is 4 years.

12. Compute the quarterly payment and the total interest paid for the loan in the previous exercise if the term is 5 years. Is more or less total interest paid for this term than for a 4-year term?

13. You wish to buy a new home for $107,000, and you can make a down payment of $15,000. You obtain a 25-year mortgage for the balance at 8.4% interest with monthly payments. Determine the first three lines of the amortization schedule.

14. William wants to borrow $19,500 for a new swimming pool. He is considering a 4-year loan with quarterly payments at 10.8% interest. Determine the first three lines of the amortization schedule.

15. Vladi wishes to purchase a $28,000 vehicle with a 5-year loan at 7.3% interest with monthly payments.

    (a) Determine the first two lines of the amortization schedule.
    (b) What is the unpaid balance after the first year?
    (c) How much interest would be paid during the first year of the loan?

16. For the loan in the previous exercise, compute the amount paid toward the principal during each of the first 3 years. How do these three amounts change?

17. Gabriella wants to buy a house costing $95,000, and she has $12,000 for the down payment. She can arrange a 25-year mortgage at $6\frac{3}{8}$% interest with monthly payments.

    (a) Determine the first three lines of the amortization schedule.
    (b) What is the unpaid balance after the fifth year?

18. Find the interest that Gabriella would pay during the first and fifth years of the loan in the previous exercise.

19. If the loan in Exercise 15 is paid with bimonthly (every 2 months) payments, what will be the amount of each payment? How much interest will be paid during the first year of such a loan?

20. Kalveccio purchased a car 3 years ago for $18,900. At that time, he assumed a loan at 4.9% interest for the entire price. The loan required monthly payments for 4 years. How much money is required for Kalveccio to pay off the loan now?

21. Roberta purchases a piece of property for $73,000 and obtains a 15-year mortgage at 4.8% interest with bimonthly payments for the entire amount. In 5 years, she sells that same property for $96,500. How much money will Roberta have left after she pays off the remaining balance on her mortgage?

22. Consider the mortgage from the previous exercise.

    (a) What is the total amount paid to the bank by Roberta over the 5 years she was making payments?
    (b) Subtract this amount from the net proceeds that Roberta had when she sold the property.
    (c) What additional factor should be considered when deciding whether this financial venture was profitable?

23. Martin is considering the purchase of a small apartment complex whose units typically rent for a total of $3,600 per month. He estimates that expenses such as taxes and maintenance will average about $750 per month. Using only this net income to make the monthly payments of a 30-year mortgage at 9% interest, determine how much money he can borrow.

24. What must be the value of the unpaid balance B_k of a loan for $k = n$, where n is the total number of payment periods?

25. There is another way to find the expression for the present value P of a simple ordinary annuity that pays n payments of R dollars each back to the investor. We find the sum of the n separate amounts that need to be deposited initially to generate each of these payments. For instance, if i is the interest rate per period, then $P_1 = R(1 + i)^{-1}$ needs to be deposited to generate the first payment, because $P_1(1 + i) = R$. Similarly, $P_2 = R(1 + i)^{-2}$ is needed to produce the second payment, because $P_2(1 + i)^2 = R$. Continuing in this fashion, we see that the total initial amount needed is

$$P = P_1 + P_2 + \cdots + P_n$$
$$= R(1+i)^{-1} + R(1+i)^{-2} + \cdots + R(1+i)^{-n}.$$

Finish this argument by factoring out $R(1 + i)^{-1}$ and using the formula from Exercise 29 in the last section for a geometric sum with a common ratio of $(1 + i)^{-1}$.

26. Often, a bank charges a premium for the borrower to receive a lower interest rate. Describe a method by which you could determine whether it would be worth paying $1,000 to reduce the interest rate on a mortgage by 0.5% given the loan amount, interest rate, and term.

27. A TV advertisement offered the possibility of buying a new $18,000 car according to two options: (a) a $1,200 immediate rebate, but an annual interest rate of $r = 4\%$, or (b) $r = 0.9\%$. Assume that the term of the loan will be 5 years and that payments are made monthly. Determine your total cost for both options. (Remember to subtract $1,200 from the cost of your loan in the first option.) Which option would cost you the least amount of money?

Name _____

use the extra space
to show your work

Chapter Review Test **2** ◯◯◯ ←

All interest rates are annual rates. Every annuity is a simple, ordinary annuity.

1. Susan has a credit card account that uses a 9¾% interest rate. Her balance at the beginning of a 30-day month was $235. If she charged $48 on the 7th day of the month, $105 on the 17th day, and $179 on the 23rd day, what was her finance fee for the month?

2. Suppose you deposit $2,670 in a savings account that offers a 5.8% interest rate. How much money will you have in 4 years if the interest is compounded annually? Monthly? Daily?

3. Two different banks are offering certificates of deposit. Central Bank is advertising 6.5% compounded quarterly, and National Trust is giving 6.25% compounded daily. Compute the effective yield offered by each bank. Which bank would you choose?

4. At his first job after college, Oscar is offered two different choices to begin a retirement annuity. The first one requires a monthly rent payment of $100 and earns 6% interest. The second one requires a quarterly rent payment of $300 and earns 6.2% interest. Determine the future value of each of these annuities in 25 years.

5. Paxton County wishes to generate the money needed to build a $4.5 million library in 5 years. How much rent must the county deposit semi-annually into a sinking fund if the fund earns 8.4% annual interest? What would be the aggregate contribution made by the county at the end of 5 years?

6. Geralyn just hit it big in the lottery and won $425,000. She wants to establish an annuity from which she will receive a biweekly (every 2 weeks) payment for the next 20 years. If the interest rate is 7.8%, what will be the amount of each payment?

7. Tyler decides to borrow $23,500 to buy a grand piano. He obtains a 4-year loan at a 6% interest rate with monthly payments.

   (a) Determine the first three lines of the amortization schedule.
   (b) What is the unpaid balance after the first year?
   (c) How much interest will be paid during the first year of the loan?

Chapter Objectives

check off when you've completed an objective

- ☐ Know how to form symbolic statements.
- ☐ Determine the logical value of a statement.
- ☐ Evaluate a truth table for a statement form.
- ☐ Identify logically equivalent statement forms.
- ☐ Know De Morgan's laws for statements.
- ☐ Evaluate a syllogism with a truth table or Venn diagram.
- ☐ Know how to find a logic circuit corresponding to a statement form.
- ☐ Learn the pseudocode for input/output, decision structures, and looping.
- ☐ Write the output of a short program.
- ☐ Know how to write pseudocode to solve basic problems.
- ☐ Learn about the lives of Bertrand Russell and George Boole.

Navigate Companion Website

go.jblearning.com/johnson

Visit go.jblearning.com/Johnson to access a Laboratory Manual, Student Solutions Manual, Interactive Glossary, and Interactive Flashcards.

chapter

3   Logic and Computer Science

3.1 Statements and Connectives

3.2 Truth and Consequences

3.3 Tautologies and Syllogisms

3.4 Computer Programming Structures

> And this shows that intuition can sometimes get things wrong. And intuition is what people use in life to make decisions. But logic can help you work out the right answer.
>
> —Mark Haddon, *The Curious Incident of the Dog in the Night-Time.* © 2004, Vintage Books.

Alice in Wonderland and *Through the Looking Glass* by Lewis Carroll are generally considered to be wonderful works of fantasy written for the amusement of children. Few people realize that Carroll was the pen name of a professor of mathematics and logic at Oxford, a man named Charles Dodgson (1832–1898). Although Dodgson's primary purpose in both of these books was entertainment, many of the adventures of the main character, Alice, are enlivened with notions involving *logic*. In addition to his many humorous works, Dodgson wrote several serious books in his field, including *Symbolic Logic*, in which he wrote

> It [symbolic logic] will give you clearness of thought—the ability to see your way through a puzzle—the habit of arranging your ideas in an orderly and get-at-able form—and, more valuable than all, the power to detect fallacies, and to tear to pieces the flimsy illogical arguments which you will so continually encounter in books, in newspapers, in speeches, and even in sermons. . . .

Logic—the study of correct reasoning—is the glue that binds together any mathematical framework and, in turn, any science or applications supported by that mathematics. Logic dictates the validity of arguments and therefore must be the essential tool not only of scientific realms, but also of philosophy, law, and any other arena that requires clearness of thought and precise reasoning. We begin by introducing the basic mathematical building blocks that we will later combine with the rules of logic to systematically evaluate the truth of an argument. We will then show how logic is used in the construction of a computer program by featuring certain types of algorithms and decision-making mechanisms.

3.1 Statements and Connectives

One of the grandest examples of pure logic is the 2,000-page book *Principia Mathematica*, which was written by the British mathematicians **Alfred North Whitehead** (1861–1947) and **Bertrand Russell** (1872–1970) over a period of 10 years and was published between 1910 and 1913. This book begins by defining a few terms and listing a small number of axioms or assumed facts and then proceeds to *deductively* reach a large number of conclusions concerning the foundations of mathematics. Many consider this towering work to be the greatest contribution to the science of logic since the time of Aristotle. As **deduction** is the process of correctly reaching a conclusion from a given set of initial axioms or statements, it is no accident that both Russell and Whitehead were also two of the twentieth century's premier philosophers. After losing his son in combat in World War I, Whitehead turned his attention in earnest to philosophical musings with the publication of *An Enquiry Concerning the Principles of Natural Knowledge* (1919) and *The Concept of Nature* (1920). After a successful 30-year stay as a mathematician at Trinity College, Cambridge, Whitehead finished his career as professor of philosophy at Harvard, where he wrote *Science and the Modern World* (1925), a highly acclaimed study of the effects of modern science on Western culture. Russell, who was a student of Whitehead at Cambridge, was one of history's great pure thinkers. In addition to mathematics, he wrote extensively on scientific, philosophical, political, and social concerns. Among these works are *The Analysis of Mind* (1921), *The Prospects of Industrial Civilization* (1923), *The ABC of Relativity* (1925), *An Outline of Philosophy* (1927), *The Scientific Outlook* (1931),

deduction Process used to reach a conclusion from a given set of premises.

Education and the Social Order (1932), *Power* (1938), *Human Knowledge* (1948), and various other books. He received a multitude of distinctions for his achievements, including the Nobel Prize in Literature in 1950.

FIGURE 3.1.1   Bertrand Russell (1872–1970).

In disciplines such as mathematics and philosophy, you often begin with a group of given *objects*, define *operations* on those objects, and then obtain conclusions by rigorously applying those operations to those objects. You have been utilizing this concept since childhood. In the game checkers, for instance, the objects are the red and black pieces, and the operations are the rules for the movement of those pieces. Every game concludes after a finite sequence of moves. In logic, our objects are *statements* for which we will define four operations, known as **connectives**, for combining statements. Our conclusions about these combinations will then stem from their *logical value*. A **statement** is a declarative sentence that has exactly one **logical value**, either true or false.

? Example 1

Which of the following are statements?

(a)   Go to the end of the line.
(b)   Tomorrow is June 21.
(c)   Are you going to the concert tonight?
(d)   That snowmobile can go faster than 45 miles per hour (mi/h).
(e)   Lions are beautiful animals.
(f)   This sentence is false.

⚙ Solution

(a)   This is a command and is neither true nor false. It is not a statement.
(b)   This is a statement—tomorrow either is or is not June 21.
(c)   This is a question and is not a statement.

(d)   This is a statement—the snowmobile either can or cannot go faster than 45 mi/h.

(e)   It is a matter of opinion whether lions are beautiful. Therefore, this is not a statement.

(f)   This is not a statement because it cannot be given a consistent logical value. If you consider it to be true, then it claims itself to be false; but if you consider it to be false, then it must be true! (Such an undecidable sentence is sometimes called a **paradox**. See Exercise 15.)   ◆

> **paradox** Sentence or set of conditions leading to a sentence with an undecidable logical value.

Next, we define a function L, which assigns to any statement p its logical value. Note that this is a well-defined function because a statement must be either true or false, but it cannot be both. The domain of L consists of the set S of all statements, and the range is the two-element set {True, False}.

p	$L(p)$
$6 + 7 = 13$.	True
Mount Everest is the tallest mountain on Earth.	True
All bears are black.	False
$10^3 = 1,000$.	True
George Bush was a U.S. senator.	False
$39 + 56 < 41$.	False

Sometimes, the logical value of a statement can change. The value of "Today is Friday" depends on the current day of the week. If one or more real number variables are used in the declaration of a statement, then we call it an **algebraic statement**. An algebraic statement depends on the current value(s) of the variable(s) involved, and hence so does its logical value.

> **algebraic statement** Statement containing one or more variables.

 Example 2

Let p represent $x + 100 > 200$. Then p changes precisely when x changes.

x	p	$L(p)$
27	$127 > 200$	False
75	$175 > 200$	False
100	$200 > 200$	False
101	$201 > 200$	True
116	$216 > 200$	True ◆

> **compound statements** Statement consisting of two or more prime statements.
>
> **prime statements** Statement that does not contain a connective.

Compound statements are formed by combining a pair of given statements using a key word called a *connective*. (Statements that contain no connectives, as in the earlier examples, are known as **prime statements**.) For instance, "Jim is driving the car and Becky is a passenger" is a compound statement formed by linking two prime statements—"Jim is driving the car" and "Becky is a passenger"—by using the connective *and*. The four main connectives are *and*

conjunction The connective *and.*

disjunction The connective *or.*

conditional The connective *if . . . then.*

biconditional The connective *if and only if.*

negation Reversal of the logical value of a statement with some form of *not.*

(**conjunction**), *or* (**disjunction**), *if . . . then* (**conditional**), and *if and only if* (**biconditional**). In addition, one may reverse the logical value (from true to false, or from false to true) of any statement by using the word *not*. This is called the **negation** of the original statement. The following are examples of compound statements.

"Jim is driving the car, or Becky is a passenger."

"Jim is driving the car, and Becky is not a passenger."

"It is not the case that Jim is driving the car or Becky is a passenger."

"If Jim is driving the car, then Becky is a passenger."

"If Jim is not driving the car, then Becky is a passenger."

"If Becky is a passenger, then Jim is driving the car."

"Jim is driving the car if and only if Becky is a passenger."

"Jim is driving the car if and only if Becky is not a passenger."

It is important to note the difference between the conditional and the *bi*conditional. The biconditional stands for two conditionals at the same time. In the statements here, the biconditional "Jim is driving the car if and only if Becky is a passenger" is completely equivalent to the combination of this pair of conditionals: "If Jim is driving the car, then Becky is a passenger" and "If Becky is a passenger, then Jim is driving the car." The notion of equivalent statements will be discussed further in the next section.

To streamline the logical analysis of compound statements, letters (such as p, q, and r) are used to represent the prime statements, while connectives and negation are denoted by the symbols shown in **Figure 3.1.2**.

Name	Word	Symbol
Conjunction	and	$\wedge$
Disjunction	or	$\vee$
Conditional	If . . . then	$\Rightarrow$
Biconditional	If and only if	$\Leftrightarrow$
Negation	not	$\sim$

FIGURE 3.1.2   The symbols used for connectives and negation.

❓ Example 3

Let p represent "I mow the lawn on Friday."

Let q represent "I play golf on Saturday."

Express the following symbols as English statements.

(a) $p \wedge q$ (b) $p \vee q$ (c) $\sim p$ (d) $p \Rightarrow q$

(e) $p \Leftrightarrow q$ (f) $\sim p \wedge \sim q$ (g) $\sim(p \wedge q)$ (h) $\sim q \Rightarrow \sim p$

⚙ Solution

(a) I mow the lawn on Friday, and I play golf on Saturday.

(b) I mow the lawn on Friday, or I play golf on Saturday.

(c) I do not mow the lawn on Friday.

(d) If I mow the lawn on Friday, then I play golf on Saturday.

(e) I mow the lawn on Friday if and only if I play golf on Saturday.

(f)   I do not mow the lawn on Friday, and I do not play golf on Saturday.

(g)   It is not the case that I mow the lawn on Friday and I play golf on Saturday.

(h)   If I do not play golf on Saturday, then I do not mow the lawn on Friday.   ◆

The English language often allows for several ways to express the same thought. Care must be used when translating a given English compound statement into symbols.

❓ Example 4

Translate each of the following statements into symbols.

(a)   If a boy is Swedish, then he is blond.

(b)   Jennifer ran fast, yet she lost the race.

(c)   All redwood trees grow to be over 100 feet (ft) tall.

(d)   Neither Mike nor Alfreda plays poker.

(e)   Dark clouds are present whenever you see lightning.

(f)   On Sundays, Albert plays either bridge or chess.

(g)   Barbara likes psychology, but she does not like statistics.

(h)   That shirt is clean if and only if Alex washed it.

⚙ Solution

(a)   Using p for "A boy is Swedish" and q for "He is blond," we write $p \Rightarrow q$.

(b)   This statement is equivalent to "Jennifer ran fast, and she lost the race." Using p for "Jennifer ran fast" and q for "She lost the race," we write $p \wedge q$.

(c)   This statement can be rewritten as "If a tree is a redwood, then it grows to be over 100 ft tall." Using p for "A tree is a redwood" and q for "A tree grows to be over 100 ft tall," we write $p \Rightarrow q$.

(d)   This statement can be rewritten as "Mike does not play poker, and Alfreda does not play poker." Using p for "Mike plays poker" and q for "Alfreda plays poker," we write $\sim p \wedge \sim q$. Observe that this is different from "Either Mike or Alfreda does not play poker." This would be translated as $\sim p \vee \sim q$.

(e)   This statement is equivalent to "If you see lightning, then dark clouds are present." Using p for "You see lightning" and q for "Dark clouds are present," we write $p \Rightarrow q$.

(f)   Using p for "Albert plays bridge on Sundays" and q for "Albert plays chess on Sundays," we write $p \vee q$.

(g)   The connective *but* is equivalent for our purposes to *and*. Using p for "Barbara likes psychology" and q for "Barbara likes statistics," we write $p \wedge \sim q$.

(h)   Using p for "That shirt is clean" and q for "Alex washed the shirt," we write $p \Leftrightarrow q$.   ◆

---

antecedent In the conditional statement $p \Rightarrow q$, p is the antecedent.

consequent In the conditional statement $p \Rightarrow q$, the consequent is q.

In the conditional statement $p \Rightarrow q$, p is known as the **antecedent** and q as the **consequent**. The conditional has a plurality of forms of English expression. In part (c) of Example 4, the statement had to be rewritten using our typical conditional language. This will be true for such categorical statements beginning with the word *all*. In part (e), the antecedent did not come first in the sentence; instead, it followed the word *whenever*. Another variety is a sentence such as "The Dow Jones average increases every time interest rates drop." The antecedent here is "Interest rates drop," while the consequent is "The Dow Jones average increases."

We can continue this process by combining compound statements with other statements, but we must be careful to include parentheses to indicate which pairs of statements are to be combined first. As in algebra, an expression contained within a pair of parentheses is considered to be a single unit possessing a meaning independent of the rest of the expression.

? Example 5

Let p represent "Philadelphia wins its last game."

Let q represent "Pittsburgh wins its last game."

Let r represent "Chicago wins the pennant."

Translate each of the following into symbols. (Teams cannot tie.)

(a) Either Philadelphia wins its last game and Pittsburgh loses its last game, or Chicago wins the pennant.
(b) It is not the case that Chicago wins the pennant if Philadelphia wins its last game.
(c) If Philadelphia loses its last game and Pittsburgh wins its last game, Chicago wins the pennant.
(d) Chicago wins the pennant if either Philadelphia wins its last game or Pittsburgh loses its last game.
(e) If Pittsburgh loses its last game, then Chicago does not win the pennant.
(f) If either Philadelphia wins its last game and Pittsburgh wins its last game, or if Philadelphia loses its last game, then Chicago wins the pennant.
(g) Chicago wins the pennant if and only if Philadelphia or Pittsburgh loses its last game.

⚙ Solution

(a) $(p \wedge \sim q) \vee r$
(b) $\sim(p \Rightarrow r)$
(c) $(\sim p \wedge q) \Rightarrow r$
(d) $(p \vee \sim q) \Rightarrow r$
(e) $\sim q \Rightarrow \sim r$
(f) $((p \wedge q) \vee \sim p) \Rightarrow r$
(g) $r \Leftrightarrow (\sim p \vee \sim q)$ ◆

Example 6

Let p represent the algebraic expression "$x + y < 21$," and let q represent "$m = 65$." Translate each of the following into symbolic language.

(a) $x + y < 21$ and $m \neq 65$.
(b) $(x + y \geq 21$ and $m = 65)$ or $m \neq 65$.
(c) Not $(x + y < 21$ or $m = 65)$.
(d) $x + y < 21$ or $(x + y \geq 21$ and $m \neq 65)$.

Solution

(a) $p \wedge {\sim} q$
(b) $({\sim} p \wedge q) \vee {\sim} q$
(c) ${\sim}(p \vee q)$
(d) $p \vee ({\sim} p \wedge {\sim} q)$ ◆

Name _____

Exercise Set **3.1** ◯◯◯←——————————

1. Determine which of the following are statements.
   (a) Alice owns a restaurant.
   (b) Billy Joel is the greatest rock 'n' roll musician alive.
   (c) Are you going to the dentist on Tuesday?
   (d) $12 + 15 = 30$.
   (e) The sun is composed of more than 90% hydrogen.
   (f) The meat in the hamburgers at Burger Universe tastes as if it comes from horses.

2. Determine which of the following are statements.
   (a) France is the most beautiful country in Europe.
   (b) France is the largest country in Europe.
   (c) $3x + 5x = 8x$.
   (d) Baseball is full of action and excitement.
   (e) Presidential elections occur every 5 years.
   (f) When are you going to visit your parents?

3. Let p represent "$2x + 40 \leq 50$." Fill in the empty columns below.

x	p	$L(p)$
0		
3		
6		
9		
12		

4. Let p represent "$r + s = 12$." Fill in the empty columns below.

r	s	p	$L(p)$
−3	28		
−2.5	14.5		
0	12		
4.2	6.8		
6.1	5.9		
20	−8		
22	−12		

5. Let p represent "$m^2 + n^2 = 100$." Fill in the empty columns below.

m	n	p	$L(p)$
0	10		
3	5		
6	6		
6	8		
7	7		
8	6		
9	1		
10	0		

6. Let p represent "$x^2 - 15 \geq 1$." Fill in the empty columns below.

x	p	$L(p)$
1		
2		
3		
4		
5		

7. Let p represent "I ride my bike every day."
Let q represent "I weigh less than 200 pounds (lb)."
Express the following symbols as English statements:
(a) $\sim p$ (b) $p \vee q$ (c) $p \Rightarrow q$ (d) $q \Rightarrow p$
(e) $\sim p \wedge q$ (f) $\sim(p \wedge q)$ (g) $p \Leftrightarrow \sim q$ (h) $\sim(p \Leftrightarrow q)$

8. Let p represent "Babe Ruth played baseball for the Yankees."
Let q represent "The Yankees won the 1927 World Series."
Express the following symbols as English statements:
(a) $p \wedge q$ (b) $p \vee q$ (c) $\sim p$ (d) $p \wedge \sim q$ (e) $\sim p \vee (p \wedge q)$
(f) $p \Rightarrow q$ (g) $\sim q \Rightarrow \sim p$ (h) $p \Leftrightarrow q$ (i) $\sim p \vee q$ (j) $\sim(p \Leftrightarrow q)$

9. Let p represent "It is summer on Mars."
Let q represent "Dust storms are occurring on Mars."
Let r represent "The brightness of Mars changes."
Translate each of the following English statements into symbolic language.
(a) The brightness of Mars changes whenever dust storms are occurring.
(b) Dust storms are occurring on Mars, and it is not summer on Mars, but its brightness is not changing.
(c) Neither is it summer on Mars, nor are dust storms occurring.
(d) Dust storms are not occurring on Mars whenever its brightness is not changing.
(e) If it is not summer on Mars, then either dust storms are occurring or its brightness is changing.
(f) It is summer on Mars if and only if dust storms are not occurring.
(g) The brightness of Mars changes if and only if it is not summer and dust storms are occurring.
(h) It is not the case that either it is summer on Mars or dust storms are occurring while its brightness changes.

10. Let p represent "Scott plays soccer."
Let q represent "Scott plays the violin."
Let r represent "Scott does well in school."
Translate each of the following English statements into symbolic language.
(a) Scott plays soccer but not the violin.
(b) Either Scott does not play soccer, or he does not play the violin.
(c) Scott plays either soccer or the violin, but he does not do well in school.
(d) Scott plays soccer if and only if he does not play the violin.
(e) If Scott plays soccer, he does well in school.
(f) Whenever Scott does well in school, he plays either soccer or the violin.
(g) Scott plays the violin but not soccer whenever he does not do well in school.
(h) If Scott either does well in school or plays the violin, then he plays soccer.

11. Identify the antecedent and the consequent in the following conditional statements.
(a) All square polygons are rectangles.
(b) Every rational number is a real number.
(c) A magnetic field is created whenever an electric current flows through a wire.
(d) His movies make a profit whenever they gross $50 million or are seen by 10 million people.
(e) Much debate occurs in Congress every time the topic is gun control or welfare.
(f) We go fishing on the weekend provided that it doesn't rain or snow.

12. Identify the antecedent and the consequent in the following conditional statements.
(a) Whenever it rains, it pours.
(b) All humans are mammals.
(c) If you plunge into the stock market, you either sink or swim.
(d) Neither Ashley nor Betty goes to the beach if the temperature is not above 75°.
(e) You receive a driver's license provided you are 16 years old and you pass the exam.
(f) All minerals that are diamonds are composed of carbon.

13. Let p represent "$x \geq 7$."
Let q represent "$5x < y$."
Let r represent "$y = 350$."
Translate each of the following into symbolic language.
(a) $x \geq 7$ or $y \neq 350$.
(b) ($x < 7$ and $5x < y$) or $y = 350$.
(c) $x \geq 7$ and $5x \geq y$ and $y \neq 350$.
(d) $x < 7$ or not ($5x < y$ and $y = 350$).
(e) ($x \geq 7$ and $5x < y$) or ($5x \geq y$ and $y = 350$).

14. Let p represent "$x + y = 2{,}500$."
Let q represent "$8w > z$."
Let r represent "$2x + 5y - 11z \leq w$."

Translate each of the following statements into symbolic language.
(a) $x + y = 2{,}500$ or $8w \leq z$.
(b) $(x + y \neq 2{,}500$ and $2x + 5y - 11z \leq w)$ or $8w > z$.
(c) Not $(8w > z$ or $2x + 5y - 11z > w)$ and $x + y = 2{,}500$.
(d) $8w \leq z$ and not $(2x + 5y - 11z \leq w$ and $x + y = 2{,}500)$.
(e) Not $((x + y \neq 2{,}500$ and $8w > z)$ or $(8w > z$ and $2x + 5y - 11z > w))$.

15. A **set** is defined as any collection of objects, regardless of whether those objects are numbers, cars, galaxies, or hairs on your head. Bertrand Russell once defined a specific set R whose objects were themselves sets by

> **set** A collection of objects.

$$R = \{\, S \mid S \text{ is a set that is not a member of itself} \,\}.$$

Is the sentence "R is a member of R" a statement? (This example is known as *Russell's paradox.*)

16. Scott makes this statement: "I am the Diabolical Disagreer. I disagree with all claims made by anyone." Nick says to Scott, "I claim you disagree with me." Is this a paradox?

17. Korlin lives in a home with several other people of all ages. Each person has a special job, and Korlin's job is to shine *only* the shoes of every person who does not shine his or her own shoes. The question is this: Does Korlin shine her own shoes? Is this a paradox?

3.2 Truth and Consequences

Every compound statement has a logical value that is a function of the values of its prime statements. For instance, if you are currently attending college, then this statement is true:

p: I am a college student.

On the other hand, this statement is (hopefully) false:

q: I am an ax murderer.

You would agree, then, that the compound statement "I am a college student and an ax murderer" is also false. In other words, any pair of statements *p* and *q* that are joined by the conjunction *and* form a false compound statement unless both *p* is true and *q* is true. Using the logical function *L*, we say that the only time that $L(p \wedge q)$ is true is when $L(p)$ is true and $L(q)$ is true.

truth table A table listing the logical value of a statement form for every combination of values for the prime statements.

A chart known as a **truth table** places T for the value True and F for the value False in every box to indicate the appropriate logical value for a compound statement given the various possible true and false combinations for the prime statements. The truth table for conjunction is given in **Figure 3.2.1**. Although the columns should be headed by $L(p)$, $L(q)$, and $L(p \wedge q)$, we just use *p*, *q*, and $p \wedge q$ for the sake of convenience. This convention will be followed in all ensuing truth tables.

p	*q*	$p \wedge q$
T	T	T
T	F	F
F	T	F
F	F	F

FIGURE 3.2.1   Truth table for conjunction.

? Example 1

Determine the truth of this statement: "George Washington was the first U.S. president, and Abraham Lincoln was the second."

⚙ Solution

If *p* represents "George Washington was the first U.S. president" and *q* represents "Abraham Lincoln was the second U.S. president," then *p* is true, *q* is false, and the given statement—represented by $p \wedge q$—is false.   ◆

? Example 2

Under what conditions would the following statement be true? "Deep space is cold and dark."

⚙ Solution

Both prime statements—"Deep space is cold" and "Deep space is dark"—must be true.   ◆

FIGURE 3.2.2   George Boole (1815–1864).

symbolic logic
The science of using formal abstractions and procedures to systemize the principles of deduction and valid reasoning. Also called *logical calculus.*

Note that the table in Figure 3.2.1 does not depend in any way on the statements represented by *p* and *q*. This is the essential feature of classical **symbolic logic**, also called *logical calculus*, which uses formal abstractions to systematize and codify the principles of deduction and valid reasoning. We are concerned with not the truth of the prime statements in a given argument, but rather whether the argument itself is valid. One of the pioneers of this endeavor was the famous British mathematician **George Boole** (1815–1864). Born into extreme poverty, Boole struggled valiantly as a child to raise his standing in society by learning Latin and Greek on his own. His success at translating the classics led to a job as an elementary school teacher while yet a teenager, which, in turn, inspired him to study higher mathematics. Although he was only 20, his natural gift for the discipline blossomed as he took up the study of the great works of Lagrange and Laplace, the successors to Isaac Newton. In 1848, Boole published *The Mathematical Analysis of Logic*, in which he wrote:

> They who are acquainted with the present state of Symbolical Algebra, are aware that the validity of the processes of analysis does not depend upon the interpretation of the symbols which are employed, but solely on the laws of their combination.

This book was the precursor to Boole's masterpiece, *The Laws of Thought*, about which Bertrand Russell said, "Pure Mathematics was discovered by Boole in [this] work." The use of truth tables for the logical examination of argument began in part from this highly original book. The tables for negation and disjunction are given in **Figure 3.2.3**.

p	~*p*
T	F
F	T

p	*q*	*p* ∨ *q*
T	T	T
T	F	T
F	T	T
F	F	F

FIGURE 3.2.3   Truth tables for negation and disjunction.

The negation of a true statement is false, and the negation of a false statement is true. For instance, if "Harry is a swimmer" is false, then "Harry is a nonswimmer" (or "Harry is not a swimmer" or "Harry does not swim") is true.

In everyday English, one often uses *or* in the exclusive sense, such as in this statement: "I am going to the opera, or I am going to the concert." Here the speaker is usually implying that she will not attend both. However, in symbolic logic, the *or* used in the disjunction of two statements is always considered to be *inclusive*. Therefore, it is always a possibility that both statements p and q can be true, and in that case, we see in Figure 3.2.3 that $p \vee q$ is true.

? Example 3

Determine the logical value of this sentence: "Earth's surface consists of more water than land, and Elvis Presley is alive." Determine the logical value of this sentence: "Earth's surface consists of more water than land, or Elvis Presley is alive."

⚙ Solution

Let p represent the first prime statement and q the second. Because p is true and q is false, the first compound statement, $p \wedge q$, is false, and the second compound statement, $p \vee q$, is true.   ◆

? Example 4

Consider these algebraic statements: p: $x \geq 13$ and q: $x \leq 29$. Here, x is a real number variable. Because p and q are expressions involving x, their values change as x changes, and so we use the L function to indicate their current logical value.

x	p	q	$L(p)$	$L(q)$	$L(p \wedge q)$	$L(p \vee q)$
10	$10 \geq 13$	$10 \leq 29$	F	T	F	T
13	$13 \geq 13$	$13 \leq 29$	T	T	T	T
20	$20 \geq 13$	$20 \leq 29$	T	T	T	T
30	$30 \geq 13$	$30 \leq 29$	T	F	F	T ◆

We now examine the possible logical configurations for this conditional statement:

"If Debra does your taxes, then your taxes are done correctly."

If both the antecedent ("Debra does your taxes") and the consequent ("Your taxes are done correctly") are true, then certainly the entire conditional is true. Furthermore, if Debra does not do your taxes, then the fact of whether they are done correctly does not affect the truth of the conditional, and so it is true regardless of the outcome. It is only if Debra does your taxes and they are done incorrectly that we see that the original conditional is false. **Figure 3.2.4** displays the truth table for the conditional statement.

p	q	$p \Rightarrow q$
T	T	T
T	F	F
F	T	T
F	F	T

FIGURE 3.2.4   Truth table for the conditional.

? Example 5

We see that some rather bizarre conditional statements can be true. "If it rains milk, then zebras have polka dots" is a true statement, because it is impossible for the antecedent to be true. The same can be said for this statement: "If $6 + 6 = 13$, then every day is a holiday." It is just as important to realize that a true consequent guarantees a true conditional. "If your middle name is Bozo, then Harrisburg is the capital of Pennsylvania" is always true.   ♦

To see the difference between a conditional and a biconditional, suppose the following two statements are *true*.

1. "If Jack receives a flu shot in October, then he does not get the flu in winter."

2. "Jack does not get the flu in winter if and only if he receives a flu shot in October."

The conditional statement 1 tells us that a true antecedent—Jack receives a flu shot in October—means that the consequent must also be true: Jack does not come down with the flu in winter. However, we know that a false antecedent does not guarantee a false consequent even though statement 1 is true. Jack may get lucky and not get the flu even if he does not get the appropriate shot.

A true biconditional statement 2 is stronger than statement 1. It states that Jack does not get the flu in winter *if and only if* he has a flu shot in October. This gives more information than statement 1 by stating that when the second part of the biconditional ("He receives a flu shot in October") is false, then the first part of the biconditional ("Jack does not get the flu") is also false. In other words, Jack's immune system is such that the only way he avoids the flu is to receive the shot. Either both prime statements are true, or both are false. *In a biconditional that is true, the logical values of the prime statements are the same.* Similarly, it follows that if the logical values of the two primes are different, then the resulting biconditional is false. These facts are summarized in **Figure 3.2.5**.

p	q	$p \Leftrightarrow q$
T	T	T
T	F	F
F	T	F
F	F	T

FIGURE 3.2.5   Truth table for the biconditional.

© Liga Lauzuma/ShutterStock, Inc.

❓ Example 6

Determine the logical value of this statement: "The water in the pond turns to ice if and only if the temperature is above 32°F."

⚙ Solution

We know this to be a false statement from common experience, because if the first prime statement is false, then the second one is true. Also, if the first prime is true, then the second is false. Because the only situation that exists is the one in which the p and q of this biconditional statement have opposite logical values, the value of the biconditional is false.   ◆

We now return to our characterization of symbolic logic as *objects* and *operations*. The statement symbols are the objects, and the tables given in the earlier figures can be considered to be definitions of the operations ($\sim$, $\wedge$, $\vee$, $\Rightarrow$, and $\Leftrightarrow$) on those objects. With these definitions, we can determine a logical value (T or F) for *any* expression of objects and operations without regard or need for the objects to have an English meaning, but based solely on the values of the prime component symbols. Such expressions—having logical value but no meaning—are called **statement forms**. Any substitution of English words for the object and operation symbols in a form is then called an **interpretation** of the form. (Hence, all the translations in the previous section went back and forth from statement forms to interpretations.) For simplicity's sake, we will continue to call the values T "true" and F "false" when referring to the values of statement symbols and forms. The *standard order* in which operations are performed is $\sim$, $\wedge$, $\vee$, $\Rightarrow$, and $\Leftrightarrow$. There is one exception to this sequence—when parentheses are used. As in algebra, any operation within a pair of parentheses is to be done first. Even when they do not alter the standard order, parentheses are often used for clarification, and in the case of nested parentheses, you begin with the operation inside the innermost pair.

> **statement forms** Any expression of statement symbols and connectives.
>
> **interpretation** A substitution of English words for the objects and symbols in a statement form.

❓ Example 7

Let p, q, r, and s be statement symbols having values T, F, T, and F, respectively. Find the values of the following statement forms.

(a) $\sim p \wedge q \vee r$

(b) $\sim p \wedge (q \vee r)$

(c) $\sim p \wedge q \Rightarrow r \wedge \sim s$

(d) $\sim(p \wedge q \vee r) \Rightarrow r \wedge s$

(e) $(p \Rightarrow q) \Leftrightarrow (r \Rightarrow s)$

(f) $\sim(p \wedge q \Leftrightarrow r) \wedge (\sim r \Rightarrow s)$

⚙ Solution

(a) Negation is done first, followed by conjunction and then disjunction.

~p	∧	q	∨	r	
F		F		T	~p has value F
	F			T	~p ∧ q has value F
			T		~p ∧ q ∨ r has value T

(b) This time, we do disjunction after negation because of the parentheses.

~p	∧	(q	∨	r)	
F		F		T	~p has value F
F			T		q ∨ r has value T
	F				~p ∧ (q ∨ r) has value F

(c)

~p	∧	q	⇒	r	∧	~s
F		F		T		T
	F				T	
			T			

(d)

~(p	∧	q	∨	r)	⇒	r	∧	s
T		F		T		T		F
	F			T			F	
			T				F	
			F				F	
					T			

(e)

(p	⇒	q)	⇔	(r	⇒	s)
T		F		T		F
	F				F	
			T			

(f)

~(p	∧	q	⇔	r)	∧	(~r	⇒	s)
T		F		T		F		F
	F			T			T	
			F				T	
			T				T	
					T		◆	

The truth table for any statement form must contain a value for every combination of values for the prime statements. Consider the form $p \wedge \sim q$. In **Figure 3.2.6**, you see that first the values for p are repeated and $\sim q$ is evaluated. Then the values for the entire form are computed in the boxed column headed by the $\wedge$ by operating on each pair of the listed values for p and $\sim q$.

p	q	p ∧ ~q	
T	T	T	F
T	F	T	T
F	T	F	F
F	F	F	T

p	q	p ∧ ~q		
T	T	T	**F**	F
T	F	T	**T**	T
F	T	F	**T**	F
F	F	F	**F**	T

FIGURE 3.2.6   Truth table for p ∧ ~q.

? Example 8

Consider this longer form: $(p \lor {\sim}q) \Rightarrow {\sim}(p \land q)$. The order of operations is important and is indicated by the numbering at the bottom of the columns in the sequence of tables in **Figure 3.2.7**. First, we list the values for p, q, and ~q. Second, we determine the values for each connective inside the parentheses. Third, we reverse the values of $(p \land q)$ because of the negation symbol ~. The fourth and final boxed column of boldface values contains the logical values resulting from connecting the values in the oval enclosed columns using the conditional ($\Rightarrow$).

p	q	(p ∨ ~q)	⇒	~(p ∧ q)	
T	T	T F		T	T
T	F	T T		T	F
F	T	F F		F	T
F	F	F T		F	F
		1 1		1	1

p	q	(p ∨ ~q)	⇒	~(p ∧ q)	
T	T	T T F		T T T	
T	F	T T T		T F F	
F	T	F F F		F F T	
F	F	F T T		F F F	
		1 2 1		1 2 1	

p	q	(p ∨ ~q)	⇒	~ (p ∧ q)	
T	T	T T F		F T T T	
T	F	T T T		T T F F	
F	T	F F F		T F F T	
F	F	F T T		T F F F	
		1 2 1		3 1 2 1	

p	q	(p ∨ ~q)	⇒	~ (p ∧ q)	
T	T	T (T) F	**F**	(F) T T T	
T	F	T (T) T	**T**	(T) T F F	
F	T	F (F) F	**T**	(T) F F T	
F	F	F (T) T	**T**	(T) F F F	
		1 2 1	4	3 1 2 1	♦

FIGURE 3.2.7   Truth table for (p ∨ ~q) ⇒ ~(p ∧ q).

? Example 9

We now evaluate a truth table for a form containing three statement symbols: p, q, and r. Because each of three symbols can have one of two logical values, we must examine $2^3 = 8$ combinations of these values, as shown in **Figure 3.2.8**. In general, one must consider 2^n combinations for a statement form composed of n statement symbols. In Figure 3.2.8, we show the results for this form: $(p \lor q) \land r$. Disjunction ($\lor$) is computed before conjunction ($\land$) because of the parentheses.

p	q	r	(p	∨	q)	∧	r
T	T	T	T	T	T	**T**	T
T	T	F	T	T	T	**F**	F
T	F	T	T	T	F	**T**	T
T	F	F	T	T	F	**F**	F
F	T	T	F	T	T	**T**	T
F	T	F	F	T	T	**F**	F
F	F	T	F	F	F	**F**	T
F	F	F	F	F	F	**F**	F
			1	2	1	3	1	◆

FIGURE 3.2.8   Truth table for a form with three prime statement symbols.

Sometimes you can deduce the logical value of a given statement form if you know the value of a related form.

Example 10

If the value of $p \Leftrightarrow q$ is true, what is the value of $p \Rightarrow q$?

Solution

From Figure 3.2.5, we see that if $p \Leftrightarrow q$ is true, then p and q are either both true or both false. In either case, the value of $p \Rightarrow q$ must be true.

Note that $p \Rightarrow q$ does not have a unique value if $p \Leftrightarrow q$ is given to be false. In that event, either p is true and q is false (in which case $p \Rightarrow q$ is false), or p is false and q is true (in which case $p \Rightarrow q$ is true).   ◆

Example 11

Based on the given information, determine whether the statement form has a unique logical value; if so, find it.

(a) $(p \wedge {\sim}q) \Rightarrow (r \vee {\sim}s)$
          T

(b) $p \wedge (q \Leftrightarrow r)$
        T

(c) $(p \wedge q) \Leftrightarrow ({\sim}p \vee r)$
   F

Solution

(a)   Because the value of r is T, the value of $r \vee {\sim}s$ must be T. The value of the conditional must then be T, because the consequent has a value of T.

(b)   Knowing the value of the biconditional does not give a unique value to the entire conjunction. If p is true, then $p \wedge (q \Leftrightarrow r)$ is true. If p is false, then $p \wedge (q \Leftrightarrow r)$ is false.

(c)   Here p has value F and so $\sim p$ has value T. Therefore, $p \wedge q$ has value F, $\sim p \vee r$ has value T, and $(p \wedge q) \Leftrightarrow (\sim p \vee r)$ has value F.   ◆

    Suppose a friend on the phone tells you it is hot and sunny outside, but you look out the window and see that it is cloudy even though the temperature is 98°. So you disagree with your friend based on the lone fact that it is cloudy. In other words, you negated the entire compound statement ("It is hot and sunny") because the negation conveys the same meaning as "It is either not hot or it is not sunny." It often happens that two different statements have the same meaning. Formally, we say that two statement forms are **logically equivalent**, or just **equivalent**, if their truth tables have identical final columns. We denote the phrase *is equivalent to* by the symbol $\equiv$. For instance, it is clear that $\sim(\sim p) \equiv p$. Also, we have just seen an interpretation of the equivalence

logically equivalent
See **equivalent**.

equivalent Descriptive of two statement forms whose truth tables have identical final columns. Also called *logically equivalent*.

$$\sim(p \wedge q) \equiv \sim p \vee \sim q$$

where p is "It is hot" and q is "It is sunny." This is verified by the truth table in **Figure 3.2.9**.

p	q	$\sim (p \wedge q)$	$\sim p \vee \sim q$
T	T	**F** T T T	F **F** F
T	F	**T** T F F	F **T** T
F	T	**T** F F T	T **T** F
F	F	**T** F F F	T **T** T

FIGURE 3.2.9   Equivalent statement forms.

    For simplicity, we also refer to interpretations of equivalent forms as *equivalent interpretations*.

❓ Example 12

Computer scientists utilize the equivalence in Figure 3.2.9 to implement negations of compound algebraic statements in computer programs. Find equivalent statements to the negation of each of the following statements.

(a) $x = 18.5$ and $y > 31.5$.
(b) $x \geq 70$ and $x \leq 75$.
(c) $a^2 \neq 1{,}000$ and $a + b < 250$.

⚙ Solution

(a) $x \neq 18.5$ or $y \leq 31.5$.
(b) $x < 70$ or $x > 75$.
(c) $a^2 = 1{,}000$ or $a + b \geq 250$.   ◆

? Example 13

"If Marlowe located Jessica Vanderkamp, then he earned \$5,000" is an interpretation of the conditional $p \Rightarrow q$. Show that this form is equivalent to $\sim q \Rightarrow \sim p$, and write out the corresponding interpretation.

⚙ Solution

The equivalence is verified by the truth table in **Figure 3.2.10**.

p	q	$p \Rightarrow q$			$\sim q \Rightarrow \sim p$		
T	T	T	**T**	T	F	**T**	F
T	F	T	**F**	F	T	**F**	F
F	T	F	**T**	T	F	**T**	T
F	F	F	**T**	F	T	**T**	T

FIGURE 3.2.10   Equivalent statement forms.

This is the corresponding interpretation: "If Marlowe did not earn \$5,000, then he did not locate Jessica Vanderkamp."   ◆

The statements shown to be logically equivalent in Figure 3.2.10 are often used interchangeably in various types of arguments. Formally, we define $\sim q \Rightarrow \sim p$ as the **contrapositive** of the conditional $p \Rightarrow q$. On the other hand, the statement "If Marlowe earned \$5,000, then he located Jessica Vanderkamp" would be an interpretation of $q \Rightarrow p$, which does *not* have the same final truth values as $p \Rightarrow q$ and thus is *not* equivalent to it. These related conditional forms, $p \Rightarrow q$ and $q \Rightarrow p$, are called **converses** of each other.

contrapositive The contrapositive of $p \Rightarrow q$ is $\sim q \Rightarrow \sim p$.

converses The converse of $p \Rightarrow q$ is $q \Rightarrow p$.

? Example 14

Determine the converse and contrapositive of this conditional statement:

"If $f(x) = 4x + 10$, then $f(2) = 18$."

⚙ Solution

The converse is obtained by simply exchanging the antecedent and consequent. In this example, this is "If $f(2) = 18$, then $f(x) = 4x + 10$." We know that many functions exist for which $f(2) = 18$, such as $f(x) = 5x + 8$ or $f(x) = 12x - 6$. Therefore, this converse is not a true statement. However, the contrapositive is "If $f(2) \neq 18$, then $f(x) \neq 4x + 10$." This is certainly true and exemplifies the fact that *any conditional statement is logically equivalent to its contrapositive.*   ◆

Translating two given compound statements into forms and then comparing truth tables is an excellent method for determining equivalence. This strikes at the heart of the social and scientific utility of the logical abstraction of complex statements. The symbols we use for statements are nothing more than indifferent variables whose combinations (forms) have values determined according to our defined set of operations. The replacement of English by logical symbolism allows for a dispassionate comparison of seemingly similar sentences and, moreover, effective analysis of long, involved proofs of mathematical facts as well as philosophical or political arguments. This will be explored more thoroughly in the next section.

Name _____

Exercise Set **3.2** ◯◯◯ ←

1. Under what conditions would the following statement be true? "Nick enjoys baseball and spinach."

2. Under what conditions would the following statement be false? "Andre lives in Spain and Asif lives in Pakistan."

3. Determine the logical value of this statement: "289 > 147 and 35 < 11." Determine the logical value of "289 > 147 or 35 < 11."

4. Determine the logical value of "The U.S. Senate consists of 100 senators or the U.S. House of Representatives consists of 100 representatives." Determine the logical value of "The U.S. Senate consists of 100 senators and the U.S. House of Representatives consists of 100 representatives."

5. Determine the logical value of "If $6 + 8 = 10$, then Earth is made of sushi." Determine the logical value of "If $6 + 8 \neq 10$, then Earth is made of sushi."

6. Determine the logical value of "Your math teacher can lift 500 pounds whenever that teacher can run 100 meters in 3.5 seconds." Determine the logical value of "Your math teacher can lift 500 pounds whenever that teacher can run 100 meters in less than 10 minutes."

7. Let p represent "$x > 1{,}000$," and let q represent "$x < 1{,}500$." Fill in the following table.

x	p	q	$L(p)$	$L(q)$	$L(p \wedge q)$	$L(p \vee q)$
500						
1,000						
1,250						
1,499						
1,500						
1,600						

8. Let p represent "$x > 0$," and let q represent "$2x + y \leq 26.3$." Fill in the following table.

x	y	p	q	$L(p)$	$L(q)$	$L(p \wedge q)$	$L(p \vee q)$
−5	45						
−1	28						
2	20						
10	5						
12	3						

9. Let p represent "$x \geq 75$," and let q represent "$x + y = 100$." Fill in the following table.

x	y	p	q	$L(p)$	$L(q)$	$L(p \wedge q)$	$L(p \vee q)$
62	15						
70	30						
75	25						
80	20						
90	11						

Let the statement symbols p, q, r, and s have values T, F, T, *and* F, *respectively. Find the values of the following statement forms.*

10. $p \wedge q \vee r \wedge s$

11. $(p \vee q \wedge r) \vee s$

12. $\sim(p \wedge r) \Rightarrow q \vee s$

13. $(q \Leftrightarrow s) \Rightarrow (\sim r \Leftrightarrow \sim s)$

14. $((p \vee r) \wedge s) \Leftrightarrow (q \vee r)$

15. $((q \Leftrightarrow s) \wedge p) \wedge \sim(p \wedge r \Rightarrow s)$

Evaluate the truth table for each of the following statement forms.

16. $p \vee \sim q$

17. $\sim p \Rightarrow q$

18. $\sim p \wedge q$

19. $\sim(p \wedge q)$

20. $\sim q \Rightarrow \sim p$

21. $p \wedge \sim p$

22. $(p \wedge q) \Rightarrow \sim(p \vee q)$

23. $(p \Rightarrow q) \wedge (q \Rightarrow p)$

24. $(p \wedge \sim q) \Rightarrow \sim r$

25. $(p \wedge q) \vee r$

26. $(p \Rightarrow q) \wedge (p \Rightarrow \sim q)$

27. $p \Rightarrow (q \vee \sim r)$

Based on the given information, determine whether the statement form has a unique logical value; if so, find it.

28. $p \Rightarrow (q \vee \sim r)$
   F

29. $p \vee (q \Rightarrow r)$
   T

30. $(p \vee q) \Leftrightarrow (r \wedge s)$
   T   F

31. $(p \vee q) \Leftrightarrow (r \wedge s)$
   F   T

32. $(p \Rightarrow q) \Rightarrow (\sim q \Rightarrow s)$
   T

33. $(p \wedge \sim p) \Leftrightarrow (q \vee \sim q)$

34. If the value of $p \wedge q$ is F, what is the value of $\sim p \vee \sim q$?

35. If the value of $p \vee q$ is F, what is the value of $p \Leftrightarrow q$?

36. If the value of $p \Leftrightarrow q$ is T, what is the value of $p \Leftrightarrow \sim q$?

37. If the value of $p \Rightarrow q$ is T, what is the value of $p \wedge \sim q$?

Verify the following equivalences.

38. $p \Leftrightarrow q \equiv \sim p \Leftrightarrow \sim q$

39. $p \Leftrightarrow q \equiv (p \Rightarrow q) \wedge (q \Rightarrow p)$

40. $\sim(p \Rightarrow q) \equiv p \wedge \sim q$

41. $p \Rightarrow q \equiv \sim p \vee q$

42. Two pairs of equivalent forms known as *De Morgan's laws* are as follows:

$$\sim(p \wedge q) \equiv \sim p \vee \sim q$$
$$\sim(p \wedge q) \equiv \sim p \vee \sim q$$

The first equivalence was verified in Figure 3.2.9. Use a truth table to verify the second equivalence.

43. Use De Morgan's laws to find a statement equivalent to this one: "It is not the case that Senator Farnsworth voted for the tax cut and for the new spending bill."

44. Use De Morgan's laws to find a statement equivalent to this one: "Last year, average family income did not increase, and average family spending did not increase."

45. Use De Morgan's laws to find a statement equivalent to this one: "It is not the case that Erika received either a B in math or a C in English."

46. Use De Morgan's laws to find a statement equivalent to this one: "In the soccer game, either Jorge scored a goal or he made an assist."

Use De Morgan's laws to find the negation of each of the following algebraic statements.

47. $x > 2$ and $y = 35$.   **48.** $x \neq 60$ or $x + y \leq 320$.

49. $x + y = 4$ or $x < 13.5$.   **50.** $x \geq 56$ and $x \leq 57$.

51. $y < 85$ or $y > 90$.   **52.** $r > 6$ and $r^2 + s^2 = 100$.

53. $(x < 10$ and $y < 10)$ or $2x + 3y = 50$.   **54.** $(a = 25$ or $b = 25)$ and $a + b > 600$.

55. Determine the converse and the contrapositive of this conditional statement: "If the Earth's core is made principally of iron, then the temperature of the core is between 4,000 and 5,000°C." Which one is equivalent to the original?

56. Determine the converse and the contrapositive of this conditional statement: "If $m = 8$, then $m^2 = 64$." Which one is equivalent to the original?

57. Determine the converse and the contrapositive of this conditional statement: "If $f(x) = 8x + 3$, then $f(2) = 19$." Which one is equivalent to the original?

58. Determine the converse and the contrapositive of this conditional statement: "If the length and width of a rectangle are 5 and 10 centimeters, then its area is 50 square centimeters." Which one is equivalent to the original?

59. Suppose we are allowed to use only the connectives ~ and ∧. We could implement the other connectives by use of equivalences. For instance, we could write ~(~p ∧ ~q) for p ∨ q according to De Morgan's laws. What would you write for $p \Rightarrow q$, using only ~ and ∧? [*Hint*: See Exercise 40.]

60. What would you write for $p \Leftrightarrow q$, using only ~ and ∧? [*Hint*: See Exercise 39.]

61. In computer science programming, the function NOR(p, q) assumes the logical values defined by NOR$(p, q) = {\sim}p \vee {\sim}q$. All the connectives can be implemented via this one function, called a **logic gate**, by varying the arguments of the function. For instance, ${\sim}p \equiv$ NOR(p, p) and so $p \vee q \equiv$ NOR(NOR(p, p), NOR(q, q)). How could you express $p \wedge q$, using NOR?

logic gate Computer mechanism that simulates a logical connective.

62. Another special function used by computer programmers is XOR, the "exclusive or." The function XOR(p, q) assumes the value true if either p is true (and q is false) or q is true (and p is false), but it is false when p and q have the same logical value. In other words, it is the same as $p \vee q$ except when p and q are both true. Create a statement form that is equivalent to this function.

3.3 Tautologies and Syllogisms

> I thought of mathematics with reverence, and suffered when Wittgenstein led me to regard it as nothing but tautologies.
>
> —Bertrand Russell

We have discussed the impact that Bertrand Russell had on modern thought, but we should mention that during his lifetime, his views were not always held in high esteem by everyone. In particular, he was an outspoken peace activist, and his support of the No-Conscription Fellowship during World War I embroiled him in a controversy that led to his removal from his lectureship in mathematics at Trinity College, Cambridge, in 1916. Apparently, you can get yourself into no small amount of trouble by being too logical.

In fact, Russell spent several productive months in jail in 1918 on charges resulting from the authorship of inflammatory antiwar literature. Prison, however, seemed not to dampen his creative spirit, for he wrote much of his *Introduction to Mathematical Philosophy* (from which a quote appears at the beginning of this section) during this time. This work contained certain results stemming from the framework for symbolic logic constructed by George Boole, and it outlined the methodology of the theory of deduction. Deduction is the process of reaching a **conclusion** from a given set of premises. A **premise** may be a prime statement or a compound statement, and an entire sequence of premises leading to a conclusion is called an **argument**. We are primarily concerned here with the determination of the *validity* of an argument. Simply stated, an argument is considered to be **valid** if it is impossible for the conclusion to be false when the premises are true. The use of a valid argument to obtain a conclusion is called *valid reasoning*. It is important to realize that this does *not* state that the validity of an argument is a property equivalent to the truth of its conclusion. Before we introduce a process for determining validity, we define an associated characteristic of statements in general.

conclusion Final statement in an argument.

premise Any of the statements of an argument except for the conclusion.

argument Collection of a sequence of statements.

valid Descriptive of an argument whose truth table corresponding to its statement form is a tautology.

Consider the statement form $(p \wedge q) \Rightarrow p$ and its associated truth table. (See **Figure 3.3.1**.)

p	q	$(p \wedge q) \Rightarrow p$
T	T	T T T **T** T
T	F	T F F **T** T
F	T	F F T **T** F
F	F	F F F **T** F

FIGURE 3.3.1 Example of a tautology.

tautology A statement form (or its interpretation) that is true for every possible combination of logical values for its prime statements.

Note that the logical value of this compound statement is true under all possible combinations of values for the prime statements. Such a statement form is called a **tautology**. Realize that tautologies exist independently of the truth of each prime statement. They simply guarantee that the statement taken *as a whole* is always true. These are four interpretations of the form that represents each row of the table in Figure 3.3.1:

If horses are mammals and sharks are predators, then horses are mammals.
If horses are mammals and Ben Franklin is alive, then horses are mammals.
If horses are reptiles and sharks are predators, then horses are reptiles.
If horses are reptiles and Ben Franklin is alive, then horses are reptiles.

All of these interpretations are true statements.

? Example 1

Determine whether the following statement forms are tautologies.

(a) $(p \Leftrightarrow \sim q) \wedge p \Rightarrow \sim q$

(b) $p \wedge (q \vee r) \Rightarrow p \wedge r$

⚙ Solution

(a)

p	q	$(p$	$\Leftrightarrow$	$\sim q)$	$\wedge$	p	$\Rightarrow$	$\sim q$
T	T	T	F	F	F	T	**T**	F
T	F	T	T	T	T	T	**T**	T
F	T	F	T	F	F	F	**T**	F
F	F	F	F	T	F	F	**T**	T
		1	2	1	3	1	4	1

The parentheses dictate that we evaluate the biconditional first. By order of operations, we evaluate the conjunction next, followed by the conditional. Because all the final values are true, this statement form is a tautology.

(b)

p	q	r	p	$\wedge$	$(q$	$\vee$	$r)$	$\Rightarrow$	p	$\wedge$	r
T	T	T	T	T	T	T	T	**T**	T	T	T
T	T	F	T	T	T	T	F	**F**	T	F	F
T	F	T	T	T	F	T	T	**T**	T	T	T
T	F	F	T	F	F	F	F	**T**	T	F	F
F	T	T	F	F	T	T	T	**T**	F	F	T
F	T	F	F	F	T	T	F	**T**	F	F	F
F	F	T	F	F	F	T	T	**T**	F	F	T
F	F	F	F	F	F	F	F	**T**	F	F	F
			1	3	1	2	1	4	1	3	1

The parentheses dictate that we evaluate the disjunction before the conjunctions. The conditional ($\Rightarrow$) is used to operate on pairs of values occurring in the two columns under the conjunctions in step 3. The occurrence of a false in the second row of the final column prevents this statement form from being a tautology. ◆

When somebody is programming a computer, algebraic statements are used to provide direction for the flow of the execution of the program. The purpose of such statements might be considered to be the control of traffic at a two-pronged fork in the road. Program execution continues along one path if the value of the expression is true and along the other path if the value of the expression is false. So the expression serves as a sort of traffic light. An algebraic statement that has a constant logical value for all possible values of the variables channels the flow into just one branch and essentially blocks off the remaining branch. Although not a tautology in the strict sense of the word, it is a related type of phenomenon in computer programming, one that creates a type of mistake called a *logic error*, since it is certainly not what the programmer intended. It serves no purpose—it is analogous to putting in a traffic light, one that always remains green at an intersection!

? Example 2

Although "$x + y \geq 30$" could be true or false depending on the values of x and y, the compound statement "$x + y \geq 30$ or $x^2 \geq 0$" will *always* be true, because the square of any number is never less than 0. Because "$x^2 \geq 0$" is always true, so is "$x + y \geq 30$ or $x^2 \geq 0$." **Figure 3.3.2** displays the two situations with a device known in computer science as a *flowchart*. Note that the algebraic expression is written inside a diamond.

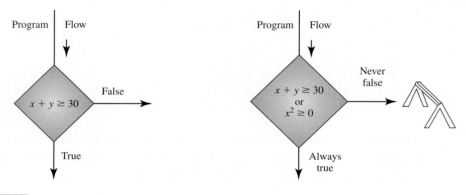

FIGURE 3.3.2   The perpetually true algebraic expression on the right is a programming error.

Other examples are

$$x > 25 \text{ or } x \leq 25, \quad 4x + 9y = 17 \text{ or } 4x + 7y \neq 17.$$

Similarly, a logic error occurs when an algebraic statement is always false. Examples are

$$2x + y \geq 100 \text{ and } 2x + y < 100, \ (x > 3 \text{ and } y > 3) \text{ and } x + y = 2.$$

Extreme care must be taken to avoid these types of mistakes when writing computer programs. ◆

In the history of logic, it was Aristotle (384–322 BC) who first initiated the systemization of logical argument. In particular, he defined a **syllogism** as an argument consisting of two premises and a conclusion, each pair of which has a connection, such as a common object, description, or action. One common type is exemplified by the following:

syllogism An argument consisting of two premises and a conclusion.

FIGURE 3.3.3   Statue of Aristotle in his birthplace, Stageira, Greece.

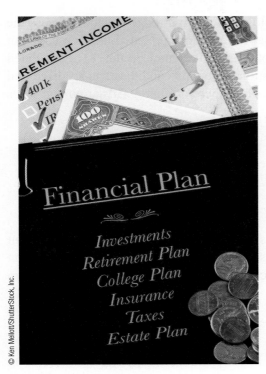

	Premise:	If consumer prices increase, then purchasing power decreases.
	Premise:	Consumer prices are increasing.
	Conclusion:	∴ Purchasing power is decreasing.

Syllogisms are often structured according to this format. The two premises are listed above a horizontal line, with the conclusion stated below it. The symbol "∴" means *therefore*.

For an accurate logical analysis, first we translate this argument into symbolic language. Let p represent "consumer prices increase," and let q represent "purchasing power decreases." Then we can join the premises of the syllogism with a conjunction to create this compound statement: $(p \Rightarrow q) \wedge p$. (Liberties may be taken with the expression and tense of the verbs.) The conclusion of this argument is then written as the consequent of the conditional statement form

$$[(p \Rightarrow q) \wedge p] \Rightarrow q,$$

which we see in **Figure 3.3.4** is a tautology.

p	q	$[(p$	$\Rightarrow$	$q)$	$\wedge$	$p]$	$\Rightarrow$	q
T	T	T	T	T	T	T	**T**	T
T	F	T	F	F	F	T	**T**	F
F	T	F	T	T	F	F	**T**	T
F	F	F	T	F	F	F	**T**	F
		1	2	1	3	1	4	1

FIGURE 3.3.4   This syllogism is a valid argument known as *modus ponens*.

This means that it is impossible for the conclusion to be false if the premises are true, a fact that is incorporated into our definition of the validity of any argument.

Whenever the truth table of the corresponding statement form of an argument is a tautology, the argument is said to be *valid*.

So we see that any syllogism having the form $[(p \Rightarrow q) \wedge p] \Rightarrow q$ is a valid argument. Because this particular reasoning scheme is so common, it has a special name. It is known both as *direct reasoning* and by the label **modus ponens**, which we will abbreviate as MP.

modus ponens
Valid reasoning scheme with form $[(p \Rightarrow q) \wedge p] \Rightarrow q$. Abbreviated as MP. Also called *direct reasoning*.

Another common type of syllogism, which falls into the category of *indirect reasoning*, can be illustrated by a review of Aristotle's proof that Earth is a sphere. Aristotle rejected the concept of a flat Earth as a result of his observations of lunar eclipses. See **Figure 3.3.5**.

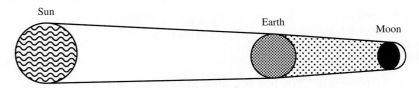

FIGURE 3.3.5   Observation leading to the rejection by Aristotle of a flat Earth.

He reasoned that if Earth were flat, it would cast a straight-edged shadow across the face of the moon during all lunar eclipses. Actual observations revealed no such edge but rather a shadow with a curved boundary. Therefore, Earth cannot be flat. (The extended conclusion of a spherical Earth comes by elimination—the only candidate shapes discussed seriously by the ancient philosophers were those of a sphere or plate.) This argument takes on this syllogistic form:

> **modus tollens**
> Valid reasoning scheme with form $[(p \Rightarrow q) \land \sim q] \Rightarrow \sim p$. Abbreviated as MT. Also called *indirect reasoning*.

Premise: If Earth is flat, then a straight-edged shadow appears on the moon during a lunar eclipse.

Premise: A straight-edged shadow does not appear on the moon during an eclipse.

Conclusion: ∴ Earth is not flat.

To analyze this by using symbolic logic, we can let p represent "Earth is flat" and q represent "A straight-edged shadow appears on the moon during a lunar eclipse." The corresponding statement form for the syllogism is then

$$\left[(p \Rightarrow q) \land \sim q\right] \Rightarrow \sim p,$$

which is shown in **Figure 3.3.6** to be a tautology. Indirect reasoning is also known by the label **modus tollens**, which we will abbreviate as MT.

p	q	$[(p$	$\Rightarrow$	$q)$	$\land$	$\sim q]$	$\Rightarrow$	$\sim p$
T	T	T	T	T	F	F	**T**	F
T	F	T	F	F	F	T	**T**	F
F	T	F	T	T	F	F	**T**	T
F	F	F	T	F	T	T	**T**	T
		1	2	1	3	1	4	1

FIGURE 3.3.6   This syllogism is a valid argument known as *modus tollens*.

Not all syllogisms are valid arguments, as shown in Example 3.

? Example 3

Construct a truth table to determine the validity of this syllogism:

Premise: If you live in Cuba, you are governed by Communists.
Premise: You are governed by Communists.
Conclusion: ∴ You live in Cuba.

⚙ Solution

Let p represent "You live in Cuba" and q represent "You are governed by Communists." Then the statement form for this argument is $[(p \Rightarrow q) \land q] \Rightarrow p$. The truth table is displayed in **Figure 3.3.7** and shows that the form is not a tautology. Hence the argument is invalid. ◆

p	q	\[(p	$\Rightarrow$	q)	$\wedge$	q\]	$\Rightarrow$	p
T	T	T	T	T	T	T	**T**	T
T	F	T	F	F	F	F	**T**	T
F	T	F	T	T	T	T	**F**	F
F	F	F	T	F	F	F	**T**	F
		1	2	1	3	1	4	1

FIGURE 3.3.7   Truth table for an invalid syllogism.

Venn diagrams
Pictorial diagram using circles for sets, initiated by John Venn and used to analyze the validity of arguments.

Many syllogisms have a categorical nature that allows them to be readily analyzed by pictorial diagrams known as **Venn diagrams**, in honor of their originator, **John Venn** (1834–1923). Venn was a professor of logic at Cambridge for four decades who made significant contributions to the formal structuring of symbolic logic initiated by George Boole. We illustrate with an example of an MP syllogism.

❓ Example 4

Premise: All flying animals have wings.
Premise: A robin flies.
Conclusion: ∴ A robin has wings.

Here the statement "All flying animals have wings" states that the category or set of flying animals is contained in the larger set of winged animals as a subset. This can be expressed visually by means of a Venn diagram. Based on the *first* premise, we draw a small circle to represent the set of all flying animals and place it inside a larger circle, which represents the set of all winged animals. (See **Figure 3.3.8**.) The placement of the specific animal, the robin, is then based on the *second* premise. Because a robin can fly, it is contained in the smaller set, which forces it to be a member of the larger set as well. The conclusion that a robin has wings is then consistent with the Venn diagram, *which has been drawn based on only the premises*. We, therefore, conclude that this syllogism is a *valid* argument.

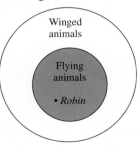

FIGURE 3.3.8   Venn diagram of an MP syllogism.

We can identify this as an MP syllogism by first rewording it as follows:

Premise: If an animal flies, then it has wings.
Premise: A robin flies.
Conclusion: ∴ A robin has wings.

We let p represent "An animal flies" and q represent "An animal has wings." Note that the use of the word *animal* serves as a variable in this case. Once *robin* has been substituted for *animal*, we may write the corresponding statement form as $[(p \Rightarrow q) \land p] \Rightarrow q$, which is the characteristic form of a type MP syllogism.   ♦

? Example 5

Draw a Venn diagram that corresponds to the following MT syllogism.

> All NFL linemen are at least 6'3" tall.
> Patrick is less than 6'3" tall.
> ∴ Patrick is not an NFL lineman.

⚙ Solution

The first premise implies that the set of NFL linemen must be contained in the set of all people who are at least 6'3" tall. (See **Figure 3.3.9**.) The second premise then requires that Patrick be placed outside the larger set, which in turn places him outside the interior set as well. Only after the diagram has been drawn do we consider the conclusion. Because the conclusion is indeed supported by the Venn diagram, we may conclude that this is a valid argument.

The corresponding statement form, $[(p \Rightarrow q) \land \sim q]$, is obtained by letting p represent "A person is an NFL lineman," letting q represent "A person is at least 6'3" tall," and substituting *Patrick* for *person*. This identifies the syllogism type as MT.   ♦

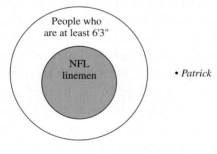

FIGURE 3.3.9   Venn diagram of an MT syllogism.

? Example 6

Draw a Venn diagram of the following syllogism, and determine its validity.

> All squares have four right angles.
> A rectangle has four right angles.
> ∴ A rectangle is a square.

⚙ Solution

Using only the premises, you find that it is possible in the Venn diagram to place the point associated with the rectangle inside the larger set yet outside the smaller one.

(See **Figure 3.3.10**.) Because the conclusion is not supported by a diagram that is consistent with the premises, this syllogism must be declared invalid. The conclusion is thus not guaranteed by the premises. If you wish to translate this syllogism to symbolic language, first you observe that it could be reworded as follows:

If a polygon is a square, then it has four right angles.

A rectangle has four right angles.

∴ A rectangle is a square.

Let p represent "A polygon is a square" and q represent "A polygon has four right angles." Once you substitute *rectangle* for *polygon*, you obtain the form $[(p \Rightarrow q) \wedge q] \Rightarrow p$, which has already been demonstrated in Figure 3.3.7 to not be a tautology. Both techniques verify that possession of four right angles does not ensure that a polygonal figure is a square. Both procedures demonstrate that the given argument is invalid. ◆

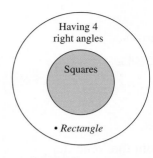

FIGURE 3.3.10 Venn diagram of an invalid syllogism.

It is important to be able to distinguish between argument validity and the truth values of the constituent statements. It is similar to our previous observations of true conditional statements having antecedents and consequents that were both false. Consider these next examples.

❓ Example 7

All winged animals can fly.   All senators are currently over 36 years old.

An ostrich has wings.   Everett is not over 36.

∴ An ostrich can fly.   ∴ Everett is not a senator.

⚙ Solution

These examples illustrate the irrelevance of the truth of either a premise or the conclusion—taken individually—when determining the *validity* of an argument. It is a fact, for instance, that an ostrich cannot fly. Likewise, any truth about the ages of senators varies with time. However, given that the premises of each argument are true, the truth of each conclusion is then inescapable according to the rules of logic. The first syllogism is of type MP, and the second is of type MT. ◆

We close with a final type of categorical example conducive to analysis by Venn diagrams.

❓ Example 8

The use of the word *not* in the first premise of each of the following syllogisms implies a separation of sets in the associated Venn diagrams. Determine the validity of the argument produced by each of the four variations of the second premise.

(a) Texans do not eat carrots.
     Erika is a Texan.
     ∴ Erika does not eat carrots.

(b) Texans do not eat carrots.
     Erika eats carrots.
     ∴ Erika is not a Texan.

(c) Texans do not eat carrots.
     Erika is not a Texan.
     ∴ Erika eats carrots.

(d) Texans do not eat carrots.
     Erika does not eat carrots.
     ∴ Erika is a Texan.

⚙ Solution

In both of the first two syllogisms, the given conclusions follow from the Venn diagrams produced by the premises. Both of these arguments are valid. However, no such thing can be said for the next two cases, because the second premise only directs us to place Erika *outside* a specific set but not necessarily *inside* the other set. Hence, both of these arguments are invalid.

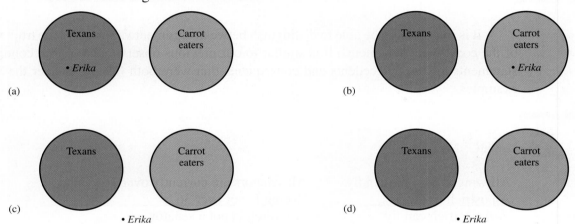

❓ Example 9

One of the most common mistakes of logic made by many people is to conclude that *all* members of one particular set are also members of some second set only because that is the case for one specific member. This mistake is the source of much prejudice throughout human history. So, for instance, just because one person of some faith or occupation or cultural background possesses some characteristic does not imply that all people of that same group share that characteristic. Is the following syllogism valid?

Jorgenson was born in Denmark.
Jorgenson dislikes rock 'n' roll music.
∴ All natives of Denmark dislike rock 'n' roll music.

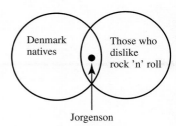

Jorgenson

 Solution

Clearly, this is an invalid argument. The premises only imply that the intersection of these two sets must be nonempty. They do not imply that one set is a subset of the other. So we can draw a diagram that is consistent with the premises but does not support the conclusion. ◆

We close this section by quoting a poignant passage from Leo Tolstoy's novella *The Death of Ivan Ilyich*, in which a man is first confronting his inner feelings about the hard reality of his imminent death. We include this here to acknowledge the simple fact that all too often logic is not enough to tackle all of life's issues.

In the depth of his heart he knew he was dying, but not only was he unaccustomed to such an idea, he simply could not grasp it, could not grasp it at all. The syllogism he had learned from Kiesewetter's logic—"Caius is a man, men are mortal, therefore Caius is mortal"—had always seemed to him correct as applied to Caius, but by no means to himself. That man Caius represented man in the abstract, and so the reasoning was perfectly sound; but he was not Caius, not an abstract man; he had always been a creature quite, quite distinct from all the others. . . . Caius really was mortal, and it was only right that he should die, but for him, Vanya, Ivan Ilyich, with all his thoughts and feelings, it was something else again.

—Leo Tolstoy

Name _____

Exercise Set **3.3** ◯◯◯←

Determine whether each of the following statement forms is a tautology.

1. $p \vee \sim p$

2. $p \vee q \Leftrightarrow p$

3. $(p \wedge q) \vee (p \vee q)$

4. $(p \wedge q) \Rightarrow p$

5. $[(p \Rightarrow q) \wedge q] \Rightarrow p$

6. $[(p \wedge q) \wedge \sim p] \Rightarrow \sim q$

7. $\sim(p \wedge q) \Rightarrow \sim p \vee \sim q$

8. $\sim(p \vee q) \Rightarrow \sim p \wedge \sim q$

9. $\sim(p \wedge q) \Rightarrow \sim p \wedge \sim q$

10. $\sim(p \vee q) \Rightarrow \sim p \vee \sim q$

11. $[(p \vee q) \wedge \sim p] \Rightarrow q$

12. $[(p \vee q) \wedge p] \Rightarrow \sim q$

13. $[(p \Rightarrow q) \vee (p \Rightarrow r)] \vee (q \Rightarrow r)$

14. $(p \wedge q) \vee (\sim p \wedge \sim q)$

15. $[(p \Rightarrow q) \wedge r] \Rightarrow [p \Rightarrow (q \wedge r)]$

16. $(p \Rightarrow q) \wedge (q \Rightarrow p) \Leftrightarrow (p \Leftrightarrow q)$

Each of the following algebraic expressions has a constant logical value. What is it?

17. $x \geq 3$ or $x < 3$

18. $7x + 2y = 10$ or $7x + 2y \neq 10$

19. $(x + 4)^2 \geq 0$

20. $x > 35$ and $x < 30$

21. $x + 3y > 6$ and $x + 3y \leq 6$

22. $(x > 25$ or $x \leq 25)$ and $|x| < 0$

23. $(x > 1$ and $y > 1)$ and $x + y = 2$

24. $a = 18$ or $(a - 18)^2 > 0$

🏆**25.** $\sin^2 \theta + \cos^2 \theta = 1$ or $\sin \theta = 0.5$

🏆**26.** $\sin \theta > 0$ and $\cos \theta = 2$

Create a statement form corresponding to each of the following syllogisms, and use it to determine its validity. When appropriate, classify the syllogism as **modus ponens (MP)** *or* **modus tollens (MT)**.

27. If it snows, then Amy builds a fort.

   It is snowing._____

   ∴. Amy is building a fort.

28. If Frederich practices his violin, then he gets dessert after supper.

   Frederich is not getting dessert._____

   ∴. He did not practice his violin.

29. If you do not watch television, you read books.

   You read books._____

   ∴. You do not watch television.

30. If the United States supports NATO, then it commits military support to Bosnia.

The United States does not support NATO.

∴ The United States does not commit military support to Bosnia.

31. Consumption increases if disposable income increases.

Consumption is not increasing.

∴ Disposable income is not increasing.

32. Every time Bert goes fishing, it rains.

It is not raining.

∴ Bert is not fishing.

33. Whenever a dog eats dirt, it gets sick.

Andy's dog does not eat dirt.

∴ Andy's dog does not get sick.

34. Whenever white bugs appear, a tomato spoils.

A tomato is spoiling.

∴ White bugs are appearing.

35. Mold forms whenever the humidity increases.

The humidity is decreasing.

∴ Mold is not forming.

36. New-home construction decreases whenever interest rates increase.

New-home construction is decreasing.

∴ Interest rates are increasing.

Draw a Venn diagram for each of the following syllogisms, and determine whether the syllogism constitutes a valid or invalid argument.

37. All men are mortal.

Caius is a man.

∴ Caius is mortal.

38. All parallelograms have two pairs of equal sides.

A rectangle is a parallelogram.

∴ A rectangle has two pairs of equal sides.

39. All men are mortal.

Caius is mortal.

∴ Caius is a man.

40. All rectangles have four right angles.

A triangle does not have four right angles.

∴ A triangle is not a rectangle.

41. All stars contain hydrogen.

The moon does not contain hydrogen.

∴ The moon is not a star.

42. Every planet orbits the sun.

Halley's comet orbits the sun.

∴ Halley's comet is a planet.

43. Everyone who smokes cigarettes has a faster than average heart rate.

Shelly does not smoke cigarettes.

∴ Shelly does not have a faster than average heart rate.

44. Humans are all capable of rational thought.

A chimpanzee is not human.

∴ A chimpanzee is not capable of rational thought.

45. Dinosaurs did not have fur.

A woolly mammoth had fur.

∴ A woolly mammoth was not a dinosaur.

46. Accountants never work fewer than 40 hours per week (h/wk).

Tracy works 45 h/wk.

∴ Tracy is an accountant.

47. No resident of Cleveland is a fan of the Pittsburgh Steelers.

Rick does not live in Cleveland.

∴ Rick is a fan of the Steelers.

48. Nobody over age 70 goes skiing.

Georgia is a skier.

∴ Georgia is not over age 70.

49. Thomas is from Wisconsin.

Thomas plays the clarinet.

∴ Everyone from Wisconsin plays the clarinet.

50. Omar is from Syria.

Omar is a farmer.

∴ All Syrians are farmers.

51. Ishtak is a lawyer.

Ishtak plays tennis.

∴ All lawyers play tennis.

52. Vito has brown eyes.

Vito is Catholic.

∴ All Catholics have brown eyes.

If possible, deduce a conclusion from the given premises (other than repeating a premise), using valid reasoning. If this cannot be done, then write, "No conclusion is possible."

53. If you study mathematics more than 6 h/wk, you will not fail math class. Alexander studies mathematics 8 h/wk. Therefore, _____.

54. If Karolyn plays poker during the week, she borrows money on Friday. Karolyn did not borrow money on Friday. Therefore, _____.

55. If the defendant is innocent, then he is not convicted. The defendant is guilty. Therefore, _____.

56. All Finns have blue eyes. Leonard has blue eyes. Therefore, _____.

57. If the density of a substance exceeds 1.0 gram per cubic centimeter (g/cm³), then the substance sinks. The cork from that wine bottle floats. Therefore, _____.

58. All Republican senators support budget amendment 22. Senator Bureaucrat does not support budget amendment 22. Therefore, _____.

59. All Republican senators support budget amendment 22. Senator Snort is not a Republican. Therefore, _____.

60. All piano players are musicians. Constantine plays the guitar. Therefore, _____.

61. Everyone who eats spinach is strong. Popeye is not strong. Therefore, _____.

62. No one from Luzerne County likes jazz music. Gilford is not from Luzerne County. Therefore, _____.

63. History majors never take more than one mathematics class. Khantana has taken three mathematics classes. Therefore, _____.

3.4 Computer Programming Structures

> Emotions are illogical.
>
> —Mr. Spock, *Star Trek*

When Mr. Spock spoke in the original *Star Trek* television series, his voice usually contained none of the inflections commonly associated with human emotion—a trait shared by the humanoid Data in the follow-up series, *Star Trek: The Next Generation*. This tendency endowed both characters with a personality that most of us would say was, at least in part, rather inhuman. At the same time, the essence of both of these characters centered on their ability to use pure logic in their problem-solving methods. This correspondence of logic and cold detachment stems from most people's observations of **computers**—machines designed to perform numerical computations and to process information. Why are these machines so good at employing logic?

computers Machine that performs numerical computations and processes information.

The answer is connected to electricity. Electricity empowers the logic center of a computer because of the simple fact that, at any given time, an electrical circuit is in one of two states: *on* or *off.* Similarly, every storage unit in a computer's memory, called a **byte**, is typically composed of 8 magnetized **bits**, each of which must be either *positive* or *negative.* We say such objects have a **binary** (two-valued) existence, or are binary-valued, and it is this quality that makes orchestrated collections of electrical circuits (as in a computer) so appropriate for implementing logical processes. We have already seen that effective logical analysis is rooted in the fact that statements have a binary quality—they are either true or false. Additionally, each connective is really a binary-valued *function* whose values may be synthesized in a computer by the use of an appropriate mechanism called a **logic gate**.

FIGURE 3.4.1 Electricity runs a computer.

© Christina Richards/ShutterStock, Inc.

The main gates used for logic control in a computer are for conjunction (AND), disjunction (OR), and negation (NOT). The AND and OR gates each have two input terminals and one output terminal, while the NOT gate has just one input terminal and one output terminal. By identifying an abstract statement with a physical electrical conduit, we can associate the value of the statement as true with a pulse being sent through the conduit (i.e., *on*) and the value of false with the absence of a pulse (i.e., *off*). To conform to the truth table for conjunction, the AND gate must be constructed according to the specifications depicted in **Figure 3.4.2**. Notice how this correlates to the values that $p \wedge q$ acquires as a result of the four combinations of logical values for p and q.

byte Computer memory storage unit, usually composed of 8 magnetized bits.

bits The smallest computer memory unit. It is binary-valued.

binary Having exactly two possible values.

logic gate Computer mechanism that simulates a logical connective.

Similarly, the truth tables for disjunction and negation mandate that the OR and NOT gates be constructed according to the diagrams in **Figure 3.4.3**.

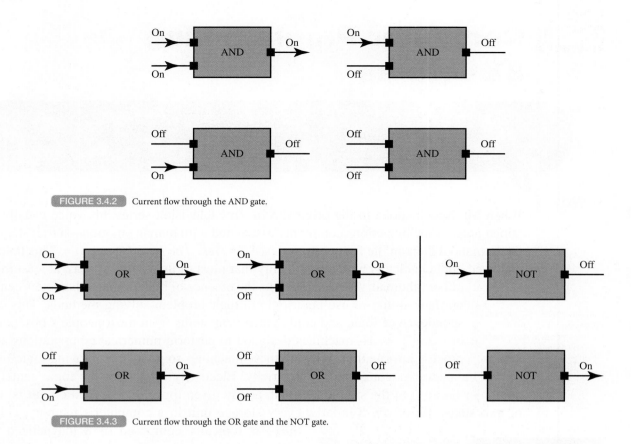

FIGURE 3.4.2 Current flow through the AND gate.

FIGURE 3.4.3 Current flow through the OR gate and the NOT gate.

A *circuit* is a closed loop of electrical conduit running between a power source and an appliance such as a lightbulb. If a switch is inserted in the loop, then the light will be *on* if the switch allows a pulse through and *off* if it does not allow a pulse. For our purposes, we define a **logic circuit** to be a combination of electrical conduits and connective gates that corresponds to a single statement form and acts as one giant switch. If the output terminal of the logic circuit is on, then this corresponds to the form being true. If it is off, then the form is false. Example 1 illustrates this process.

logic circuit
Combination of logic gates (i.e., electrical circuits and connective gates) that simulates a statement form.

❓ Example 1

Diagram a logic circuit that corresponds to the statement form $\sim(p \vee q) \wedge \sim r$, given that p and q both have value T and r has value F. Is the circuit on or off?

⚙ Solution

Naturally, the order of evaluation of the connectives is crucial in setting up the circuit in **Figure 3.4.4**. We see that this circuit will be off if p, q, and r are initially T, T, and F (on, on, and off, respectively). This means that the corresponding statement form, $\sim(p \vee q) \wedge \sim r$, is false. Clearly, changing the values of the prime statements corresponds to a change in the initial states of the pulses into the circuit, which in turn can affect the final state of the circuit. ◆

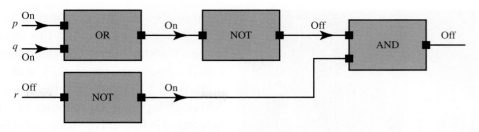

FIGURE 3.4.4   Logic circuit corresponding to the form $\sim(p \lor q) \land \sim r$.

FIGURE 3.4.5   A portrait of John von Neumann (1903–1957).

© INTERFOTO/Alamy Images

central processing unit The brain of a computer that performs the arithmetic and logical operations. Abbreviated CPU.

input device Computer component that transmits information to the memory.

output device Computer component that displays results from the performance of a computer program.

memory The component of a computer where information is electronically stored in locations called cells.

cells Computer memory storage unit composed of 1 or more bytes.

program List of instructions for a computer to translate to machine language and then perform.

This process gives us a bit of the flavor of how logical operations are done by a computer. A rapid expansion in the sophistication and plurality of uses of computers has occurred in recent years. The original idea for the design of the computing machine is due, in large part, to **John von Neumann** (1903–1957), one of the foremost mathematicians and visionary thinkers of the twentieth century.

Born in Budapest, Hungary, and educated in Europe, von Neumann moved to America in 1930 to teach mathematical physics at Princeton University. In 1933, he accepted an offer to be a professor at the Institute for Advanced Study, where he produced highly original work in quantum theory, nuclear physics, economics, logic, set theory, and the theory of high-speed computing devices. His memory, reasoning abilities, and computational speed were legendary. He received many prestigious awards in his lifetime, including the Medal of Merit (for his service to the military as a consultant), the Enrico Fermi Award, the Medal of Freedom, and the Albert Einstein Award. The image, however, of mathematicians as a somber, serious group lacking in social graces certainly did not fit John von Neumann. Witty, charming, and easygoing, "Good-time" Johnny was the host of innumerable entertaining parties throughout his illustrious career. The premature death of this popular genius from cancer in 1957 saddened the hearts of many. Among his greatest and most enduring gifts to society are his contributions to the development of the computer.

In von Neumann's early machines, the computations were done with large, bulky vacuum tubes. (A 1950s version equivalent to a current desktop personal computer would have filled a sizable garage!) In the basic architecture of today's modern computing machine, the logical and arithmetic operations are performed in an elegant weave of minute circuitry known as the **central processing unit** (CPU). The CPU is the so-called brain that orchestrates the other main components, such as the *input device, memory,* and *output device.* (See **Figure 3.4.6**.) The **input device** (usually a keyboard) transmits information from the user to the memory, while the **output device** (usually a monitor) provides a medium for transmitting the results of the computer operations to the user. The **memory** contains a large number of storage locations called **cells**, which, in turn, are each composed of a fixed number of bytes. In order for the computer to perform some task, a list of instructions or commands, called a **program**, must be provided by the user. Most programs are written in a high-level language such as BASIC,

FIGURE 3.4.6   John von Neumann with one of the first computers.

---

machine language
The program instructions in a special language that the computer understands.

algorithm A finite, ordered list of steps for solving a problem.

pseudocode
Description of an algorithm that the programmer designs before writing the program.

Java, or C++ that uses recognizable English terms. Once read into memory through the input device, a program is then translated by a compiler into a set of instructions called **machine language** that is written in eight-digit, base-2 numbers that the computer can understand. Although learning the specifics of either of these types of languages is beyond the scope of this book, we will sneak a brief look at how a programmer at least begins the process of designing a program to perform certain tasks with a computer.

An **algorithm** is a finite, ordered list of steps for solving a problem. To organize our thoughts about constructing an appropriate algorithm for a specified task, we need a bridge from the fuzziness and ambiguities of everyday English language to the rigid program rules required by the computer. This bridge is basically a generic description of the algorithm designed and written by a programmer in easily understandable instructions called **pseudocode**. We will use a version of the standard pseudocode that has been slightly modified for our purposes.

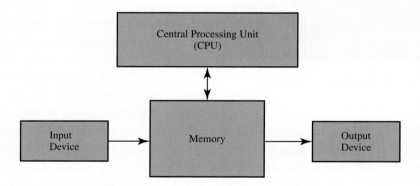

Suppose we want the computer to add any two given numbers and to print the sum. This task requires four distinct steps to be performed. First, the user needs to be notified that the computer is ready to add two numbers. Second, the computer must accept the numbers from the user and store them in two memory cells. Every cell in the computer's memory has an address that is accessed by the use of a *variable* in a program. We shall use variables as needed in our pseudocode without worrying about specific details concerning the type of information they store. Third, the CPU must add the numbers and store the result in a third separate cell; fourth, the content of this cell must be printed on the screen. Numbering our steps, we write

1. **print** "Input two numbers to be added."
2. **read** x, y
3. $x + y \rightarrow S$
4. **print** "The sum of", x, "and", y, "is", S
5. **stop**

We make several important observations about this pseudocode.

- Each word in boldface indicates a computer command. These are similar to the actual commands used in the high-level language version.
- Each italicized letter is a variable.
- The **read** command accepts whatever numbers the user types on the keyboard and places them in memory into two cells addressed by x and y. For example, if the user types in 9 and 15, then we might imagine their storage in memory as

$$\boxed{9} \quad \boxed{15}$$
$$x \qquad y$$

- The arrow, $\rightarrow$, indicates an assignment or placement. The two numbers stored in x and y are added by the CPU, and the sum is placed into the memory cell addressed by the variable S. The situation in memory takes on this form:

$$\boxed{9} \quad \boxed{15} \quad \boxed{24}$$
$$x \qquad y \qquad S$$

- The **print** command prints verbatim on the screen those words that are placed between quotes. These provide communication with the user and are called I/O (for input/output) messages.

Any program written from such pseudocode would terminate after one execution. Suppose we wished the computer to repeatedly ask for two positive numbers that it would add until the user supplied a nonpositive input. We would have to modify our algorithm by allowing for task repetition while also incorporating a decision-making ability to tell the computer when to stop. These are two separate capabilities found in all major programming languages.

> **looping** Program structure for repeating a set of steps.

The ability of a computer to rapidly repeat the same task or set of steps again and again is a process known as **looping**, and it is one of its most powerful applications. Looping is one type of control structure implemented in various ways depending on the language. We will indicate that a particular set of steps is to be repeated by nesting it (via indentation) inside the pseudocode words **loop** and **endloop.** A decision must be made to terminate a loop based on some condition being satisfied. A very powerful **decision structure** employed by all major languages is the **if-then-else-endif** statement, which allows for program flow to be directed along one of two main branches. The indentations in this pseudocode are for clarity.

> **decision structure** Programming method for making a decision.

1. **if** $(a > 1$ and $b \geq 3)$ **then**
2. $\quad a/2 + 5 * b \rightarrow r)$ (The symbol "$*$" is used for multiplication.)
3. **else**
4. $\quad 5 * a + b/2 \rightarrow r$
5. **endif**

> **Boolean expression** Computer programming name for an algebraic statement. It is either true or false.

You should recognize the *algebraic statement* following the **if** and recall that its logical value (true or false) depends on the current values of a and b. Computer scientists call such true or false statements **Boolean expressions** (in honor of the father of symbolic logic, George Boole). A Boolean expression must always appear after the **if** in our pseudocode. If the expression is true, the action stated after the **then** is performed, and the action following the **else** is skipped.

On the other hand, if the expression is false, the action stated after the **then** is skipped, and the action following the **else** is carried out. For example, the following values are placed in r as a result of this pseudocode and correspond to the stated values of a and b.

6	7	38
a	*b*	*r*

2	4	21
a	*b*	*r*

0	110	55
a	*b*	*r*

9	2	46
a	*b*	*r*

After the action is performed, program execution then proceeds to the line immediately after the **endif**. In the event that we do not wish any alternative action to occur when the Boolean expression is false, the **else** portion may be omitted.

Usually, there exists more than one way to write a decision structure with an **if-then-else-endif**. You should check to see that the example here could also have been written as

1. **if** $(a \leq 1$ or $b < 3)$ **then**
2. $5 * a + b/2 \to r$
3. **else**
4. $a/2 + 5 * b \to r$
5. **endif**

We note that the Boolean expression used following the **if** in this case is the negation of the previous one. The choice of pseudocode generally is a matter of style, efficiency, and the individual programmer.

Example 2

Write pseudocode for a program that repeatedly adds a pair of numbers and prints the sum until either of the inputted values is negative or zero. Include I/O messages.

Solution

A loop structure is now needed to allow for repetition. To exit the loop when the required condition is met, we also need an appropriate decision structure.

1. **loop**
2.   **print** "Input two numbers to be added."
3.   **read** x, y
4.   **if** $(x > 0$ and $y > 0)$ **then**
5. $x + y \to S$
6.   **print** "The sum of", x, "and", y, "is", S, "."
7.   **else**
8.   **go to** 11
9.   **endif**
10. **endloop**
11. **print** "Have a good day!"
12. **stop** ◆

Steps 2 through 9 constitute the *body* of the loop and are to be repeated as long as the inputs are both positive. Note the action command "**go to** 11," following the **else**. This provides an exit from the loop as the program directs the computer to immediately branch to the line numbered 11. Usually, it is preferable to avoid the use of the **go to** command in the high-level language version because of the multiple branching, program entanglements, and confusion it can create. Most languages have loop structures that use a Boolean expression to terminate the loop without the use of a **go to** command. The expression is located at the beginning or end of the loop and creates a more streamlined top-down format. Unfortunately, a more detailed discussion would get quite involved and would divert us from our main goals here.

The above algorithm could also have been written as follows:

1. **loop**
2.   **print** "Input two numbers to be added."
3.   **read** x, y
4.   **if** $(x \leq 0 \text{ or } y \leq 0)$ **then go to** 8 **endif**
5. $x + y \to S$
6.   **print** "The sum of", x, "and", y, "is", S, "."
7.   **endloop**
8.   **print** "Have a good day!"
9.   **stop**

Note that the **else** is unnecessary here and that we have collapsed the exit condition to a single line for easier reading of the pseudocode. Here is a sample run of a program written from either of the above pseudocode designs.

```
Input two numbers to be added.
9  15
The sum of 9 and 15 is 24.

Input two numbers to be added.
54  27
The sum of 54 and 27 is 81.

Input two numbers to be added.
60  -3
Have a good day!
```

? Example 3

Suppose you have a job at which you earn $6.50 per hour for the first 40 h in a week and $9.75 (or 1½ times the normal rate) for every hour after 40 h. Write pseudocode for a program that determines your weekly earnings, given the number of hours you worked that week.

Solution

1. **print** "How many hours did you work this week?"
2. **read** h
3. **if** $(h \leq 40)$ **then**
4. $6.5 * h \rightarrow m$
5. **else**
6. $6.5 * 40 + 9.75 * (h - 40) \rightarrow m$
7. **endif**
8. **print** "Your earnings this week are", m
9. **stop** ◆

---

incrementation
Repeated increase of the value of a variable by a fixed amount.

One final programming technique we examine is the notion of **incrementation**, or the repeated increase of the value of a variable by a fixed amount. Consider the following algorithm.

1. $19 \rightarrow N$
2. **loop**
3. $N + 1 \rightarrow N$
4.   **print** N
5.   **if** $(N = 25)$ **then go to 7 endif**
6. **endloop**
7. **stop**

initialization
Assignment of an initial value to a variable.

In line 1, the value 19 is stored in the variable N. (This is called an **initialization** of a variable. Without it, some random unknown value could initially be stored in N.) In line 3, the current value stored in N is *replaced* by a value that is 1 larger than N, that is, $N + 1$. We say that N has been *incremented* by 1. The output of this algorithm would resemble the following column of numbers.

$$20$$
$$21$$
$$22$$
$$23$$
$$24$$
$$25$$

Observe that your output depends on the location of the print command within the algorithm. If we exchanged lines 4 and 5 of this example, then the number 25 would not get printed. In that case, the output would be changed to the following list of numbers.

$$20$$
$$21$$
$$22$$
$$23$$
$$24$$

pixels Smallest element of a television or computer screen.

Incrementation is used in the drawing of graphics on your computer screen. The smallest elements that make up your television or computer screen are called **pixels**, which are arranged in a rectangular fashion and labeled in a manner similar to a Cartesian coordinate system. (More pixels result in better resolution of the picture.) The number of bits associated with each pixel determines the pixel's color and brightness. In the simplest case, only one bit is used, which can be turned either on (black) or off (white). We will consider this case.

To create graphical images on a computer screen, instructions must be given to the computer to tell it which pixels are to be turned on or off. Consider the (enlarged) grid of pixels arranged according to the format in **Figure 3.4.7**. Starting with column 1 on the left and row 1 on the bottom, each pixel is labeled by an ordered pair of positive integers (x, y), where the x-coordinate marks the column number and the y-coordinate marks the row number. The pixels (3, 5) and (11, 8) have been marked.

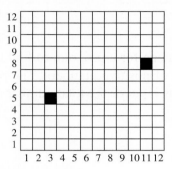

FIGURE 3.4.7   A grid of pixels. The pixels (3, 5) and (11, 8) have been marked.

? Example 4

Create an algorithm that would draw the line shown in **Figure 3.4.8**.

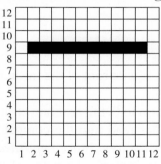

FIGURE 3.4.8   Drawing a line on a pixel grid.

⚙ Solution

We will use the command **flip**(x, y) to instruct the computer to "flip" the pixel at position (x, y) from off to on. The line begins at position (2, 9) and ends at (11, 9). Our algorithm must tell the computer to darken each pixel between and including these two points.

1. $2 \rightarrow x$
2. $9 \rightarrow y$
3. **loop**
4.   **flip**(x, y)
5. $x + 1 \rightarrow x$
6.   **if** $(x = 12)$ **then go to** 8 **endif**
7. **endloop**
8. **stop** ◆

Functions play a large role in computer science, just as they do in all areas of mathematics. In the design and creation of any program, they help to modularize, or compartmentalize, the tasks that the program is supposed to solve. All high-level languages have a set of intrinsic functions built into them, which the programmer may employ by using the correct function name and the proper arguments. Examples would include many of the function buttons on your calculator. Example 5 uses the function $sqrt(x)$, which takes the square root of a nonnegative number x.

? Example 5

Write the pseudocode for a program to find the square root of the mean (arithmetic average) of a list of positive numbers. The program should end if either a negative number is given as input or the total number of values in the list exceeds 10. Include I/O messages.

⚙ Solution

1. $0 \rightarrow s$
2. $0 \rightarrow n$
3. **print** "Input the next number."
4. **read** x
5. $n + 1 \rightarrow n$
6. **if** $(x \geq 0$ and $n \leq 10)$ **then**
7. $s + x \rightarrow s$
8.   **go to** 3
9. **endif**
10. $sqrt(s/n) \rightarrow r$
11. **print** "The square root of the mean is", r
12. **stop**

By initializing n to 0, the program ensures that the assignment $n + 1 \rightarrow n$ counts the number of values read into the computer. (In such cases, n is called a *counter* variable.) *Note how the loop is accomplished here.* If both $x \geq 0$ and $n \leq 10$, then the new input is used to increase the value of the sum s, and we go back to statement 3 and read in the next value. However, if either of those conditions is false, then we skip statements 7 and 8 and finish up. In this case, the **loop-endloop** process is not needed, because that procedure has been accomplished with the **go to** inside the **if-then-endif.** ♦

Programming on a Calculator

Many calculators have the capacity to be programmed using a language specific to the model. On the Texas Instruments (TI) graphing calculators, for example, the button **PGRM** accesses this option. We demonstrate how the pseudocode in Example 5 can be implemented on the TI-83+. The code is given first, followed by an explanation. The companion guidebook gives much more detailed programming instructions.

```
PROGRAM: EX5
: 0 → S
: 0 → N
: Lbl 1
    : Disp "NEXT NUMBER?"
    : Prompt X
    : N + 1 → N
    : If (X ≥ 0 and N ≤ 10)
    : Then
        : S + X → S
        : Goto 1
    : End
: √(S/N) → R
: Disp "THE SQRT IS", R
: Stop
```

1. Press **PGRM**, and scroll to select **NEW**. Select **Create New**, and enter a name for the program. You will now automatically be in the program editor. Note that each statement must follow a colon sign.

2. Type "$0 \to S$" using the **STO** button for the storage arrow and **ENTER**. Note that the purpose of this arrow is the same as in our pseudocode version. Type "$0 \to N$".

3. Several means are available to implement a loop, including **For, While,** and **Repeat** control statements. However, we will use **Goto,** a previous statement label, to avoid introducing another new command. (**While** is explored in Exercises 52–57.) We must therefore insert a label number as a marker for the program to reenter the top of our loop. (Label numbers are only used to provide reentry points to the program.) While in the editor, press **PGRM**, select **CTL,** and then **Lbl.** Type "1" as your label. Note that all the control statements are accessed under **CTL,** which appears after pressing **PGRM** while in the editor.

4. Our **print** pseudocode is implemented by the TI's **Disp** (for display) statement. This is listed under **I/O**, which you can select with the cursor, after pressing **PGRM**.

5. Our **read** pseudocode is implemented by **Prompt**. **Prompt** X will ask for a value for X by printing "X = ?" on the screen when you run the program. **Prompt** is also found in the **I/O** list.

6. The relational operators ≥, **and**, and ≤ in the Boolean expression after the **If** are listed under **TEST** and **LOGIC** after pressing **2nd** and **MATH**. If, Then, **End**, and **Goto** are all control statements.

7. All programs must terminate with **Stop**.

We conclude our brief look at programming by using the computer to encrypt messages.

Example 6

The computer is used extensively in *cryptology,* the study of code encryption. A *Caesar cipher* is a scheme for writing a secret message by shifting every letter to the right in the standard alphabet by a constant number k of positions. (The encryption of CAT to FDW is a Caesar encryption with shift constant $k = 3$.) Write pseudocode to shift a given letter up by 5 positions. Include I/O messages.

⚙ Solution

Every computer has a fixed enumeration that it associates with the alphabet, called its *collating sequence*. We will assume that letters A to Z are numbered 0 to 25 (we do not concern ourselves with lowercase letters). We need two intrinsic functions, called *chr(n)* and *ord(x)*. The first function returns a letter, given its number in the collating sequence. For instance, *chr*(2) is C and *chr*(24) is Y. The second does the opposite— it returns the number corresponding to a given letter. For instance, *ord*(F) is 5 and *ord*(W) is 22. To implement a Caesar cipher, we need to be able to shift letters at the end of the alphabet to those at the beginning. This is done using what is known as the *modulo* function, **mod**. The output of x **mod** 26 is the remainder obtained when x is divided by 26; for example, 29 **mod** 26 is equal to 3. Note that the variables a and b used here hold character—not numerical—values.

1. **print** "Please input an uppercase letter."
2. **read** a
3. $ord(a) + 5 \rightarrow x$
4. x **mod** $26 \rightarrow y$
5. $chr(y) \rightarrow b$
6. **print** "The letter shifted by 5 positions is", b
7. **stop** ◆

Pseudocode Word List

print	read	stop	if-then-else-endof
loop	endloop	goto	flip

Name _____

Exercise Set **3.4** ⃝⃝⃝◀──────────────

Diagram a logic circuit that corresponds to each of the following statement forms. Given that all the prime statements are true, determine whether the circuit would be on or off.

1. $\sim p \vee q$

2. $\sim(p \vee q)$

3. $p \wedge \sim(q \wedge r)$

4. $(p \wedge \sim q) \wedge \sim r$

5. $(\sim p \vee \sim q) \vee r$

6. $(p \wedge q) \wedge \sim(r \wedge s)$

7. $p \wedge \sim q \vee \sim(r \wedge \sim s)$

8. $\sim p \wedge (q \vee r) \wedge \sim s$

9. $(\sim p \vee \sim q \wedge r) \wedge s$

10. $\sim(p \vee q) \wedge \sim(r \vee s)$

Write an equivalent decision structure, using the negation of the given Boolean expression.

11. **if** $(x \geq y)$ **then**

   $x + 60 \to m$

   else

   $y + 60 \to m$

   endif

12. **if** $(N = 500)$ **then**

   $7 * N \to a$

   else

   $6 * N + 8 * (N + 1) \to a$

   endif

13. **if** $(r \geq 75 \text{ and } r \leq 100)$ **then**

   $2 * r + 1 \to s$

   else

   $3 * r + 1 \to s$

   endif

14. **if** $(w > 0 \text{ and } t > 0)$ **then**

   $2 * (w + t) \to D$

   else

   $2 * (w - t) \to D$

   endif

15. **if** $(M \neq N \text{ or } M + N > 1{,}000)$ **then**

   $sqrt(M * N) \to P$

   else

   $sqrt(M + N) \to P$

   endif

16. **if** $(S > 32 \text{ or } S \neq T + 10)$ **then**

   $S \bmod 2 \to E$

   else

   $T \bmod 3 \to E$

   endif

In Exercises 17 and 18, compute the value to be assigned to the empty memory cell, using the given values for the other cells.

17. (a) For the pseudocode in Exercise 11, find m.

18	13	☐
x	y	m

(b) For the pseudocode in Exercise 13, find s.

$$\boxed{45} \qquad \boxed{}$$
$$r \qquad\qquad s$$

(c) For the pseudocode in Exercise 15, find P.

$$\boxed{72} \qquad \boxed{72} \qquad \boxed{}$$
$$M \qquad\quad N \qquad\quad P$$

18. (a) For the pseudocode in Exercise 12, find a.

$$\boxed{500} \qquad \boxed{}$$
$$N \qquad\qquad a$$

(b) For the pseudocode in Exercise 14, find D.

$$\boxed{537} \qquad \boxed{-463} \qquad \boxed{}$$
$$w \qquad\qquad t \qquad\qquad D$$

(c) For the pseudocode in Exercise 16, find E.

$$\boxed{30} \qquad \boxed{20} \qquad \boxed{}$$
$$S \qquad\quad T \qquad\quad E$$

Write the output of a program written from the pseudocode given in each of the following problems.

19. 1. $0 \to n$

   2. loop

   3. $n + 2 \to n$

   4.   print n

   5.   if $(n > 11)$ then go to 7 endif

   6. endloop

   7. stop

20. 1. $5 \to x$

   2. $7 \to y$

   3. if $(x \geq 0$ or $y \geq 10)$ then

   4. $x + y \to s$

   5. else

   6. $x * y \to s$

   7. endif

   8. print s

   9. stop

21. 1. $256 \to n$

    2. loop

    3. $sqrt(n) \to n$

    4.   print n

    5.   if ($n < 4$) then go to 7 endif

    6. endloop

    7. stop

22. 1. $5 \to x$

    2. $7 \to y$

    3. if ($x \geq 0$ and $y \geq 10$) then

    4. $x + y \to s$

    5. else

    6. $x * y \to s$

    7. endif

    8. print s

    9. stop

If a loop is never exited, it results in a programming nightmare known as an infinite loop. This usually happens when the Boolean expression used to exit the loop has a constant logical value, as discussed in the previous section. Explain why each of the following decision structures would result in an infinite loop.

23. 1. $0 \to n$

    2. loop

    3. $n + 1 \to n$

    4.   print n

    5.   if ($n < 1$) then go to 7 endif

    6. endloop

    7. stop

24. 1. $5 \to x$

    2. loop

    3. $2 * x \to x$

    4.   print x

    5.   if ($|x| < 0$) then go to 7 endif

    6. endloop

    7. stop

25. 1. $6 \to k$

    2. $12 \to m$

    3. loop

    4. $k + 3 \to k$

    5. $m + k \to S$

    6.   if ($S > 40$ and $m < 12$) then go to 8 endif

    7. endloop

    8. stop

26. 1. $7 \to A$

    2. $-7 \to B$

    3. loop

    4. $3 * A \to A$

    5. $(-5) * B \to B$

    6. $A * A + B * B \to R$

    7.   if ($R \leq 0$) then go to 7 endif

    8. endloop

    9. stop

Using incrementation, write pseudocode for an algorithm that draws each of the graphic images on the given pixel grid.

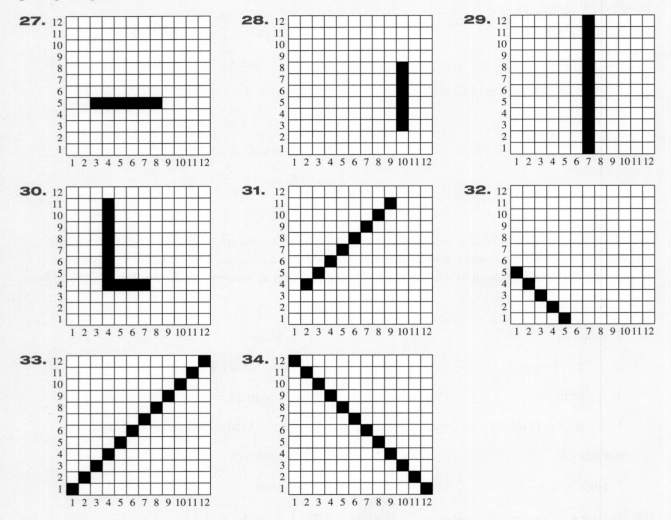

Write the pseudocode for a program that performs each of the following tasks. Include I/O messages. Use a loop if repetition is requested.

35. The property tax in Greenborough County is 1% of the worth of a person's house unless the property is worth $80,000 or more. In that case, the owner is charged $800 plus 2% of the amount in excess of $80,000. Determine the tax on a property, given its value.

36. The income tax in FantasyLand is 5% of gross income for people who earn less than $32,000 annually and 8% otherwise. Determine the net (after-tax) income, given the gross income as input.

37. Arthur's Art Gallery is having a sale this month. A total purchase of less than $100 will be discounted 10%, while any purchase of $100 or more will be reduced by 15%. Determine the final charge to the customer (including the sales tax of 6%), given the total cost of the purchases.

38. Determine the area of a rectangle, given any positive-valued length and width, and allow for repetition. The program should stop if an input is given that is a negative number or zero.

39. Determine the perimeter of a rectangle, given any positive-valued length and width, and allow for repetition. The program should stop if an input is given that is a negative number or zero.

40. Determine the area and circumference of any circle, given any positive-valued radius, and allow for repetition. The program should stop if an input is given that is a negative number or zero. (Use 3.1416 as an approximation for π.)

41. The roots of the quadratic equation $ax^2 + bx + c = 0$ are real numbers if $b^2 - 4ac \geq 0$, and they are complex numbers if $b^2 - 4ac < 0$. Determine whether the roots of this quadratic equation are real or complex, given any coefficients a, b, and c.

42. Determine the length of the hypotenuse of a right triangle given the lengths of both legs.

43. In Example 6, the Caesar encryption scheme is defined. Devise an algorithm to shift a letter by k positions, given any letter and any shift constant k.

44. In Example 6, the Caesar encryption scheme is defined. Devise an algorithm to shift any given letter by 12 positions.

Programming with a Calculator

In Exercises 45–50, use the pseudocode from the stated previous exercise to write the corresponding program on a programmable calculator. Use the program to do the requested computation.

45. Exercise 35: Determine the tax for a property worth (a) $61,000; (b) $95,800; (c) $146,500.

46. Exercise 36: Determine the net income for a person who earned an annual income of (a) $21,500; (b) $30,700; (c) $54,900; (d) $93,150.

47. Exercise 37: Determine the final charge of a purchase of (a) $78.00; (b) $96.15; (c) $137.25; (d) $210.98.

48. Exercise 38: Determine the area of a rectangle having (a) width 12 and length 20; (b) width 17.5 and length 87.7; (c) width 23.11 and length 49.88.

49. Exercise 41: Find the nature of the roots (real or complex) for (a) $a = 2$, $b = -7$, $c = 3$; (b) $a = 4$, $b = 3$, $c = 5$; (c) $a = -1$, $b = 2$, $c = 11$; (d) $a = 3$, $b = 6$, $c = 3$.

50. Exercise 42: Find the length of the hypotenuse, given that the lengths of the legs are (a) 3 and 4; (b) 5 and 12; (c) 6 and 13; (d) 29 and 42.

51. An alternative to the -83 code that performs the loop in Example 5 is given here. Does this accomplish the same task? Why do you think it seems less appealing to use than the code in Example 5?

```
: Lbl 1
    : Disp "NEXT NUMBER?"
    : Prompt X
    : N + 1 → N
    : If (X < 0 or N > 15)
    : Then: Goto 2
    : End
    : S + X → S
: Goto 1
: Lbl 2
```

*Write the pseudocode that performs the tasks in the remaining exercises. Code them on a programmable calculator, using a **While** command, as explained in Exercise 52. Use the program to do the requested computations.*

52. Use of the **While** command to implement a loop facilitates a "top-down" programming approach. On a TI-83, the statements between **While** (Boolean expression) and **End** are executed as long as the Boolean expression remains true. For example, the following code finds the mean (average) of 15 numbers given by the user.

```
: 0 → S
: 1 → N
: While (N ≤ 15)
    : Prompt X
    : S + X → S
    : N + 1 → N
: End
: S/15 → A
: Disp "The mean is", A, "."
: Stop
```

The three statements between **While** and **End** will be repeatedly performed until N increases to 26. At that point, the expression "$N \leq 15$" becomes false, and execution skips past the **End** to "$S/15 \rightarrow A$." Code this program on a calculator and run it.

53. Find the square root of the mean of the squares of a list of 10 numbers. Find this value for (a) the first 10 even positive integers; (b) 10, 20, 30, . . . , 100.

54. Every term (except the first one) in an *arithmetic sequence* with common difference d is equal to the sum of d and the preceding term. Add the first 25 terms of an arithmetic sequence, given any first term a and any common difference d. Find this sum for (a) $a = 7$ and $d = 3$; (b) $a = 10$ and $d = 2.8$.

55. Every term (except the first one) in a *geometric sequence* with common ratio r is equal to r times the preceding term. Add the first 50 terms of a geometric sequence, given any first term and any common ratio. Find this sum for (a) $a = 7$ and $r = 0.5$; (b) $a = 10$ and $r = 0.75$; (c) $a = 3$ and $r = 2$.

56. An approximation to e^x is given by this finite series:

$$1 + x + \frac{x^2}{2!} + \frac{x^3}{3!} + \frac{x^4}{4!} + \frac{x^5}{5!} + \frac{x^6}{6!}.$$

Use this series to approximate e^x for $x = 1, 5, 12$.

57. The first few terms in the Fibonacci sequence are 1, 1, 2, 3, 5, 8, 13, 21, 34, 55, 89, . . . with every number (except for 1) equaling the sum of the previous two numbers. Devise an algorithm to print out the first 30 terms of the Fibonacci sequence. What is the 30th term?

Foxtrot by Bill Amend.

Name _____

use the extra space to show your work

— Chapter Review Test **3** ☐☐☐ ←——————————————————

1. Which of the following sentences are statements?

   (a) Please mow the front lawn.
   (b) Russia is the largest country in the world.
   (c) Are you going to Europe this summer?
   (d) Jack Nicholson is the greatest actor ever.
   (e) The U.S. inflation rate was less than 4% in 1992.
   (f) Jupiter has more than 12 moons.
   (g) This sentence is true.
   (h) Mathematics is wonderful!

2. Let p represent "The high school graduation rate exceeds 90%." Let q represent "The disciplinary code is strictly enforced." Translate the following statements into symbolic form.

   (a) The high school graduation rate exceeds 90%, and the disciplinary code is strictly enforced.
   (b) It is not the case that either the high school graduation rate exceeds 90% or the disciplinary code is strictly enforced.
   (c) The high school graduation rate does not exceed 90% whenever the disciplinary code is not strictly enforced.

3. Determine the logical value of the following statements.

   (a) Abraham Lincoln was the 16th U.S. president, or Christmas is on June 25.
   (b) Abraham Lincoln was the 16th U.S. president, and Christmas is on June 25.
   (c) If $45 + 30 = 75$, then whales can fly.
   (d) If there are 50 seconds in 1 minute, then this chapter is more than 1,000 pages long.
   (e) Canada is south of the United States if and only if the Mississippi River is in California.

4. Translate the following forms into interpretations, given these representations.

   p: Michael studies 30 hours per week.
   q: Michael gets good grades.
   r: Michael goes to a party on Saturday.

   (a) $p \vee {\sim}q$
   (b) $p \Rightarrow (q \wedge r)$
   (c) ${\sim}(p \wedge r)$

5. Let *p* represent the algebraic expression "$x > 0$," and let *q* represent "$x + y = 10$." Fill in the following table. Use T for true and F for false.

x	*y*	*p*	*q*	*L(p)*	*L(q)*	*L(p ∧ q)*	*L(p ∨ q)*
−8	10						
−1	11						
1	7						
5	5						

6. Construct the truth table for this statement form: $(p \wedge q) \Leftrightarrow {\sim}(p \vee q)$.

7. Determine whether this statement form is a tautology: $(p \wedge q) \Rightarrow (p \vee q)$.

8. Given that *p* is true, determine whether each statement form has a unique logical value and, if so, find it.

   (a) $p \vee q$
   (b) $p \Leftrightarrow ({\sim}p \wedge q)$

9. Use De Morgan's laws to find the negation of each of the following algebraic statements.

   (a) $x + y = 17$ and $5z > 21$
   (b) $A \neq 100$ or $B \leq 200$

10. Create a statement form that corresponds to the following syllogism, and use that form to determine its validity.

     If LeBron James scores at least 40 points, then the Miami Heat win.
     The Miami Heat lost.
     ∴ LeBron James scored less than 40 points.

11. Draw a Venn diagram to determine the validity of the following syllogism.

     All humans are sentient creatures.
     An orangutan is not human.
     ∴ An orangutan is not sentient.

12. If possible, deduce the correct conclusion, using valid reasoning. If you cannot reach a conclusion by using valid reasoning, then write, "No conclusion possible."

    (a) All of the planets orbit the sun in an elliptical path. Mercury is a planet. Therefore, _____.
    (b) Usain Bolt runs the 100-meter sprint in less than 10 seconds whenever he is healthy. Usain Bolt ran the 100-meter sprint in 9.8 seconds. Therefore, _____.

13. Diagram a logic circuit that corresponds to this statement form: $(p \lor \sim q) \land r$. Given that all the prime statements are true, determine whether the circuit would be on or off.

14. Write the output of a program written from this pseudocode:

1. $2 \rightarrow k$
2. **loop**
3.   **print** k
4. $k + 3 \rightarrow k$
5.   **if** $(k > 17)$ **then go to** 7 **endif**
6. **endloop**
7. **stop**

15. The sales tax rate in Brackets County is 2% if the amount purchased is less than $50 and is 3% if it costs $50 or more. Write pseudocode for a program that would compute the final amount charged for any purchase. Include I/O messages.

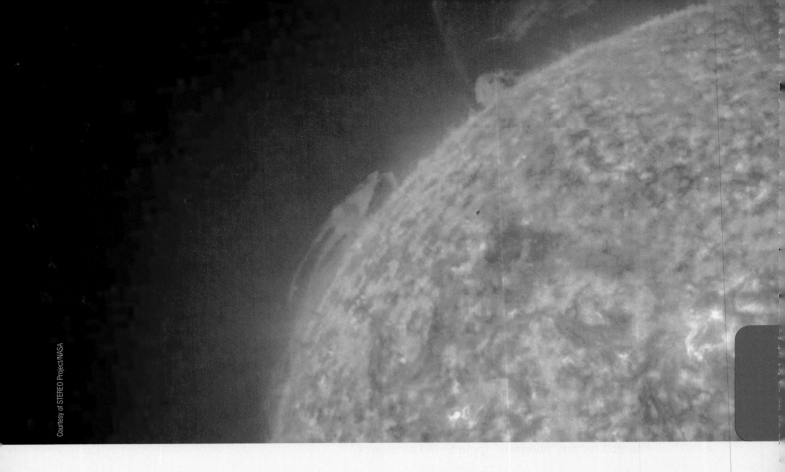

Chapter Objectives

check off when you've completed an objective

☐ Know the scientific method.

☐ Learn units of distance and angular measure for celestial objects.

☐ Use the small-angle formula.

☐ Know the planetary systems of Ptolemy and Copernicus.

☐ Know and apply Kepler's three laws of planetary motion.

☐ Understand the concepts of velocity, acceleration, mass, weight, and force.

☐ Use the functions for the distance and velocity of a dropped object.

☐ Learn about the lives of Galileo and Isaac Newton.

☐ Apply the inverse-square law.

☐ Learn the importance of the *Principia* and the law of universal gravitation.

☐ Find the velocity and acceleration of a body in a circular orbit.

Navigate Companion Website

go.jblearning.com/johnson

Visit go.jblearning.com/Johnson to access a Laboratory Manual, Student Solutions Manual, Interactive Glossary, and Interactive Flashcards.

4 Astronomy and the Methods of Science

4.1 Ancient Milestones

4.2 The Two Great Systems

4.3 The Defense of Copernicanism

4.4 And All Was Light

Philosophy is written in this grand book, the universe, which stands continually open to our gaze. But the book cannot be understood unless one first learns to comprehend the language. . . . It is written in the language of mathematics. . . . Without these one wanders about in a dark labyrinth.

—Galileo Galilei, *The Philosophy of the Sixteenth and Seventeenth Centuries*, translated by Richard Henry Popkin, p. 65. Copyright (c) 1966, Free Press.

FIGURE 4.1.1   The night sky.

astronomy The science of the observation and study of the universe.

natural philosophy A set of principles or ideas concerning the workings of nature. The addition of mathematics to formalize such ideas led to the recasting of this phrase as *science*.

Few activities have enjoyed such universal appeal through the ages as gazing up at a starry sky on a pleasant summer's night (see **Figure 4.1.1**). We have all spent an idle evening wistfully musing as those faraway diamonds sparkled against the magnificent black backdrop of the cosmos. Some evenings we have taken the next step and pondered the inevitable cascade of questions that seem to hang in the air right next to the stars and planets. What are they made of? How long have they been there? How far away are they? Along what path do they move, and how fast? The search for the answers to such curiosities comprises the beautiful subject of **astronomy**. See **Figure 4.1.2**.

A brief look at the history of humankind's search to understand its place in the universe provides us with a classic example of the interweaving of the development of mathematics with the genesis of modern science and of its impact on the evolution of human culture. We shall see that mathematics is not just numbers and equations, but rather an extremely effective language for *deductive* reasoning and for defining the terms and concepts with which to debate the great questions of the world. Indeed, mathematics often has to be *created* to explain newly discovered relationships and to sort and separate facts. It has been very successful in rendering the tenets of **natural philosophy**, a phrase from an earlier era that today we would call science.

Science can be reasonably defined as systemized knowledge logically deduced from observations. Certainly a claim can be made declaring astronomy to be the *first* science; the quest to understand the celestial theater that surrounds us began several millennia ago (and continues vigorously today). This amazing odyssey provides us a rich tapestry of lives and achievements well worth our effort to examine.

FIGURE 4.1.2   The astronomer reaches for truth. (Camille Flammarion, *L'atmosphère: météorologie populaire*, Paris, 1888.)

Replete with more failures than successes, the history of astronomy is fraught with the full range of emotions and hardships that are always the handmaidens to any human experience. Therefore, the style and tone of this chapter will be slightly different than found elsewhere in the text. A story is going to be told, and as it unfolds, we will intersperse our study of specific mathematical and astronomical principles with important events in the lives of the key contributors.

4.1 Ancient Milestones

We begin by realizing that the nocturnal scenes witnessed in the evening skies by people 2,500 years ago are pretty much the same ones you can see today in your own backyard on any clear night. Back in the days before indoor distractions such as televisions, people used to regularly gather outside to while away the evening in pleasant conversation. Attentive individuals would notice over a period of months that there were some things in the arrangement of the swarm of sparkling lights above them that stayed the same and others that changed. For example, compare the two pictures shown in **Figure 4.1.3**, giving the same view in the same direction but separated chronologically by several weeks (or possibly months).

We note that, except for the brighter object, all the stars have remained in the same locations *relative to one another*. In the right-hand picture, however, the apparent motion of the bright object through the background stars marks it as different somehow, and any model of the universe we might propose must account for the meanderings of this so-called "wanderer." Indeed, *planet* is the Greek word for wanderer, and the ancient Greeks observed five such planets (Mercury, Venus, Mars, Jupiter, and Saturn) gallivanting across the heavens.

These are observations of occurrences happening over a period of time. Likewise, no one can escape noticing the daily movements of the sun and moon rising in the east and setting in the west. Most of the stars also follow this same pattern, although if we direct our gaze to the north, we see that one particular star (Polaris) seems to not move much at all during the night while nearby stars follow circular paths around it, never dipping below the horizon. Such a stable beacon of the night has been observed for centuries both by sailors,

FIGURE 4.1.3 The view on the right is of the same group of stars but several weeks later.

who used it for navigating, and by bards, who sang of its permanence. In *Julius Caesar*, Shakespeare wrote:

> If I could pray to move, prayers would move me;
> But I am constant as the northern star,
> Of whose true-fix'd and resting quality
> There is no fellow in the firmament.

What model of the sun, moon, planets, and stars might you invent that would fit these facts? Incidentally, that's quite important—"fitting the facts." One key to understanding a history of the important discoveries in any scientific arena is an examination of the development and refinement of the **scientific method**, a process involving three main steps:

scientific method
A formalized procedure for exploring natural phenomena involving:

1. Careful observation and the recording of data.
2. Objective analysis of the data and the creation of a model to fit the data.
3. Use of the model to make predictions that can be tested against new observations.

axioms A statement universally accepted as true.

1. **Careful observation of a phenomenon and the recording of data**
2. **Objective analysis of the data and the creation of a model to fit the data**
3. **Use of the model to make predictions that can be tested against new observations**

Inherent in the construction of any model of a natural phenomenon is the concept of the underlying *axioms*. Essentially, **axioms** are simply stated assumptions whose truth seems so self-evident that most people accept them as true without proof. Just as any sturdy edifice needs cornerstones laid as a foundation for building, so does every field of science or mathematics adopt a set of axioms as a starting point from which to proceed. The Greeks of the classical Athenian period (c. 600–350 BC) felt that such basic principles were so primordial that the knowledge of them resided within everyone, and so their realization needed no further explanation. This is akin to the notion that you need not prove the existence of your own soul—you just feel it. The problem with this approach is that the history of science and mathematics is full of instances where one or more of the foundational axioms of a particular theory have been shown to be unreliable. This can then lead to a dramatic change in the consequences of those axioms, perhaps to the point of completely abandoning the entire theory. Such a happening is called a

paradigm shift An altering of the standard model for a phenomenon.

paradigm shift—an altering of the standard model for a particular phenomenon; in fact, our study of astronomy will highlight one of the classic paradigm shifts in the history of science. The lesson of these revisions has led to a modern attitude in which we adopt our axioms with a bit less certainty and a bit more flexibility.

We begin our exploration of the development of our current model of the planetary system with that ancient, mysterious character **Pythagoras** (c. 580–500 BC) (see **Figure 4.1.4**) and his school of followers. To the Pythagoreans, numbers were very important tools to solving the riddles of the natural world around them, an idea that grew increasingly strong as these thinkers discovered the roles that numbers played in establishing relationships in geometry and music. Furthermore, because the universe was a pure and harmonious place, they believed that these relationships extended to the movement of the planets. The "music of the spheres" was that perfect blend of tones produced by the planets moving in cosmic harmony in the same way as the proper strings of the right lengths vibrate on a musical instrument.

quadrivium A group of studies in medieval universities consisting of arithmetic, music, geometry, and astronomy.

Establishing connections between such seemingly unrelated disciplines led to the famous set of grouped studies historically known as the **quadrivium** of knowledge, consisting of arithmetic, music, geometry, and astronomy. The study of these subjects led to a degree in many medieval universities, and they were all considered to be branches of the same field, mathematics.

Although the so-called music of the spheres was pure fantasy, Pythagoras and his followers also believed that because circles and spheres were the perfect forms of geometry, they must be the paths and shapes that the gods had supplied to the universe. These ideas tilled the soil for the growth of the concept of the sun, planets, and stars moving in continuous circular paths around a spherical Earth. In fact, these two primary axioms influenced most of the planetary models produced from 500 BC to the 1500s:

1. Earth is the stationary center of the universe (called a **geocentric** universe).

2. Every celestial body travels in a circular path.

In light of the state of knowledge of the natural world at that time, such axioms are quite reasonable. Surely, thought the ancients, if Earth were sailing through empty space, wouldn't we all be sent hurtling off its surface, much like ants off a thrown apple? Circles, in fact, have had such a special place in the geometry of the world—the most graceful curve in nature—that their use remained ingrained in any depiction of the planetary orbits well into the sixteenth century, as seen in **Figure 4.1.5** in the illustration of the geocentric system done by Portuguese cosmographer Bartolomeu Velho in 1568 (Bibliothèque Nationale, Paris).

Although we know today that the planets do not move in circles centered on Earth, the notions of a round Earth and planets in repeating orbits were giant steps in the right direction. Bent on using mathematics to provide answers to philosophical questions, the Pythagoreans dispensed with many previous mythologies and planted the seed for the growth of the scientific method by attempting to describe the phenomena of nature based on observation.

For example, the lack of observational *parallax* supports the notion of a stationary Earth. You already are familiar with this optical notion: Close one eye, extend your arm with your thumb held vertically, and use your other eye to line up your thumb with some object several meters distant. Now close your open eye and open the closed one. What happens? Your thumb appears to have moved with respect to the distant object. This is known as **parallax**. Well, the same should be true for our view from Earth if it is truly a vehicle that is carrying us through space. Why do the closer stars not change their orientation with respect to the more distant stars over time? The answer, of

© Offscreen/ShutterStock, Inc.

FIGURE 4.1.4   Bust of Pythagoras.

geocentric Earth-centered.

parallax The change of position of a close object with respect to a more distant background when viewed from two different locations.

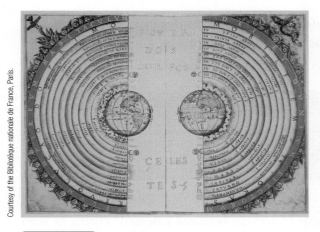

Courtesy of the Bibliothèque nationale de France, Paris.

FIGURE 4.1.5   The geocentric system.

course, is that they do! It's just that they are *all* so far away (the closest star after the sun, Alpha Centuri, is more than 41 trillion kilometers (km) from Earth) that stellar parallax is not observable with the naked eye. See **Figure 4.1.6**.

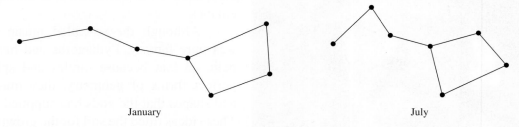

January   July

FIGURE 4.1.6   Fictitious scenario for how stars in the Big Dipper constellation might change as a result of Earth's motion if they were closer to us.

Example 1

Miles and kilometers are units of distance that are useful for terrestrial measurement but quickly become cumbersome when used to describe the vastness of space. Modern-day astronomers use the **astronomical unit** (abbreviated as AU) in computing distances among the planets, asteroids, comets, and other members of our solar system. One astronomical unit (1 AU) is equal to the mean distance of Earth from the sun—93,000,000 miles (mi) or 150,000,000 km. Incredibly, stars and galaxies are so much farther away that even astronomical units become unwieldy, and *light-years* have become the standard unit of measurement. One **light-year** (ly) is the distance light travels in 1 year (yr) at its speed of 300,000 kilometers per second (km/s). If we multiply this speed by the number of seconds in a year, we get the distance:

astronomical unit
The mean distance from Earth to the sun, equal to about 93 million mi or 150 million km. Abbreviated AU.

light-year Distance traveled by light in 1 yr, equal to about 9.5 trillion km.

$$\left(300,000 \ \frac{km}{s}\right)\left(3,600 \ \frac{s}{h}\right)\left(24 \ \frac{h}{day}\right)(365 \ days) \approx 9.5 \times 10^{12} \ km.$$

So 1 ly is equal to about 9.5 trillion km. Because Alpha Centauri is 41 trillion km away, this translates to $41/9.5 \approx 4.3$ ly. ◆

Modern literature contains many references to the great distances that all celestial objects are from Earth. The following poetic passage is from *The Risk Pool* by Richard Russo.

Here was a wish from another lifetime, granted twenty-five years too late, as if God were in a place so distant that it took almost forever for wishes to travel there, like pale starlight from a distant galaxy, eons old and all worn out even as we look at it.

The only reasonable way to gauge the separation between celestial objects as seen from Earth is by measuring an angle. We say, for example, that star A and star B have

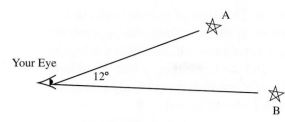

FIGURE 4.1.7 Angular separation.

angular separation
The measure of the angle formed at the observer's position by the two lines of sight to two separate objects.

arcminutes
One-sixtieth of a degree. Used as a unit for angular separation.

arcseconds
One-sixtieth of an arcminute. Used as a unit for angular separation.

an **angular separation** of 12° if the two lines of sight from the observer to those stars form an angle of 12°. See **Figure 4.1.7**. Note that this tells us nothing about the actual distance in space *between* the two stars because their separate distances from Earth could be very different. It follows that the motion of an object is often given in a certain number of degrees or radians per unit of time. Smaller units are often needed, and so 1 degree (1°) is defined to be 60 **arcminutes** (abbreviated by 60′) while 1 minute (1′) of arc is equal to 60 **arcseconds** (abbreviated by 60″).

In the following passage from *Cold Mountain* by Charles Frazier, references are made to the motion of stars in the night sky. Knowing basic astronomy can augment your enjoyment of literature.

Orion had fully risen and stood at the eastern horizon, and from that Inman made the time to be long past midnight. Orion was girded about tight, his weapon ready to strike. Traveling due west every night and making unfailing good time.

Example 2

As the sun moves across the sky each day, you could easily compute that it moves at a rate of 15° per hour. The ancients thought the sun was circling Earth, but of course today we know that Earth rotates on its axis once every 24 h (360° divided by 24 h yields the required rate). Naturally, the stars at night display this same apparent motion and at the same angular rate, but you will notice that the path each individual star appears to follow is actually along a circle centered at Polaris, which itself never moves. This is so because the northern end of our planet's rotational axis points (almost) directly at Polaris (the end star in the handle of the Little Dipper). Stars in the northern part of the sky, such as those in the constellations Cassiopeia and the Big Dipper, move in tighter circles. Stars that have a greater angular separation from Polaris, such as those in Orion, move in circles large enough to give the appearance of rising in the east and setting in the west. (In the southern hemisphere, an analogous arrangement prevails with circles of apparent motion centered at a spot in the southern part of the sky.) ◆

Example 3

If you were to look directly east at whatever group of stars is peeking over the horizon every night at suppertime, you would easily ascertain after a few weeks that those stars were appearing about 1° higher each night. (Looking at the stars at the same time every night eliminates the problem of accounting for the daily rotation of Earth.) Six months from now, that same group is setting in the west at suppertime, and in a year's time, it has returned to its original position. This type of motion is physically distinct from that described in Example 2, but as in that example, the movements are along Polaris-centered circles. Earth's annual revolution around the sun accounts for this transport through 360° over a period of 365 days, yielding a star "marching speed" of about 1° per day toward the west. Careful measurement of any constellation would show a movement of about 30° over a 1-month period.   ◆

As we mentioned earlier, the mysterious travels of the five visible planets had inspired the wonder of habitual stargazers for centuries. Why do these five orbs appear to move among the background of an ocean of stars? Well, for one thing, they must be much closer than the stars. Moreover, there was little argument as to their relative proximities because of the assumption that the more slowly the planet moved, the farther it must be from Earth. So observations of the planets' relative speeds put them in the order of Mercury, Venus, Mars, Jupiter, and Saturn.

However, two major problems could not be explained by simple circular motion around the Earth. The motions of all the planets are restricted to a rather narrow ribbon of the sky called the **zodiac** that also contains the apparent path of the sun. (We know today that this is the result of all the planets orbiting the sun in pretty much the same plane.) Although Mars, Jupiter, and Saturn typically travel in eastward paths through the zodiac, one of them occasionally slows to a complete stop, reverses its motion, and goes westward for a few weeks or months until it stops and reverses again to resume its original direction (**Figure 4.1.8**). This reversal is referred to as **retrograde motion**, and it was a mystery that all the early cosmologists struggled to resolve. Accounting for retrograde motion under the axiom system described earlier eventually led to the **Ptolemaic system**, a model we will examine closely in the next section.

The other big question concerned the fact that something altogether different was going on with Mercury and Venus. These two planets could be seen only in the early morning just before sunrise or in the evening just after sunset, but never in the middle of the night. They stuck close to the sun with Venus never separated by more than 48° and Mercury never by more than 28°. It was this riddle that **Heracleides** (c. 388 BC) solved by being the first to suggest the incredibly novel idea of having Mercury and Venus in orbit *around the sun*. (See **Figure 4.1.9**.)

zodiac A narrow, beltlike region around the sky containing the apparent paths of the sun and planets.

retrograde motion The apparent reversal of the movement of a planet.

Ptolemaic system Model of the planetary system with Earth at the center and employing epicycles and deferents to explain retrograde motion of the planets.

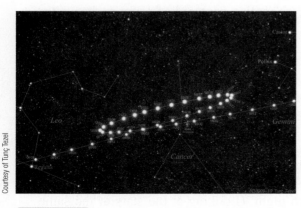

FIGURE 4.1.8   The position of Mars taken at 10-day intervals.

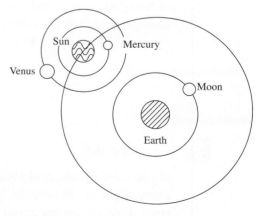

FIGURE 4.1.9   The model of Heracleides.

Now here was a thought of far-reaching significance indeed! True, he still had the sun circling the Earth, but no one had ever offered up the possibility of anything moving around the *sun*. We now use this revised picture to compute some relative distances.

Although undetectable by the naked eye, the width of the planetary disks varies as their distance from us changes. Modern telescopes can determine the angular width of the visible disk of a celestial body with great precision. The **angular diameter** α of any object (as opposed to its actual or linear diameter) is defined as the angular separation from one side of the object to the other side. By measuring the angular diameter, you can find the distance to the object if you know the linear diameter, or vice versa. The distance to most celestial objects is great enough to allow us to think of the linear diameter pictured in **Figure 4.1.10** as a portion of a large circle whose radius is the distance. If we use radians to measure α, then geometry tells us that

> **angular diameter**
> The angular separation of opposite sides of an observed object.

$$\alpha = \frac{\text{linear diameter}}{\text{distance}}.$$

However, α is generally measured more conveniently in arcseconds because of its small size, and so we may utilize the fact that 1 radian (rad) consists of 206,265 arcseconds to obtain the proportion

$$\frac{\alpha}{206,265} = \frac{\text{linear diameter}}{\text{distance}},$$

where α is now measured in arcseconds. The above relationship is often referred to by astronomers as the *small-angle formula*.

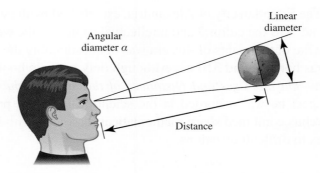

FIGURE 4.1.10 Angular and linear diameters.

❓ Example 4

The moon's diameter is 3,500 km. We can sometimes assume a constant Earth–moon distance, even though, in reality, it is continuously changing. If the distance varies in one month from a minimum of 363,000 km to a maximum of 405,000 km, what are its minimum and maximum angular diameters that month?

⚙ Solution

Substituting the appropriate numbers from above into the small-angle formula, we get

$$\frac{\alpha}{206,265} = \frac{3,500}{363,000},$$

and this gives us

$$\alpha = 0.00964(206,265)$$
$$= 1,990''$$
$$= 33'$$

for the maximum angular diameter. Likewise,

$$\frac{\alpha}{206,265} = \frac{3,500}{405,000}$$
$$\alpha = 0.00864(206,265)$$
$$= 1,780''$$
$$= 30' \text{ for the minimum.}$$

Although this fluctuation of 3' is probably not noticeable to the casual observer, it is apparent in the differences it causes in total solar eclipses. At maximum distance, the smaller lunar disk fails to cover the entire sun, producing a thin brilliant ring of fire surrounding the disk. It is known as an *annular eclipse*. ◆

The Egyptian city of Alexandria, established by the warrior-king Alexander the Great in 322 BC, was the cultural and intellectual capital of the western world for several hundred years. One of the giants of this era was unquestionably the brilliant **Aristarchus** of Samos, who was born around 310 BC. Although only a single book of his writings survived to the present—*On the Sizes and Distances of the Sun and the Moon*—his fame is secured by the respect he was accorded in the scientific literature produced by his contemporaries. Aristarchus combined sharp mathematical reasoning with keen fact gathering to formulate answers to difficult questions.

❓ Example 5

Aristarchus invented an ingenious method for determining the ratio of the Earth–sun distance to the Earth–moon distance and, in the process, greatly enlarged the Greek estimates of the size of the universe. His technique rested on examining the right triangle formed by the sun, moon, and Earth when the moon is at the position in its orbit that yields the view of a half-illuminated disk (first-quarter or third-quarter phase).

Somehow Aristarchus estimated the angle θ in **Figure 4.1.11** to be about 87°, and this then implies that the small angle must be 3°. Because each degree equals 3,600″, the small-angle formula can be applied to give us

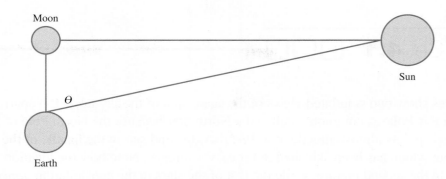

FIGURE 4.1.11 The method used by Aristarchus to compare the distances to the sun and moon.

$$\frac{10,800}{206,265} = \frac{EM}{ES},$$

where EM = the Earth–moon distance and ES = Earth–sun distance. Therefore,

$$0.0524 = \frac{EM}{ES}$$

$$ES = \frac{EM}{0.0524}$$

$$= 19\,EM.$$

Note that a small change in the measurement of the angle induces a large change in this factor (see the exercises). Although the actual ratio of the Earth–sun distance to the Earth–moon distance is about 390, this was nonetheless a truly astounding discovery for that era. ◆

Among his many accomplishments, Aristarchus is commonly credited with being the first man to place *the sun at the center* of the planetary system with *all* the planets, including Earth, in circular orbits around it. This is known as a **heliocentric** system (the Greek word *Helios* means sun god). He felt the scale of the universe to be so grand and the stars so enormously distant that any parallax caused by the movement of our planet would not be measurable. These ideas gained little acceptance in Aristarchus's lifetime. Perhaps Aristarchus himself was not fully convinced of the truth of his conjecture because there existed no strong observational evidence to favor this scheme over any other. It was simply a philosophical alternative that seemed, somehow, less secure than that of a geocentric universe. In fact, it would be almost 1,800 years before a sun-centered planetary model would be seriously considered again.

heliocentric
Sun-centered.

> The scanty conceptions to which we can attain of celestial things give us, from their excellence, more pleasure than all our knowledge of the world in which we live.
>
> —Aristotle (*Parts of Animals* I, 5)

Name _____

Exercise Set **4.1** ◯◯◯←

1. These figures show two simulated views of the same area of the sky taken 6 h apart. The star labeled P is Polaris, commonly called the North Star because the North Pole of Earth's rotational axis points almost directly at it. Polaris is the end star in the handle of the Little Dipper constellation, which has been sketched in for easy reference. Note how the location of Polaris is unchanged in the second picture, while the rest of the stars in the constellation *appear* to have rotated around Polaris in circular paths. What phenomenon regarding the daily motion of Earth actually accounts for this observation? The entire constellation has rotated counterclockwise about 90°. Thus, we see that each star moves at an angular rate of 90° every 6 h, or 15° per hour. Is this consistent with your answer to the above question? Why?

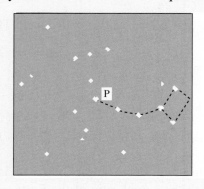 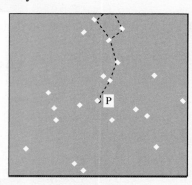

2. Suppose you go out tonight at 9:00 P.M. and pick out a nice bright star on the eastern horizon. If you view that same star 1 month from now at 9:00 P.M., what angular displacement (on a large circle centered at Polaris) will that star have undergone?

3. Explain why the assumption of a stationary Earth was so prevalent for such a long time.

4. What observational facts led Heracleides to propose that Mercury and Venus orbited the sun instead of Earth?

5. Name the two axioms for building any planetary model that were believed by most of the Greek astronomers of antiquity.

6. How many degrees does the sun move across the sky between 10:00 A.M. and 3:00 P.M. due to the rotation of the Earth?

7. What is the angular rate at which the stars rotate around Polaris in units of degrees per hour? Arcminutes per minute? Arcseconds per second?

8. One *light-second* is the distance light travels in 1 s. What is the mean distance of Earth from the sun in light-seconds? In light-minutes? [Recall that the speed of light is 3×10^5 km/s and the distance from Earth to the sun is 1.5×10^8 km.]

9. The mean distance of Pluto from the sun is about 39 AU. What is that distance in light-seconds?

10. How many astronomical units are in 1 ly?

11. The moon takes roughly 30 days to complete one trip around Earth. Compute the angular rate (in degrees per hour) at which the moon moves through its orbital path.

12. Phobos, one of the two moons of Mars, orbits the planet at a distance of about 5,980 km. It is considerably smaller than our own moon, having a diameter of only 20 km. What angular diameter would Phobos present to an observer on Mars? Would Phobos appear larger or smaller in the Martian sky than our moon appears to us here on Earth?

13. The maximum angular diameter of Jupiter as seen from Earth is about 50″ (arcseconds). If the linear diameter of Jupiter is 144,000 km, determine the minimum distance (in kilometers) of Jupiter from Earth.

14. The planet Saturn is well known for its colorful rings. The diameter across the outermost ring is 340,000 km. When Saturn is 9.0 AU from Earth, what is the angular diameter of the outermost ring?

15. In the previous exercise, what is the angular diameter of the outermost ring of Saturn when the planet is 10.0 AU from Earth?

16. If d and α are the linear and angular diameters, respectively, of a celestial object and r is the distance to the object, then, for any particular object (cf. Example 5), α clearly changes as r changes. Write α as a function of r. Does α achieve a maximum or a minimum when r is a minimum? Does α achieve a maximum or a minimum when r is a maximum?

17. If the angular separation between two objects is less than 5°, for our purposes we may continue to assume that the triangle formed by those two objects and Earth is a long, narrow right triangle. Megrez and Phad are two of the stars forming the bowl of the constellation commonly called the Big Dipper. Both of these stars happen to be about 80 ly from Earth and have an angular separation of 4.5°. How far apart are they in space? (1° = 3,600″.)

© Volkova Anna/ShutterStock, Inc.

18. Many of the points of light in the night sky that appear to be single stars are actually composed of two stars revolving around a common point between them. These are called *binary systems*. A binary system appears to be a single star because it is too far away for the human eye to resolve into two separate light sources. About the closest star separation that the eye can distinguish is 4′. Mizar, a binary system in the crook of the handle of the Big Dipper (see above picture), is about 73 ly from Earth and can only be resolved by the keenest human eye. What must be the maximum distance between the two stars this system?

19. Alpha Centauri, our sun's closest stellar neighbor at 4.3 ly, is actually a triple star system. The two biggest, brightest stars are known as Alpha Centauri A and B. They orbit each other every 80 yr, having minimum and maximum separations of 11 and 35 AU, respectively. What are the corresponding minimum and maximum angular separations of these two stars as seen from Earth? Would you need a telescope to identify these as two different stars? (From the previous exercise, we see that the answer is yes if the angular separation is always less than 4′.)

20. Aristarchus computed a distance to the moon of about 80 Earth radii (the actual figure is closer to 60) by using an ingenious technique that involved clocking the time it took the moon to pass through Earth's shadow during a lunar eclipse. Refer to Example 5 to determine what Aristarchus concluded must be the distance to the sun.

21. A more accurate measurement of the sun–Earth–moon angle θ in the Aristarchus triangle in Figure 4.1.11 is 89.853°. If the mean distance to the moon is about 384,000 km, what is the mean distance to the sun?

22. A *solar eclipse* occurs when the moon interposes itself between the sun and Earth. It is a dynamic demonstration that the moon is closer to us than the sun and also reveals that the average angular diameters of these two bodies are the same—about 30′. Similar triangles then allow us to conclude that the *ratio* of the diameters of the sun and moon is the same as the *ratio* of their distances from Earth:

$$\frac{\text{Distance to sun}}{\text{Distance to moon}} = \frac{\text{diameter of sun}}{\text{diameter of moon}}.$$

The work of Aristarchus in Example 5 gave a value of 19 for the ratio on the left-hand side of the equation in the figure. If the diameter of the moon is given to be 3,500 km, what would be the diameter of the sun?

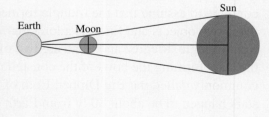

23. In Exercise 21, you computed a different value for the ratio of the Earth–sun distance to the Earth–moon distance than the one computed by Aristarchus. Use this value to answer the same question as in Exercise 22: If the diameter of the moon is given to be 3,500 km, what is the diameter of the sun?

24. One of the greatest scholars of the Alexandrian Greek world was **Eratosthenes** (275–194 BC). He was the first to compute an extremely good estimate of the circumference of Earth. Eratosthenes noticed that at summer solstice, the sun was directly overhead in the city of Syene (S in the accompanying picture) and at an angle of 7.5° from straight overhead in Alexandria (A), which was located 500 mi to the north. The angle at the center E of Earth is formed by radii extended through both points. Because we can think of sunlight rays as parallel, the measure of this angle must also be 7.5°. Because $\dfrac{7.5}{360} = \dfrac{1}{48}$, this means that the portion of the circle from A to S is $\dfrac{1}{48}$ of the circumference. What is the circumference?

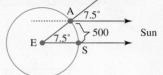

25. Suppose Eratosthenes measured the angle at E in the accompanying picture to be 6°. What would have been his value for the circumference of Earth?

26. If you were to travel to an exotic land and wished to locate the North Star some evening, there is a wonderful little fact you should know. The elevation angle from the horizon to Polaris is related to the latitude of your location. For instance, where would Polaris be in the sky if you were situated at the North Pole? At the equator? Your *latitude angle L* is defined to be the angle between radii drawn to the equator and to your location. In the picture here, your line of sight to Polaris and the line from Earth's center to the North Pole are parallel. Therefore, the two angles marked by θ are equal. Prove that the angle of elevation α is equal to L.

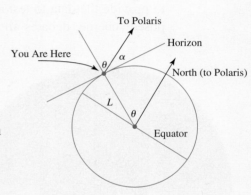

27. In the previous exercise, would Polaris appear higher in the sky as viewed from Dallas, Texas, or Chicago, Illinois?

28. Elongation of a planet is the angular separation of the planet from the sun as seen from Earth. The angle *SEM* in the figure here is the angle of elongation of the planet Mercury for two different positions of Mercury in its orbit. The maximum elongation angle for Mercury is about 28°, giving the configuration in the diagram on the right. We see that the line *EM* is tangent to the orbit of Mercury and therefore is perpendicular to the radius *MS* of the orbit. Hence, the three bodies form a right triangle with hypotenuse equal to the Earth–sun distance *ES*. Estimate the distance from Mercury to the sun and the minimum distance to Mercury from Earth in astronomical units. (Trigonometry is needed.)

elongation The angular separation between the sun and a planet.

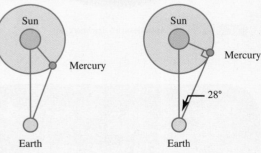

29. The maximum elongation angle for Venus is about 48°. As in the previous exercise, find the distance from Venus to the sun in both astronomical units and kilometers. At this elongation, how far apart are Earth and Venus? If the linear diameter of Venus is about 12,000 km, what would be its angular diameter at this point in its orbit?

30. Use the results of the previous exercise to find the minimum distance Venus can ever be from Earth. Give your answer in both astronomical units and kilometers. If the linear diameter of Venus is about 12,000 km, what would be its angular diameter at this point in its orbit?

4.2 The Two Great Systems

> To be accepted as a paradigm, a theory must seem better than its competitors, but it need not, and in fact never does, explain all the facts with which it can be confronted.
>
> —Thomas S. Kuhn

The creative thinking of people like Heracleides and Aristarchus drives the progression of science, but a recurring problem in those olden days was the lack of sophisticated instrumentation to provide the data to test a new model. Without hard numerical data to corroborate a proposed scheme, debates about the best model of the universe continued unabated, and the set of axioms used to build any theory played a large role. Of course, in the fourth century BC, one large barrier to real knowledge was a lack of physical concepts that could be measured. Notions such as weight, mass, force, velocity, acceleration, and so on were either partially or wholly undeveloped quantitatively. One might speak of heavy or light, but never of 175 pounds (lb). In such an environment, the people who became revered authorities of learning were those who could articulate their philosophies about the nature of the world with the best rhetorical polish. Chief among these were **Plato** (429–348 BC) and his esteemed pupil **Aristotle** (384–322 BC), a man whose clarity of insight empowered his writings on philosophy, economics, history, politics, poetry, drama, and the sciences (see **Figure 4.2.1**). His main work detailing his thoughts about the structure of the cosmos is *On the Heavens*. The high esteem awarded to the circle and sphere is evident in this book. He wrote:

© marekuliasz/ShutterStock, Inc.

FIGURE 4.2.1   Aristotle on a 5 drachma coin.

> Let us consider generally which shape is primary among planes and solids alike. Every plane figure must be either rectilinear or curvilinear. Now the rectilinear is bounded by more than one line, the curvilinear by one only. But since in any kind the one is naturally prior to the many and the simple to the complex, the circle will be the first of plane figures.... And the sphere holds the same position among the solids. For it alone is embraced by a single surface, while rectilinear solids have several. The sphere is among solids what the circle is among plane figures.... The shape of the heaven is of necessity spherical; for that is the shape most appropriate to its substance and also by nature primary.

We see these arguments as favoring not only the axiom of circular motion, but also the axiom of a spherical Earth. In fact, in a departure from the approach of Plato, who disdained experimentation, Aristotle included some observations in his studies on this point, stating:

The evidence of the senses further corroborates this. How else would eclipses of the moon show segments shaped as we see them? ... In eclipses the outline is always curved: and, since it is the interposition of the earth that makes the eclipse, the form of this line will be caused by the form of the earth's surface, which is therefore spherical. Indeed there are some stars seen in Egypt and in the neighborhood of Cyprus which are not seen in the northerly regions ... all of which goes to show not only that the earth is circular in shape, but also that it is a sphere of no great size. ...

See **Figure 4.2.2**.

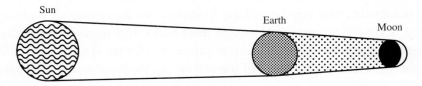

Sun   Earth   Moon

FIGURE 4.2.2   Observations of lunar eclipses led Aristotle to believe Earth was a sphere.

FIGURE 4.2.3   A medieval painting of Ptolemy.

epicycle Imaginary circle around which a planet moved in the Ptolemaic system. The center of the epicycle moved along the deferent.

deferent Imaginary circle centered on Earth, invented by Hipparchus, around which the center of a planet's epicycle moved.

Plato was convinced that God had designed a perfect mathematical pattern for the motions of all the celestial bodies. Aristotle's view of the planetary model was one that wholeheartedly endorsed Platonic geocentricity but labored to explain retrograde motion. Aristotle elaborated on a model proposed by the mathematician **Eudoxus** (408–355 BC)—one that accounted for the strange reversals of the planets by having them attached to a complicated nest of interlocking ethereal spheres, each turning around Earth at a different yet uniform speed. These teachings of Aristotle led the mainstream of Greek astronomy to later embrace a geocentric, or Earth-centered, model that was to become known as the Ptolemaic system. See **Figure 4.2.3**.

Preserving both geocentricity and circular motion while simultaneously accounting for observed retrograde was accomplished through a mathematical creation inspired originally by **Hipparchus** (c. 150 BC). His idea allowed for an object in orbit around Earth to follow a path that was a geometric combination of two circles and still yielded observations of apparent back-and-forth motion. Further refinements and applications of this scheme to all the planets were made 300 years later by the last of the renowned Alexandrian astronomers, **Claudius Ptolemy** (c. 150 AD), for whom the entire model is named. The essential element is that each planet P travels in a small circle (see **Figure 4.2.4(a)**) called an **epicycle** at the same time as the center Q of the epicycle moves along another larger (imaginary) circle surrounding Earth called a **deferent**. Imagine a small ball (the planet) attached to a string 1 foot (ft) long that is tied to the end of a 5-ft stick, which you are holding. If you (Earth) turn in place while you gyrate the stick so as to twirl the ball in the same plane as the stick, the resulting motion of the ball is what Hipparchus had in mind. It moves in a curve that mathematicians call a *cycloid*, shown in **Figure 4.2.4(b)**.

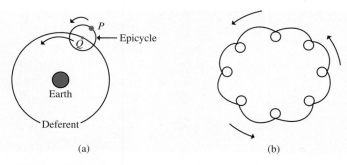

(a)   (b)

FIGURE 4.2.4   An epicycle combined with a deferent gives a geocentric explanation of retrograde motion.

Although the complete plan of Ptolemy contained some other modifications to this basic framework, note how beautifully it accounts for the retrograde problem. Each loop in the cycloid is viewed as retrograde motion from Earth. It is produced by choosing a speed for *P* greater than that for *Q*, along with appropriate radii for the epicycle and deferent. By tinkering with the sizes and angular rates of the pair of circles associated with a particular planet, you could predict its position with enough accuracy to match the observations available in that era.

The entire model and the accompanying mathematics are contained in Ptolemy's *Almagest*, one of the most significant books of antiquity. In the final configuration (**Figure 4.2.5**), Ptolemy ignored the suggestion of Heracleides concerning Mercury and Venus (as Aristotle had) and placed the centers of the epicycles containing these two bodies on an imaginary line connecting Earth to the sun, whose own orbit was placed beyond that of Venus. This very neatly explained the perpetual proximity of the two innermost planets to the sun within the constructs of the model.

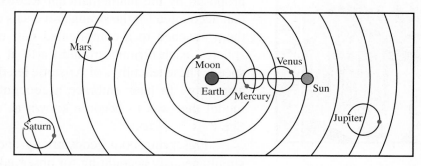

FIGURE 4.2.5   The Ptolemaic system.

So here we have a geocentric model of the universe in which all bodies moved in circles yet gave fairly good matches of their positions with observational data. It was an achievement of monumental proportions. If the primary criterion by which to judge the success of a model is the duration of acceptance by the learned community, then the Ptolemaic system—which remained the standard for more than 1,500 years—is among the most successful in the history of science. Morris Kline, one of the most respected mathematics writers of the twentieth century, called the Ptolemaic system the "supreme achievement of all Greek efforts" and wrote in his book *Mathematics and the Search for Knowledge* that

> No other product of the entire Greek era rivals the *Almagest* in the profound influence it exerted on the conceptions of the universe.

During the next 15 centuries, the development of new mathematics in the West and the associated inquiries of natural philosophy came to a stop. The penetration to Mediterranean lands by fierce barbarian tribes from the north, culminating in the sack of Rome in 410 AD, began a lengthy period of chaos and suffering. The teachings of Aristotle were considered the foundations of education, and because religious doctrine also endorsed a geocentric view of the universe, no serious disagreement with the Ptolemaic system emerged for quite some time.

Then a series of events in the late medieval period began to create a new atmosphere for learning. The compass; the printing press; easier access to books; lenses, corrective glasses, and improved vision; and later the invention of the microscope and telescope all fueled an invigorating spirit of investigation. A new guiding principle of observation combined with reason began to replace a reliance on authority. As the accuracy of the data increased, the errors inherent in the geocentric model became more noticeable and prompted a reexamination of the grand system of Ptolemy.

Nicholas Copernicus (1473–1543) (see **Figure 4.2.6**) never had any intention of discrediting the astronomy of Ptolemy, for whom he had the utmost respect. Born in Poland, Copernicus was adopted and raised by a wealthy and powerful uncle, a bishop who valued science and education as highly as religion. He sent his nephew to study in Italy, where the young man probably first hatched his seminal ideas that led to a new view of the cosmos. However, cautious by nature and fearful of condemnation by the Church as well as scorn from contemporary scientists, Copernicus did not publish his classic astronomical work, *De Revolutionibus Orbium Caelestium* (meaning *On the Revolutions of the Celestial Orbs*), until just prior to his death. (It is interesting to note that this book led to the meaning of *revolutionary* as being radically new.)

The two main claims of his treatise were that Earth was just one of six planets in circular orbit around a stationary sun (**Figure 4.2.7**) and that the daily passages of the sun, moon, and stars across the sky were an illusion created by Earth's rotation on its axis. *De Revolutionibus* began the reevaluation process that eventually dethroned the Ptolemaic system. In particular, it lent such momentum to the heliocentric model of the planets that the entire theory is often referred to as the **Copernican system**. It is unknown whether Copernicus obtained his ideas from Aristarchus of ancient Greece, but in the preliminary version of his manuscript, he wrote:

© Brendan Howard/ShutterStock, Inc.

FIGURE 4.2.6   Nicholas Copernicus. This 1973 stamp was issued in honor of the 500th anniversary of his birth.

Copernican system Model of the planetary system with the sun at the center and Earth rotating on its own axis.

> Philolaus believed in the mobility of the Earth, and some even say that Aristarchus of Samos was of that opinion.

We saw how to estimate the solar distances of Mercury and Venus in the last section. Copernicus devised a clever geometric technique to also compute the distances (relative to Earth's distance) of the planets lying beyond Earth's orbit. These distances are given in

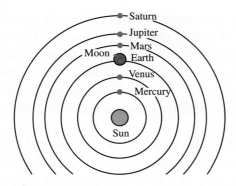

FIGURE 4.2.7   The Copernican system.

Figure 4.2.8, along with the modern astronomical unit value and the orbital period (time of 1 revolution around the sun).

Planet	Copernicus (AU)	Modern (AU)	Period (yr)
Mercury	0.38	0.387	0.24
Venus	0.72	0.723	0.62
Earth	1.00	1.00	1.00
Mars	1.52	1.52	1.88
Jupiter	5.22	5.20	11.87
Saturn	9.17	9.54	29.46

FIGURE 4.2.8   Distances and periods of the planets known at the time of Copernicus.

As we mentioned earlier, even in ancient times, the order of the planets was well established and was based on the assumption that the more distant the planet, the slower it moved. By computing each planet's solar distance, Copernicus was now in a position to estimate its velocity.

? Example 1

If we suppose each planet to move at a constant rate in a perfect circle of radius a centered at the sun, then it travels a distance $2\pi a$ (the circumference of the circle) during one period p. Therefore, its velocity must be $v = \dfrac{2\pi a}{p}$ AU/yr. These units are not very meaningful to us because we are more accustomed to terrestrial terms, and so we convert it to kilometers per second (km/s). There are 1.5×10^8 km in 1 AU and

$$365 \times 24 \times 60 \times 60 = 3.1536 \times 10^7 \text{ s in 1 year. So, 1 AU/yr} = \frac{1.5 \times 10^8}{3.1536 \times 10^7} \text{ km/s} \approx 4.76 \text{ km/s}.$$

Multiplying the above formula by 4.76 gives

$$v = \frac{9.52\pi a}{p}.$$

For instance, the orbital speed of Mercury would be $v = \dfrac{9.52\pi (0.38)}{0.24} \approx 47.4$ km/s.

Copernican velocity Speed of a planet assuming it traveled in a circular orbit.

Today, we know the planets move along near-circular orbits at nonconstant speeds. Hence, this is only a good estimate of the average speed of Mercury, which we shall call the **Copernican velocity**. ◆

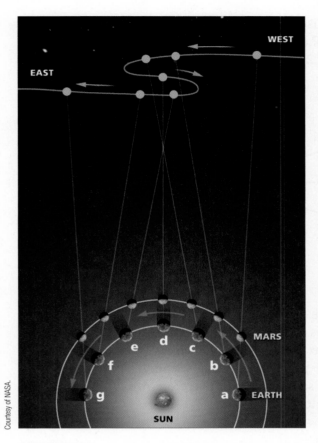

FIGURE 4.2.9 Retrograde motion explained by the Copernican system.

What prompted Copernicus to examine the time-honored Ptolemaic system? One primary reason was that putting Earth in motion around an immobile sun provided a much simpler and more elegant explanation of retrograde motion. Imagine two horses running at the same speed on a racetrack. Although they may be racing dead-even on the straightaway, the one on the inside lane will edge ahead on the curve. To the jockey on the inside horse, the outside horse appears to be going backward in spite of its forward motion. Now consider the six known planets as the horses, with Earth in the third lane from the inside. In the heliocentric system, this phenomenon, in combination with the different speeds of the planets, produces the optical illusion of retrograde motion. **Figure 4.2.9** displays the geometry of the situation for the planet Mars. Note that the reversals do not occur along exactly the same path, resulting in an S shape. This is so because the planetary orbits lie in slightly different planes.

Recall that the Ptolemaic system had been the *paradigm*, or standard model, of the planets for 1½ millennia. By replacing the first axiom (geocentricity) with a new one (heliocentricity), Copernicus pulled out the cornerstone on which the system of Ptolemy had been erected. Such an extreme change in an axiomatic system and resultant model is called a paradigm shift. At the same time, we see that he retained the second axiom and continued to use circles to describe the paths of the planets. As we shall see, each planet actually travels in an elliptical path.

One problem with the Copernican system was that the concept of a moving Earth reintroduced the mystery of a lack of visible stellar parallax. The stars appeared to be permanently attached to some rigid unchanging latticework—an impossibility for close stars unless Earth was forever immobile. Copernicus claimed, correctly as it turned out, that the stars were too distant to exhibit parallax. Determining distances to celestial objects continues to be one of the prime challenges in astronomy. In fact, parallax caused by Earth's motion is so subtle that even 200 years after Galileo first turned a telescope skyward, the first reliable stellar distance was still a mystery. Finally, in 1838, **Friedrich Wilhelm Bessel** determined a stellar parallax angle of a star. From this angle, we can then obtain a distance in the following manner.

❓ Example 2

Because Earth does, in fact, move in orbit around the sun, two pictures can be taken of a nearby star 6 months apart, and an angle of a right triangle can sometimes be inferred from the pictures by measuring the parallax shift of the star in question among the background stars. (Recall the analogous situation, discussed in the last section, created by extending your arm and siting your thumb against a more distant object with each eye separately. A picture is being taken by your brain from two vantage points, and each one displays your thumb differently among the background objects.) The actual angle that astronomers call **stellar parallax** is the angle p in **Figure 4.2.10**.

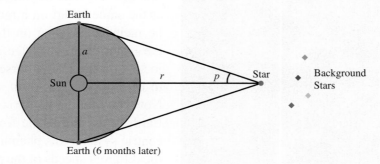

FIGURE 4.2.10   Finding the parallax angle.

The distance r to the star under consideration is quite large in comparison to the Earth–sun distance a, and therefore p is extremely small. For instance, the nearest star, Alpha Centauri, has a stellar parallax of only 0.76″, much less than the angle subtended by the thickness of a piece of paper held at arm's length. Convenience thus dictates that we measure p by using seconds of arc. Therefore, we can use the small-angle formula from the last section to write

$$\frac{p}{206,265} = \frac{a}{r}.$$

Because $a = 1$ AU, we can isolate r in the above equation to get distance as a function of parallax:

$$r = \frac{206,265}{p}.$$

This computes distance in terms of astronomical units and yields large, unwieldly values for r. Instead, astronomers define 1 **parsec** as the distance at which a star possesses a parallax of 1 second of arc. This forces 1 parsec (abbreviated pc) to be equal to 206,265 AU. With these convenient units, we may determine the distance to a star in parsecs simply by inverting its stellar parallax, measured in arcseconds:

$$r = \frac{1}{p}.$$

The above-mentioned Alpha Centauri, for example, is located at a distance

$$r = \frac{1}{0.76} = 1.31 \text{ pc.}$$

Because 1 pc = 3.26 ly, this translates to 1.31(3.26) = 4.3 ly. The insertion of the Hubble Telescope into Earth's orbit increased the number of stars with measurable parallax and so extended the utility of this once rather limited method. ♦

We pause at this point to observe that one of the driving forces behind the lifelong motivation of Copernicus was the further glorification of God. He felt that the establishment of a simpler system for the universe was, in fact, bold evidence of a divine creation. It is interesting at this point to note that neither Ptolemy nor Copernicus made any definite claims in their defining works—the *Almagest* and *De Revolutionibus*—concerning the physical reality of the systems they set forth. Each of these books reveals a mathematical structure for simulating the planetary motions and making predictions, but does not address whether these structures, in truth, describe the nature and arrangement of celestial spheres. It is an example of a long-standing debate that continues today: Does every piece of mathematics have an existent representation somewhere in the universe, or is most of it simply the creation of the human mind? One person who was very much a realist—a believer that his mathematics provided an accurate description of a real situation—was born shortly after Copernicus died.

Few are the people who persevere in the face of constant tragedy to achieve significant accomplishments, but **Johannes Kepler** (1571–1630) is an example of Shakespearean proportions. (See **Figure 4.2.11**.) Kepler was born into poverty in southwestern Germany. His father was a brutal man who disappeared entirely while Johannes was still quite young. Kepler's mother was no jewel either. By his own account, she was "swarthy, gossiping, and quarrelsome, of a bad disposition." Later in her life, she required a defense by her son in a 3-year trial on charges of witchcraft. She barely escaped being burned at the stake.

FIGURE 4.2.11 Johannes Kepler on a Polish stamp.

Kepler himself was a frail and sickly child whose life was fraught with health problems. His intellectual prowess at the local public school, however, led to young Johannes being sent to the seminary in 1584 and to the University of Tübingen 4 years later. It was during his time at this famous school that Kepler began to resonate with an energy and a curiosity to understand the universe. He became convinced that the world and its attendant mysteries were knowable in a form intricately woven in mathematics by the hand of God. He wrote:

> Geometry existed before the Creation. It is co-eternal with the mind of God. . . . Geometry provided God with a model for the Creation. . . . Geometry is God Himself.

At Tübingen, the astronomer Michael Maestlin privately introduced his eager star pupil to the principles of Copernicism. Although Kepler was reluctant to leave his theological studies, his immersion into mathematics and astronomy qualified him for a mathematics teaching position at Graz in Austria in 1594, which he accepted out of financial need. Six years later, he was hired as the assistant to the Imperial Mathematician at the court of Emperor Rudolf II in Prague. Kepler's first assignment was to analyze a large set of observational data of the planets and to determine, once and for all, the elusive orbit of Mars. It was during his 4-year effort to accomplish this task that he made two of the three great discoveries later to be known as **Kepler's laws of planetary motion**.

The path of nearby Mars had always invoked speculation, partly because it displayed a relatively large deviation from circular motion. From the data at Prague, Kepler determined a shape for Mars's orbit that indicated it to be noncircular. Additionally, as both a mystic and a realist, Kepler had long felt that the sun played a key role in causing the movements of all the planets through some type of attraction. This firm belief, along with the approximate curve he had produced, prodded him into the eventual realization that Mars orbited the sun not along a circle but along another of the conic sections—an *ellipse*. This startling conclusion is now known as **Kepler's first law:**

> **The orbit of each planet is an ellipse with the sun at one focus.**

Led by his intuition, Kepler had replaced one noble curve of geometry with another of equal standing. He had discovered a true law of nature and, in the process, had accomplished a most cherished goal. The significance of this event is hard to overestimate. Owen Gingerich, the renowned historian of astronomy at the Harvard-Smithsonian Institute for Astrophysics, said in *Circle of the Gods: Copernicus, Kepler, and the Ellipse* that

> As Newton later claimed, Kepler had guessed—but it surely was an inspired guess based on the latest observations of previously unavailable quality and quantity. The circles of the gods had finally failed.

Ellipses and their beautiful properties had been known since the days of Hipparchus. Any school child can draw one by simply taking up the slack with a pencil in a piece of string whose endpoints F_1 and F_2 have been fixed. One then keeps the string tight while tracing out a closed curve with the pencil. The resulting curve is called an **ellipse**, and each of the points F_1 and F_2 is called a **focus**. (See **Figure 4.2.12**.) Note that this implies that we can define an ellipse as the set of all points whose sum of distances from two fixed foci is constant (i.e., the length of the string).

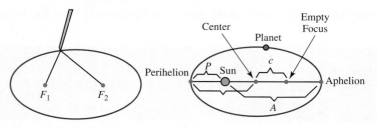

FIGURE 4.2.12   Kepler's first law.

? Example 3

eccentricity Numerical value between 0 and 1 indicating the extent to which an ellipse departs from a circle.

perihelion The point in the orbit of an object in our solar system where the object is at a minimum distance from the sun.

aphelion The point in the orbit of an object in our solar system where the object is at a maximum distance from the sun.

The extent to which an ellipse deviates from being a true circle is measured by a parameter known as its **eccentricity**. (I'm sure you all know someone who deviates enough from normal to be labeled an eccentric.) The *center* of an ellipse is the midpoint of the line that goes through the two foci, known as the *major axis*. If c represents the distance from the center to either focus and a represents one-half the length of the major axis, then the eccentricity e is defined by the ratio $e = \dfrac{c}{a}$. Note immediately that it must be true that $0 \leq e < 1$ and also that the closer the foci are to each other, the smaller the values of c and e. An ellipse having $e = 0$ is a circle. On the other hand, values for e close to 1 are indicative of stretched out, cigar-shaped ellipses. Because all the planets travel in near-circular paths, their orbits have relatively low eccentricity, and this contributed to the cosmic mystery for some time. (On the other hand, many comets, such as Halley's comet, have quite elongated elliptical orbits around the sun, with eccentricities typically greater than 0.9.)

It turns out that eccentricity is more easily determined by using the distances P and A to the points in a planet's orbit that are closest to the sun (labeled the **perihelion** by Kepler) and farthest from the sun (**aphelion**). Considering Figure 4.2.12, we see that

$$e = \frac{c}{a} = \frac{2c}{2a} = \frac{A - P}{A + P}.$$

The often-quoted figure of 150 million km for the distance from Earth to the sun is really the value of the parameter a, referred to as the *mean distance*. In reality, Earth achieves aphelion 152 million km from the sun and perihelion at 147 million km, yielding an eccentricity for our home planet's orbit of

$$e = \frac{152 - 147}{152 + 147} = \frac{5}{299} = 0.0167. \quad \blacklozenge$$

Kepler was not finished with his contributions. In 1609, he published his results in a book called *Astronomia Nova* (*The New Astronomy*), which gives highly mathematical demonstrations for his claim of elliptical orbits. One famous result is now referred to as **Kepler's second law.**

The line joining a planet and the sun sweeps out equal areas in equal amounts of time.

Most astronomers up to this point in history had always felt that the speed of each planet was uniform. (Ptolemy was a notable exception. He used varying speeds for his deferents.) Note how Kepler's second law implies that the speed of any planet is continuously changing. In **Figure 4.2.13**, each of the shaded regions has the same area, and each of the arclengths *AB*, *CD*, and *EF* are traveled by the planet in the same amount of time. The planet travels slowest along the path *AB* because that has the shortest of the three lengths; it goes faster along *CD* and faster still along *EF*. I think you can

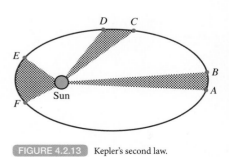

FIGURE 4.2.13 Kepler's second law.

imagine the body accelerating from aphelion to perihelion, achieving a maximum speed there, and then slowing down after it rounds the curve and returns to aphelion, where it achieves its minimum speed before starting the whole trip over again.

Kepler was not quite done. In his final hurrah, he published *Harmonice Mundi* (*The Harmony of the World*) in 1619, a grand monument to the labors of his life in which he attempted to unify astronomy, astrology, music, and geometry in the ultimate explanation of the universe. Although most of it was fantastical musings, one final major planetary rule, **Kepler's third law**, was revealed.

> The square of the period of a planet's revolution is proportional to the cube of its mean distance from the sun.

In equation form, this is written.

$$p^2 = ka^3,$$

where p is the period and a represents the same value as in our definition of eccentricity—one-half the length of the major axis of the planet's orbit. If we choose years for our time measurement and astronomical units for distance, our constant of proportionality has a convenient value of 1. So we get simply

$$p^2 = a^3.$$

? Example 4

Consulting Figure 4.2.8, we find that the mean distance a of Mercury from the sun is 0.387 AU. Thus,

$$p^2 = (0.387)^3 = 0.0580 \Rightarrow p = \sqrt{0.0580} \approx 0.24 \text{ yr.}$$

This value matches up with that in the table. Also, because it is true that $p = a^{2/3}$ and $a = p^{3/2}$, note that we can now write the Copernican velocity v as a function of a.

$$v = \frac{2\pi a}{p} = \frac{2\pi a}{a^{3/2}} = \frac{2\pi}{\sqrt{a}} \text{ AU/yr.}$$

Recall that this formula is based on the assumptions of the planet moving at a constant speed in a circular motion. Because Kepler revealed both of these assumptions to be false, the Copernican velocity now becomes simply a good estimate of the *average velocity* of the planet. If we wish to convert the units to kilometers per second, recall that we need to multiply by the conversion factor of 4.76.

$$v = \frac{2\pi}{\sqrt{a}}(4.76) = \frac{9.52\pi}{\sqrt{a}} \text{ km/s.} \quad \blacklozenge$$

So the penetrating insight and imagination of Johannes Kepler had arrived at three of the most famous results in the search to understand our universe. In so doing, he provided substantial momentum to the revolution begun by Copernicus. Kepler himself, never one to shy away from the dramatic statement, proclaimed in *The Harmonies of the World* that

With this symphony of voices man can play through the eternity of time in less than an hour, and can taste in small measure the delight of God, the Supreme Artist. . . . I yield freely to the sacred frenzy. . . the die is cast, and I am writing the book—to be read either now or by posterity, it matters not.

Name _____

Exercise Set **4.2** ⬡⬡⬡⟵

Note: Answers should be given using the appropriate number of significant digits (cf. Appendix III).

1. What student of Plato used lunar eclipses to demonstrate that Earth was a sphere?

2. The epicycles and deferents used by Ptolemy in his model accounted for what type of unusual motion displayed by the planets?

3. Approximately how long was the Ptolemaic system used as a model for our planetary system?

4. What major paradigm shift was achieved by Nicholas Copernicus in his book *De Revolutionibus?*

5. Kepler correctly postulated in his first law that each planet travels around the sun according to what type of orbital curve? What position does the sun occupy relative to this curve?

6. What is Kepler's second law?

Use the following table for computing the Copernican velocities (in kilometers per second) of the following four planets.

Planet	Mean Distance (AU)	Period (yr)
Venus	0.723	0.62
Earth	1.00	1.00
Jupiter	5.20	11.87
Saturn	9.54	29.46

7. Earth

8. Venus

9. Jupiter

10. Saturn

11. Altair is the brightest star in the constellation Aquila. If it has a parallax of 0.20″, what is its distance in parsecs? In light-years? In astronomical units?

12. The brightest star in the sky is Sirius, and it is easily located in the early evenings of January and February. If it has a parallax of 0.38″, what is its distance in parsecs? In light-years? In astronomical units?

13. If a star is located 23 ly from Earth, what is its parallax angle?

14. If a star is located 48.5 ly from Earth, what is its parallax angle?

15. How many kilometers equal the distance of 1 pc?

16. The Andromeda galaxy is about 2 million ly from Earth. Do you think this distance estimate was found by the parallax method? Explain.

17. Take a piece of string 8 inches (in.) long, and use it to draw three ellipses. First, fix the endpoints of the string 6 in. apart, and call those points *A* and *B*. Take up the slack in the string with your pencil and, keeping the string tight, let it guide your hand as you sketch an ellipse. Repeat this procedure with A and B only 4 in. apart and then again with a 1-in. spread. Which ellipse most closely resembles a circle? Which one has the smallest eccentricity? The greatest? Which would be most likely to represent the path of a planet orbiting the sun?

Use the following table for computing the eccentricities of the following four planets.

Planet	Perihelion (AU)	Aphelion (AU)
Mercury	0.3060	0.4670
Venus	0.7184	0.7282
Mars	1.381	1.666
Jupiter	4.951	5.455

18. Mercury

19. Venus

20. Mars

21. Jupiter

22. Neptune has an orbital eccentricity of 0.0100. If its aphelion distance from the sun is 30.4 AU, determine its perihelion distance.

23. One of the reasons Pluto lost its classification as a planet is that its orbital eccentricity of 0.2484 is larger than that of all the remaining eight planets. (Pluto is now called a dwarf planet.) If its aphelion distance from the sun is 49.2 AU, determine its perihelion distance. Compare these values to those of Neptune in the previous exercise. What conclusion do you reach?

24. Many comets orbit the sun in a highly eccentric elliptical path, revealing themselves only periodically to observation from Earth. Comet Halley has perihelion and aphelion distances of 0.587 and 35.3 AU, respectively. Find the eccentricity of its orbit.

25. Comet Encke has a perihelion distance of 0.332 AU and an eccentricity of 0.8499. Find its aphelion distance.

26. Explain why we may conclude from Kepler's second law that the speed of each planet varies, reaching a maximum at perihelion and a minimum at aphelion.

27. Assuming that the mass of a moon revolving around a planet is significantly less than that of the planet, the moon's orbital path is an ellipse with the center of the planet at one focus. In this case, the point of greatest distance is known as the *apogee*, and the point of least distance is called the *perigee*, usually measured from the center of the planet to the center of its moon. The eccentricity of the orbit is again given by $e = \dfrac{A - P}{A + P}$, where A and P are the apogee and perigee distances, respectively. In Example 5 from the first section, the distances to perigee and apogee of our moon were 363,000 and 405,000 km, respectively. What is the eccentricity of our moon's orbit?

←

28. Kepler's laws of planetary motion apply to any body in our solar system—all the asteroids, comets, and planets discovered since Kepler's time. Use the third law to fill in the blanks below. Do the values for Jupiter and Saturn match up with Figure 4.2.8?

Planet	Mean Distance (AU)	Period (yr)
Jupiter	5.20	
Saturn		29.5
Uranus	19.19	
Neptune		164.8
Pluto	39.44	
Comet Halley		76.0
Comet Encke	2.21	

29. Express the Copernican velocity (in kilometers per second) as a function of the period p (yr) of a planet. Use this function and the periods computed in the previous problem to find the velocities for Uranus, Neptune, and the dwarf planet Pluto.

30. In Figure 4.2.12, we can see that the sum of the distances to aphelion and perihelion must be equal to twice the mean distance. Symbolically, $2a = A + P$. If the mean distance to Uranus is 19.2 AU and the perihelion distance is 18.3 AU, find the aphelion distance and the eccentricity of its orbit.

31. The comet Hyakutake was sighted for the first time in January 1996 and was bright enough in the evening sky by March to be seen with the naked eye. In fact, this brightness caused it to receive a great deal of media attention even though about a dozen comets are sighted each year. Its period was determined to be about 9,100 yr. Use Kepler's third law to find its mean solar distance.

32. Perihelion distance P to Comet Hyakutake was observed to be about 0.20 AU. Use the mean distance a computed in the previous exercise to find the aphelion distance. (*Hint:* From Exercise 30, we know that $2a = A + P$.)

33. Although each planet varies in brightness over time as a result of its changing distance from Earth, this is a phenomenon that cannot be perceived by the naked eye, and so pretelescopic astronomers were unaware of it. Modern telescopes reveal that the angular diameters of Mercury range from a minimum of 4.7″ to a maximum of 13″. Given that Mercury has a linear diameter of 4,900 km and approximating its orbit by a circle, we can estimate the (mean) diameter of the orbit. Find the minimum distance r_{min} (corresponding to 13″) and the maximum distance r_{max} (corresponding to 4.7″) from Earth to Mercury as in the accompanying figure, and subtract them to get the orbital diameter. (Recall that the angular the diameter α and distance r are related by $\dfrac{\alpha}{206,265} = \dfrac{4,900}{r}$ when α is measured in arcseconds.)

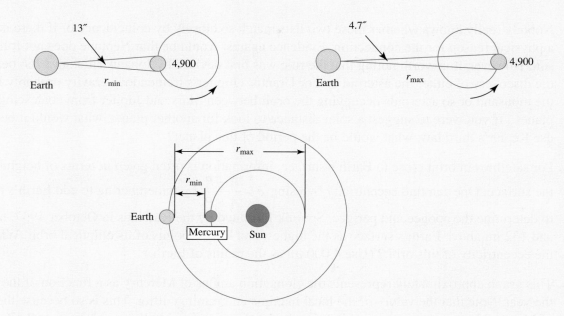

34. The angular diameter of Mars ranges from a minimum of 4″ to a maximum of 24″. If Mars has a linear diameter of 6,800 km, estimate the mean diameter of its orbit.

35. Venus presents a disk of 10″ at full phase up to 60″ at crescent phase. If it has a linear diameter of 12,000 km, estimate the mean orbital diameter of Venus.

36. The minimum and maximum angular diameters of Jupiter are about 30″ and 50″, respectively. If Jupiter has a linear diameter of 144,000 km, estimate the mean orbital diameter.

37. The Titius–Bode rule is an algorithm for producing a sequence of numbers that eerily coincide with the average distances from the sun of the first seven planets (but not for Neptune). The sequence is generated by first writing 0, 3, 6, 12, 24, and so on, doubling each successive number. Next, 4 is added to each number, and then the result is divided by 10. This gives a predicted distance in astronomical units that matches up with the actual distance with remarkable accuracy. The dwarf planet Pluto and the asteroids are also included in the following list.

Planet	Titius–Bode Prediction (AU)	Actual (AU)
Mercury	(0 + 4)/10 = 0.4	0.387
Venus	(3 + 4)/10 = 0.7	0.723
Earth	(6 + 4)/10 = 1.0	1
Mars	(12 + 4)/10 = 1.6	1.524
Asteroid belt	(24 + 4)/10 = 2.8	2.77 average
Jupiter	(48 + 4)/10 = 5.2	5.203
Saturn	(96 + 4)/10 = 10.0	9.539
Uranus	(192 + 4)/10 = 19.6	19.18
Neptune		30.06
Pluto	(384 + 4)/10 = 38.8	39.44

Nobody really knows whether these two lists match so closely by coincidence or if there is really a physical reason for the connection. Evidence against would be that Neptune does not follow suit. However, it is worth noting that the rule was first devised by Johann Titius in 1766 before the discovery of either the asteroid belt or Uranus. (Jupiter's tremendous gravity probably kept the thousand or so asteroids occupying the orbit between Mars and Jupiter from coalescing into a planet.) If you were to suggest a solar distance to look for another planet, what would it be? If you use Kepler's third law, what would be the period of this planet?

38. For satellites in orbit close to Earth's surface, information is often given in terms of heights above the surface. One can find eccentricity by using $e = \dfrac{A - P}{A + P}$ by remembering to add Earth's radius to determine the apogee and perigee. Sputnik I, orbited by the Russians in October 1957, had 583 and 132 mi above Earth's surface as the highest and lowest points of its elliptical orbit. What is the eccentricity of this orbit? (Use 4,000 mi as the radius of Earth.)

39. This graph approximately represents the elongation angles of Mercury as a function of the day of the year. Note that the values of the local maxima and minima differ. This is so because the orbit of Mercury has an eccentricity of 0.2056, the largest of the eight planets. Mercury is 0.47 AU from the sun at aphelion and 0.31 AU at perihelion. If Mercury happened to be at aphelion coincidentally with maximum elongation as seen from Earth, what would be the elongation angle? What would be the angle at perihelion? (See Section 4.1, Exercise 28.)

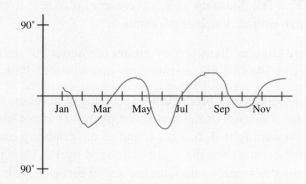

40. Kepler needed to know the formula for the area of an ellipse when studying the data that led to his second law. In his book *Calculus Gems*, George F. Simmons points out that a procedure similar to that of Kepler was used by his contemporary **Bonaventura Cavalieri** (1598–1647). When centered at the origin of a Cartesian coordinate system, an ellipse with semi-axes a and b and a circle of radius a have the respective equations

$$\frac{x^2}{a^2} + \frac{y^2}{b^2} = 1 \quad \text{and} \quad x^2 + y^2 = 1.$$

Solving both of these equations for y, we obtain

$$y = \pm \frac{b}{a}\sqrt{a^2 - x^2} \quad \text{and} \quad y = \pm\sqrt{a^2 - x^2}.$$

So we see that the *y*-coordinate on any point of the ellipse is *b/a* times the corresponding *y*-coordinate of the point on the circle having the same *x*-coordinate.

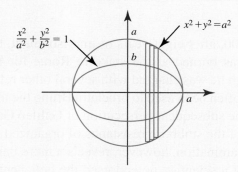

If we think of the areas as being approximated by a set of thin rectangles, then each rectangle used to compute the area of the ellipse has length *b/a* times the corresponding rectangle used to compute the area of the circle. Hence

$$\text{Area of ellipse} = \frac{b}{a}\left(\text{area of circle}\right) = \frac{b}{a}\left(\pi a^2\right) = \pi ab.$$

What is the area of the ellipse with semi-axes of lengths 8 and 12? What is the area of an ellipse with semi-axes of lengths 15 and 28?

41. If the orbit of a planet has eccentricity *e* and mean distance (major semi-axis length) *a*, then the length of the minor semi-axis is $b = a\sqrt{1-e^2}$. Use the formula in the previous excercise to find an expression for the area of the ellipse in terms of *a* and *e*. For orbits with low eccentricity, does this differ much from the area of a circle of radius *a*?

4.3   The Defense of Copernicanism

> In medicine it was sufficient to quote Galen, as to quote Aristotle in practically everything else. For Galileo, to quote was not sufficient; he turned to mathematics.
>
> —George Polya

On February 17, 1600, after eight years of imprisonment, the Dominican monk **Giordano Bruno** (b. 1548) was burned at the stake in Rome for heresy, a victim of the Roman Inquisition. Although he was charged with several other religious offenses, his advocacy of Copernicanism has often been used to proclaim Bruno the first martyr of the new astronomy. His execution and the subsequent persecution of Galileo Galilei have been commonly cited as prime examples of the stubborn resistance of organized religion to the noble advance of science. A closer examination, however, reveals a more balanced perspective.

Bruno was not a scientist; he endorsed the heliocentric theory only as one of many arguments to be used in his efforts to reconcile Protestants and Catholics in a time of violent religious warfare. He had a poor technical understanding of *De Revolutionibus*, yet he seized its ideas as a grand philosophical metaphor. The important thing to Bruno was that Earth had been shown not to be at the center of the universe. This centerless universe not only must be of infinite extent, but also must contain an infinite number of other worlds because Earth now had no claim to uniqueness.

Unfortunately, the condemnation of Bruno set the stage for the persecution of **Galileo Galilei** (1564–1642), one of the first true scientists and the valiant champion of Copernicanism. (See **Figure 4.3.1**.) His goals in life were completely different. Galileo vigorously supported the heliocentric theory based solely on years of scientific investigation and attached to it no religious implications. Bruno's shadowy rhetorical arguments lacked what Galileo accrued in abundance—conclusions based on solid experimental evidence. Historians of science credit Galileo and his contemporary, **Réné Déscartes**, with the reformulation of the composition of scientific activity. By molding a new methodology of experimentation and analysis, they forged a permanent bond between science and mathematics, one begun by Copernicus and Kepler. Galileo, in particular, insisted that

FIGURE 4.3.1   Galileo on an Italian banknote.

mathematical relationships, derived from experimentation and embodied by equations and functions, were the cornerstone to any theory worth postulating. His work paved the way for the astonishing successes of modern science in the last three centuries.

Prior to the seventeenth century, the approach to understanding nature had never varied much from that taken by the ancient Aristotle and his disciples—one in which explanations for natural phenomena were qualitative rather than quantitative. Actual experimentation either to initiate exploration of a new process or to corroborate the predictions of an existing theory was never seriously done. Aristotle claimed, for example, that a body twice as heavy as a second body must fall twice as fast, a conjecture no one ever bothered to check prior to Galileo.

By 1609, Galileo had been teaching mathematics at the University of Padua for 18 years. During this time, his research of the properties of pendulums and his important discoveries in mechanics—the physics of motion—had already earned him a glowing reputation. From his experiments, he formulated two laws giving the velocity and distance of a dropped object as functions of time alone, *regardless of the size or mass of the object*.

Strictly speaking, the **velocity** of a moving object is specified by both the speed at which it is moving *and* the direction of movement. Going 30 miles per hour (mi/h) east on Main Street is a different velocity than going 30 mi/h west on Main Street. **Acceleration** of a moving object measures how the velocity changes per unit of time. This can be a change in speed, a change in direction, or both. If a moving object is subject to no acceleration, then the velocity is a constant value V, and the distance traveled in time T is just the product VT. Let $v = V$ be the horizontal line in the coordinate system on the left in **Figure 4.3.2**. Then the distance traveled in time T is equal to the area of the rectangle under that line and over the interval $[0, T]$.

velocity The rate of change of position of a moving body, i.e., its speed and its direction of motion.

acceleration Rate of change of velocity with respect to time. It has both a numerical value and a direction.

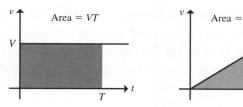

FIGURE 4.3.2   Distances traveled by a moving object can be represented as areas. The area on the left corresponds to an object moving at constant speed $v = V$. The area on the right corresponds to an object moving at speed $v = gt$.

Galileo observed that falling bodies gain speed as they fall. He speculated that the acceleration of a body dropped from rest was constant in both numerical value and direction (toward the center of Earth). He then experimented to verify this relationship by rolling balls down inclined planes—a related phenomenon of motion that facilitated the necessary measurements. Today, we represent this constant by the letter g for acceleration due to gravity near the surface of Earth (or whatever planet where you may be). If v is the velocity of a dropped object after a time t from the moment of release, then

$$v = gt.$$

The graph of this linear function is shown on the coordinate system on the right in Figure 4.3.2. Because the velocity is continuously changing, we cannot just multiply a velocity by a time to obtain a distance. But notice that after a specific elapsed time T, the velocity is $v = gT$. By again equating the distance d traveled as the area under the curve as before, we compute d as the area $\frac{1}{2}gT^2$ of the right triangle with legs T and gT. Writing d as function of any arbitrary time t, we have

$$d = \frac{1}{2}gt^2.$$

These two functions give the velocity and distance of a dropped object as a function of time after release near the surface of any planet (neglecting the effects of air resistance). Of course, the constant g (sometimes referred to simply as *surface gravity*) depends on the planet. A complete list of values is given in **Figure 4.3.3**.

Planet	Acceleration g (m/s²)
Mercury	3.8
Venus	8.9
Earth	9.8
Mars	3.7
Jupiter	24.9
Saturn	10.4
Uranus	8.8
Neptune	12.0
Pluto	0.7

FIGURE 4.3.3   Acceleration due to gravity near the surfaces of the eight planets plus Pluto.

? Example 1

The value of g for Earth has been determined to be about 9.8 meters per second (m/s) every second. (These units are abbreviated as m/s².) If a baseball takes 4 s to drop from the top of a building, how tall is the building, and how fast is the ball traveling when it hits the ground? (This is referred to as the *impact velocity*.)

⚙ Solution

The ball will have an impact velocity directed straight down of $v = 9.8(4) = 39.2$ m/s. The height of the building must be equal to the distance traveled by the ball: $d = \frac{1}{2}(9.8)(4)^2 = 78.4$ m. Note the intermediate values for v and d in **Figure 4.3.4**.

t (s)	w (m/s)	d (m)
0	0	0
1	9.8	4.9
2	19.6	19.6
3	29.4	44.1
4	39.2	78.4

FIGURE 4.3.4   Velocity and distance of a falling object.   ◆

mass Property of an object that is a measure of the amount of matter in the object. It can be thought of as a measure of its inertia.

grams The basic measure of mass in the metric system.

kilogram 1,000 grams.

weight Amount of force that a body exerts on the surface of a planet or moon as a result of gravity.

newton Unit of force in the metric system.

It is important to note that neither of Galileo's functions for velocity or distance incorporate the *mass* of the falling body. The **mass** of an object is the measure of the amount of matter of which it is composed. This value never varies with respect to the location of the object—it is the same on Earth, on the moon, or in the vacuum of space. Mass is measured in **grams** (g) in the metric system. One **kilogram** (1 kg) is 1,000 g.

We must be careful to never confuse mass with the separate concept of *weight*. An object's **weight** is the force with which it is attracted to the body on which it is resting. You weigh less on the moon, for example, because it attracts you with less force than Earth does. Force (to be discussed further in the next section) is measured using pounds in the English system or newtons in the metric system. One pound is the equivalent of about 4.45 newtons (N). One **newton** is defined as the amount of force required to give an acceleration of 1 m/s² to a mass of 1 kg. Because gravity on the surface of Earth is 9.8 m/s², it gives a weight (force) of 9.8 N to any object having a mass of 1 kg, a weight of 19.6 N to a mass of 2 kg, and so on. In general, we can write the relationship between the mass m and weight W of any object on a planet with surface gravity constant g as

$$W = mg.$$

? Example 2

If you weigh 160 lb on Earth, what is your mass? What would your weight be on Mars?

⚙ Solution

Because mass is almost always expressed with metric units, we first determine your metric weight to be (160 lb)(4.45 N/lb) = 712 N. We substitute this for W in $W = m(9.8)$ to obtain

$$712 = m(9.8)$$

$$m = \frac{712}{9.8} = 72.7 \text{kg}.$$

(Note that this says that a 160-lb object has a mass of 72.7 kg. Pounds are often mistakenly used as units of mass with a so-called conversion factor of 2.2 lb/kg. See Exercise 10.)

Your mass, of course, remains the same everywhere (assuming you've laid off the pasta), but the acceleration due to gravity on the smaller Mars is 3.7 m/s². Thus, your weight there would be $W = mg = 72.7(3.7) = 269.0$ N. Converting back to English units, we divide 269.0 N by 4.45 N/lb to get the equivalent weight of 60.4 lb. ◆

Because the surface gravity on Mars exerts just $\frac{3.7}{9.8}$ of the pull that it exerts on Earth, we also could have found your Martian weight in the above example by simply multiplying: $\left(\frac{3.7}{9.8}\right)160 = 60.4$ lb. In general, if your weight is W_E on Earth and W_P on planet P with surface gravitational acceleration g, then

$$\frac{W_P}{W_E} = \frac{mg}{m(9.8)}$$

$$W_P = \left(\frac{g}{9.8}\right)W_E.$$

GARFIELD, YOU WEIGH TOO MUCH

I HEAR YOU, JON

I WANT YOU TO LOSE WEIGHT, AND I MEAN NOW!

YESSIR

WHERE ARE YOU GOING?

TO A PLANET WITH A WEAKER GRAVITATIONAL PULL

JIM DAVIS 5-24

Example 3

Garfield weighs 28 lb. What would he weigh on Venus? On Jupiter?

Solution

Consulting Figure 4.3.3, we see that Garfield would weigh

$$W_{\text{Venus}} = \left(\frac{8.9}{9.8}\right)(28) = 25 \text{ lb on Venus}$$

and

$$W_{\text{Jupiter}} = \left(\frac{24.9}{9.8}\right)(28) = 71 \text{ lb on Jupiter.} \quad \blacklozenge$$

As significant as his achievements were in the field of physics, Galileo acquired his greatest fame (and got into the most trouble) for his defense of the Copernican system. At the stately age of 45, his rather orderly life as a university professor took a dramatic turn when he learned of a wondrous new invention, the telescope. Immediately, he constructed one of his own and began a systematic study of the night skies, which was the first use of this scientific instrument in astronomy. The impact was dramatic. *Everything* in the heavens upon which Galileo turned his scope was being observed for the first time. In 1610, Galileo published a book of his studies, *Sidereus Nuncius*, which rocked the scientific and religious worlds. As a title, he wrote

> ASTRONOMICAL MESSAGE
> Containing and Explaining Observations Recently Made,
> With the Benefit of a New Spyglass, About the
> Face of the Moon, the Milky Way, and Nebulous
> Stars, about Innumerable Fixed Stars and also Four
> Planets hitherto never seen, and named
> MEDICEAN STARS

Galileo continued to give more detailed descriptions of these findings, two of which are of particular interest to us here. Early in the book he commented on the craggy surface of the moon:

> . . . we certainly see the surface of the Moon to be not smooth, even, and perfectly spherical, as the great crowd of philosophers have believed about this and other heavenly bodies, but, on the contrary, to be uneven, rough, and crowded with depressions and bulges. And it is like the face of the Earth itself. . . .

This description flew in the face of the standard Aristotelean view that all the heavenly bodies must, by nature and God, be perfectly round and smooth. To Galileo, the moon offered convincing evidence that Earth was not unique but rather just one of the collection of rocky orbs that circled the sun. His newly discovered objects supported that notion as well. On January 7, 1610, he had noticed three peculiar stars along a straight line through Jupiter, two near to the planet on the east and one on the west. When he looked again the next night, he noticed the same three stars but in a *different* arrangement. Intrigued, he began making nightly sketches of this mysterious behavior and soon realized from their varying patterns that these were not stars at all, but four new bodies (he later sighted a fourth) in orbit *around Jupiter*. This conclusion was a stunning discovery, for it proved the existence of a second center of motion, thereby providing a disclaimer to Earth's special status in that role. This belief was falsified by the finding of a planet that clearly retained its satellites as it traveled through space.

Galileo had privately long believed in the heliocentric theory, but now, as the evidence piled up, he became a fervent torchbearer for its validity. Any lingering doubts were forever banished when he turned his telescope on Venus and observed that the planet went through a full series of phases just as our own moon does each month. This is possible in a Copernican scheme, because Venus would appear more than half-illuminated (called a

gibbous A phase of an illuminated body in which more than one-half of the illuminated side is visible from Earth.

crescent A phase of an illuminated body in which less than one-half of the illuminated side is visible from Earth.

gibbous phase) when it was farther from us than the sun. It is impossible if Venus is riding on an epicycle between the sun and Earth, because most of the side illuminated by the sun would always face away from Earth, revealing *only* a **crescent** phase. (See **Figure 4.3.5**.) It dealt the Ptolemaic system another shattering blow.

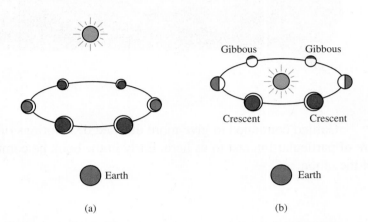

FIGURE 4.3.5   The phases of Venus as they would appear in (a) the Ptolemaic system and (b) the Copernican system.

? Example 4

We must stop for a moment and consider the logic involved with using these observations in support of a particular planetary system. Galileo offered them as "proof" of the Copernican model. But is this a valid conclusion? An analogous situation would be the following syllogism.

> Fact: If a car is a Porsche, then it is fast.
> Hypothesis: Sabrina owns a Porsche.
> Conclusion: Therefore, Sabrina owns a fast car.

In other words, being fast is a necessary consequence of being a Porsche. Is the reverse true, however? Is being fast sufficient to ensure being a Porsche?

> Hypothesis: Rusty owns a fast car.
> Conclusion(?):   Therefore, Rusty owns a Porsche.

converse For the conditional statement "if P, then Q," the converse is "if Q, then P." Its truth value is not related to that of the original statement.

Any student of logic (and cars) will tell you that this is a false conclusion. A logician would say that the **converse** to a given true statement need not also be true. The converse to a statement such as "if *P*, then *Q*" is "if *Q*, then *P*." The converse to our above fact is false:

Converse: If a car is fast, then it is a Porsche.

For our present purposes, we are concerned with the validity of a planetary model. So we examine this statement:

> Fact: If the model is heliocentric, then Venus will show many phases.

Galileo has observed the phases of Venus. Can he justifiably conclude that the system is heliocentric? No, he cannot. The converse to this fact is not true. Just as other types

> **contrapositive** For the conditional statement "if *P*, then *Q*," the contrapositive is "if not *Q*, then not *P*." Its truth value is the same as that of the original statement.
>
> **hypothetico-deductive method** A method of inquiry in which a theory gains increasing acceptance as more evidence for its validity is observed. This is *not* the same as a deductive method, by which a definitive conclusion is reached.

of cars may also be fast, there were other postulated schemes for the planetary structure that also predicted Venusian phases.

On the other hand, the **contrapositive** of a given fact must always be true. The contrapositive of the statement "if *P*, then *Q*" is "if *not Q*, then *not P*." In the above example, this would take the following form.

Contrapositive: If a car is *not* fast, then it is *not* a Porsche.

Galileo was quite correct in using the phases of Venus to eliminate the Ptolemaic system as a candidate, for it is the contrapositive of this true statement:

Fact: If the system is Ptolemaic, then Venus cannot show many phases. ◆

Hermann Weyl, the famous twentieth-century mathematician and physicist, once said, "Logic is the hygiene the mathematician practices to keep his ideas healthy and strong." Galileo was much too good a mathematician to not realize that his observations did not allow him to definitively *conclude* that Earth moves around a centralized sun; yet he was convinced this was the case. He was using what is now referred to as the **hypothetico-deductive method**—one in which a given hypothesis gains steadily increasing acceptance as it continues to pass a sequence of tests. Although Galileo was convinced of the truth of Copernicanism, the science of today usually does not profess to discover such absolutes. The phrase *current model* is the typical claim, that is, a mathematical framework that can be used for predictive purposes. A model is pronounced good if it gives accurate predictions. (Recall that neither Ptolemy nor Copernicus made any claims in their main astronomical works that their models represented the *real* structure of the cosmos.) The irony of Galileo's celebrated troubles with the Catholic Church is that it made only one substantial request that he declare his beliefs to be just that—a hypothetical model, convenient for use by mathematicians but *not* a description of physical reality. In 1616, the powerful Cardinal Roberto Bellarmino evaluated one of Galileo's analytical treatises in a letter to a friend:

> For to say that assuming the earth moves and the sun stands still saves all the appearances better than eccentrics and epicycles is to speak well. This has no danger in it, and it suffices for mathematicians.
>
> But to wish to affirm that the sun is really fixed in the center of the heavens and that the earth is situated in the third sphere and revolves very swiftly around the sun is a very dangerous thing, not only by irritating all the theologians and scholastic philosophers, but also by injuring our holy faith and making the sacred Scripture false.

This passage is indicative of the sympathies accorded Galileo by many leading theologians of the day, including Cardinal Maffeo Barberini, with whom he had many friendly and enlightening cosmological discussions. However, as in Example 3, they had legitimate doubts stemming from the potential fallacies in Galileo's logic. His results were simply not conclusive enough to force a reinterpretation of a centuries-old belief system. The stubborn Galileo did not agree. Even though he was a devout Roman Catholic, he was always fond of saying, "The Bible teaches how to go to heaven, not how the heavens go."

Unlike his shy predecessor Copernicus, Galileo was an egotistical and, at times, abrasive individual who loved to be at the center of controversy. As evidenced by the fate of

FIGURE 4.3.6   Frontispiece and first page of the *Dialogue*.

Bruno, however, the religious upheavals of the 1600s had created a tense atmosphere for the debate of the heliocentric theory in Italy. In 1616, *De Revolutionibus* was placed on the Index of Forbidden Books, and Galileo was formally warned by Cardinal Bellarmino against publicly defending the Copernican system. However, in 1623, his friend Cardinal Barberini was elected Pope Urban VIII. Filled with an increased sense of security, the headstrong Galileo overplayed his hand in 1632 when he published his most famous work, *Dialogue Concerning the Two Chief World Systems*. (See **Figure 4.3.6**.) Although it was billed by the author as an objective debate of the Copernican and Ptolemaic models, even the most casual reader realized the *Dialogue* to be an extensive argument in favor of Copericanism. Even worse, the enemies of Galileo seized upon the fact that one of the main characters, Simplicio, whom every reader knew also meant "simpleton," had essentially espoused the opinions of the pope himself. Such clear disrespect forced the hand of Pope Urban VIII, and in February 1633, Galileo was summoned to Rome at the age of 70 to face the Inquisition. Fourth among the stated charges listed against him was the following:

> **Dialogue** The book written by Galileo, that defends the Copernican system through an extended conversation among three men.

> The author claims to discuss a mathematical hypothesis, but he gives it physical reality, which mathematicians never do. Moreover, if defendant had not adhered firmly to the Copernican opinion and believed it physically true, he would not have fought for it with such asperity, nor . . . would he have held up to ridicule those who maintain the accepted opinion, and as if they were dumb mooncalves [*hebetes et pene stolidos*] described them as hardly deserving to be called human beings.
>
> Indeed, if he had attacked some individual thinker for his inadequate arguments in favor of the stability of the Earth, we might still put a favorable construction on his text; but, as he holds all to be mental pygmies [*homunciones*] who are not Pythagorean or Copernican, it is clear enough what he has in mind. . . .

This leads us to believe that the trial was probably motivated primarily by the desire for public submission by the arrogant Galileo. Extreme punishment was not necessary, and Galileo probably understood from the beginning that a complete admission of his error was the inevitable conclusion. At the final point in the proceedings, the head of the inquisitors refrained from putting specific passages of the *Dialogue*, one at a time, to Galileo and demanding a refutation. This could have forced the issue of heresy back to the forefront, leading to a far more dangerous situation and a potentially crueler punishment. As it was, the next day his sentence read, in part,

> We pronounce ... the said Galileo ... have rendered yourself in the judgment of this Holy Office vehemently suspected of heresy, namely, of having believed and held the doctrine—which is false and contrary to the sacred and divine Scriptures—that the Sun is the center of the world and does not move from east to west and that the Earth moves and is not the center of the world....
>
> And, in order that this your grave and pernicious error and transgression may not altogether go unpunished ... we ordain that the book of the "Dialogue of Galileo Galilei" be prohibited by public edict.
>
> We condemn you to the formal prison of this Holy Office during our pleasure....

Upon hearing his sentence, Galileo dutifully knelt before the 10 judges and recanted his sin of promoting the Copernican system. One intriguing legend among the tales of scientific lore holds that, on rising from this humbling experience, the feisty old man flashed his still unbroken spirit by whispering, *"Eppur si muove"* ("Still it moves"), meaning, of course, Earth. It seems doubtful that one would take such a risk under the circumstances, but it certainly would have been in character for Italy's most famous pioneer of mathematics, physics, and astronomy.

> According to historic record, the *segno* was revealed in a mode the Illuminati called *lingua pura.*
>
> The pure language?
>
> Yes.
>
> Mathematics?
>
> That's my guess. Galileo was a scientist after all, and he was writing *for* scientists. Math would be a logical language in which to lay out the clue.
>
> —Dan Brown, *Angels & Demons*

FIGURE 4.3.7 Statue of Galileo in Florence, Italy.

Name _____

Exercise Set **4.3** ⭘⭘⭘←—————————————————————————————————

1. A ball dropped from a tall building takes 6.0 s to reach the ground (on Earth). What is the impact velocity of the ball, and how tall is the building?

2. A B1 bomber drops a bomb from a height of 6,000 m. Neglecting the effects of air resistance, how long does it take for the bomb to hit the ground, and what is its impact velocity?

3. Using Galileo's formulas for the velocity and distance of a dropped object on Earth, write the impact velocity v of the object as a function of the distance d it drops. (*Hint*: Solve the distance formula for t in terms of d, and substitute into the velocity formula.) What is the impact velocity of an object dropped from 100, 500, 1,000, and 4,000 m? Is this an increasing function?

Use these values to answer Exercises 4–9.

	Acceleration g (m/s²)
Mercury	3.8
Venus	8.9
Earth	9.8
Mars	3.7
Jupiter	24.9
Saturn	10.4
Uranus	8.8
Neptune	12.0
Pluto	0.7
Moon	1.6

4. In 10 s, how fast would an object dropped from rest be traveling on Venus? Mars? Neptune? How far would it drop on those three planets?

5. How long would it take a body to drop from a height of 500 m on Mercury? Earth? Saturn? What would the impact velocity be on those three planets? On which of these three planets would it take the least time to fall a specific distance? The greatest time?

6. Write the velocity v of the object as a function of the dropped distance d and the acceleration g. What is the impact velocity of a rock dropped from 100 m on Venus? Jupiter? Pluto?

7. Estimate a height from which you feel you could safely jump on Earth, and determine your impact velocity for that height. If you use that value as a safe impact velocity, from what height could you jump on Pluto and not be injured?

8. Lars weighs 890 N on Earth. What is his weight in pounds? What would he weigh (in pounds) on Venus? Mars? On which of the eight planets would his weight be a maximum?

9. Maria has a weight on Earth of 120 lb. What is her mass? What would her weight be (in pounds) on Mercury? Jupiter? Uranus?

10. Pounds are sometimes mistakenly used as units of mass. Use $W = 9.8m$ to show why the so-called conversion factor from kilograms to pounds is 2.2 lb/kg. Remember that first you must change pounds to newtons before you use this formula.

11. Luke Skywalker lands on the planet Xantok and measures the weight of a large igneous rock to be 533 N. Luke drops his light saber from the top of a cliff and determines the gravitational acceleration to be 6.5 m/s². What is the mass of the rock? What would it weigh on Earth in newtons? In pounds?

Assume each of the following statements is true. Write the converse and contrapositive of each statement, and classify each one as true or false.

12. If it is a cloudy day, then you do not catch fish on Fence Lake.

13. If someone is watching television, then the television is on.

14. If the light switch is up, then the light is on.

15. If a polygon is a square, then it is a rhombus.

16. If the wind is blowing, then the windmill is turning.

17. If a house is on Paradise Lane, then it costs at least $180,000.

18. If $x = 10$, then $x^2 = 100$.

19. If $x = 12$, then $2x + 15 = 39$.

20. If it does not rain, then we do not cancel soccer practice.

21. If my shirt is solid red, then it is not blue.

22. If the temperature is above 45°F, then it does not snow.

Determine whether each of the following five statements is true or false. Write the converse and contrapositive of each statement, and identify each one as true or false.

23. If $f(x)$ is an increasing function over $[a, b]$, then $f(a) < f(b)$.

24. If $f(x)$ is a decreasing function over $[a, b]$, then $f(a) < f(b)$.

25. If $f(x)$ is an increasing function, then $f(x)$ is a linear function with positive slope.

26. If $f(\theta) = \sin \theta$, then $f(30°) = 0.5$.

27. If $f(x) = 5x + 13$, then $f(x) = 23$.

28. Explain why Venus can be viewed going through a set of phases in the Copernican system but not in the Ptolemaic system. Do you think an outer planet (such as Mars) would exhibit two identifiably different sets of phases in the two systems? Explain.

29. Why didn't the observation of the phases of Venus prove the validity of the Copernican system? Why did it falsify the Ptolemaic system?

←——

30. Galileo openly held a stronger opinion of the heliocentric model than did Copernicus. In what way? How did this position lead to conflict with the Catholic Church?

31. What book written by Galileo led to his being summoned to Rome to face the Inquisition?

32. According to legend, what did Galileo say upon recantation of his beliefs to the judges of the Inquisition?

33. Galileo offered a clever argument for why his formulas for the velocity and distance of a falling object were independent of mass. Suppose two objects x and y have different masses, with x being the lighter one. Aristotle claimed that y must fall faster than x. Galileo proposed, "What if we tie the two objects together and drop them?" Then the lighter object should retard the velocity of the combined $x + y$, and so the velocity of $x + y$ should be less than that of y alone. On the other hand, $x + y$ is more massive than y and so should fall faster, according to Aristotle. What is your conclusion to this thought experiment?

4.4 And All Was Light

> Nature and Nature's laws lay hid in night:
> God said, "Let Newton be!" and all was light.
>
> —Alexander Pope

FIGURE 4.4.1   Isaac Newton.

Galileo's sentence of life imprisonment was commuted to permanent confinement to his villa, where he spent the last nine years of his life studying physics and writing his finest book on the subject. As befits a good drama, 1642 saw both the death of Galileo and the birth of the man whom many declare to have possessed the most profound intellect in the history of humankind. **Sir Isaac Newton** (1642–1727) was a towering icon to the conquest of the mind over the mysterious universe (see **Figure 4.4.1**). He was described by biographer Gale E. Christianson as "a mutant, seeming more a phenomenon than a man . . . the incarnation of the abstracted thinking machine" and one who, in the words of Albert Einstein, "stands before us, strong, certain, and alone." An only child, born shortly after his father's death on a farm at Woolsthorpe, England, Newton experienced an unusually lonely childhood. By his third birthday, his mother had remarried an elderly rector to better her financial position, leaving young Isaac in the care of his aged grandmother at Woolsthorpe. As a youth, he spent many solitary hours constructing an array of mechanical devices such as water clocks, sundials, and various types of kites. He was not reunited with his mother until the age of 11, and the acute sense of abandonment felt by the young Isaac had a profound effect on his adult personality. A lifelong bachelor and recluse, he was penurious and sternly disciplined in his personal habits. Shunning the usual entertainments of civilized society, Newton was always happiest when left to himself to indulge his relentless curiosity. When asked later in life by one of his few friends, the astronomer **Edmond Halley** (1656–1743), how he had made his great discoveries, he responded, "By thinking about the problem unceasingly."

In 1665, Newton received his B.A. from Trinity College, Cambridge, where he had studied mathematics and physics. That same summer, the college was evacuated due to the spread of the bubonic plague that was devastating England and Europe, and so Newton returned to Woolsthorpe, where he spent two years in solitary contemplation of mathematics, optics, astronomy, and mechanics. This was a period in his life often referred to by historians as the *miracle years* because of the many discoveries he made at that time. In a 1718 letter, Newton wrote:

> In the beginning of the year 1665 I found the Method of approximating series . . . in May I found the method of Tangents . . . & in November had the direct method of fluxions . . . in January the Theory of Colours. . . . I began to think of gravity extending to y⁰ orb of the Moon & . . . from Kepler's rule . . . I deduced that the forces [which] keep the Planets in their Orbs must [be] reciprocally as the squares of their distances from the centers about [which] they revolve: & thereby compared the force requisite to keep the Moon in her orb with the force of gravity at the surface of the earth, & found them answer pretty nearly. All this was in the two plague years of 1665 & 1666. For in those days I was in the prime of my age for invention & minded Mathematics & Philosophy [science] more then at any time since.

Here we see a mind overflowing with ideas. His "method of fluxions" led directly to his later development of the calculus, his "theory of colours" was the first comprehensive treatment of the properties of light, and his deductions about the force of gravity were the roots of the monumental treatise on the physics of motion he produced 20 years later. Newton states that the forces that "keep the Planets in their Orbs must [be] reciprocally as the squares of their distances from the centers about which they revolve." This means that the gravitational force of the sun on any planet varies inversely as the square of the distance of the planet from the sun. Recall that if x and y are two variables and y *varies inversely as x,* or *y is inversely proportional to x,* then $y = k/x$, where k is a constant. Therefore, in this case, we may write $F = k/r^2$, where F is the force of gravity due to the sun and r is the planet–sun distance. Several other natural phenomena that involve varying quantities are known to behave according to this famous **inverse-square law**.

> **inverse-square law** A general principle by which the effect of a phenomenon on an object varies inversely as the square of its distance to the source.

Think of a slowly inflating blue balloon. The dye giving the balloon its color keeps fading to lighter shades of blue as the balloon gets bigger. (See **Figure 4.4.2**.) This is so because a fixed amount of the dye keeps getting spread over a steadily increasing surface area. The surface area of a sphere of radius r is $4\pi r^2$. Assume the balloon to be a sphere painted evenly with b units of blue dye. There would be a concentration $C_0 = \dfrac{b}{4\pi R^2}$ units per square inch (in.²) on the balloon's surface when it is inflated to a radius of $r = R$ in. If you blow up the balloon and double its radius to $r = 2R$, the concentration must decrease to $\dfrac{b}{4\pi(2R)^2} = \dfrac{1}{4}\left(\dfrac{b}{4\pi R^2}\right) = \dfrac{1}{4}C_0$, or one-fourth the original concentration. Similarly, if you tripled the original radius to $r = 3R$, the concentration would decrease to $\dfrac{b}{4\pi(3R)^2} = \dfrac{1}{9}\left(\dfrac{b}{4\pi R^2}\right) = \dfrac{1}{9}C_0$, and so on. We see that b and 4π are constants that play no part in the basic relationship between the concentration C and the radius r, which we can express as $C = k/r^2$, where $k = b/4\pi$.

If we think of light from a star as radiating from a single point source, we may imagine it to be traveling in spherical waves much as the ripples in a pond caused by a thrown stone emanate in circular waves from the point where the stone entered the water. Other phenomena such as gravity or radio waves can be visualized in the same way. As the light

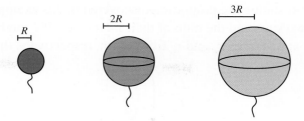

FIGURE 4.4.2   Demonstration of the inverse-square law.

gets farther from its source, it is spread out over a larger sphere. Hence its intensity I, like the concentration of dye on the balloon, also decreases inversely as the square of its distance r from the star. We may write

$$I = \frac{k}{r^2}.$$

❓ Example 1

Suppose two stars, call them A and B, each have the same intrinsic brightness (like two lightbulbs of the same wattage). If star A is 20 pc away and the light intensity from star A is 25 times greater than that from star B, then how distant is star B? As with the balloon analogy, the sphere of light reaching us from B must have a larger radius than that of the sphere of light from A. The question is, by what factor? If I_A and I_B are the intensities of light from stars A and B, respectively, then $I_A = 25I_B$ and so

$$25 = \frac{I_A}{I_B} = \frac{k/20^2}{k/r^2} = \left(\frac{k}{20^2}\right)\left(\frac{r^2}{k}\right) = \frac{r^2}{20^2} = \left(\frac{r}{20}\right)^2.$$

The constant k cancels, and we equate the first and last terms to get

$$\left(\frac{r}{20}\right)^2 = 25$$

$$\frac{r}{20} = 5$$

$$r = 5(20) = 100\,\text{pc}.$$

Note that the desired factor turned out to be $\sqrt{25} = 5$. If, instead, we had received 25 times more light from star B than star A, then it would have to be *closer* by a factor of 5. This is so because $I_B = 25I_A$ in this case, and so $\dfrac{I_A}{I_B} = \dfrac{1}{25}$. Then the distance to star B would be found by

$$\left(\frac{r}{20}\right)^2 = \frac{1}{25}$$

$$\frac{r}{20} = \frac{1}{5}$$

$$r = 4\,\text{pc}. \quad \blacklozenge$$

It is worth stating the proportion discovered in this example involving the ratio of the distances of two stars and the ratio of their intensities. If I_A and I_B are the intensities of two stars A and B at distances r_A and r_B from Earth, respectively, then

$$\frac{I_A}{I_B} = \left(\frac{r_B}{r_A}\right)^2.$$

It was at Woolsthorpe where the famous tale of the falling apple allegedly occurred. In his old age, Newton was fond of reminiscing about it, yet no one is sure whether it actually happened or was a bit of mischief he concocted for the sake of a good story. Supposedly, he was musing in his garden one day, thinking deeply about the force that held the moon in its orbit, when he saw an apple fall from a tree. In a creative flash, he linked these two seemingly unrelated occurrences—the *same* power of gravity attracted the apple and kept the moon in its orbit—and unified the theories of terrestrial and celestial motion (See **Figure 4.4.3**.). Furthermore, he realized that *any* body exerts a gravitational force on any other body, which is a function of the distance between their centers, according to the inverse-square law. Key to this understanding was his formulation of the *central force law*. (See Exercise 26.)

FIGURE 4.4.3   Newton under the apple tree.

central force The force on a moving body directed toward a central fixed point.

centripetal acceleration The acceleration on a moving body directed toward the center of its path of motion that results from a central force.

Suppose you are holding a rope with a tether ball attached to the end, and you are rotating in a circle. You are providing a **central force** on the ball that imparts an acceleration on the ball that is directed toward you, the center. (Remember that acceleration has a directional component as well as a numerical component.) This force is necessary to keep the ball in a circular orbit around you. If the rope were to break, for example, the ball would go flying off in a straight line tangent to its previous circle of motion. Newton's central force law states that the acceleration a directed toward the center—known as **centripetal acceleration**—needed to keep an object in a circular orbit is

$$a = \frac{v^2}{r},$$

where v is the velocity of the object and r is its distance from the center. Note that this indicates that the acceleration increases as the distance decreases. You are familiar with this from riding in the passenger side of a car. If you are making a gentle left turn (large r) on the freeway, then little force (hence acceleration) is exerted by the seat and side of the car to keep you from going straight forward. However, if you try a very sharp turn (small r) at the same speed, you experience a much greater force—and you hope your car door is shut tight!

? Example 2

Like the planets, the orbit of the moon has a low eccentricity, and so we may assume that it is circular for the sake of approximation. The radius of its orbit is $r = 3.84 \times 10^8$ m, and its period is $p = 27.3$ days $= (27.3 \text{ days})(24 \text{ h/day})(60 \text{ min/h})(60 \text{ s/min}) = 2.36 \times 10^6$ s. Recall that the Copernican velocity v is the circumference of its orbit divided by its period:

$$v = \frac{2\pi r}{p} = \frac{2\pi \left(3.84 \times 10^8 \text{ m}\right)}{2.36 \times 10^6 \text{ s}} = 1,020 \text{ m/s}.$$

The moon is under the influence of a central force—Earth's gravity. Its centripetal acceleration is

$$a = \frac{v^2}{r} = \frac{\left(1,020 \text{ m/s}\right)^2}{3.84 \times 10^8 \text{ m}} = 0.0027 \text{ m/s}^2. \quad \blacklozenge$$

When Newton returned to Trinity at the age of 26, the results of his independent studies earned him the appointment to the prestigious position of Lucasian Professor of Mathematics at Cambridge. This position afforded Newton the freedom necessary for a life devoted to investigation and discovery. We can learn much from his methods. True, his genius is undeniable, but the many successes of Isaac Newton can also be attributed to his unwavering devotion to learning, his respect for the works of his predecessors, and, most importantly, his ability to learn from his mistakes.

Interestingly enough, Newton's greatest contribution may never have surfaced if not for a chance visit from a friend. In 1684, Edmond Halley came to Cambridge to see Newton and to pose a question that had been plaguing many of London's scientists: If the sun attracted each planet with a force inversely proportional to the square of the distance, what type of path would be traced out by the planet? Newton immediately replied that he had solved the problem long ago—it must be an ellipse. Astonished, Halley urged his friend to overcome his reluctance to publish and share this grand discovery with the rest of the world! So for the next 18 months, Newton brought his prodigious powers of

Principia The grand treatise of Isaac Newton that established the foundations of physics and astronomy. Its explanations for natural phenomena are mathematically derived from a core set of three laws of motion.

reasoning and insight to bear on finishing his greatest masterpiece, *Philosophiae Naturalis Principia Mathematica* (*The Mathematical Principles of Natural Philosophy*). The magnificent **Principia**, as it is known, is considered to be one of the supreme achievements of the human mind (see **Figure 4.4.4**). In the beginning, he states his goal:

> For the whole burden of philosophy seems to consist of this—from the phenomena of motions to investigate the forces of nature and then from these forces to demonstrate all other phenomena.

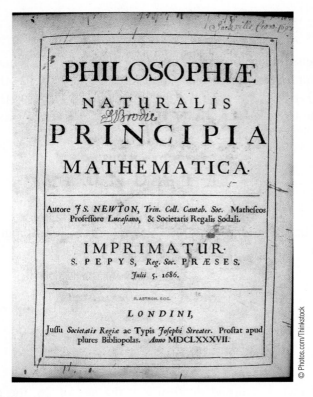

FIGURE 4.4.4   The *Principia*.

laws of motion The three basic principles of motion, which Newton states in the beginning of the *Principia*, and from which he derives the causes for many celestial and terrestrial phenomena. See Figure 4.4.5.

force An entity applied to an object, that changes its velocity. It is equal to the product of the mass of the object and its acceleration.

inertia The ability of a body to resist a change in its state of motion.

He then states his famous three **laws of motion** (**Figure 4.4.5**), axioms that establish the foundations for a wealth of results concerning mathematics, physics, and astronomy. The second law is a means of recognizing when a **force** has been applied to an object: either the speed or the direction of motion changes (i.e., an acceleration has been added). **Inertia** is the natural resistance to such changes, and so this rule is often referred to as the *law of inertia*. The path of a thrown baseball is a long, graceful arc because the force of gravity continuously acts on the ball to overcome its inertia, serving to decrease its speed and pull it downward. If it were thrown in the negligible gravity of outer space, the ball would continue in a straight line forever. This explains why a space vehicle launched from Earth can travel such long distances with so little fuel. Once free of the gravitational field of Earth, it travels in a near-linear path (affected only by the sun's gravity) until the firing of a small steering rocket redirects it or it enters the gravitational field of another planet.

NEWTON'S LAWS OF MOTION

1. The Law of Inertia
   A body remains at rest or continues in a straight line at a constant speed unless acted upon by an external force.

2. The Law of Force
   The total force on a body is equal to the product of the mass of the body and its acceleration.

3. The Law of Equilibrium
   If one body exerts a force on a second body, then second body must exert an equal and opposite force on the first body.

FIGURE 4.4.5   The simplicity of these natural laws is beautiful.

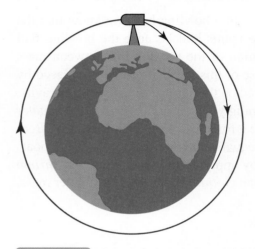

FIGURE 4.4.6   Newton's idea for putting a cannonball into orbit.

The first law allowed Newton to understand how Earth's gravity keeps the moon in orbit around it and, similarly, how the sun's gravity is responsible for the orbits of the planets. He compared the process to the firing of a cannonball off the top of a tall mountain (see **Figure 4.4.6**). As with a thrown baseball, Earth's gravitational force imparts an acceleration that continuously pulls the cannonball toward the surface. The greater its initial velocity, the farther it travels before impact. Surely, then, if it is endowed with a large enough initial velocity, its curved path keeps "missing" Earth, and so it enters a perpetual orbit as the pull of gravity and the inertia of the cannonball achieve a sort of balance.

In Example 2, we saw that the centripetal acceleration of the moon is 0.0027 m/s^2. Just as the twirling tetherball is pulled by the rope toward you, the moon is pulled by Earth's gravity with enough force every second to impart to it a speed of 0.00271 m/s toward Earth's center. Combined with its existing inertial velocity, this is just enough pull to keep it in orbit, as we see in **Figure 4.4.7(a)**. A schematic diagram in part **(b)** of the figure shows how the moon moves to the corner of the rectangle formed by the inertial velocity and the velocity change induced each second by the centripetal acceleration toward Earth. We hasten to add that because this is a continuous process, the end result is a smooth curve, not a sequence of jumps.

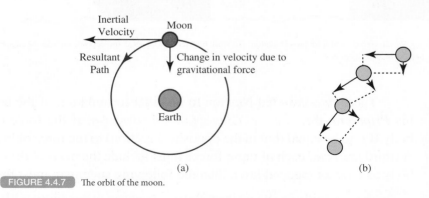

(a)

(b)

FIGURE 4.4.7   The orbit of the moon.

Newton was now in a position to test his theory from Woolsthorpe concerning gravitational force. Galileo had shown the value of acceleration due to gravity at Earth's surface (1 radius from its center) to be 9.8 m/s^2. The distance to the moon had been refined since the days of Aristarchus to be about 60 Earth radii. If the force of gravity (and its imparted acceleration to an object) *did* reach out into space as far as the moon and was inversely proportional to the square of the distance, its value there should be $\left(\dfrac{1}{60}\right)^2 = \dfrac{1}{3,600}$ of that at the surface. Lo and behold, $\dfrac{9.8}{3,600} = 0.0027$ m/s², beautifully matching his previous computation or, in Newton's words, "found them answer pretty nearly." This was the validation he needed to confirm the inverse-square relationship.

Newton's second law of motion is formulated simply as

$$F = ma.$$

Perhaps an easier way to grasp its meaning is to write it as $a = \dfrac{F}{m}$. If we deliver a push to a tennis ball on a flat table with the same force as to a bowling ball, we know that the bowling ball will acquire less acceleration than the tennis ball because the bowling ball has the greater mass. Recall our previous examination of the relationship between mass and weight. This is seen now simply as a special case of this law, with force F replaced by weight W and a replaced by the acceleration g induced on the surface of a planet by gravity.

The third law can best be illustrated by imagining yourself in *Le Shrimp*, the smaller of two boats next to each other on a lake. If you give a forceful shove to *Goliath*, the larger boat, you will simultaneously experience a force of equal value returned to you. Both boats will accelerate—but in opposite directions. Again by applying the second law, *Le Shrimp* will achieve a greater velocity a few seconds later than *Goliath* because of its smaller mass. (See **Figure 4.4.8**.)

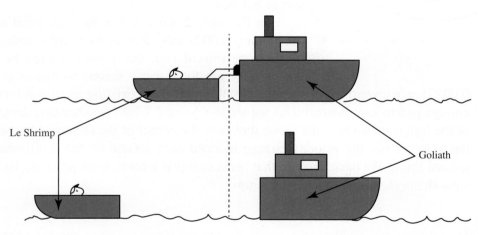

Le Shrimp

Goliath

FIGURE 4.4.8   The second and third laws of motion. The small boat receives an equal and opposite force to the large boat but with a greater acceleration because of its smaller mass.

Law of Universal Gravitation Any body attracts another body with a force directed along the line joining their centers, which is proportional to the product of the masses of the bodies and inversely proportional to the square of the distance between them.

These two laws led Newton to his final formulation of the crown jewel of his *Principia*, the **Law of Universal Gravitation**. If the force of body A on body B is to be equal (but in the opposite direction) to the force of body B on body A (third law) *and* each of these forces must include the mass of the corresponding body as a factor (second law), then the following statement must be true:

Every body in the universe attracts every other body with a force directed along the line joining their centers, which is proportional to the product of the masses of the bodies and inversely proportional to the square of the distance between them.

In symbols, the force F is computed by

$$F = \frac{GMm}{r^2},$$

where M and m are the masses of the two bodies, r is the distance between their centers, and G is the constant of variation known as the *gravitational constant*. In 1798, English physicist Henry Cavendish made the first fairly accurate determination of this constant. In the metric system, its value is $G = 6.67 \times 10^{-11}$ newton-square meters per square kilogram (N·m²/kg²).

Note that the force of gravity is a decreasing function of distance and an increasing function of mass. Although a space vehicle leaving our home planet frees itself of Earth's gravity after a relatively short distance, it continues to be under the influence of the sun for a considerably longer interval because the mass of the sun is about 330,000 times the mass of Earth.

So both mass and distance are important when computing the force of gravity of one body on another. If the mass of the moon were to somehow double, for instance, the resulting force between the moon and Earth would also double. On the other hand, because gravity follows the inverse-square law, if the distance from Earth's center to the moon or to any other orbiting satellite were to double, the gravitational force would decrease to one-fourth of the original value.

Next we observe that although gravitational force depends on mass, Galileo demonstrated that all bodies, regardless of mass, are attracted with the same acceleration to the surface of Earth. Newton's laws bear this out. Let M be the mass of Earth; m, the mass of any other body; and r, the distance between their centers. Because the force felt by the other body must be given by both of the force laws, we may equate the two expressions for F to get

$$ma = \frac{GMm}{r^2}$$

$$a = \frac{GM}{r^2}.$$

This formula for the acceleration of the body depends not on its own mass, but only on the mass of Earth and the other body's distance from Earth's center. Recall that this was precisely as Galileo had originally asserted!

? Example 3

The previous argument shows us that the acceleration imparted by any central mass M to any body at a distance r is dependent *only* on r, according to the inverse-square relationship. Moreover, we need not know the value of M to compare the accelerations imparted at different distances. We proceed with the same method we used in comparing the intensity of light from two sources. For example, let a_E and a_J be the mean accelerations imparted by the sun to Earth and Jupiter, respectively, at their mean distances r_E and r_J from the sun. Then we have

$$\frac{a_J}{a_E} = \frac{GM/r_J^2}{GM/r_E^2} = \left(\frac{GM}{r_J^2}\right)\left(\frac{r_E^2}{GM}\right) = \frac{r_E^2}{r_J^2} = \left(\frac{r_E}{r_J}\right)^2.$$

Notice the similarity to our previous formula for the ratio of light intensities from two different stars. Because Jupiter is about 5.2 AU from the sun, $r_J = 5.2r_E$ and so

$$\frac{a_J}{a_E} = \left(\frac{r_E}{5.2r_E}\right)^2 = \left(\frac{1}{5.2}\right)^2 = 0.037.$$

Thus,

$$a_J = 0.037a_E.$$

So the mean acceleration imparted by the sun to Jupiter is 0.037 of that which it gives to Earth. If we conveniently define one gravitational unit (GU) to be the mean acceleration allotted to Earth by the sun, then Jupiter receives 0.037 GU. By the same token, at the closer mean distance of 0.39 AU, Mercury feels a greater pull from the sun than Earth by a factor of $1/(0.39)^2 \approx 6.6$. So we say that Mercury receives 6.6 GU from the sun.   ◆

? Example 4

When rocket scientists plan the launch of a satellite to orbit Earth, they make use of the relationship for acceleration a as a decreasing function of distance r from Earth's center in order to compute a at increasing distances. They already know by experiment that gravitational acceleration must equal 9.8 m/s² on the surface of Earth. This must therefore be the value for a at a distance from Earth's center equal to its radius, call it R. In other words, $a = 9.8$ for $r = R$. So we have

$$9.8 = \frac{GM}{R^2}.$$

Without substituting for any other letters, we can determine the acceleration induced by Earth at a distance of 2, 3, or 10 Earth radii. We simply let $r = 2R$ to get

$$a = \frac{GM}{(2R)^2} = \left(\frac{1}{2^2}\right)\left(\frac{GM}{R^2}\right) = \left(\frac{1}{4}\right)(9.8) = 2.45\,\text{m/s}^2.$$

Similarly, for $r = 3R$,

$$a = \frac{GM}{(3R)^2} = \left(\frac{1}{3^2}\right)\left(\frac{GM}{R^2}\right) = \left(\frac{1}{9}\right)(9.8) = 1.09\,\text{m/s}^2,$$

and for $r = 10R$,

$$a = \frac{GM}{(10R)^2} = \left(\frac{1}{10^2}\right)\left(\frac{GM}{R^2}\right) = \left(\frac{1}{100}\right)(9.8) = 0.098\,\text{m/s}^2.\quad ◆$$

? Example 5

If we fix the distance in the prior formula, acceleration now becomes a function of mass. By substituting the radius of Earth (6,400 km), for r, 9.8 m/s² for a, and the value given earlier for the gravitational constant G, we can actually determine the mass of the Earth!

$$9.8 = \frac{\left(6.67 \times 10^{-11}\right)M}{\left(6.4 \times 10^{6}\right)^{2}}$$

$$M = \frac{(9.8)(6.4)^{2} \times 10^{12}}{6.67 \times 10^{-11}}$$

$$\approx 60.2 \times 10^{23} \text{ kg} = 6.02 \times 10^{24} \text{ kg}.$$

(*Note*: Because the units for G are in terms of meters rather than kilometers, we needed to express the radius in terms of meters. Likewise, the units of mass are necessarily kilograms.) ◆

Because the acceleration induced by the sun is a decreasing function of distance, we must also note that this is consistent with our knowledge from a prior section that the maximum velocity of a planet occurs at perihelion and the minimum velocity at aphelion. And finally we consider how the special case of a circular orbit (constant r) of an object around the sun or any central body of mass M leads to the velocity function well known to every aerospace engineer. We have seen that the centripetal acceleration that causes the object to move in a circular motion of constant velocity v at a distance r is given by $\frac{v^{2}}{r}$. Equating this to the acceleration that the central body induces on the object, we get

$$\frac{v_{c}^{2}}{r} = \frac{GM}{r^{2}}$$

or

circular velocity
Constant speed at which a body under the influence of a central gravitational force must move to maintain a circular path.

$$v_{c} = \sqrt{\frac{GM}{r}}.$$

This value is often referred to as the **circular velocity** and is denoted by v_{c}.

? Example 6

The first human-made satellite to be successfully placed in orbit around our planet was *Sputnik I*, launched by the Soviet Union on October 4, 1957. With what velocity did it move through its circular orbit? Because its orbit only carried it a few hundred miles above the surface, we use a value for r that is approximately equal to Earth's radius. Using a slightly more accurate value for the mass of Earth than computed in Example 5, we get

$$v_c = \sqrt{\frac{\left(6.67 \times 10^{-11}\right)\left(5.98 \times 10^{24}\right)}{6.4 \times 10^6}}$$

$$= 7{,}900 \, \text{m/s} = 7.9 \, \text{km/s}.$$

(Getting beat into space by our cold war opponents had an electrifying effect on the United States. Fear of being "bombed from above" by Soviet space vehicles triggered a funneling of funds into space research and spurred reforms in mathematics and science education.) ◆

? Example 7

Because the mass of the sun is 2.0×10^{30} kg, the speed of any object in a circular orbit at a distance of 1.5×10^{11} m (Earth's distance) is

$$v_c = \sqrt{\frac{\left(6.67 \times 10^{-11}\right)\left(2.0 \times 10^{30}\right)}{1.5 \times 10^{11}}} = 3.0 \times 10^4 \, \text{m/s} = 30 \, \text{km/s}.$$

You should compare this to the Copernican velocity you computed for Earth in Exercise Set 4.3. They match to two significant figures. The difference is that you need to have *observed* the movement of a known body in order to compute the Copernican velocity. For a given central mass, v_c is a function of r only. ◆

---

natural laws
Foundational principles that provide a basis of explanation of other relationships found in nature.

It is difficult to overestimate the impact of Newton and his *Principia* on the development of science and on the shape of modern times. His laws of motion and gravitation were the first set of **natural laws**—underpinning principles that provide a basis for the derivation of a host of other relationships. Following the statement of the laws in the book, he demonstrates how all three of Kepler's laws are immediate mathematical consequences. He then goes on to give the first mathematical treatment of wave motion; explains the orbits of comets; calculates the masses of the sun, Earth, and the planets with satellites; accounts for the equatorial bulge of Earth and how it causes the precession of the equinoxes; and shows how the gravitational pulls of the moon and sun are responsible for the daily high and low tides. Kepler and Galileo had obtained formulas in an empirical fashion; that is, they noticed mathematical patterns in the data that they encoded in equations. Although these equations were far better descriptions of nature than those that had been rendered by Aristotle, they failed to provide *reasons* for why they were true. Newton supplied this crucial missing piece of the puzzle. His laws were important because

1. They unified the laws of terrestrial and celestial motion.
2. They threw out the wordy, dead-end descriptions favored by Aristotelians and replaced them with crisp, simple axioms from which many observable phenomena could be deduced.
3. They were predictive. Positions of planets (and the paths of future space vehicles) could be calculated with great precision for the first time.

The last statement deserves a special mention. The predictive power of any new theory is one of the key tests of its acceptance by the scientific community. The many dramatic successes of Newton's calculus and natural laws brought quick and unanimous acknowledgment of the brilliance of his work. Mathematics became the new tool for the study of nature. Edmond Halley, for example, was able to employ the new mathematics to accurately forecast the return of the comet that bears his name. (See **Figure 4.4.9**.) It was not long before all serious doubters of the heliocentric theory disappeared. The revolution begun by Copernicus was complete.

FIGURE 4.4.9   Halley's Comet on a 1986 stamp from Laos.

Newton was lionized in his day in recognition of his accomplishments. He reigned as president of the Royal Society for 20 years and was given the lucrative position of Master of the Mint of England, a service he performed so admirably that he was knighted by the queen. He did, however, suffer his share of personal misfortune. In 1693, he had an acute mental breakdown, probably due to the poisoning of his system from the chemicals he routinely tasted in his alchemy experiments. He recovered from this episode, but he was never able to master his lifelong paranoia and eccentric, reclusive behavior, as evidenced by his bitter 25-year quarrel with the great German mathematician Gottfried Leibniz over who first created the calculus.

So yes, Newton was a strange bird to be sure, but we can be sure of one thing—he revealed a grand order in the universe. Every time one of America's splendid space vehicles is launched or a new space communications satellite is put into orbit, the memories of Isaac Newton and those who preceded him are recalled. While reflecting on his career late in life, he credited his predecessors by remarking that

> If I have seen farther than other men,
> it is because I stood on the shoulders of giants.

The stories of these giants—Copernicus, Kepler, Galileo, and Newton—offer a rich historical lesson. Cosmological conjecturing has occupied, and always will occupy, a special niche in the human quest for intellectual fulfillment. Although the search for solutions to the many puzzles offered up by Nature cannot be wholly separated from the surrounding social structure, it is, at its core, an individual endeavor. Any *absolute truth* as to the fabric of the universe can never be fully realized, only glimpsed from afar. It is those glimpses, those momentary clear visions in the mind of a single person—of you—that add an essential element to your life. It is incumbent upon you, therefore, to garner as much information as you can in order to experience the sharpest glimpses possible.

> Where the statue stood
> Of Newton with his prism and silent face,
> The marble index of a mind for ever
> Voyaging through strange seas of Thought, alone.
>
> —William Wordsworth

Name _____

Exercise Set **4.4** ◯◯◯←

1. Where was Isaac Newton born? What prolonged event prompted him to return there in 1665 for two years of isolated study?

2. Who prompted Newton to publish his results in what eventually became the masterpiece, *Philosophiae Naturalis Principia Mathematica*?

3. Star A and star B have the same intrinsic brightness. If star A is 15 pc away and we receive 9 times as much light from A as from B, then how far away is star B? If we instead receive 9 times as much light from star B as from star A, how far away is it?

4. Neptune lies at an average distance from the sun of 30 AU. What is the intensity of light from the sun on Neptune as a portion of the intensity on Earth? Mercury lies at an average distance from the sun of 0.4 AU. By what factor is the intensity of light from the sun on Mercury greater than the intensity on Earth?

5. Stars A_1 and A_2 have the same intrinsic brightness and lie at distances of 14 and 56 ly, respectively. How much more light do we receive from A_1 than from A_2?

6. Which of the following is representative of the graph of gravitational force F between two given bodies as a function of distance r? Note that specific units are not necessary. Just think of one force unit existing at one distance unit. Is F an increasing or a decreasing function of r?

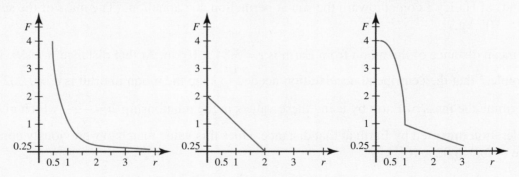

7. If the mass of Earth were magically doubled, how would the resulting gravitational force on any orbiting satellite change?

8. If the distance between Earth and the moon were doubled, what fraction of the original gravitational force between them would remain in effect?

9. If we define 1 gravitational unit (1 GU) to be the mean acceleration due to gravity that the sun imparts to Earth, how many gravitational units does the sun exert on Venus? Mars? Uranus? Pluto? The asteroids in the asteroid belt?

Body	Mean Distance from Sun (AU)
Venus	0.723
Mars	1.524
Asteroids	2.77 average
Uranus	19.18
Pluto	39.44

10. The diameters of Venus and Earth are almost identical. What must account for the fact that the surface gravity acceleration of Venus is less than that of Earth?

11. The acceleration on the surface of Mars is 3.7 m/s². What acceleration does it induce at a distance from its center of 2 Mars radii? 5 radii? 8 radii?

The gravitational constant $G = 6.67 \times 10^{-11}$ N·m²/kg² will be needed for Exercises 12–17.

12. You land on the moon and you wish to determine its mass. You already know its radius is 1,740 km, and you measure the gravitational acceleration near its surface to be 1.6 m/s². What is the mass (in kilograms) of the moon?

13. The mass of Uranus is 8.68×10^{25} kg, about 20 times greater than the mass of Venus at 4.87×10^{24} kg. However, the gravitational accelerations at the surface of both planets are about the same. Why? Compute the radius of each planet, given the surface accelerations of Venus and Uranus of 8.9 and 8.8 m/s², respectively. Are the values for the radii consistent with your response?

14. A comet is a mixture of ice and carbonaceous dust often referred to as a dirty snowball. Comets in our solar system typically have very eccentric orbits. Halley's comet achieves perihelion at 0.53 AU from the sun and aphelion at 35.1 AU. Determine the acceleration (in meters per square second) of Halley's comet toward the sun at perihelion and aphelion. (The mass of the sun is 1.99×10^{30} kg.)

15. The mean distance of the moon from Earth is $r = 3.84 \times 10^8$ m. At that distance, we saw in Example 2 that the centripetal acceleration needed to keep the moon in orbit is $a = 0.0027$ m/s².

    Determine the mass of Earth by using these values in the relationship $a = \dfrac{GM}{r^2}$, which gives the acceleration imparted by Earth at that distance. Does this value match our previously computed value of $M = 6.02 \times 10^{24}$ kg?

16. We can determine the mass M of the sun in a manner similar to the previous exercise. We equate the centripetal acceleration needed to keep Earth in its orbit ($r = 1.5 \times 10^{11}$ m) to the acceleration imparted by the sun at that distance and solve for M.

$$\frac{v^2}{r} = \frac{GM}{r^2},$$

$$M = \frac{v^2 r}{G}.$$

Use the Copernican velocity of Earth $v = 3.0 \times 10^4$ m/s, which we already computed in problem 7 of Exercise Set 4.3 to find the mass of the sun.

17. The immense gravitational hold of the massive Jupiter (1.9×10^{27} kg) on the four Galilean moons has forced them into almost perfect circular orbits. Because any satellite of a planet must also behave according to Newton's laws, use this table to answer the following questions.

Moon	Distance from Jupiter (km)
Io	422,000
Europa	671,000
Ganymede	1,070,000
Callisto	1,883,000

(a) Which moon must have the greatest circular speed?

(b) Which moon must have the smallest circular speed?

(c) What is the circular speed of Europa? (Remember to use meters as the units for r.)

18. When M stands for the mass of Earth, $GM = 4.0 \times 10^{14}$ and so the circular velocity v_c of a satellite orbiting Earth reduces to

$$v_c = \sqrt{\frac{GM}{r}} = \frac{2.0 \times 10^7}{\sqrt{r}}. \quad \text{(Remember, } r \text{ must be in meters.)}$$

How fast must a satellite travel to remain in a circular orbit around Earth with an orbital radius of 9,000 km from Earth's center?

19. One of the most practical applications of Newton's discoveries in modern times is the use of communication satellites. These satellites are placed in strategic orbits to relay radio signals between two distant locations on Earth. A *geostationary* satellite is one that always remains above the same place on Earth's surface, and so its orbital period must be 24 h. The previous exercise gives the velocity v_c necessary for a circular orbit around Earth as a function of r alone. Determine the radius r necessary for the orbit of a geostationary satellite by equating this formula to $\dfrac{2\pi r}{24(60)(60)}$ and solving for r.

20. The electrostatic force acting between two charges also follows the inverse-square law. In 1780, Charles Augustin de Coulomb showed that the force is proportional to the product of the two charges and inversely proportional to the square of the distance between them. Letting E stand for the electrostatic force, Q and q for the two charges, and r for the distance, write the function for E in symbols. (See the statement of the Law of Universal Gravitation.)

21. The force between two charges is 0.0045 N. Using the function for E given in the previous exercise, what is the new force if the distance between these two charges is increased by a factor of 3?

22. Suppose an object is at a distance r from the center of a body of mass M. The speed required for that object to escape from that central body's gravitational hold is called the *escape velocity* v_e and is given by the function

$$v_e = \sqrt{\frac{2GM}{r}}.$$

Note the similarity to the function that gives the circular velocity v_c. If an object were already in a circular orbit around a central mass, by what factor would it have to increase its velocity in order to escape?

23. What are the advantages of Newton's laws of motion as a set of axioms to describe nature over the explanations offered by Aristotle?

24. For a body in a circular orbit around the sun, we have two algebraic expressions that give us the speed of the body: $v_c = \sqrt{\dfrac{GM}{r}}$ and the Copernican velocity $v = \dfrac{2\pi a}{p}$ from the last section. Replace r with a, equate these two expressions, and deduce Kepler's third law.

25. In Example 5, we determined the mass of Earth by knowing the value of the acceleration it produced on a body near its surface. In other words, we knew the value of acceleration that Earth imparts to an object at a specific distance. Describe a procedure by which we could determine the mass of a planet possessing moons by observing the radius and period of the orbit of one moon.

26. Kepler's third law was the key to Newton's realization that the gravitational force of the sun on the planets followed an inverse-square law. The *central force law* states that the centripetal acceleration a imparted by the sun on a body in a circular orbit of radius r moving with velocity v is given by

$$a = \frac{v^2}{r}.$$

According to Kepler, $p^2 = kr^3$, where p is the period of revolution of a body. This means that the ratio $\dfrac{r^3}{p^2}$ is always constant. Substitute the Copernican velocity $v = \dfrac{2\pi r}{p}$ for v to show that the acceleration a (and hence the force) varies inversely as the square of the distance r.

27. We think of the orbit of the moon as resulting solely from Earth's gravitational pull, but in fact the sun also exerts a significant influence because of its enormous mass. This is true for a satellite of any planet. In the May 1963 issue of *The Magazine of Fantasy and Science Fiction*, Isaac Azimov computed what he called the "tug-of-war" value. For any satellite of mass m, he defined this to be the ratio of F_p, the gravitational force of the planet, to F_s, the force due to the sun. Specifically,

$$\frac{F_p}{F_s} = \frac{GM_p m/r_p^2}{GM_s m/r_s^2} = \left(\frac{M_p}{M_s}\right)\left(\frac{r_s}{r_p}\right)^2,$$

where M_p and M_s are the masses of the central planet and sun, respectively, and r_p and r_s are the distances of the satellite from the planet and sun, respectively. The Galilean moons of Jupiter, for instance, have ratios ranging from a high of 3,260 for nearby Io to a low of 160 for the more distant Callisto. It turns out that our moon is the only major satellite in the solar system with a tug-of-war ratio less than 1. What does this mean? Rather than a satellite circling Earth, what might be a better description of the orbital path of the moon? Compute the tug-of-war ratio for the moon. (Recall from the first section that the sun is 390 times as far from Earth as the moon. Also, the mass of Earth is 0.000003 of the sun's mass.)

28. A *black hole* is an object in space containing a very large amount of mass in a volume so small that the resulting gravitational field is strong enough to prevent even light from escaping. Explain how this is consistent with the roles of force F and distance r as they are related in Newton's Law of Universal Gravitation.

Name _____

*use the extra space
to show your work*

— Chapter Review Test **4** ☐☐☐ ←────────────────────

1. What new idea was introduced by Heracleides to the planetary model concerning the orbits of Mercury and Venus?

2. Define an axiom, and explain the role of a set of axioms in creating a model of a natural phenomenon.

3. What star in the sky remains practically motionless throughout the night? Does it appear higher above the horizon as viewed from Scranton, Pennsylvania, or Atlanta, Georgia?

4. What is the mean distance of Earth from the sun, measured in light-seconds?

5. If you go outside and look at the constellation Orion at 10 P.M. one winter night and then again a month later at 10 P.M., about how many degrees has Orion shifted across the sky?

6. Saturn has an orbital period of 29.5 yr. Compute its mean distance (in astronomical units) from the sun and its Copernican velocity (in kilometers per second).

7. How far away is a star having a parallax angle of 0.28″? Give your answer in parsecs and light-years.

8. Halley's comet orbits the sun in an elliptical orbit, with a perihelion distance of 0.53 AU and an aphelion distance of 35.1 AU. Compute the eccentricity of its orbit. Would you say the shape of this orbit resembles a circle or a cigar?

9. Given that the eccentricity of Mercury is 0.2056 and its maximum distance from the sun is 0.467 AU, find its minimum distance from the sun.

10. Consider the statement "If it is a July day in Phoenix, then the temperature is over 100°." Write the converse and contrapositive of this statement. Assuming the original statement is true, is the converse necessarily true? Is the contrapostive necessarily true?

11. List two observations made by Galileo that convinced him of the validity of the Copernican system.

12. The acceleration due to gravity near the surface of Mars is 3.7 m/s^2. How much time would it take for an object to drop from 1,250.6 m on Mars? What would be the impact velocity?

13. Stars Alpha and Beta have the same intrinsic brightness. If Alpha lies at a distance of 50 pc from Earth and we receive $\dfrac{1}{49}$ as much light from Beta as from Alpha, then how far away is Beta?

14. Mercury and Mars have about the same surface acceleration due to gravity, yet Mars has a greater radius than Mercury. Use the law of universal gravitation to explain what must account for this fact.

15. The planet Uranus lies at a mean distance of about 19 AU from the sun. If 1 GU is the gravitational acceleration exerted by the sun on Earth, how many gravitational units does the sun exert on Uranus?

16. The asteroid belt is a region between the orbits of Mars and Jupiter in which thousands of chunks of rock known as asteroids revolve around the sun at an average distance of about 2.77 AU. Even though they have many different masses, is the acceleration imparted to them by the sun about the same for all of them? Why? What is it? (The mass of the sun is 1.99×10^{30} kg.)

17. You land on the moon, and you wish to determine its mass. You already know its radius is 1,740 km, and you measure the gravitational acceleration near its surface to be 1.6 m/s^2. What is the mass (in kilograms) of the moon?

Chapter Objectives

check off when you've completed an objective

☐ Know the definition of a graph and how to construct one as a model.

☐ Find paths and circuits in a graph.

☐ Determine when two graphs are isomorphic.

☐ Compute the number of edges in a complete graph.

☐ Know and apply Euler's theorem.

☐ Find an Eulerization of a graph.

☐ Learn about the lives of Leonhard Euler and Rowan Hamilton.

☐ Find a Hamilton circuit in a graph.

☐ Compute the number of Hamilton circuits in a complete graph.

☐ Know and apply the brute-force and nearest-neighbor algorithms.

Navigate Companion Website

go.jblearning.com/johnson

Visit go.jblearning.com/Johnson to access a Laboratory Manual, Student Solutions Manual, Interactive Glossary, and Interactive Flashcards.

5

Graph Theory

5.1 Graphs and Paths

5.2 Euler Circuits

5.3 Hamilton Circuits

Many tasks of the modern marketplace require traveling among a large number of places that have a variety of paths connecting them. Delivery of mail and packages, removal of snow from streets, and traveling by bus, car, or airplane are but a few of the examples that involve orchestrating the visitation of many places in a manner that is time- and cost-efficient. Many of you, for instance, probably took one or more campus tours when visiting colleges as part of your decision-making process for your plans after graduation from high school. Tours of this sort are mapped out by workers in admissions offices and are constructed to allow the potential student to see as many important campus sights as possible within a particular time frame. The problems involved in constructing such a tour are representative of several types of similar issues that face workers who plan the schedules and routes of everything from airplane flights to telephone traffic. The solution to a problem of this type consists of a path that satisfies the needs of the traveler and is optimal in some way, usually meaning that it minimizes time or cost. This chapter studies some of the basic tools used in finding these solutions, all of which lie in the broad area of mathematics known as **graph theory**.

graph theory The study of the properties and applications of graphs.

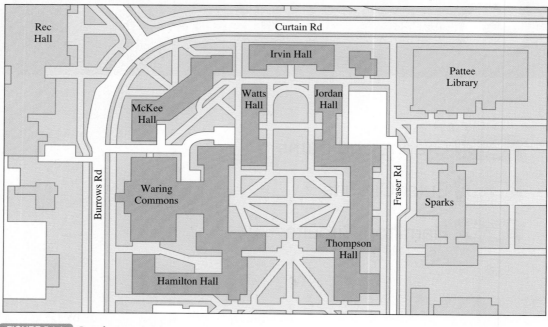

FIGURE 5.1.1   Part of a state university campus.

5.1 Graphs and Paths

An aerial photograph of a portion of the main campus of a state university is shown in **Figure 5.1.1**. Look at a small subset of sidewalks in the upper left of this picture (**Figure 5.1.2**), and suppose an historical plaque is located at each intersection. We wish to view each plaque without walking on any sidewalk more than once. To think about this problem, it is useful to draw a diagram that represents the possible paths to be taken. Because we are concerned not with distances or orientation but only with the sequence of connecting sidewalks to be traversed, **Figure 5.1.3(a)** or **(b)** will serve our purpose. The capital letters represent the plaques, and the lines connecting them represent the sidewalks.

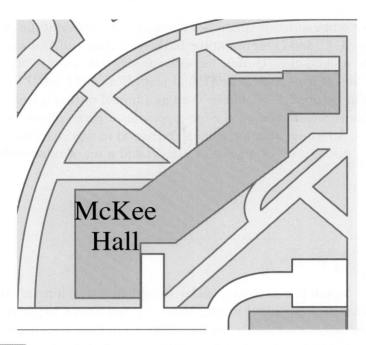

FIGURE 5.1.2 A subset of sidewalks on campus.

graph A collection of vertices and edges.

network Another name for a graph.

vertices One of a set of points that all represent a similar type of object. The vertex set of a graph G is denoted by $V(G)$.

nodes Another name for a vertex.

edges A line segment drawn between two vertices that represents some relationship between them.

isomorphism A one-to-one correspondence between the vertices and edges of two graphs. The degrees of corresponding vertices must be equal.

Each of these diagrams is called a **graph** (or **network**), and every graph consists of two essential features:

- A set of points, known as the **vertices** or **nodes**
- A set of **edges**, each of which joins a pair of vertices

The lengths and positioning of the edges differ in the two diagrams, but the essential connecting features of each layout are the same. Therefore, either graph can be used for our purposes. Two graphs are said to be *isomorphic* if their vertices and edges match up in a one-to-one correspondence in such a way that an edge joining two vertices in one graph always corresponds to an edge joining the corresponding vertices in the other graph. When the actual correspondence is constructed as a function between the graphs, it is called an **isomorphism**. If graphs G and G_1 are isomorphic, we will denote this by $G \cong G_1$. This is the case for the graphs in Figure 5.1.3.

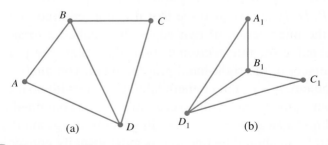

FIGURE 5.1.3 (a) Graph G. (b) Graph $G_1 \cong G$.

What other situation might be represented by either of the graphs in Figure 5.1.3? Perhaps A, B, C, and D represent the basketball teams at Alabama, Buffalo, Creighton, and Drake universities. During the season, suppose team A played B and D, team B played C and D (in addition to A), and team C also played D (in addition to B). Then the graph in Figure 5.1.3 could be used as a model of that schedule, in which each vertex represents a team and each edge represents a game played between the two teams it connects. So we see that a graph can be useful in representing any construct in which there are a discrete set of objects (vertices) and a set of relationships (edges) that connect them.

? Example 1

Sashif and Morah live in Seattle and wish to map out an itinerary to visit Portland, Oregon; Denver, Colorado; Lincoln, Nebraska; and Iowa City, Iowa, by plane. They wish to take only direct flights and have discovered that direct flights exist between:

- Seattle and each of the other cities.
- Portland and Denver as well as Portland and Iowa City.
- Denver and Lincoln.

Draw a graph that represents the possible routes taken by Sashif and Morah.

⚙ Solution

In this case, the set of vertices stands for the cities and can be denoted by $\{S, P, D, L, I\}$. A suitable graph does not have to correspond to the actual geography of the cities involved. Any one of the following graphs can be used, and each pair of them is isomorphic.

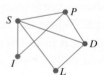 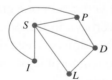 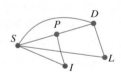

The vertex set of a given graph G is denoted by $V(G)$. In this example, $V(G) = \{S, P, D, L, I\}$. We also see that it is not required that a vertex be located at the intersection of two edges. It is called a **crossing** if no vertex is located at the intersection of two edges (e.g., the point where SD and PI intersect in the right-hand graph). When you analyze a graph to solve a problem, it is usually beneficial to use one with no crossings. If a graph is isomorphic to one with no crossings and is contained in a plane surface, it is known as a **planar graph**. In general, the graph that you use to model a problem should be one that is most visually convenient for you.   ◆

crossing An intersection of two edges that is not considered to be a vertex.

planar graph A graph that is isomorphic to a graph without crossings that is contained in a plane surface.

Example 2

Which of the following graphs are isomorphic?

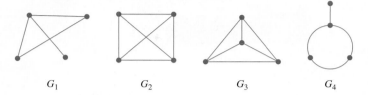

G_1 G_2 G_3 G_4

Solution

When you are exploring the possibility of an isomorphism, it is helpful to think intuitively of the edges of a graph as being "made of rubber." It is permissible to stretch or distort any edge as long as it is not broken or torn. Similarly, the vertices can be moved in any direction as long as they remain in the plane of the graph and no vertices are added or removed. Then a graph G will be isomorphic to another graph H if you can deform the edges and vertices of G until its shape resembles H. In this manner, you can conclude for the graphs here in Example 2 that $G_1 \cong G_4$ and $G_2 \cong G_3$. Note that this implies that both G_1 and G_2 are planar. ◆

degree of a vertex
The number of edges that meet at vertex v.

An important characteristic of a vertex in any graph is the number of edges that are adjoined to the vertex. The **degree of a vertex** v, denoted by $\delta(v)$, is defined to be equal to the number of edges that meet at v. Clearly, δ is a function whose domain is the vertex set of any graph and whose range is the set of nonnegative integers. In the graph in **Figure 5.1.4**, $\delta(a) = 2$, $\delta(b) = 3$, $\delta(c) = 1$, $\delta(d) = 2$, and $\delta(e) = 0$.

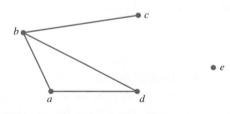

FIGURE 5.1.4   Vertices having various degrees.

Example 3

The map here features most of the counties of the state of Massachusetts. Create a graph G in which each vertex represents a county and two vertices are joined by an edge if the corresponding counties share a border. (Ignore the eastern part of Norfolk County and the two small counties by Boston.) Define $V(G)$ for this graph. What are the degrees of the vertices? What does it mean for a vertex to have degree 0?

Solution

In the graph G that corresponds to the map, the vertex set is $V(G) = \{B, F, H, HA, W, M, E, S, N, BR, P, BA, D, NA\}$, if we denote each county by the first letter of its name (or the first two letters when necessary to avoid duplication). The degrees of the vertices are as follows: $\delta(B) = 3$, $\delta(F) = 3$, $\delta(H) = 4$, $\delta(HA) = 3$, $\delta(W) = 5$, $\delta(M) = 4$, $\delta(E) = 2$,

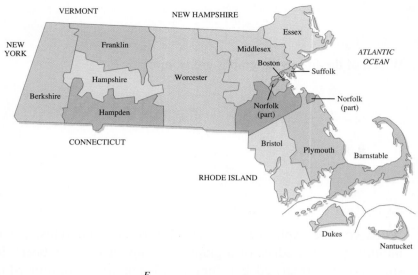

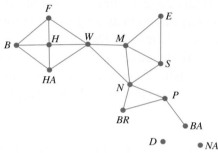

$\delta(S) = 3, \delta(N) = 5, \delta(BR) = 2, \delta(P) = 3, \delta(BA) = 1, \delta(D) = 0, \delta(NA) = 0.$ If the degree of a vertex is 0, it means that the county does not share a border with any other county. In this particular case, this is because Dukes and Nantucket are islands.   ◆

---

odd vertex A vertex whose degree is an odd integer.

even vertex A vertex whose degree is an even integer.

adjacent vertices Two vertices joined by an edge.

adjacent edges Two edges that meet at a common vertex.

path A sequence of consecutive adjacent vertices.

simple path A path in which no vertex occurs more than once.

circuit A path in a graph whose terminal vertex is the same as its initial vertex.

base The initial and terminal vertex of a circuit.

simple circuit A circuit in which the base is the only vertex used more than once.

A vertex v in a graph is called an **odd vertex** if $\delta(v)$ is an odd integer and an **even vertex** if $\delta(v)$ is an even integer. Two vertices are called **adjacent vertices** if they are joined by an edge, and two edges are called **adjacent edges** if they meet at a common vertex. Any finite sequence of edges joining two given vertices is a called a **path**. A path from vertex a to vertex b is typically represented by a finite, ordered sequence of vertices beginning with a and ending with b such that every pair of consecutive vertices is adjacent. For example, three paths from a to c in Figure 5.1.4 are given by the sequences abc, $adbc$, and $abdabc$. If no vertices are repeated in a path, it is called a **simple path**. Therefore, abc and $adbc$ in Figure 5.1.4 are simple paths. A **circuit** is a path that begins and ends at the same vertex, called the **base**. If the base is the only repeated vertex in a circuit, the circuit is also referred to as a **simple circuit**. (Two sequences consisting of the same list of vertices but in reverse order will be considered to represent the same circuit.)

If graphs G and G' are isomorphic, then any pair of corresponding vertices must have the same degree. Similarly, isomorphic graphs must have equal numbers of paths, circuits, and odd and even vertices. However, no converse to any of these statements is true. Two graphs can have the same number of vertices and equal numbers of vertices with the same degrees but still not be isomorphic. An example of this is given in **Figure 5.1.5**.

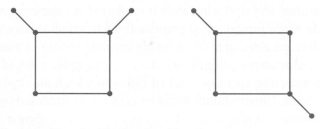

FIGURE 5.1.5   Non-isomorphic graphs with vertices of equal degrees.

? Example 4

In the graph in **Figure 5.1.6**, list several:

(a)   simple paths from vertex *s* to vertex *x*.
(b)   simple paths from vertex *t* to vertex *w*.
(c)   simple circuits based at *t*.
(d)   simple circuits based at *x*.

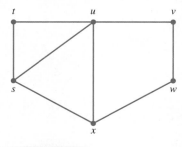

FIGURE 5.1.6

⚙ Solution

(a)   Five simple paths from *s* to *x* are *sx*, *sux*, *stux*, *suvwx*, and *stuvwx*.
(b)   Six simple paths from *t* to *w* are *tuvw*, *tuxw*, *tusxw*, *tsxw*, *tsuvw*, and *tsuxw*.
(c)   Three simple circuits based at *t* are *tust*, *tuxst*, and *tuvwxst*.
(d)   Four simple circuits based at *x* are *xsux*, *xstux*, *xuvwx*, and *xstuvwx*.   ♦

connected graph
A graph in which every vertex is connected to at least one other vertex.

disconnected graph A graph that is not connected.

bridge An edge whose removal changes a connected graph into a disconnected graph.

If every vertex in a graph is connected by a path to every other vertex, then the graph is said to be a **connected graph**. A graph that is not connected is called a **disconnected graph**. Any graph containing a vertex of degree 0 must be disconnected. If the removal of a single edge changes a connected graph to a disconnected one, then that edge is called a **bridge**. Examples of these are shown in **Figure 5.1.7**. Clearly, a connected graph and a disconnected graph cannot be isomorphic to each other. Most of the applications we will examine primarily involve connected graphs.

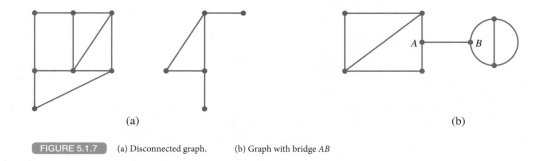

(a)   (b)

FIGURE 5.1.7   (a) Disconnected graph.   (b) Graph with bridge *AB*

complete graph
A graph in which every pair of vertices is joined by an edge. The complete graph on n vertices is denoted by K_n.

A graph in which every vertex is joined to every other vertex by an edge is called a **complete graph**. The complete graph on n vertices is usually denoted by K_n. The K here is used in honor of the brilliant Polish mathematician **Kazimierz Kuratowski** (1896–1980). Kuratowski was instrumental in maintaining and nurturing Polish mathematics during and following World War II. He made fundamental contributions to graph theory, set theory, and topology.

Complete graphs can arise as an appropriate model in many contexts involving communication and transportation. Consider, for example, a set of towns in which each pair has a road connecting them or a set of cities in which an airplane is flown daily between each pair of cities. Either situation can be effectively modeled by a complete graph. Several examples of complete graphs are shown in **Figure 5.1.8**. Some exploration of these graphs may convince you that K_n cannot be planar if $n > 4$.

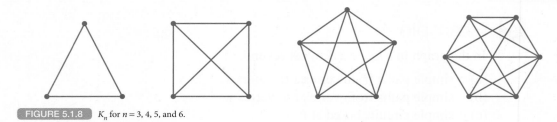

FIGURE 5.1.8 K_n for $n = 3, 4, 5,$ and 6.

The graphs in Figure 5.1.8 suggest that we could devise a formula for the number of edges in a complete graph if we knew the number of vertices. Because $n - 1$ edges are connected to each of n vertices in K_n and each edge joins two vertices, the total number of edges must be one-half of $n(n - 1)$. For example, K_4 has $(4 \cdot 3)/2 = 6$ edges and K_5 has $(5 \cdot 4)/2 = 10$ edges, as you can check for yourself. In general, the total number of edges in K_n is given by $\dfrac{n(n-1)}{2}$.

This section has introduced some of the terminology and several of the main concepts associated with graph theory. The next two sections will explore two particular types of problems that are best analyzed by graph-theoretic techniques. They involve searching for certain types of circuits, known as Euler and Hamilton circuits, that provide optimal paths through a network representing some real-life situation. The term *optimal circuit* here means that traversing this circuit requires a minimum amount of time or cost or reduces replication of a task. Such circuits are excellent examples of practical applications of an area of mathematics that consists of quite abstract and sophisticated ideas.

Name _____

Exercise Set **5.1** ◯◯◯←──────────────

1. Betty, Gretchen, and Harold are college students who all took History 101 at the same time. Furthermore, Betty, Harold, and Perry all took Math 120 together. Draw a graph in which vertices represent students and an edge represents a shared class. Define the vertex set for your graph.

2. You go to a meeting and shake hands with each of the other five people in attendance. If everyone has shaken hands with everyone else, draw a graph that represents this situation. How many handshakes occurred?

3. The mayors from Los Angeles, Chicago, Boston, Philadelphia, and Miami all have direct-line phone hookups with one another. Draw a graph that shows these connections, and define the associated vertex set $V(G)$. Is this a complete graph? How many edges are there?

4. Let the vertices of a graph G represent a set of five movie actors: A, B, C, D, and E. Two vertices are joined by an edge if the two corresponding actors have appeared in the same movie. Draw the appropriate graph if actor A has appeared with B and D; actor B has appeared with C, D, and E; actor C has appeared with D; and actor D has appeared with E. What are the degrees of each vertex of this graph?

5. The map of Eastern Europe features several countries. Create a graph in which each vertex represents a labeled country and two vertices are joined by an edge if the corresponding countries share a border. Define $V(G)$ for your graph.

Which of the pairs of graphs in Exercises 6–11 are isomorphic to each other? Which graphs contain bridges?

6.

7.

8.

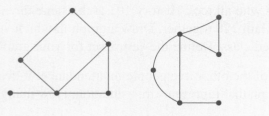

9.

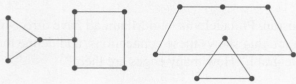

10.

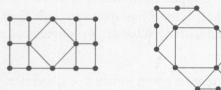

11.

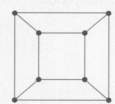
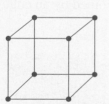

12. Draw a graph that is isomorphic to the given graph but does not contain any crossings.

(a)   (b)

13. Draw a graph that is isomorphic to the given graph but does not contain any crossings.

(a)   (b)

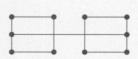

14. In the graph shown here, find five paths starting at a and ending at e. Find five paths starting at c and ending at g. Find four circuits based at b and at k.

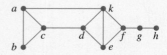

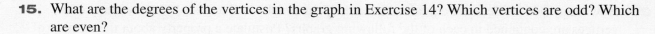

15. What are the degrees of the vertices in the graph in Exercise 14? Which vertices are odd? Which are even?

16. Which edge in the graph shown here is a bridge? Find three paths starting at vertex a and ending at g. Any path beginning at a, b, c, or d and ending at e, f, g, or h must contain what edge? Find three circuits based at a, all of which contain g. Find three circuits based at e, all of which contain d. Are these simple circuits?

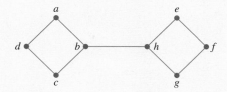

17. Identify two simple circuits based at a in the graph on the left and two simple paths from d to f. Identify four simple circuits based at h in the graph on the right and five simple paths from n to k.

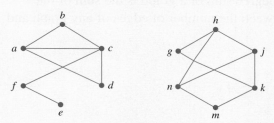

18. What are the degrees of the vertices in both graphs in Exercise 17? Which vertices are odd? Which are even? Are these graphs isomorphic?

19. Three of the graphs shown here are isomorphic to one another, and two different ones are also isomorphic to each other. Identify them.

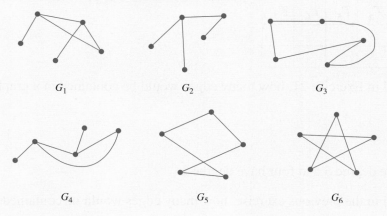

G_1 G_2 G_3

G_4 G_5 G_6

←

20. How many odd vertices are contained in each of the graphs in Exercise 19? How many odd vertices are contained in each of the following graphs? Postulate a property about the number of odd vertices contained in any graph.

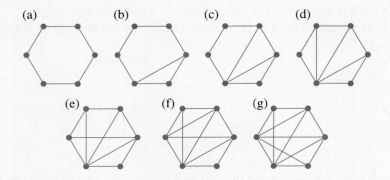

21. Complete the table for the following graphs. The degree sum of a graph is the sum of the degrees of its vertices. Postulate a relationship between the number of edges of any graph and its degree sum.

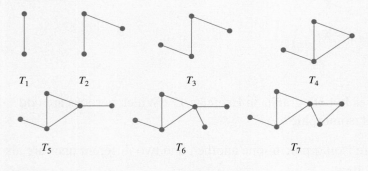

Graph	T_1	T_2	T_3	T_4	T_5	T_6	T_7
Edges							
Degree Sum							

22. Based on the formula developed in Exercise 21, how many edges would be contained in a graph that has:

(a) 10 vertices, each of degree 3.
(b) 7 vertices, each of degree 2.
(c) 8 vertices, of which four have degree 5 and four have degree 2.

23. Based on the formula developed in the previous exercise, how many edges would be contained in a graph that has:

(a) 4 vertices, each of degree 3.
(b) 6 vertices, each of degree 4.
(c) 7 vertices of degrees 2, 2, 3, 3, 4, 5, and 5.

24. What is the degree sum of a graph with 35 edges? 186 edges? 250,000 edges?

25. How many edges are contained in the complete graphs K_7, K_{11}, K_{20}?

26. Graph (a) is isomorphic to which of the six graphs, (b) to (g)?

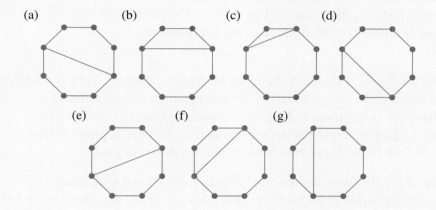

(a)   (b)   (c)   (d)

(e)   (f)   (g)

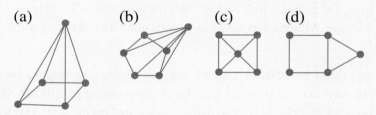

27. The graph on the left, (a), is isomorphic to which of the three graphs on the right, (b) to (d)?

(a)   (b)   (c)   (d)

28. Three houses are to be hooked up to three utilities. Gas, water, and electricity lines all need to be established, and no one of these lines can cross any of the others. Draw a graph that shows all nine of these connections. Is it possible to draw a representative graph without any crossings? Is this a planar graph?

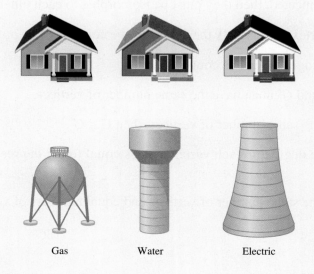

Gas   Water   Electric

←

29. A **bipartite graph** is one in which all the edges join a vertex in one group *R* to a vertex in a second group *S*. These graphs are useful in modeling what are called *matching problems*. For example, suppose three people are applying for one of a set of four jobs (*W*, *X*, *Y*, and *Z*) at a particular business. Arthur is qualified for jobs *W*, *X*, and *Z*; Bridget is qualified for jobs *X* and *Y*; and Chin is qualified for jobs *W* and *Y*. Draw a bipartite graph that represents this situation. Is it possible for all three of the applicants to each obtain a different job?

> **bipartite graph**
> A graph whose only edges consist of those that join every vertex in one group of vertices to every vertex in another distinct group.

30. Four people are applying for four jobs at Franklin Premium Insurance Company. Arin is qualified to be a programmer and an actuary, Brett is qualified to be an accountant and an actuary, Casey is qualified to be a hardware technician and a programmer, and Darcy is qualified to be an accountant. Draw a bipartite graph (defined in the preceding exercise) that represents this situation. Is it possible for all four of the applicants to each obtain a different job?

31. A *digraph* is a graph whose edges also have an indicated direction associated with them. For instance, Web page *A* that contains a link from itself to site *B* does not imply that site *B* has a link back to *A*. The direction of an edge in a graph is usually denoted by drawing an arrowhead in the middle of the edge. Suppose Tom has links on his homepage to the Web pages of his friends Omar, Beatrice, and Maggie. Likewise, Omar has a link on his Web page to that of Beatrice, and Beatrice has links to Tom and Maggie. Maggie has no links to anyone. Draw a digraph that represents this network of links.

32. Define the vertices of *V(G)* to correspond to 10 different people who each have a homepage on Facebook. Define an edge between two vertices to indicate that the corresponding people are in each other's friends' lists. If each member of *V(G)* is friends with every other member of *V(G)*, then what is the total number of edges of *G*?

Identify each statement as true or false. If the statement is false, give an example that demonstrates it is false. Let G and G' be graphs.

33. If *G* and *G'* are connected, then they must be isomorphic to each other.

34. If a graph has a vertex of degree 0, then it cannot be connected.

35. If a graph has a bridge, it must be connected.

36. If $G \cong G'$, then *G* and *G'* must have the same number of vertices.

37. If *G* and *G'* have the same number of vertices, then $G \cong G'$.

38. If $G \cong G'$, then the degree of each vertex of *G* is equal to the degree of the corresponding vertex of *G'*.

39. If *G* and *G'* have the same number of vertices and equal numbers of vertices with the same degree, then $G \cong G'$.

40. If neither G nor G' has a circuit, then $G \cong G'$.

41. If the number of odd vertices of G is not equal to the number of odd vertices of G', then G cannot be isomorphic to G'.

42. If G and G' have the same number of paths, then $G \cong G'$.

43. The degree sum of any graph is always even.

44. The Orbiter Camera on the Mars Global Surveyor took the pictures shown here of the surface of Mars in 2003. Similar surface features can be found on Earth. In what ways do you think graph theory might be helpful in comparing them to the Martian structures shown here?

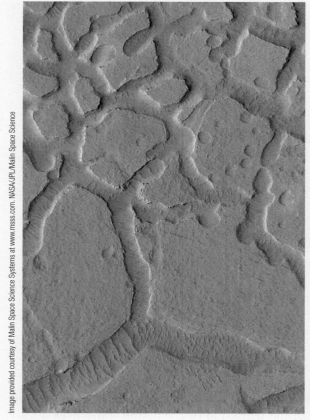

Image provided courtesy of Malin Space Science Systems at www.msss.com. NASA/JPL/Malin Space Science

Mesas and troughs.

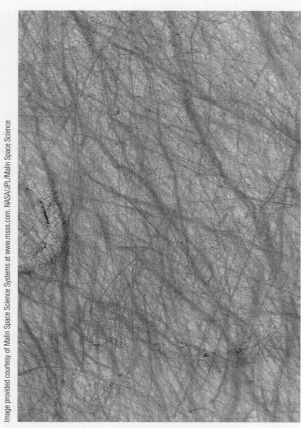

Image provided courtesy of Malin Space Science Systems at www.msss.com. NASA/JPL/Malin Space Science

Dust devils.

5.2 Euler Circuits

One of the primary modern-day uses of graph theory is for the analysis of routing problems. A *routing problem* is concerned with choosing an efficient path that takes a traveler along a required route or that leads him or her to visit a specific set of destinations. Examples of such problems include finding an optimal way to plow snow without repetition of streets, visiting a list of cities with a minimum of travel cost, or channeling a telephone call through a chain of relays in the shortest amount of time. The use of graph theory as a tool to solve such problems traces its beginnings to a recreational puzzle involving Sunday strollers in the city of Koenigsburg in eighteenth-century Prussia (currently Kalingrad, Russia). A river with several islands flowed through the center of this town, and seven bridges connected two of the islands with each other and with opposite banks of the river. A picture of the situation is given in **Figure 5.2.1**. Through the years, people had become convinced that it was not possible to walk a path that traveled across each of the seven bridges and returned you to your starting point unless you had crossed at least one bridge more than once. In order to find a definitive proof that such a circuit was indeed impossible, Leonhard Euler (1707–1783) created what was perhaps the first graph as we have defined it. He thought of each landmass (separated by water from other land) as a single destination, and so it could be represented by a single point or vertex. In this way, each bridge between two landmasses could correspond to an edge joining two vertices. The graph representing this clever idea is shown in Figure 5.2.1.

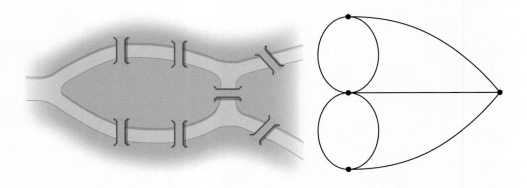

FIGURE 5.2.1   The bridges of Koenigsburg and the associated graph.

You may have heard of the influence of Leonhard Euler (**Figure 5.2.2**) on other areas of mathematics, such as the development of the definition of a function. Euler was a truly remarkable individual who made many deep and lasting contributions to mathematics and science in more than 500 published papers. His memory and computing skills were unmatched. A friend once remarked that Euler could "calculate as easily as other men breathe." It is said that Euler, a caring father, was able to work even with small children seated in his lap. Astonishingly, he continued to work and produce deep results for many years, even after he was overtaken by complete blindness. In fact, when he first became blind in his right eye, his comment was, "Now I will have less distraction." His numerous mathematical accomplishments provided a critical step forward in providing clarity and precision to the concept of proof and to the construction of abstract mathematical models that could be used to attempt to solve difficult practical problems.

FIGURE 5.2.2   Portrait of Euler.

> **Euler circuit** A closed path through a graph that contains every edge exactly once.

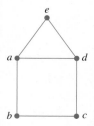

Named in honor of Leonhard Euler, an **Euler circuit** is defined to be a path through a graph that traverses every edge exactly one time and terminates at the same vertex from which it began. As you might expect, not every graph will possess an Euler circuit. However, by examining a sequence of graphs, we will develop a condition for determining precisely when an Euler circuit exists. We begin with the simple observation that each edge in the rectangular graph shown here is used once in forming a circuit. Each vertex has one edge that can be used to approach the vertex and another one to leave it. In other words, the degree of each vertex is 2.

Now note that if any corner has an additional edge added to it, it becomes impossible to begin a circuit at that point and return to it more than once. This is the case at vertex a or d in the graph shown here. Likewise, a circuit beginning elsewhere can approach such a vertex twice but only leave it once. So no Euler circuit can be found in this situation. We note that the degree of vertex a (and d also) is 3. In fact, we observe that whenever the degree of a vertex is odd, the number of edges in any circuit that approach the vertex will differ by 1 from the number of edges that leave the vertex. So no Euler circuit can exist in a graph that has one or more vertices of odd degree. This fact is stated in the following rule.

> **Every vertex of a connected graph G has even degree if and only if G possesses an Euler circuit.**

? Example 1

Do either of these graphs contain an Euler circuit? If so, find one.

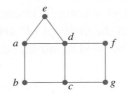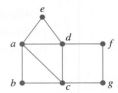

⚙ Solution

In the graph on the left, vertices a and c both have odd degree. So no Euler circuit can be found. All the vertices in the graph on the right have even degree, and so the graph must have an Euler circuit. While many such circuits could be found, two examples based at a are $aedfgcdacba$ and $abcdeacgfda$. ◆

Eulerization
Procedure in which edges are duplicated to create a new graph containing no odd vertices.

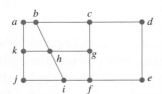

Graph theory is useful in the efficient operation of city services. Optimal routes for such tasks as the removal of snow, delivery of mail, and checking of telephone and electric lines can be found by searching for Euler circuits in graphs that represent networks of city streets. Clearly, the mileage accumulated by a snowplow is minimized by driving a route that traverses each street in a particular network with as little duplication as possible. Consider the map of a sector of Manhattan, New York, shown in **Figure 5.2.3**. Suppose a driver needs to clear snow along each street between and including 5th and 7th avenues and 40th and 42nd streets. If we assume that only one trip down each street is needed to clear away the snow, can the driver do this without having to travel a street twice? When we examine the representative graph that corresponds to the map, we see that vertices b, c, g, f, i, and k all have degree 3. Because this means no Euler circuit can be found, the next question would be, What circuit could be taken that reuses a minimum number of edges? Solving this related problem uses a technique known as the **Eulerization** of a graph. This is a procedure in which we add edges in order to create a new graph with no vertex of odd degree. The idea is to duplicate existing edges in a manner that adds as few new edges as possible.

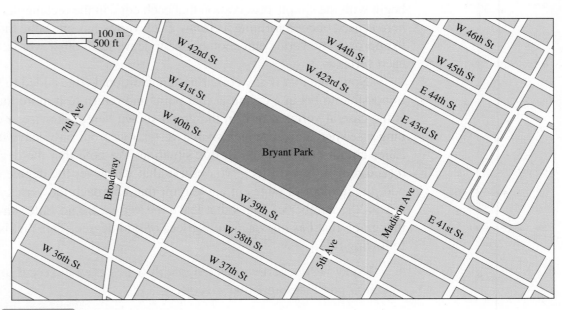

FIGURE 5.2.3   A Manhattan neighborhood.

For example, in the Eulerization of the Manhattan graph shown in **Figure 5.2.4**, we note that each vertex originally of odd degree has had at least one edge joined to it in order to change the degree to an even number. Observe also that although both vertices *h* and *j* initially had even degree, edges were attached to them to alter the degree of adjacent vertices. Moreover, we must be careful to duplicate only existing edges and not create any new connections, because creating a new edge would correspond

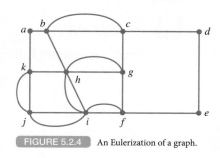

FIGURE 5.2.4   An Eulerization of a graph.

to a street that doesn't actually exist. For example, a new edge cannot be added between vertices *b* and *g*, or between *c* and *e*.

Now consider what the addition of an edge represents to the snowplow's task. It implies that any street corresponding to a duplicated edge will be traveled twice by the truck along any possible circuit constructed from the new graph. Because we are not currently accounting for the lengths of the streets represented by the edges, the main criterion for an optimal Eulerization is the addition of a minimum number of duplications. A graph satisfying this criterion is called a **minimum Eulerization**. Consider the two examples in **Figure 5.2.5**. Both of these required only four duplicated edges, as compared to the six edges added in Figure 5.2.4. A quick bit of experimentation will endorse the fact that no other modification exists that requires fewer than four new edges, and so both of these are minimum Eulerizations.

> **minimum Eulerization**
> Eulerization of a graph that uses a minimum number of duplicated edges.

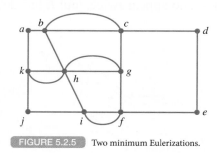 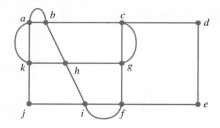

FIGURE 5.2.5   Two minimum Eulerizations.

Because the degrees of all the vertices in each of these graphs are now even, each graph possesses an Euler circuit. Once the circuit is identified, it is then squeezed back onto the original Manhattan graph to obtain a circuit that has twice-used edges. For example, an Euler circuit based at *a* in the left-hand graph of Figure 5.2.5 would be *abcbhgcdefgh-ifijkhka*. The edges *bc*, *hg*, *fi*, and *hk* are all used twice when this circuit is traversed in the original graph. (See **Figure 5.2.6**.)

Incidentally, note that the number of times each vertex (except the base) appears in the sequence for labeling the circuit must necessarily be equal to one-half its degree. For example, the vertex *h* has degree 6 in Figure 5.2.6 and is listed three times in our circuit.

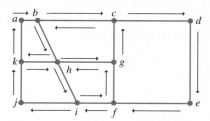

FIGURE 5.2.6   A circuit with reused edges.

? Example 2

Find a minimum Eulerization for the following graph.

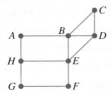

⚙ Solution

The degrees of all the vertices are even, with the exceptions of *H* and *D*, and therefore an edge must be added to each to Eulerize the graph. If *H* and *D* were adjacent, this could be done with a single edge, but this is not the case. We can see that the next best solution is to duplicate the edge connecting *H* to *E* and then another one connecting *E* to *D*.

The resulting minimum Eulerization is shown on the left in **Figure 5.2.7**. One Euler circuit in this graph would be *ABCDBEDEFGHEHA*, and squeezing it onto the original graph results in the circuit shown on the right side of Figure 5.2.7. Again we note the 2-to-1 ratio between the degree of a vertex and the number of times it appears in the sequence for the circuit. Vertices *C*, *F*, and *G* each have degree 2 and appear once, while *B*, *D*, and *H* have degree 4 and appear twice, and *E* has degree 6 and appears three times.   ◆

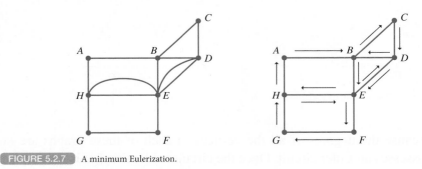

FIGURE 5.2.7   A minimum Eulerization.

In some situations, it is desirable to be able to construct a connected graph that possesses an Euler circuit that corresponds to a given sequence of vertices. For instance, one consideration in the planning of a new housing subdivision might be cost-efficient street maintenance. Therefore, it makes sense to design the street network such that it contains an Euler circuit even when faced with strange surface contours in the building site. The 2-to-1 relationship that we have observed in the circuits earlier can guide us in this construction. We conclude this section with an example.

❓ Example 3

Suppose that the path *ABCDCEFEDECA* is an Euler circuit in a particular graph.

(a)   What are the degrees of the vertices in this graph?
(b)   Draw a graph for which this path could be an Euler circuit.

⚙ Solution

(a)   With the exception of the starting point, the degree of each vertex must equal twice the number of times the vertex occurs in the circuit. Therefore, $\delta(A) = 2$, $\delta(B) = 2$, $\delta(C) = 6$, $\delta(D) = 4$, $\delta(E) = 6$, and $\delta(F) = 2$.

(b)   All graphs that have *ABCDCEFEDECA* as an Euler circuit must be isomorphic. (Why?) Both of the following graphs are examples.   ◆

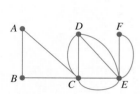 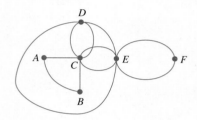

Name _____

Exercise Set **5.2** ⭘⭘⭘←

Determine the degree of the vertices in each of these graphs. If the graph has an Euler circuit, find it.

1.

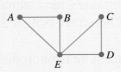

2.

3.

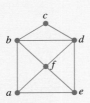

4.

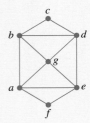

5.

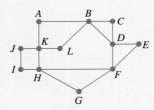

6.

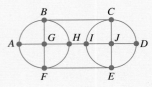

7.

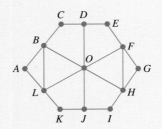

8.

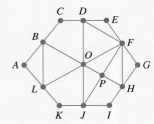

9. The Manhattan street maintenance crew wishes to check the no-passing lines for possible repainting. Suppose they wish to find the most efficient route by which to check all the streets north of 39th and south of 42nd streets and east of 7th and west of 5th avenues. Draw a graph that represents this region (including the boundary streets).

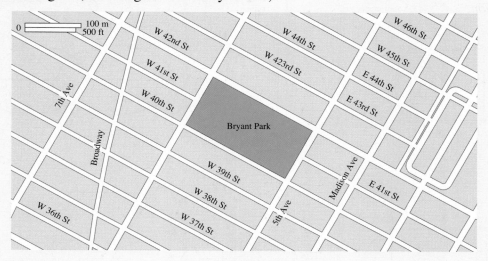

10. Does the graph in Exercise 9 have an Euler circuit? If so, find one.

11. Now draw a graph that extends the south boundary of the region in Exercise 1 to 37th Street. Does this graph have an Euler circuit?

12. This map shows a river, two islands, and five bridges. A jogger wishes to cross each of the bridges just one time as she runs a route that begins and ends on the north side of the river. Draw a graph that models the situation.

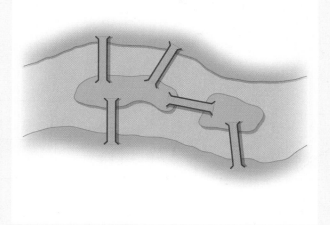

13. Does your graph in the previous exercise have an Euler circuit? If so, find one that begins on the north (top) side of the river.

Find a minimum Eulerization for each of the following graphs, and identify a circuit in that Eulerization.

14.

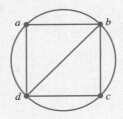

15.

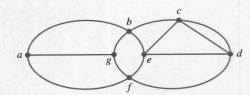

16.

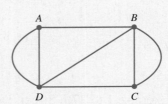

17.

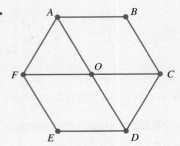

18.

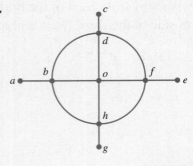

19.

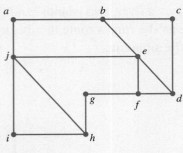

20.

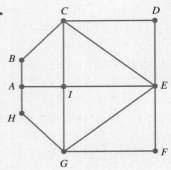

21.

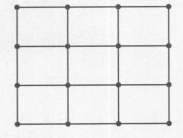

22.

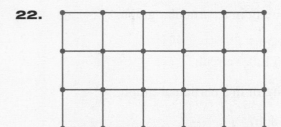

23. The map shown here displays a portion of Washington, D.C. A street inspector wants to check for holes and cracks on the streets between and including L and M streets and 17th and 19th streets. Form a graph that represents this network of streets, and find an optimal circuit.

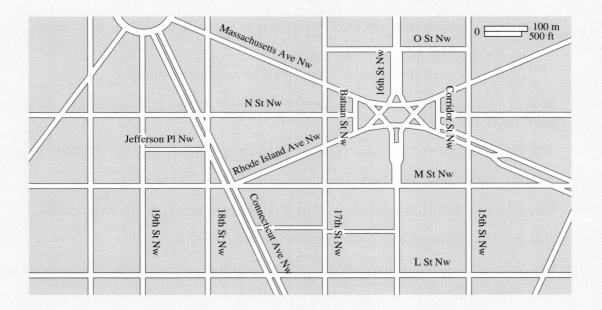

24. Draw a graph of the street network of the region between and including M and O streets and 15th and 17th streets. How many vertices of odd degree does this graph contain?

25. Draw a graph that represents the street network of the region between and including M and N streets and 17th and 19th streets. (Use a single edge to represent Connecticut Avenue.) How many vertices of odd degree does this graph contain?

26. Suppose that the path *ABCDCEBA* is an Euler circuit in a particular graph.

(a) What are the degrees of the vertices in this graph?
(b) Draw a graph for which this path would be an Euler circuit.

27. Suppose that the path *ABDEBECA* is an Euler circuit in a particular graph.

(a) What are the degrees of the vertices in this graph?
(b) Draw a graph for which this path would be an Euler circuit.

28. Suppose that the path *ABDCDEFEDFBA* is an Euler circuit in a particular graph.

(a) What are the degrees of the vertices in this graph?
(b) Draw a graph for which this path would be an Euler circuit.

29. Suppose that the path *ABCDCEFEDECA* is an Euler circuit in a particular graph.

(a) What are the degrees of the vertices in this graph?
(b) Draw a graph for which this path would be an Euler circuit.

Identify each statement as true or false. If the statement is false, give an example that demonstrates it is false. Let G and G′ be graphs.

30. If *G* is isomorphic to *G′* and *G* has an Euler circuit, then *G′* also has an Euler circuit.

31. If *G* and *G′* both have Euler circuits, then *G* is isomorphic to *G′*.

32. If an even number of vertices of *G* have odd degree, then *G* has an Euler circuit.

33. A graph cannot have an Euler circuit if it is not connected.

34. A graph with exactly two vertices of odd degree has a minimum Eulerization with just one additional edge.

35. A graph with exactly two vertices of odd degree, in which the vertices are also adjacent, has a minimum Eulerization with just one additional edge.

36. A graph has only one minimum Eulerization.

37. In an Eulerized graph, the number of times a vertex (other than the base vertex) appears in any Euler circuit must equal to one-half of its degree.

5.3 Hamilton Circuits

In this section, we shift our attention in the use of graphs from a search for a closed path that uses every edge once to looking for one that visits every vertex once. A vertex-oriented circuit does not have to use every edge in a graph and is the solution to a different kind of problem in graph theory. If the task is to plow a network of streets, then every road must be included in the chosen route, and an Euler circuit is appropriate. However, if the task is to visit just the intersections of the streets in order to check traffic patterns, then the most efficient path to travel would probably not use every street. For instance, one can envision an abundance of problems in the delivery of packages by companies such as Federal Express, UPS, and the U.S. Postal Service that require efficient paths that stop once at each of many destinations. These companies' economic survival depends on rapidly finding these paths every day. Our first example displays an elementary but representative situation in which the solution is a path that visits every desirable stop exactly once.

❓ Example 1

The students in a middle school class are going to visit the Lincoln Park Zoo in Chicago, and they wish to find an efficient route to walk in order to see their favorite animal exhibits. A map of the zoo is shown in **Figure 5.3.1**. They wish to create a graph in which the edges represent unobstructed walking paths and the vertices consist of the following:

- Penguin house (N)
- Bears (B)
- Primate house (P)
- Petting zoo (Z)
- Elephant exhibit (E)

FIGURE 5.3.1   The Lincoln Park Zoo in Chicago.

- Birds of prey (R)
- Giraffes (G)
- Sea lion pool (L)
- Swan pond (S)
- Landmark cafe (C)

Use the graph to find a closed path that visits each of the exhibits.

⚙ Solution

Any graph isomorphic to the one displayed here would suffice to model the zoo.

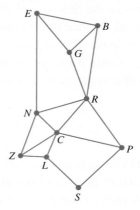

One circuit that visits each vertex once is shown by the darkened edges in **Figure 5.3.2.** If we think of it as beginning and ending at the cafe (*C*), then we could denote it by *CZNEGBRPSLC*, though the same circuit could also be designated by any of several sequences of vertices, including, for example, *ZNEGBRPSLCZ* or *NEGBRPSLCZN*. It is the *ordering* of the vertices that is the identifying characteristic as opposed to the choice of starting point.   ♦

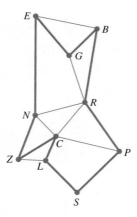

FIGURE 5.3.2   A Hamilton circuit.

Hamilton circuit A closed path through a graph that contains every vertex exactly once.

The circuit in Figure 5.3.2 is called a **Hamilton circuit** and is defined as any closed path in a connected graph that includes every vertex exactly once. A given graph may have none, one, or many Hamilton circuits. A graph with no such circuit is shown here.

Unlike the search for an Euler circuit, no precise conditions can be tested (such as the parity of the vertex degrees) in a graph that indicates when a Hamilton circuit can be found. Although exceptions to this statement for certain specialized graphs will be explored in the exercises, in general we can only identify certain helpful clues.

? Example 2

Find a Hamilton circuit in the graph shown here.

⚙ Solution

We start our path at *A* and head first to vertex *B*. At *B*, we have a choice of continuing on to *C* or to *E*. The problem with heading to *E* is that *C* could then not be included in any circuit that does not reuse a vertex because any path back through *C* would visit *B* a second time. Therefore, from *B*, we must head next to *C* and then necessarily on to *D*, because $\delta(C) = 2$. We now have a choice of *E* or *F*, either of which would lead to a Hamilton circuit. Choosing *F* would result in the circuit *ABCDFHGEIA*, and picking *E* would give *ABCDEGFHIA* as a solution. These are shown in **Figure 5.3.3**. ♦

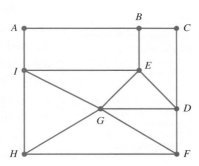

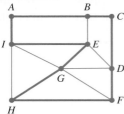 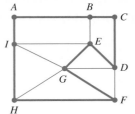

FIGURE 5.3.3   Two Hamilton circuits.

FIGURE 5.3.4   Sir William Rowan Hamilton.

This example illustrates the fact that as we construct our closed path, if we are currently resting at a vertex that is adjacent to a vertex of degree 2, then that vertex must be visited next. Otherwise, no Hamilton circuit could be completed that includes it.

The Hamilton circuit is named for **Sir William Rowan Hamilton** (1805–1865). Born and raised in Dublin, he is one of the most famous mathematicians to ever hail from Ireland. Hamilton was raised and educated by an uncle. His keen mind was apparent as a young child, and by age 11, he had mastered 13 different languages. As a teenager, his attention turned to mathematics, and by the age of 20, he had absorbed Newton's *Principia*, corrected an error in Simon Laplace's difficult *Mecanique Celeste*, and was already engaged in doing original mathematical work. He achieved remarkable success at Trinity College, Dublin, and astonishingly was appointed to the prestigious position of Professor of Astronomy at the unprecedented young age of 21. Fond of writing poetry, Hamilton formed a lifelong friendship with literary master William Wordsworth. The two men spent much time discussing science and poetry, and Wordsworth eventually was forced to advise his friend to stick to science. His poetry was not so good!

Although he accomplished much professionally, Hamilton's personal affairs were fraught with angst and heartbreak. Throughout his life, he was plagued by an unrequited love for a woman he lost because of the interference of her parents. In constant personal turmoil, he succumbed to a lifelong predilection for alcohol that, in his middle forties, led to a deterioration of his health and eventually to his death. If you ever visit Dublin, you can still see the equations he engraved on the stone bridge where they first occurred to him.

Because most graphs that occur in practice contain multiple Hamilton circuits, the next question is, Which of these is the best solution, where *best* usually is defined in terms of some numerical quantity that is a function of the circuit. Often, the relationship represented by each edge of a graph has a number value, or **weight**, associated with it, such as the length of a particular street, a distance between cities, a cost of transmission along a phone line, or a time to complete a certain task. A graph that has a weight associated with each edge is called a **weighted graph**, and the sum of these weights is called its *total weight*. A **minimal Hamilton circuit** is one whose total weight is less than that of any other Hamilton circuit.

> **weight** A numerical value assigned to an edge.
>
> **weighted graph** A graph in which every edge has a weight.
>
> **minimal Hamilton circuit** A Hamilton circuit in a weighted graph whose total weight is less than that of any other Hamilton circuit.

As an example, suppose that a school superintendent needs to visit five schools in her district and wishes to do so in the least amount of time. The roads between the schools are located as shown in **Figure 5.3.5**, with the distances (in miles) between the schools given by the indicated values. A weighted graph representing this situation is also shown, and the shaded circuit below it is one Hamilton circuit that the superintendent could use.

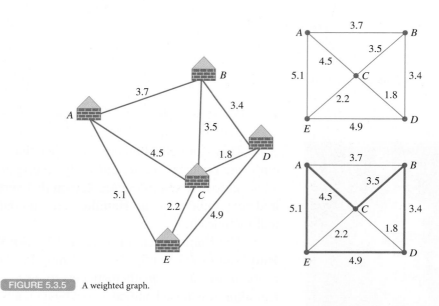

FIGURE 5.3.5   A weighted graph.

If we add up the weights along the edges of the circuit *ACBDEA,* then this total weight corresponds to the mileage associated with this particular route. But do other routes exist that would have a smaller total? Three other Hamilton circuits are shown in **Figure 5.3.6**.

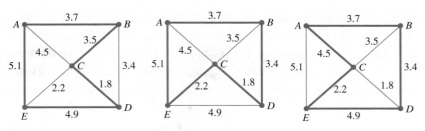

FIGURE 5.3.6   Three more Hamilton circuits.

The total weights of all four circuits are as follows:

$$ACBDEA: \; 4.5 + 3.5 + 3.4 + 4.9 + 5.1 = 21.4$$

$$ABCDEA: \; 3.7 + 3.5 + 1.8 + 4.9 + 5.1 = 19.0$$

$$ABDCEA: \; 3.7 + 3.4 + 1.8 + 2.2 + 5.1 = \mathbf{16.2}$$

$$ABDECA: \; 3.7 + 3.4 + 4.9 + 2.2 + 4.5 = 18.7$$

We conclude that the path *ABDCEA* has the shortest total weight and therefore provides the optimal solution to this problem. In Figure 5.3.6, *ABDCEA* is a minimal Hamilton circuit.

> **brute-force algorithm** A problem-solving procedure in which the total weight of every Hamilton circuit is computed to determine an optimal circuit.

The previous example demonstrates the use of the **brute-force algorithm** to find an optimal Hamilton circuit. According to this method, *every* existing Hamilton circuit is identified, and the total weight is computed. A minimal circuit is then an optimal solution. (There can be more than one.) In practice, however, this method has a serious drawback. Most graphs used in applications consist of a large number of vertices and therefore possess a very large number of Hamilton circuits. Finding all of them and calculating their total weights can be a time-consuming task even in the modern world of powerful high-speed computers. The brute-force method can take hours or days of computer time depending on the size of the graph, and therefore, it is not economically viable. As a result, other algorithms are sought by mathematicians that can determine solutions that may not necessarily be optimum but are still useful and have the advantage that can they can be found quickly. Such procedures are known as

> **heuristic algorithm** A problem-solving procedure that produces a solution quickly. However, this solution may not be optimal.

heuristic algorithms and sometimes are colloquially described as "quick and dirty." They have great value in the world of business, where the perfect solution often may not be worth the cost of finding it.

A complete graph (on *n* vertices) was defined in the previous section to be the graph K_n with *n* vertices, in which every pair of vertices is joined by an edge. We now explore the question of how many Hamilton circuits are contained in K_n by considering the example of K_4.

❓ Example 3

We wish to compute the number of circuits in K_4 that visit every vertex. For convenience, we label the vertices v_1, v_2, v_3, and v_4 and choose one of them, say, v_1, as the starting point. We then have three choices for the second node in the circuit, and after picking one, we would have two choices for the third. After selecting one of those, we have just a single node left. These would give us $3 \cdot 2 \cdot 1 = 6$ circuits. However, because we are not concerned with the

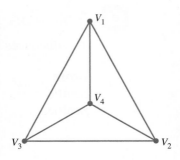

direction traveled along any circuit, we note that *both* the sequences $v_1 v_2 v_3 v_4 v_1$ and $v_1 v_4 v_3 v_2 v_1$ represent the same circuit, namely, the one in **Figure 5.3.7(a)**. Similarly, the sequences $v_1 v_2 v_4 v_3 v_1$ and $v_1 v_3 v_4 v_2 v_1$ both represent the path in **Figure 5.3.7(b)**, and **Figure 5.3.7(c)** displays both $v_1 v_4 v_2 v_3 v_1$ and $v_1 v_3 v_2 v_4 v_1$. Therefore, we must divide our previous result by 2 to get a total of three Hamilton circuits in K_4. ◆

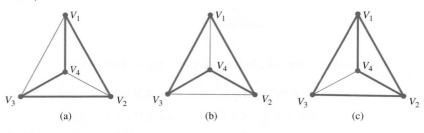

FIGURE 5.3.7   The three Hamilton circuits in K_4.

The process used in Example 3 can be generalized to computing the number of Hamilton circuits in K_n. After picking a starting point for a sequence of vertices, we would have $n - 1$ choices for the second node, $n - 2$ for the third, and so forth. Then, as in Example 3, the product $(n-1)(n-2) \cdots 3 \cdot 2 \cdot 1$ gives us the total number of sequences formed. Recalling that every sequence and the corresponding reverse-order sequence represent the same circuit, we then divide this product by 2 to obtain the final number of Hamilton circuits. Factorial notation allows us to summarize this result. The factorial of a positive integer m is denoted by $m!$ and is defined to be the product of m and all the positive integers less than m. For example, $5! = 5 \cdot 4 \cdot 3 \cdot 2 \cdot 1 = 120$ and $6! = 6 \cdot 5 \cdot 4 \cdot 3 \cdot 2 \cdot 1 = 720$.

The total number of Hamilton circuits in the complete graph K_n is $(n-1)!/2$.

Traveling Salesperson Problem (TSP) The problem of finding a minimum-weight Hamilton circuit in a complete weighted graph.

One historically well-known problem involving complete weighted graphs is the **Traveling Salesperson Problem** (TSP). This deals with the problem a salesperson faces when required to visit n different cities in the most cost-efficient manner. Assuming that transportation exists between any two cities, we see that the optimal solution to this problem concerns finding a minimal Hamilton circuit in a weighted K_n.

? Example 4

A salesperson based in Seattle needs to visit Salt Lake City, Houston, Dallas, and Memphis. What sequence of trips would consist of the least total distance traveled? In other words, find the optimal solution to the TSP for these five cities.

⚙ Solution

The distances between the given cities can be easily found on the Internet.

From	To	Distance (miles)
Seattle	Salt Lake City	700
Seattle	Houston	1,891

From	To	Distance (miles)
Seattle	Dallas	1,683
Seattle	Memphis	1,872
Salt Lake City	Houston	1,201
Salt Lake City	Dallas	1,003
Salt Lake City	Memphis	1,256
Houston	Dallas	224
Houston	Memphis	484
Dallas	Memphis	419

The complete graph on five vertices with the appropriate weights is given in **Figure 5.3.8**. There exist $4!/2 = (4 \cdot 3 \cdot 2 \cdot 1)/2 = 12$ Hamilton circuits. Each must be identified and its total weight evaluated. For example, the circuit *SLMDHS* has weight given by $700 + 1,256 + 419 + 224 + 1,891 = 4,490$. Similarly, *SLMHDS* has weight $700 + 1,256 + 484 + 224 + 1,683 = 4,347$.

The weights of all the circuits are shown in the following table.

Circuit	Total Weight
SLMDHS	4,490
SLMHDS	4,347
SLDMHS	4,497
SLDHMS	**4,283**
SLHMDS	4,487
SLHDMS	4,416
SMDLHS	6,386
SMHLDS	6,243
SMLDHS	6,246
SDMLHS	6,450
SDHLMS	6,038
SDLMHS	6,317

The optimal solution is the minimal Hamilton circuit *SLDHMS*. ◆

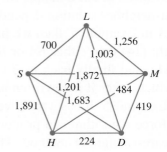

FIGURE 5.3.8 Complete graph for the TSP.

Finding all the Hamilton circuits, even for only five vertices, is an arduous task. You can imagine how much more difficult that process becomes as the number of vertices increases. Therefore, a quicker heuristic algorithm is needed for many practical applications. The **nearest-neighbor algorithm** is just such a method. It is procedure for forming a Hamilton circuit that usually has a smaller weight than most other circuits, although it may not be a minimum. Basically, you travel from vertex to vertex along a low-weight edge, being careful not to form a circuit until you return to your starting point. Specifically, it works like this:

1. Choose a starting vertex. Of the edges that meet at that vertex, choose the one of lowest weight, and travel along it to the next vertex. (In case of a tie, pick either one.)

2. Continue this process, being careful not to add an edge that completes a circuit until you return to your initial vertex.

? Example 5

A meat distributor, located at A, needs to make a delivery to three different markets every morning before they open for business. He wishes to do this in the shortest time. The time (in minutes) to travel between each pair of stores is given in the accompanying graph. Use the nearest-neighbor algorithm to find a solution.

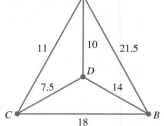

⚙ Solution

Starting at A, we proceed first to D, because edge AD has the lowest weight (10). From D, we have two choices for the next edge, and we choose DC because it has a lower weight (7.5) than DB. From C we must choose CB (18) next even though it has a greater weight than CA because if we proceed along CA, we will complete a circuit that does not visit each vertex. Finally, from B we have only BA (21.5) to select for the last edge in our circuit. The circuit $ADCBA$ has total weight $10 + 7.5 + 18 + 21.5 = 57$. ◆

We make two observations about Example 5:

1. The nearest-neighbor solution depends on the starting point. If we start at B instead of A in Example 5, then we first travel to D, because edge BD has a lower weight (14) than either BA or BC. From D we choose edge DC (7.5), then CA (11), and finally back to B via AB (21.5). This gives us the circuit $BDCAB$, which, after it is rewritten as $ABDCA$, is a different solution from the one in Example 5. It has a total weight of $14 + 7.5 + 11 + 21.5 = 54$.

2. The brute-force algorithm can produce a different solution from that found with the nearest-neighbor method. The three Hamilton circuits in Example 5 are $ADCBA$, $ADBCA$, and $ABDCA$, having total weights 57, 53, and 54, respectively. Hence, circuit $ADBCA$ is the solution found using the brute-force method and indeed is the optimal one. For small problems where the search time is not a factor, the brute-force method is preferred. However, most real-life applications rarely deal with graphs having a small number of circuits.

It is important to observe that it need not be the case that corresponding edges of isomorphic graphs have the same weights. Our definition of isomorphism in the first section did not require that edge weights be preserved. Therefore, even if G and G' are isomorphic graphs, a minimum-weight Hamilton circuit in G would not have to also be an optimal solution in G' (although it is true that a Hamilton circuit in one must certainly correspond to a Hamilton circuit in the other).

Researchers in many fields continue to discover new and powerful ways to employ graph theory in the continuing quest to understand our environment. In the article "100 Trillion Connections" in the January 2011 issue of *Scientific American,* Carl Zimmer explains how networks are used to examine the inner workings of the 100 billion neurons (vertices) and their attendant 100 trillion connections (edges) that compose the architecture of the human brain. Graphs and subgraphs are identified that become active when humans perform a particular task and show how faulty connections can lead to devastating diseases such as schizophrenia or dementia. **Figure 5.3.9** illustrates one such large-scale network.

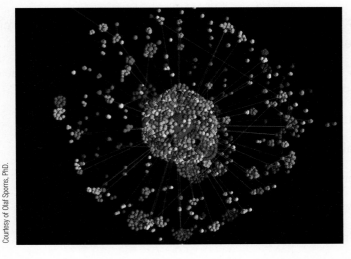

Courtesy of Olaf Sporns, PhD.

FIGURE 5.3.9   100 trillion connections: Efforts to understand the human brain's architecture require graph theory. *Scientific American,* January 2011.

It is hoped that this chapter has convinced you of the importance of graph theory. From areas such as street maintenance to the complexities of the central nervous system, the use of graphs as a mathematical model provides an effective tool for solving a wide range of difficult and practical problems. As the nature of problems confronting society continues to evolve, the flexibility and power of mathematics arm us with an effective tool for meeting new challenges.

Name _____

Exercise Set **5.3** ⃝⃝⃝←————————————————

Identify the statements in Exercises 1–11 as true or false. Let G and G′ be graphs.

1. Every connected graph must have a Hamilton circuit.

2. If G and $G′$ are isomorphic to each other and G has a Hamilton circuit, then so does $G′$.

3. If G and $G′$ both do not have a Hamilton circuit, then they must be isomorphic to each other.

4. Let G and $G′$ be isomorphic. If G has a minimum-weight Hamilton circuit H, then the corresponding Hamilton circuit $H′$ in $G′$ also has a minimum weight.

5. Let G and $G′$ be isomorphic. If H is a Hamilton circuit in G, then the corresponding closed path $H′$ in $G′$ must also be a Hamilton circuit.

6. The complete graph K_6 contains 60 Hamilton circuits.

7. A heuristic algorithm always finds the optimal solution.

8. The brute-force algorithm is heuristic.

9. The brute-force algorithm always finds an optimal solution.

10. The nearest-neighbor solution depends on the starting point.

11. If $\delta(v) = 2$ for every vertex v in a connected graph G, then G has a Hamilton circuit.

12. Draw the graph from Example 1 whose vertices correspond to the same major exhibits in the Lincoln Park Zoo shown here. Find another Hamilton circuit that is different from the one found in the text.

13. Use the graph from the previous exercise to find two closed paths that begin and end at the cafe and that stop once each at the exhibits of bears, giraffes, penguins, sea lions, and primates.

14. Use the graph from Example 1 to find two closed paths that begin and end at the cafe and that stop once each at the exhibits of penguins, bears, primates, swans, and elephants.

15. Find two Hamilton circuits in each of the following graphs.

(a)

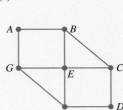

(b)

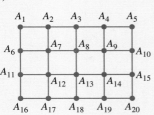

(c)

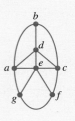

16. Find two Hamilton circuits in each of the following graphs.

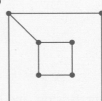

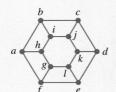

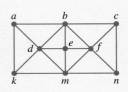

17. Do any of the following graphs have a Hamilton circuit? Why not? State a condition that would prevent a graph from having a Hamilton circuit.

(a)   (b)   (c)   (d)

18. Which of the following graphs does not have a Hamilton circuit? State a condition that would prevent a graph from having a Hamilton circuit.

(a)   (b)

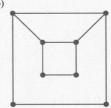

19. Find a Hamilton solution to the following graphs (starting from A), using both the brute-force algorithm and the nearest-neighbor algorithm. Do you get the same solution with both methods?

(a)

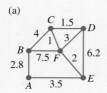

(b)

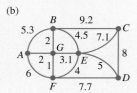

20. Find a Hamilton solution to the following graphs (starting from a), using both the brute-force algorithm and the nearest-neighbor algorithm. Do you get the same solution with both methods?

(a)

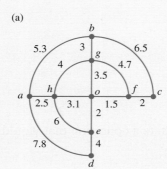

(b)

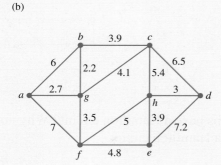

Solve the TSP in each of the following exercises, using the distances given in the following table.

From	To	Distance (miles)
Denver	Boston	1,766
Denver	Atlanta	1,204
Denver	New Orleans	1,078
Denver	Minneapolis	693
Boston	Atlanta	938
Boston	New Orleans	1,348
Boston	Minneapolis	1,126
Atlanta	New Orleans	412
Atlanta	Minneapolis	905
New Orleans	Minneapolis	1,043

21. Starting in Denver, find a solution to the TSP for the cities of Denver, Atlanta, Boston, and New Orleans, using:

(a) the brute-force algorithm.
(b) the nearest-neighbor algorithm.

Are these two solutions the same? Which one is optimal?

22. Starting in Atlanta, find a solution to the TSP for the cities of Atlanta, Boston, Denver, and Minneapolis, using:

(a) the brute-force algorithm.
(b) the nearest-neighbor algorithm.

Are these two solutions the same? Which one is optimal?

23. Starting in Boston, find the optimal solution to the TSP for the cities of Boston, Atlanta, New Orleans, and Minneapolis.

24. Starting in Denver, find a heuristic solution to the TSP for all five cities: Denver, Boston, Atlanta, New Orleans, and Minneapolis.

25. How many Hamilton circuits are contained in the complete graph K_5? In K_7? In K_{10}?

26. What is the maximum whole number n such that $n!$ cannot be computed on your calculator? What type of company might need to use a complete graph K_n for that value of n or greater?

27. A polyhedron is a three-dimensional solid geometrical object, such as a pyramid or soccer ball, that consists of vertices, edges, and plane sides or faces. Only five polyhedrons, known as regular polyhedrons, possess faces that are all congruent to one another. They are the tetrahedron (4 sides), cube (6 sides), octahedron (8 sides), dodecahedron (12 sides), and icosohedron (20 sides). Each of the following graphs represents the projection (compression) of a regular polyhedron onto the plane of the page. Note that each one contains a Hamilton circuit. In each case, find another such circuit in addition to the one indicated. (This was one of Hamilton's original problems.)

(a) (b) (c) (d) (e)

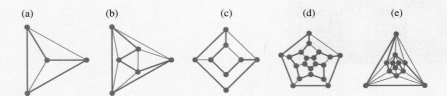

28. A **complete bipartite graph** B is one in which the vertices can be sorted into two groups, R and S, such that *every* vertex in R is connected by an edge to *every* vertex in S, and these are the only edges of B. If the number of vertices in R is m and the number of vertices in S is n, then the bipartite graph on R and S is denoted by B_{mn}. The following are graphs for B_{32}, B_{33}, and B_{34}. Which of these contain(s) a Hamilton circuit?

> **complete bipartite graph** A bipartite graph in which every vertex in one vertex group is joined by an edge to every vertex in the second group.

(a)   (b)   (c)

29. Determine a criterion for a complete bipartite graph B_{mn} in terms of m and n that will tell us precisely when B_{mn} contains a Hamilton circuit.

30. In the game of chess, the piece called the knight moves in an L-shaped path. It travels two squares nondiagonally in any direction followed by one square perpendicular to that direction. An interesting problem that dates back more than a century concerns searching for a knight's tour. A *knight's tour* is the name given to a sequence of moves that visits every square on the chessboard exactly once. A tour is also called *closed* or *reentrant* if the terminal square of the tour is one knight's move away from the initial square. The numbered figure here that corresponds to the chessboard depicts one knight's tour on the standard playing board that consists of 64 squares arranged in 8 rows and 8 columns.

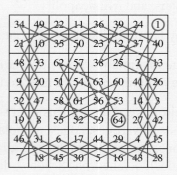

The generalized $m \times n$ knight's tour uses a nonstandard board with m rows (ranks) and n columns (files). A reentrant tour problem can be solved by searching for a Hamilton circuit in a graph in which the vertices $v_1, v_2, \ldots$ are the squares and an edge joins v_i to v_j only if a knight can be moved from v_i to v_j. A 3×4 board and the associated graph are shown here. Can you find a Hamilton circuit?

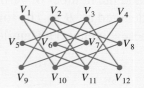

31. Draw a graph to model the 3×5 knight's tour problem. Does this have a closed tour?

32. Draw a graph to model the 4×5 knight's tour problem. Does this have a closed tour?

33. Schwenk's Theorem states that a closed knight's tour exists on an $m \times n$ graph unless:

(a) m and n are both odd.
(b) $m = 1$, 2, or 4.
(c) $m = 3$ and $n = 4$, 6, or 8.

What are the smallest values for m and n for which a closed tour exists on an $m \times n$ graph?

34. Use the map of the southeast United States given here to draw a graph in which each vertex represents a state and two vertices are joined by an edge if the corresponding states share a border. Could you visit every state in this map exactly once and return to where you started? Explain your answer.

Name _____

Chapter Review Test **5** ◯◯◯ ←

1.   True/False. Let G and G' be graphs.

   (a) If G is isomorphic to G', then the two graphs must have the same number of edges.
   (b) If G and G' have the same number of even vertices, then the two graphs must be isomorphic.
   (c) If the number of Euler circuits of G is not equal to the number of Euler circuits of G', then G must be isomorphic to G'.
   (d) A graph cannot have an Euler circuit if it is not connected.
   (e) A graph cannot have an odd number of odd vertices.
   (f) The graph shown here is a planar graph.

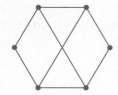

   (g) The nearest-neighbor algorithm always finds an optimal solution.
   (h) The brute-force algorithm is a heuristic method.
   (i) A graph may have none, one, or many Hamilton circuits.
   (j) A graph has only one minimum Eulerization.

2.   Use the map of the northeast United States given here to draw a graph in which each vertex represents a state and two vertices are joined by an edge if the corresponding states share a border. Could you visit every state in this map exactly once and return home? Explain your answer.

3. Fill in the blanks.

    (a) The total number of edges in the complete graph K_{12} is_____.

    (b) The total number of Hamilton circuits in K_8 is_____.

    (c) The degree sum of a graph with 150 edges is _____.

    (d) The number of edges contained in a graph having 500 vertices, each of degree 8,

       is_____.

4. Draw a graph that is isomorphic to the given graph but does not contain any crossings.

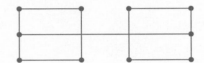

5. Which of the following graphs, (a) to (d), is isomorphic to graph (e)?

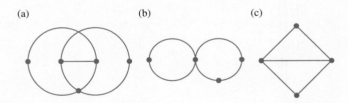

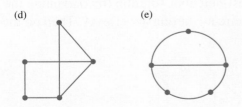

6. Does the graph here contain an Euler circuit? If so, find one, and list it as a sequence of vertices.

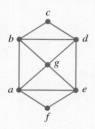

7. Suppose that the path *ABCDCEBA* is an Euler circuit in a particular graph.

(a) What are the degrees of the vertices in this graph?
(b) Draw a graph for which this path would be an Euler circuit.

8. Consider a trip that starts in Scranton, Pennsylvania. Find the optimal solution to the TSP for the following cities. Be sure to draw the complete graph.

From	To	Distance (miles)
Scranton	Philadelphia	104
Scranton	Miami	1,111
Scranton	Lexington	522
Philadelphia	Miami	1,015
Philadelphia	Lexington	509
Lexington	Miami	873

9. Use the graph shown here to find the nearest-neighbor solution (by darkening the appropriate edges) and its total weight for a Hamilton circuit starting at vertex *A*. Then do the same, starting at *B*.

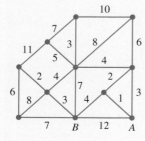
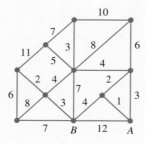

10. Define a bipartite graph. Give a problem that might need such a graph for its solution.

11. Draw an appropriate graph to model the 3×4 knight's tour problem. Be sure to give a clear definition of a vertex and an edge.

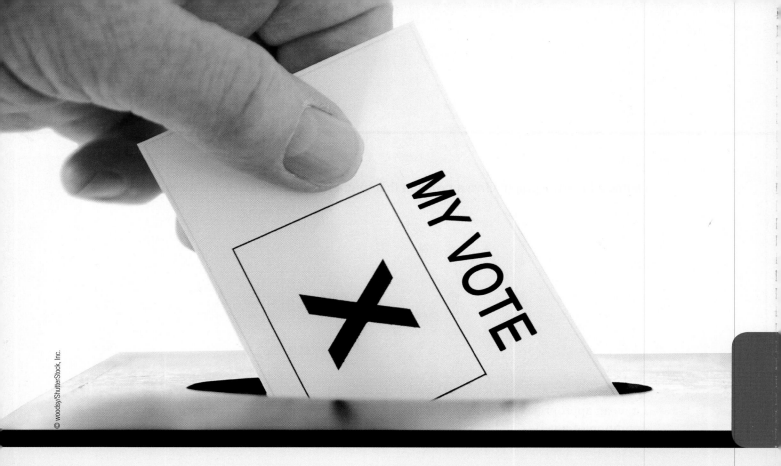

Chapter Objectives

check off when you've completed an objective

- ☐ Construct a preference table.
- ☐ Implement the plurality and Borda count methods.
- ☐ Know the fairness criteria.
- ☐ Implement the Hare system and the approval voting method.
- ☐ Know Condorcet's paradox.
- ☐ Find *n*-cycles in a preference table.
- ☐ Know Arrow's impossibility theorem.
- ☐ Construct and analyze a Saari triangle.
- ☐ Determine election results from a Saari triangle.

chapter

6

Social Choice and Voting Methods

6.1 Plurality and the Borda Count

6.2 The Hare System

6.3 Approval Voting

> Elections belong to the people. It's their decision. If they decide to turn their back on the fire and burn their behinds, then they will just have to sit on their blisters.
>
> —Abraham Lincoln

Wisely choosing a person or policy statement from a list of candidates or alternatives is one of the most important functions of a healthy society. Most people who believe in the democratic process wish to have the best candidate possible representing them and their community while making decisions on the local school board, ensuring efficient operation of their city, and enacting public policy in Congress and the White House. Propositions appear on most ballots in general elections that affect your taxes, your state and local services, your civil rights, and your natural environment. Any decision about passing such a proposition into law deserves to be made in a manner that best reflects the collective will of the voting populace. Therefore, most people would probably agree that the mechanism by which those choices are made is vitally important. Ironically, the one proven result in the centuries-old search for some form of perfect voting system is that, by most acceptable standards, no such system exists. This chapter introduces a brief list of voting methods that are in use today and those criteria that one would expect any reasonably good system to satisfy. It is both interesting and slightly disturbing to realize that, in fact, none of these systems meets all these criteria.

6.1 Plurality and the Borda Count

Voting for a political candidate in America in a general election is often viewed as picking from only two choices—a Democrat or a Republican. If this is indeed the case, then each voter marks his or her ballot for one of the two candidates, and the winner is the one receiving the majority of the votes. This is a completely reasonable system and indeed is the cornerstone of what most people view as a participatory democracy. It involves no further discussion, and so these types of elections will not be our focus here. Rather, we shall examine the issues involved with what happens when more than two choices are available. This is actually quite often the case, for example, when choosing the recipient of a major honor (the Heisman Trophy or a Nobel Prize), new members to be inducted into a society (a sports Hall of Fame or a college sorority), an officer of an organization (the president of the Teamsters Union or the president of your senior class), or a proposition that affects the public good (the regulation of smoking in restaurants). It also frequently happens in the political arena in primary or local elections and even occasionally in a national election, such as when Ross Perot garnered a significant percentage of the vote in his 1992 bid for the U.S. presidency against Bill Clinton and George H. Bush. Strong showings by third-party candidates do, in fact, occur with some regularity, and in a close election, even a relatively minor alternate candidate can have a major impact on the results. Such was the case in the 2000 U.S. presidential election in Florida when postelection surveys indicated that Ralph Nader attracted enough traditional Democratic voters away from Al Gore to tip the scales to Republican George W. Bush.

preference ballot
A ballot in which the voter ranks the candidates from most favored to least favored.

To be able to thoroughly examine the subtleties involved in selecting a winning candidate who accurately reflects the choice of a group of people, all the ballots considered in this section will be **preference ballots**. This type of ballot gives a list of candidates ranked in order of the voter's preference. You are probably more familiar with the practice of voting for a single choice

George W. Bush

Al Gore

from a set of alternatives, but this method does not reveal information about the voter's opinion of the rest of the field. If your top choice is not going to be elected, you would certainly want your opinions of the other candidates to impact the results. Consider the following example.

BALLOT PAPER
PRESIDENT OF THE GOURMET CLUB

Stoneford University

Rank your three choices from the
three canidates A, B, and C

A) Alicia ☐

B) Bert ☐

C) Carlos ☐

Number every box to make your vote count.

The 12 members of the Gourmet Club at Stoneford University recently held an election for president of their club. They each ranked their three choices from the candidates Alicia (A), Bert (B), and Carlos (C), and the individual ballots are shown below.

First	A	B	B	C	A	A	C	C	A	B	C	C
Second	C	A	C	B	B	B	A	A	B	A	A	B
Third	B	C	A	A	C	C	B	B	C	C	B	A

Because three choices can be ranked in just six different ways (*ABC*, *ACB*, *BAC*, *BCA*, *CAB*, *CBA*), it must be true that at least one sequence (and probably more) must occur more than once. Therefore, it makes sense to organize the results in a more concise form that is easier to analyze. One way to do this is by means of a **preference table** that counts the number of times each sequence occurs. In the current case, we get the preference table in **Figure 6.1.1**.

preference table
A table that shows the number of voters who have cast a preference ballot of each particular ranking of candidates.

	Number of Ballots					
	3	**1**	**2**	**1**	**3**	**2**
First	A	A	B	B	C	C
Second	B	C	A	C	A	B
Third	C	B	C	A	B	A

FIGURE 6.1.1 Preference table for the Gourmet Club election.

From this table, we can easily see that no single candidate received a majority (more than one-half) of the first-place votes. Carlos did, however, receive the most

plurality method
The voting method in which the candidate with the most votes wins.

majority criterion
If a candidate receives the majority of the first-place votes in an election, then that candidate is the winner.

first-place votes (5), and if we declare Carlos to be the winner based on this fact, then the method we have used is called the **plurality method**. Note that under this method, any candidate receiving a majority of the votes would be declared the winner. Therefore, we say it satisfies the **majority criterion**, stated as follows:

> **Majority criterion.** If a candidate receives the majority of the first-place votes in an election, then that candidate is the winner.

The main problem with the plurality method is that it does not take into account the other preferences expressed by the voters. In the current example, Carlos could have been ranked last by the other seven voters and still have won the election. One sensible way to use all the information provided by preference ballots is to assign a certain number of points to be awarded for each place in the rankings and then compute a total number of points for each candidate, using these point values. This system is known as the **Borda count method**. In a 3-rank system, for instance, we could assign 3 points for a first-place ranking, 2 points for a second-place ranking, and 1 point for a third-place ranking. Under this scheme, because Alicia had 4 first-place votes, 5 second-place votes, and 3 third-place votes, her point total, or Borda count, is $4(3$ points$) + 5(2$ points$) + 3(1$ point$) = 25$ points. The Borda counts for all three candidates are tallied below.

Borda count method The voting method using preference ballots in which points are awarded for each rank and the total number of points is computed for each candidate. The winner is the candidate with the highest point total.

	Rank			
	First	**Second**	**Third**	**Borda Count**
Alicia	4	5	3	$4(3) + 5(2) + 3(1) = \mathbf{25}$
Bert	3	5	4	$3(3) + 5(2) + 4(1) = \mathbf{23}$
Carlos	5	2	5	$5(3) + 2(2) + 5(1) = \mathbf{24}$

We see that under this voting scheme, Alicia is the winner despite the fact that she received fewer first-place votes than Carlos. So we see that these two voting methods can produce different results based on the *same* election. The election is therefore dependent on the choice of voting methods as well as the actual ballots! This is rather disconcerting especially because both methods seem, on the surface at least, to be reasonable ones. One way to help determine the strength of the Borda count method is to first find out whether it satisfies the majority criterion. Proving a statement in mathematics to be true in the general case can be a difficult task, but to show it to be false can sometimes be done by producing a single counterexample, such as that displayed in the preference table for three candidates shown in **Figure 6.1.2**.

	Number of Ballots	
	7	**5**
First	*C*	*A*
Second	*A*	*B*
Third	*B*	*C*

FIGURE 6.1.2   Proof that the Borda count method fails the majority criterion.

By the Borda count method, $7(3) + 5(1) = 26$ points are received by C, and $5(3) + 7(2) = 29$ points are received by A. So candidate A wins the election even though C received a majority of first-place votes. Therefore, the Borda method fails to satisfy the majority criterion.

? Example 1

In the last election for the congressional representative from Jacoby County, Republican Harold Chapman (C) received 38% of the vote, Democrat Sheila Forsythe (F) received 35%, and Independent Frank Kilpatrick (K) received 27%. Because U.S. political elections use the plurality method, Chapman won this election. However, exit pollsters asked voters to rank each candidate as they left the voting booth, and the preference table of their responses is shown below.

	Percentage					
	26	**12**	**29**	**6**	**20**	**7**
First	C	C	F	F	K	K
Second	F	K	K	C	F	C
Third	K	F	C	K	C	F

Who is the winner of this election, using the Borda count method?

⚙ Solution

We can think of each of the percentages as the typical number of votes allotted to a candidate from a random sample of 100 voters. Then, using the standard procedure of assigning 3 points for first place, 2 points for second place, and 1 point for third place, we get the following Borda counts.

	Rank			
	First	**Second**	**Third**	**Borda Count**
Chapman	38	13	49	$38(3) + 13(2) + 49(1) = \mathbf{189}$
Forsythe	35	46	19	$35(3) + 46(2) + 19(1) = \mathbf{216}$
Kilpatrick	27	41	32	$27(3) + 41(2) + 32(1) = \mathbf{195}$

Amazingly, not only did Forsythe beat Chapman when we used Borda counts, but so did Kilpatrick! One might easily begin to question the effectiveness of the plurality voting method based on a close election like this one. ◆

The Borda count method seems to represent the collective will of the voters better than the plurality method, but it fails more than just the majority criterion. Under this method, the results can be affected by voters who deliberately vote contrary to their beliefs in order to secure a particular outcome. One particularly intriguing example of this concerns altering the order of *irrelevant alternatives* on a preference ballot. In our prior example of the Gourmet Club, Alicia won the election when the Borda count method was used, and Carlos finished second. (See Figure 6.1.1.) Suppose two of the three voters in the first column (who liked Alicia the best) had instead listed Carlos in second place and Bert in third place. The preference table would then have be altered to look like **Figure 6.1.3**.

Number of Ballots

	1	3	2	1	3	2
First	A	A	B	B	C	C
Second	B	C	A	C	A	B
Third	C	B	C	A	B	A

FIGURE 6.1.3   Altered preference table for the Gourmet Club election.

We can see that Carlos picked up 2 more points, and so the new Borda counts would now make Carlos the winner even though the two new ballots still meant that Alicia was ranked over Carlos!

	Rank			
	First	**Second**	**Third**	**Borda Count**
Alicia	4	5	3	$4(3) + 5(2) + 3(1) = \mathbf{25}$
Bert	3	3	6	$3(3) + 3(2) + 6(1) = \mathbf{21}$
Carlos	5	4	3	$5(3) + 4(2) + 3(1) = \mathbf{26}$

In this instance, Bert is called an irrelevant alternative, because he does not win the election in either situation. If these two voters wanted to ensure that Alicia won, then they would definitely list Carlos after Bert on their ballots in spite of the fact that they might actually despise Bert. In the literature of social choice, we say that they have displayed an **insincere voting** pattern in order to manipulate the election. In other words, they have voted in a manner that does not reflect their true preferences. A voting system that is immune to this type of tampering is said to satisfy the following criterion.

insincere voting
The act of voting in a manner that does not reflect the true wishes of the voter.

independence of irrelevant alternatives (IIA)
In an election that declares candidate A to be the winner, the order on any preference ballot can be changed if it does not affect the ranking of candidate A with respect to candidate B, and B still does not win.

strategic manipulation
Submission of a ballot or block of ballots in which voters have voted insincerely.

> **Independence of Irrelevant Alternatives (IIA).** In an election that declares candidate A to be the winner, the order on any preference ballot can be changed if it does not affect the relative ranking of candidate A with respect to candidate B, and B still does not win.

If savvy voters can affect the outcome of an election by using some predetermined scheme, then that insincere voting method is said to be vulnerable to **strategic manipulation**, and such insincere voting is one way in which that weakness is exploited. By maximizing the distance on a preference ballot between the voter's top choice A and the closest competitor B, the voter reduces the number of points allotted to B. The additional points scored by the irrelevant candidates ranked between A and B have no bearing on the results. This strategy may not reflect the voter's true preferences, but it definitely increases the chances of his or her favorite candidate winning.

? Example 2

Each year, the Heisman Trophy (**Figure 6.1.4**) is awarded to the most outstanding player in college football in the United States. The winner is determined by preference ballots that are cast by a collection of sportswriters and announcers. Each ballot lists the voter's top three choices, and the winner is determined by the Borda count method. The results of the 2009 election were very close and are given in the accompanying table. Each first-place vote is worth 3 points; each second-place vote, 2 points; and each third-place vote, 1 point. Suppose 29 of the voters who placed Mark Ingram in first place also put Colt McCoy in second place and Toby Gerhart in third place. What would have been the results if these voters had switched the positions of Colt McCoy and Toby Gerhart?

© Black Russian Studio/ShutterStock, Inc.

FIGURE 6.1.4   The Heisman Trophy is awarded annually to the best player in American college football.

Player	First	Second	Third	Total
Mark Ingram, Alabama	227	236	151	1,304
Toby Gerhart, Stanford	222	225	160	1,276
Colt McCoy, Texas	203	188	160	1,145
Ndamukong Suh, Nebraska	161	105	122	815
Tim Tebow, Florida	43	70	121	390

Solution

Toby Gerhart would have received 29 more second-place votes and therefore 29 more points. These additional points would have given him a total of 1,305 and therefore ranked him first despite the fact that the altered ballots all retained Mark Ingram as the number one choice. This demonstrates that the Borda count method does not satisfy the IIA criterion.   ◆

A number of other means are available to manipulate ballots that are counted by using the Borda method. In the next section, we will examine another voting scheme along with two more standards by which to measure the validity of all methods.

Name _____

Exercise Set **6.1** ⭘⭘⭘←────────────

1. The Jefferson High School Backpackers Club schedules a 5-day excursion each summer. The 10 members have narrowed their choices for the upcoming summer to the national parks at Denali (D), Yosemite (Y), and Zion (Z), and they decide to vote to determine the final location. The individual ballots are shown here.

First	D	Y	Y	Z	D	Z	Z	Y	D	D
Second	Y	Z	Z	D	Y	Y	Y	Z	Z	Y
Third	Z	D	D	Y	Z	D	D	D	Y	Z

(a) Write the preference table for this election.
(b) Did any of the locations receive a majority of first-place votes in this election?
    If so, which one?
(c) Did any of the locations receive a plurality of the first-place votes in this election?
    If so, which one?

2. By assigning 3 points for first place, 2 points for second place, and 1 point for third place, determine the winner of the election in Exercise 1 by using the Borda count method. Is this different from the plurality winner?

3. The local amateur Astronomy Club decided to have its 12 members vote on their favorite star from the choices Altair (A), Betelgeuse (B), Castor (C), and Deneb (D). The individual ballots are given here.

First	B	B	D	C	A	B	D	C	A	B	C	B
Second	C	A	C	B	D	A	A	B	D	C	B	C
Third	D	C	A	A	B	C	B	A	B	D	A	D
Fourth	A	D	B	D	C	D	C	D	C	A	D	A

(a) How many different preference sequences are possible with four alternatives?
(b) Write the preference table for this election.
(c) Did any of the stars receive a majority of first-place votes in this election? If so, which one?
(d) Did any of the stars receive a plurality of first-place votes in this election? If so, which one?

4. By assigning 4 points for first place, 3 points for second place, and so forth, determine the winner of the election in Exercise 3 by using the Borda count method. Is this different from the plurality winner?

5. The French magazine *France Football* annually awards the Ballon d'Or for the best European footballer of the calendar year. The voters consist of one journalist from each member country of the Union of European Football Associations, and each voter ranks 5 players. The Borda count method is used, and the results for the top 10 players in 2008 are given below. How many points are given for each of the five rankings? (*Source: France Football*, no. 3269, December 2, 2008.)

			Player					
Player	**Country**	**Club**	**1**	**2**	**3**	**4**	**5**	**Total**
Cristiano Ronaldo	Portugal	Manchester	77	11	4	1	3	446
Lionel Messi	Argentina	FC Barcelona	6	33	27	14	10	281
Fernando Torres	Spain	Liverpool	5	13	24	9	12	179
Iker Casillas	Spain	Real Madrid	2	16	12	8	7	133
Xavier Hernández	Spain	FC Barcelona	3	9	4	15	4	97
Andrei Arshavin	Russia	St. Petersburg	—	3	4	10	20	64
David Villa	Spain	Valencia	1	2	4	10	10	55
Ricardo Santos	Brazil	Milan AC	—	4	2	4	1	31
Zlatan Ibrahimovic	Sweden	Inter. Milan	—	1	4	5	4	30
Steven Gerrard	England	Liverpool	—	2	2	5	4	28

6. In the table from Exercise 5, the winner, Cristiano Ronaldo, was in the top 5 on every ballot cast. How many people voted? If the 51 voters who placed Lionel Messi in third, fourth, or fifth place had instead ranked him in second, would he have won?

7. In the 1992 U.S. presidential election, Bill Clinton (C) received 43% of the popular vote; George H. W. Bush (B), 37%; and H. Ross Perot (P), 19%. Suppose that each voter had completed a preference ballot instead of just voting for one candidate and that the resulting preference table was as follows. Determine the winner of the election, using the Borda count method.

	Percentage					
	36	**7**	**31**	**6**	**14**	**5**
First	C	C	B	B	P	P
Second	P	B	P	C	B	C
Third	B	P	C	P	C	B

8. If 53 people vote for the four candidates W, X, Y, and Z, complete the following preference table so that it exhibits a violation of the majority criterion by the Borda count method.

	Number of Ballots		
	27	**16**	**10**
First	Y	X	Z
Second			
Third			
Fourth			

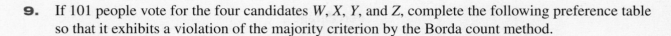

9. If 101 people vote for the four candidates W, X, Y, and Z, complete the following preference table so that it exhibits a violation of the majority criterion by the Borda count method.

	Number of Ballots			
	51	30	11	9
First	X	Y	Z	W
Second				
Third				
Fourth				

10. The Baseball Writers' Association of America uses the Borda count method to choose the Rookie of the Year award recipients in Major League Baseball in the United States. The table here lists the results of the 32 ballots cast for the National League Rookie of the Year in 2009. How many points are given for a first-place rank, second-place rank, and third-place rank? Which player was listed among the top three on every ballot?

Player, Team	First	Second	Third	Total
Coghlan, Marlins	17	6	2	105
Happ, Phillies	10	11	11	94
Hanson, Braves	2	6	9	37
McCutchen, Pirates	2	5	0	25
McGehee, Brewers	1	3	4	18

11. Determine a new allocation of first-, second-, and third-place votes in the table of the previous exercise so that Coghlan still wins the Rookie of the Year award by the plurality method but loses it using the Borda count method.

12. One hundred people are voting for four candidates using the plurality method.

    (a) What is the minimum number of votes a candidate must receive to have a chance of winning?
    (b) If each candidate knows definitively that at least 15 friends are voting for each of them, then what is the minimum number of additional votes that will guarantee a win for any candidate?

13. The town of Cranky Valley has a population of 1,000 people, all of whom vote in every election. Five people are running for mayor, and each candidate knows definitively of 100 relatives who are voting for each of them. If the plurality method is used, then what is the minimum number of additional votes that will guarantee a win for any candidate?

14. In the 1860 U.S. presidential election, Abraham Lincoln (L) received 40% of the popular vote; John Breckenridge (B), 18%; John Bell (J), 13%; and Stephen Douglas (D), 29%. Suppose that each voter had completed a preference ballot instead of just voting for one candidate and that the resulting preference table had been as follows. Determine the winner of the election using the Borda count method.

	Percentage					
	35	**5**	**10**	**8**	**13**	**21**	**8**
First	L	L	B	B	J	D	D
Second	B	D	L	D	B	L	L
Third	J	J	D	L	L	J	B
Fourth	D	B	J	J	D	B	J

15. Manipulate the preference table in Exercise 14 so that Douglas wins the election by the Borda count method, but without altering the relative order of Lincoln and Douglas. What voting criterion does this violate?

16. In the 2000 U.S. presidential election in Florida, the closeness of the contest and various procedural mistakes delayed the tabulation of the results. The final count was amazingly tight, with 48.85% for George W. Bush (*B*), 48.84% for Al Gore, Jr. (*G*), and 1.63% for Ralph Nader (*N*). Suppose an exit poll produced the following preference table. Using this table, determine who would have won the Florida (and therefore the national) election, using Borda counts.

	Percentage					
	30.15	**18.70**	**42.50**	**6.34**	**1.23**	**0.40**
First	B	B	G	G	N	N
Second	G	N	N	B	G	B
Third	N	G	B	N	B	G

17. If preference ballots had been used in the 2000 U.S. presidential election, who do you think would have been elected president? Give reasons for your answer.

18. To be elected to the American Baseball Hall of Fame, a player must appear on at least 75% of the ballots, which are cast by members of the Baseball Writers' Association of America. Is this one of the methods discussed in this chapter? What are some of the strengths and weaknesses of this method?

6.2 The Hare System

Hare system The voting method using preference ballots such that if no candidate has a majority of first-place votes, then the candidate with the fewest first-place votes is eliminated, and all the ballots are adjusted accordingly. This process continues until one candidate receives a majority. Also called *instant runoff*.

instant runoff Another name for the Hare system of voting.

The election process in many countries can actually involve 2 days at the voting booths if the voting system requires that a second runoff election be held whenever no single candidate receives a majority of votes on the first ballot. This system has two major drawbacks. Not only is it markedly more expensive, but also it is almost always the case that voter turnout drops significantly in the second vote due to the additional effort and decreased interest.

The **Hare system**, developed by Thomas Hare in 1861, uses preference ballots just as in the last section, but they are tabulated according to a different scheme. In this system, the candidate with the fewest first-place votes is eliminated on every ballot, and then the preference table is adjusted by advancing each of the candidates ranked under the deleted one. This process continues until one candidate receives a majority of votes. Because everyone presumably would vote for his or her initial second choice in a runoff election (see **Figure 6.2.1**) if the first choice had been eliminated, the Hare method is also referred to as **instant runoff**. Consider the following example.

© Smart7/ShutterStock, Inc.

FIGURE 6.2.1   Runoff elections create many problems.

? Example 1

The public information director of the regional chapter of Students Against Drunk Driving (SADD) has to retire, and the members want a reliable replacement for this important position. They decide to use the Hare system to count the votes for Asif, Barahti, Christina, and Dirk, and the 100 members produced the preference table in **Figure 6.2.2**. Who is the winner?

	Number of Ballots					
	27	**23**	**22**	**13**	**11**	**4**
First	B	D	A	C	C	B
Second	C	A	C	B	D	A
Third	D	B	D	A	A	C
Fourth	A	C	B	D	B	D

FIGURE 6.2.2   Round 1 preference table for the SADD election.

Solution

In the first round, the numbers of first-place votes for each candidate are 31 for Barahti, 24 for Christina, 23 for Dirk, and 22 for Asif. Because no one at this stage has a majority, we note that Asif has the fewest first-place votes, and so we delete his name from Figure 6.2.2. By moving up the candidates ranked below him, we produce the round 2 preference table given in **Figure 6.2.3**.

	Number of Ballots					
	27	**23**	**22**	**13**	**11**	**4**
First	*B*	*D*	*C*	*C*	*C*	*B*
Second	*C*	*B*	*D*	*B*	*D*	*C*
Third	*D*	*C*	*B*	*D*	*B*	*D*

FIGURE 6.2.3   Round 2 preference table for the SADD election.

In this table, Christina gains 22 more votes for a total of 46, but this is still not enough for a majority (51 votes in this case). Because Barahti and Dirk still have 31 and 23 votes, respectively, we delete Dirk from Figure 6.2.3 and move up the candidates ranked below him. This gives the table for round 3, shown in **Figure 6.2.4**.

	Number of Ballots					
	27	**23**	**22**	**13**	**11**	**4**
First	*B*	*B*	*C*	*C*	*C*	*B*
Second	*C*	*C*	*B*	*B*	*B*	*C*

FIGURE 6.2.4   Round 3 preference table for the SADD election.

Figure 6.2.4 shows that Barahti wins the election because he picked up 23 votes to give him a total of 54 while Christina's total remained constant at 46.   ◆

single transferable voting systems (STVS) A voting method in which votes are transferred after a candidate has been eliminated.

Because votes are transferred from one candidate to another in each round, the Hare method is one of a family of **single transferable voting systems** (STVS). (See Exercises 26 and 27 for another example.) When it was first introduced in the nineteenth century, the STVS was hailed as a major advance in voting practices. Although it has not been adopted in the United States for national political elections, it is the system of choice for several countries, including Australia, Iceland, and Ireland. (See Exercises 7–10.) Some American cities, such as San Francisco, Oakland, and Portland, Maine, also have recently adopted it for mayoral elections. In its issue of October 22, 2011, *The Economist* magazine reported that the system had "already produced a big upset" in the election for the mayor of Oakland in 2010. The second- and third-choice votes put the winner over the top even though a better-known candidate had received more first-choice votes. [She] "had knocked on doors and told people she would be happy to be their second choice. When she was their first choice, she suggested another candidate for second. Those other candidates reciprocated, thus keeping the tone unusually civil."

Clearly, the Hare system satisfies the majority criterion and does possess certain other virtues, but unfortunately, it fails one rather curious test. Suppose we return to the original preference table (Figure 6.2.2) and exchange the order of Barahti and Christina in the fourth ballot ranking, as shown in **Figure 6.2.5**. One would probably feel safe in predicting that Barahti would remain the winner, because the only change made was to *elevate* his ranking on one type of ballot.

	Number of Ballots					
	27	**23**	**22**	**13**	**11**	**4**
First	B	D	A	B	C	B
Second	C	A	C	C	D	A
Third	D	B	D	A	A	C
Fourth	A	C	B	D	B	D

FIGURE 6.2.5   Original preference table (Figure 6.2.2) altered with one rank exchange.

monotonicity criterion If candidate X is the winner in an election before her or his ranking is elevated one place on any ballot, then X remains the winner.

However, when we apply the Hare method to this new table, we first eliminate Christina (only 11 first-place votes) to get **Figure 6.2.6**. Asif would be deleted next, resulting in a third round, making Dirk the winner with a total of 56 votes! This is a very counterintuitive result, demonstrating that the Hare method fails to be monotonic. In the language of social decision making, it fails a standard known as the **monotonicity criterion**.

> **Monotonicity criterion.** If candidate X is the winner in an election before his or her ranking is elevated one place on any ballot, then X remains the winner.

	Number of Ballots					
	27	**23**	**22**	**13**	**11**	**4**
First	B	D	A	B	D	B
Second	D	A	D	A	A	A
Third	A	B	B	D	B	D

FIGURE 6.2.6   Round 2 preference table after the rank exchange.

Equally surprising are the results obtained when we apply the Borda count method to either Figure 6.2.2 or Figure 6.2.5. In *both* cases, the winner turns out to be Christina. The point totals for Figure 6.2.2 are as follows.

	Rank				
	First	**Second**	**Third**	**Fourth**	**Borda Count**
Asif	22	27	24	27	$22(4) + 27(3) + 24(2) + 27(1) = \mathbf{244}$
Barahti	31	13	23	33	$31(4) + 13(3) + 23(2) + 33(1) = \mathbf{242}$
Christina	24	49	4	23	$24(4) + 49(3) + 4(2) + 23(1) = \mathbf{274}$
Dirk	23	11	49	17	$23(4) + 11(3) + 49(2) + 17(1) = \mathbf{240}$

In Figure 6.2.5, we observe additionally that Barahti would have won that election by using the plurality method, and so we are faced with the troubling fact that three different voting methods applied to the same set of ballots can yield three different winners! Before we try to determine an optimal voting method, first we present one more common evaluation standard used for judging them.

The **Marquis de Condorcet** (1743–1794) was a French mathematician and economist, as well as a strong advocate of educational reform, who made key contributions to calculus, probability theory, and the philosophy of mathematics. His work was considered to be so significant that he was granted membership in the Académie des Sciences in Paris in 1769. In his most important paper, "Essay on the Application of Analysis to the Probability of Majority Decisions," published in 1785, he established his famous criterion by which all modern voting schemes are measured.

Condorcet (see **Figure 6.2.7**) was a vocal champion of the Republican cause during the French Revolution and, as a result, fell from favor when a more radical group came to power during his term in the Legislative Assembly. He was arrested for his views and died shortly thereafter in prison. He was described in the *Encyclopaedia Britannica*, which stated:

> Wholly a man of the Enlightenment, an advocate of economic freedom, religious toleration, legal and educational reform, and the abolition of slavery, Condorcet sought to extend the empire of reason to social affairs. Rather than elucidate human behaviour by recourse to either the moral or physical sciences, he sought to explain it by a merger of the two sciences that eventually became transmuted into the discipline of sociology.
>
> —H. B. Acton

Condorcet candidate
A candidate associated with a preference table who is preferred by a majority of voters in one-on-one comparisons with each of the other candidates.

Condorcet criterion If one candidate defeats the other candidates in one-on-one comparisons, then that candidate should be the winner of the election.

A candidate who is preferred by a majority of voters in all one-on-one comparisons is called a **Condorcet candidate** for that preference table. It turns out that a preference table exists for each of the plurality, Borda, and Hare methods in which the winner is *not* a Condorcet candidate. Therefore, none of these methods satisfies the **Condorcet criterion** in all cases. This is a serious deficiency and is demonstrated in the following example.

FIGURE 6.2.7   Portrait of Condorcet.

Courtesy of Library of Congress, Prints & Photographs Division.

 Example 2

An election of the team captain for the local soccer club yields the preference table given in **Figure 6.2.8**. Compute the winner by using the Hare system, and determine whether that person is also a Condorcet candidate.

	Number of Ballots			
	7	**6**	**3**	**2**
First	*A*	*C*	*B*	*B*
Second	*B*	*B*	*A*	*C*
Third	*C*	*A*	*C*	*A*

FIGURE 6.2.8   Preference table for the soccer captain election.

 Solution

According to the Hare system, we eliminate candidate *B* in the first round because *B* has the fewest first-place votes (5). After the votes are redistributed, *A* receives 3 more votes, for a total of 10. Because this is a majority, *A* wins the election. However, *B* is the Condorcet candidate, because in one-on-one comparisons, *B* is ranked higher than *A* on 11 out of 18 ballots (often written as $B > A$), and *B* is also ranked higher than *C* on 12 out of 18 ballots ($B > C$). ◆

This example shows that the Hare system does not satisfy the Condorcet criterion. Analogous demonstrations for the plurality and Borda methods are explored in the exercises. We also note that it is possible for a preference table to not even have a Condorcet candidate. Even more remarkable is the seemingly infeasible situation exemplified by a three-candidate election in which candidate A is preferred to B and B is preferred to C, but C is preferred to A. This is demonstrated by the simple case in **Figure 6.2.9**, which is representative of any election where the ballots are divided evenly among a symmetric set of rankings.

	Number of Ballots		
	1	1	1
First	A	B	C
Second	B	C	A
Third	C	A	B

FIGURE 6.2.9   A preference table showing the Condorcet paradox.

According to the table in Figure 6.2.9, two-thirds of the voters favor A to B, two-thirds favor B to C, and two-thirds favor C to A. This runs counter to the manner in which an individual person typically relates preferences among a set of three choices. For instance, if Samantha prefers chocolate to vanilla and vanilla to strawberry, then she almost certainly prefers chocolate to strawberry. This characteristic of individual preference is called the transitive property, and the example in Figure 6.2.9 shows that the collective preferences of a large population of voters do not have to behave transitively. In the theory of social choice, this is called **Condorcet's paradox**, and the even distribution of circular rankings displayed on the three ballots in Figure 6.2.9 is called a *cycle* in majority relationships. Notice that this cycle can be expressed as $A > B > C > A$. An **n-cycle** (n being an integer greater than 2) involves n candidates and may emerge from any subset of candidates in an election. The occurrence of an n-cycle precludes the existence of a Condorcet candidate in an election with n choices.

Condorcet's paradox An election involving n candidates in which an n-cycle occurs.

n-cycle A pattern of circular rankings of n candidates in an election.

Example 3

Find the cycles in the preference table in **Figure 6.2.10**.

	Number of Ballots			
	4	3	1	2
First	A	D	C	B
Second	B	C	D	C
Third	C	A	B	D
Fourth	D	B	A	A

FIGURE 6.2.10   A four-candidate preference table containing a 3-cycle.

Solution

The six possible pairwise comparisons result in $A > B$, $B > C$, $B > D$, $C > A$, $C > D$, and $D > A$. These yield two 3-cycles, $A > B > C > A$ and $A > B > D > A$, and one 4-cycle, $A > B > C > D > A$. This election has no Condorcet candidate.   ◆

Name _____

Exercise Set **6.2** ⬜⬜⬜←━━━━━━━━━━━━━━━━━━━

1. The Marywood University football team must choose from offers to holiday bowl games in Miami (M), Phoenix (P), and San Diego (S). The preference table is shown here. Use the plurality, Hare, and Borda methods to compute the winner. Are the answers different?

	Number of Ballots			
	19	16	12	7
First	M	S	P	S
Second	S	P	M	M
Third	P	M	S	P

2. Rearrange the preference table in Exercise 1 so that the following two conditions result: (1) The same winner is produced by either the Borda or Hare voting method, and (2) a different winner is produced by the plurality method.

3. The 20 members of the Franklin University debate team voted for Wanita (W), Xavier (X), Yoshi (Y), or Zoe (Z) for their team captain and produced the following preference table. Use the plurality, Hare, and Borda methods to compute the winner. Are the winners different?

	Number of Ballots			
	7	5	5	3
First	Y	Z	X	W
Second	X	Y	Z	X
Third	W	X	W	Z
Fourth	Z	W	Y	Y

4. Rearrange the preference table in Exercise 3 so that the following two conditions result: (1) The same winner is produced by either the Borda or Hare voting methods, and (2) a different winner is produced by the plurality method.

5. Use the Hare system to determine the winner of the election with the following preference table. Exchange the order of two candidates in one column to demonstrate that the Hare system violates the monotonicity criterion.

	Percentage of Ballots					
	18	20	22	10	20	10
First	A	A	B	B	C	C
Second	B	C	C	A	A	B
Third	C	B	A	C	B	A

6. Use the Hare system to determine the winner of the election with the following preference table. Exchange the order of two candidates in one column to demonstrate that the Hare system violates the monotonicity criterion.

	Percentage of Ballots			
	30	**28**	**27**	**15**
First	X	Y	Z	Z
Second	Y	X	Y	X
Third	Z	Z	X	Y

The Hare system of voting is used for political elections in Australia. The following commentary and tables come from the Australian Electoral Commission website (www.aec.gov.au).

To vote for a Member of the House of Representatives, an elector is required to write the number "1" in the box next to the candidate who is their first choice, and the numbers "2," "3," and so on against all the other candidates, in order of the elector's preference.

A candidate receiving more than 50% of the first preference votes is immediately elected. However, if no candidate gains 50% of first preferences, the one with the fewest votes is excluded. That candidate's votes are then transferred to the other candidates according to the preferences shown. This process continues until one candidate has more than 50% of the votes and is declared elected.

For example, the table shows the number and percentages of votes received by each candidate for the division of Namadgi (ACT) at each stage of the scrutiny for the election. In this case, the candidate who received the most first preference votes (Brendan Smyth—45.14%) actually lost when preferences were distributed.

The first count row shows the first-preference votes of each candidate, the total number of first-preference votes, and each candidate's votes as a percentage of total votes. The subsequent rows show how votes were distributed from excluded candidates, the percentage of excluded candidates' and continuing candidates' total votes after the distribution of preferences, and each candidate's votes as a percentage of total votes.

	RATTENBURY Shane (GRN)		ELLIS Annette (ALP)		ROSBOROUGH Derek (IND)		SMYTH Brendan (LP)			
Count	**Votes**	**%**	**Votes**	**%**	**Votes**	**%**	**Votes**	**%**	**Total**	**Exh.**
First	4,579	7.22	28,583	45.06	1,636	2.58	28,638	45.14	63,436	—
Second	716	44.78	399	24.95	Excluded		484	30.27	1,599	37
Total	**5,295**	**8.35**	**28,982**	**45.71**			**29,122**	**45.93**	**63,399**	**37**
Third	Excluded		3,560	70.27			1,506	29.73	5,066	229
Total			**32,542**	**51.51**			**30,628**	**48.49**	**63,170**	**266**
			Elected							

The candidate with the majority of total votes (51.51%), Annette Ellis, was elected.

7. Why was Rosborough eliminated after the first round? Of the 1,636 ballots that listed him first, what number of them listed Rattenbury second? Ellis? Smyth? What is an Exh. (exhausted) ballot?

8. Can a preference table be constructed that corresponds to the information given for the election shown in Exercise 7? Explain your answer.

9. Why was Rattenbury eliminated after the second round? Of the 5,295 ballots that listed him first or second, what number listed Ellis next? Smyth?

10. Who would have won the election shown in Exercise 7 if the plurality method had been used? Can the Borda count method be used based on the information given for the above election?

11. The faculty at Progressive University has always used the plurality method for the selection of members to its Curriculum Committee. The Elections Committtee has decided to let the faculty elect a voting method from the plurality (P), Borda (B), and Hare (H) systems. Of course, the question is, What method should the faculty use to tally the votes?! It decides to use all three methods to count the votes and then hopes that at least two of the procedures produce the same winner. The following preference table shows the percentage of the faculty that cast each type of ballot.

Percentage of Ballots					
28	12	27	9	16	8
First					
B	B	P	P	H	H
Second					
H	P	H	B	P	B
Third					
P	H	B	H	B	P

(a) Determine the winner, using the plurality method.
(b) Determine the winner, using the Borda method.
(c) Determine the winner, using the Hare method.
(d) Does this preference table have a Condorcet candidate?

12. Change the percentages in the table in Exercise 11 so that H wins, using either the plurality or Borda method.

13. Change the percentages in the table in Exercise 11 so that B wins, using either the plurality or Hare method.

14. Change the percentages in the table in Exercise 11 so that P wins, using either the Borda or Hare method.

15. Is the Hare system independent of irrelevant alternatives? Create an example that demonstrates your answer.

16. In an election with only two candidates, would all three voting methods discussed in this chapter satisfy the monotonicity criterion? Condorcet criterion?

17. Does the following election have a Condorcet candidate? Who is it? Who is the plurality winner? Does the plurality method satisfy the Condorcet criterion?

	Number of Ballots		
	10	**9**	**7**
First	A	B	D
Second	C	C	C
Third	B	D	A
Fourth	D	A	B

18. Create your own preference table for an election with three candidates, demonstrating that the plurality method violates the Condorcet criterion.

19. Suppose candidate X wins an election using the Borda method and then the preference table is altered by raising the ranking of X on any ballot.

   (a) Would the total points accumulated by X increase or decrease?
   (b) Would the points of any other candidate increase?
   (c) Does X still win the election?
   (d) Does the Borda method satisfy the monotonicity criterion?

20. This table lists the final top 10 teams of the *USA TODAY*/ESPN college men's basketball coaches' poll in the United States in 2013. The total points are based on 25 points for first place, 24 points for second place, . . ., up through 1 point for twenty-fifth place. Also, it shows that Louisville received 31 first-place votes.

Rank	School	Record	Points
1	Louisville (31)	34–5	775
2	Michigan	31–7	744
3	Syracuse	30–10	696
4	Wichita State	30–9	643
5	Duke	30–6	607
6	Ohio State	29–8	594
7	Indiana	29–7	568
8	Kansas	31–6	533
9	Florida	29–8	530
10	Miami	29–7	487

Data from *USA Today*

   (a) Did Louisville receive any votes that ranked it lower than first place?
   (b) Is it possible to determine the winning team from this information by using the Hare system?
   (c) Is it possible to determine a Condorcet candidate from this information? If so, which team is it?

21. We consider a simpler version of the ranking in the previous exercise by considering a preference table of ballots for the top five teams from Arizona (A), Duke (D), Illinois (I), Kentucky (K), and Purdue (P).

Percentage of Ballots

	35	32	23	10
First	A	P	D	D
Second	I	A	P	A
Third	K	I	A	I
Fourth	D	K	I	K
Fifth	P	D	K	P

(a) Using a Borda count, which team would be ranked first?
(b) Does this table have a Condorcet candidate?
(c) Find a 4-cycle in this table.
(d) Does this table contain a 5-cycle?

22. Construct a simple preference table that contains a 4-cycle in an election with four candidates. In every one-on-one comparison, what fraction of the voters prefers the leading candidate?

23. Construct a simple preference table that contains a 5-cycle in an election with 5 candidates. In every one-on-one comparison, what fraction of the voters prefers the leading candidate?

24. Create a preference table with 4 candidates that contains a 3-cycle but not a 4-cycle.

25. The International Olympic Committee selects the city to host the Olympics by voting according to a variation of the Hare method. Multiple rounds of voting are used, and after each round, the city with the fewest votes is eliminated. The members vote repeatedly for their favorite remaining city until one city has a majority. Rio de Janeiro was elected as the host city of Games of the XXXI Olympiad in 2016 as a result of the following voting.

Round 1		Round 2		Round 3	
Madrid	28	Rio de Janeiro	46	Rio de Janeiro	66
Rio de Janeiro	26	Madrid	29	Madrid	32
Tokyo	22	Tokyo	20		
Chicago	18				

Did the same number of members vote in each round? What could not have happened in this election if preference ballots had been used? (*Source*: www.olympic.org.)

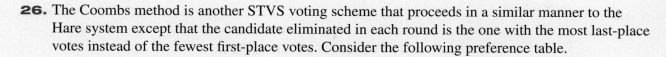

26. The Coombs method is another STVS voting scheme that proceeds in a similar manner to the Hare system except that the candidate eliminated in each round is the one with the most last-place votes instead of the fewest first-place votes. Consider the following preference table.

	Number of Ballots			
	7	**3**	**3**	**2**
First	A	B	C	C
Second	B	C	B	A
Third	C	A	A	B

(a) Determine the winner by using the Hare system.
(b) Which candidate is eliminated first by using the Coombs method?
(c) Determine the winner by using the Coombs method.

27. Does the Coombs method satisfy the majority criterion? Monotonicity criterion?

6.3 Approval Voting

In the last two sections, we described several voting methods and four standards of fairness with which to evaluate those methods. These standards are among those commonly used by social scientists to assess various types of social-choice functions and are repeated as shown here.

FAIRNESS CRITERIA

1. **Majority criterion.** If a candidate receives the majority of the first-place votes in an election, then that candidate is the winner.
2. **Independence of irrelevant alternatives.** In an election that declares candidate A to be the winner, the order on any preference ballot can be changed if it does not affect the ranking of candidate A with respect to candidate B, and B still does not win.
3. **Monotonicity criterion.** If candidate X is the winner in an election before his or her ranking is elevated one place on any ballot, then X remains the winner.
4. **Condorcet criterion.** If one candidate defeats all the other candidates in one-on-one comparisons, then that candidate should be the winner of the election.

> **Arrow's impossibility theorem** It is mathematically impossible for a democratic voting method to satisfy all the fairness criteria.

We have seen that each of the voting methods explored violates at least one of the fairness criteria in this list. It is reasonable to ask if there does, in fact, exist a democratic voting system that is perfectly fair. Remarkably, the answer to this question was mathematically proved to be no in a groundbreaking paper written by the economist Kenneth J. Arrow in 1951 (see **Figure 6.3.1**). His result, stated here, is known as **Arrow's impossibility theorem**, and it was one of the reasons he received the Nobel Prize in Economics in 1972 and the National Medal of Science in 2004.

Arrow's impossibility theorem. It is mathematically impossible for any voting method to satisfy all reasonably defined standards of fairness.

FIGURE 6.3.1 Kenneth Arrow (1921–).

Arrow's work ushered in an entire new approach to social-choice theory that combines mathematics, economics, and the political and social sciences. Although the proof of Arrow's theorem is well beyond the scope of this book, we examine one of the contradictions that could occur if we assumed that some perfectly fair voting scheme (call it V) exists that will produce at least one winner. Suppose we use V to count the votes in an election with three candidates A, B, and C. We have seen that it is quite possible that this election could produce a preference table in which V yields a winner, say A, that is a Condorcet candidate even though A did not receive a majority of first-place votes. Because V has to be independent of irrelevant alternatives, we know that if we exchange the rankings of B and C, then C cannot win even if C gains a majority. However, if C does have a majority after a rank exchange with B on any ballot, then it would have to be true that C was preferred to A in a one-on-one comparison in the original preference table. This contradicts the assumption that *A* is a Condorcet candidate. Therefore, this is a preference table for *V* in which it is impossible for all the fairness criteria to hold at the same time. So *V* cannot be perfectly fair.

Although Arrow's impossibility theorem eliminates the possibility of an ideal voting system in multicandidate elections (according to our definition of fairness), we can still search for a method that is a "best fit" to an ideal in the sense that its failure to meet all the criteria is less extreme than that of any other method. Of course, the right method for one type of election may not be appropriate for a different type. Selecting a U.S. senator or a state governor from a large voting populace, for example, would certainly have a different list of desirable requirements than voting for a CEO of a major corporation or the president of your neighborhood gardening club. For practical considerations alone, it is safe to say that implementing any type of ranked balloting in a national election with millions of voters would be a daunting task.

Regardless of the type of election, however, most social theorists would probably agree that the prime characteristic of any best voting method should be resistance to manipulation. In other words, *voting sincerely should always be the voter's optimal strategy.* We have already seen that any voting scheme utilizing preference ballots is subject to varying degrees of manipulation. (The Borda count, for instance, encourages ranking the closest competitor to your top choice at the bottom even if you prefer that competitor over other candidates.) However, one method does exist that has been shown in other texts to foster sincere voting in many common scenarios. In this method, known as **approval voting**, the voter simply casts a vote for every candidate she or he considers to be acceptable. A so-called approval vote can be made for more than one candidate, but no ranking is indicated. Hence, the procedure still allows for voter input on every candidate, but in the form of a thumbs up (approval) or a thumbs down (no approval). The winner is the candidate who receives the most approval votes. This voting system has many advocates among social theorists for a variety of reasons and has been adopted by a number of large professional organizations such as the Institute of Electrical and Electronic Engineers, the American Mathematical Society, and the National Academy of Sciences. We will consider the advantages after looking at a few examples.

approval voting
The voting method in which each person votes for as many candidates as he or she finds acceptable. The winner is the candidate with the most approval votes.

Approval Ballot	
Baker	
Dempsey	X
Foster	X
Nailon	
Peterson	X

? Example 1

The Riverdale High School band used approval voting to elect its president, and the results are displayed in the following table. Each X indicates an approval vote.

	Number of Ballots				
	6	**4**	**3**	**3**	**1**
Adena	X		X		
Mohat	X	X			X
Jong-Sei			X	X	X

(a) Who won this election?
(b) What percentage of the voters approved of each candidate?

⚙ Solution

(a) Adena received 9 votes, Mohat received 11 votes, and Jong-Sei received 7 votes. Therefore Mohat is the winner.

(b) Adena is acceptable to 53% (9 out of 17) of the voters, Mohat to 65% (11 out of 17), and Jong-Sei to 41% (7 out of 17). Clearly, the percentages do not have to sum to 100%. ◆

---

dichotomous preference
A division of candidates by a voter into two subsets of preferred and nonpreferred choices. The voter considers the members within each particular subset to be equally qualified.

We can see that the basic premise of approval voting is that each voter makes a separate judgment on each candidate as to whether that person can do an acceptable job if elected. In other words, the voter divides the list of candidates into two subsets of preferred and nonpreferred choices. If the voter considers the candidates in each of these subsets to be equally qualified, then this binary division of choices is known as a **dichotomous preference**. It has been shown that if all voters have dichotomous preferences and if every voter does indeed vote for every member in her or his subset of preferred choices, then approval voting is a system in which the best strategy by each voter is to vote sincerely.

❓ Example 2

In the preference table shown here, who would win by using the plurality method? Create an approval table by marking an approval vote for a first-place or second-place rank. Who is the winner according to this table?

	Number of Ballots			
	9	**7**	**6**	**3**
First	B	A	C	A
Second	C	B	B	C
Third	A	C	A	B

⚙ Solution

Candidate *A* would win by the plurality method even though 15 of the 25 voters ranked *A* as their third-place choice. The corresponding approval table giving votes for first-place and second-place ranks is the following.

	Number of Ballots		
	15	**7**	**3**
A		X	X
B	X	X	
C	X		X

The winner is candidate *B* with 22 approval votes. ◆

Example 2 demonstrates one of the advantages of approval voting. In the preference table, we see that candidate *B* was ranked last by only 3 votes as opposed to 7 last-place votes for *C* and 15 for *A*. So, of all the candidates, most people considered *B* to be the least unacceptable. One can interpret this to mean that the candidate most voters would be willing to support if their first choice did not get elected is candidate B. Therefore, a voter who wishes to support a minor-party candidate can do so without "wasting a vote" because he or she can still indicate equal support of a major-party candidate who has a more realistic chance of winning. This characteristic of approval voting encourages not only minor-party participation but voter turnout as well. These are both very good things in a democratic society.

Just as in ranked voting, the use of multiple votes on a ballot guarantees fuller expression of the voter's political judgment. However, in contrast to ranked voting, the fact that each approval vote is of equal value implies that a voter is not forced to choose one of several equal candidates to designate as a first choice. This can encourage voter participation and also reduces the extent to which appropriate fairness standards are violated. In particular, criteria such as the independence of irrelevant alternatives and the monotonicity criterion are strictly related to rankings and therefore are no longer applicable. In addition, the use of negative campaigning would probably be reduced because overt criticism of an opponent could cost a candidate secondary support from voters who consider that opponent a firm top choice.

One final observation is that if all voters in an election have dichotomous preferences and vote for their preferred choices, then *approval voting must elect the Condorcet candidate*. This is true because if candidate A has received the most approval votes, then A is in the preferred subset of the most voters and so would have to win a one-on-one contest with any other candidate. Satisfying the Condorcet criterion is an attribute long considered to be of primary importance in any optimal voting system.

? Example 3

Does the following preference table have a Condorcet candidate? Create an approval table by assuming that all voters have dichotomous preferences whose preferred subsets consist of their first two ranked choices. Does the new table have a Condorcet candidate?

	Number of Ballots				
	11	8	5	4	4
First	B	A	C	C	D
Second	A	C	A	D	B
Third	C	B	D	B	C
Fourth	D	D	B	A	A

⚙ Solution

Pairwise comparisons indicate no Condorcet candidate in the preference table. The corresponding approval table is as follows.

	Number of Ballots			
	11	13	4	4
A	X	X		
B	X			X
C		X	X	
D			X	X

Because A received the most approval votes (24), A is the Condorcet candidate in the approval table and the winner of the election. ◆

One additional positive feature of approval voting is that it would be extremely unlikely for a winner not to be acceptable to the majority of the electorate. For example, the only U.S. president to ever receive less than 40% of the popular vote was Abraham Lincoln (39.8%) in the election of 1860. Under approval voting, a candidate who has been ranked first by 40% of the voters would only have be acceptable to an additional 10% (one-sixth of the remaining 60%) in order to achieve at least a majority of approval. Voters would probably be quite encouraged to know that the winning choice is acceptable to most of them. We summarize the advantages of approval voting in the table in **Figure 6.3.2**. However, note that approval voting is not the favored method among all voting theorists. Like all the methods we have examined, it also has its share of flaws.

- The best strategy for every voter is to vote sincerely.
- The winner is acceptable to the largest portion of the electorate.
- Voter turnout is encouraged.
- Negative campaigning is discouraged.
- The outcome is not vulnerable to monotonicity or irrelevant candidates.
- The issues and candidates of minor parties would probably receive greater attention.
- The method is easy to understand and implement.

FIGURE 6.3.2   Advantages of approval voting.

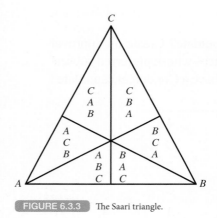

FIGURE 6.3.3   The Saari triangle.

We conclude this chapter with an introduction to a method for geometrically organizing the information in a preference table that displays the results of a ranked voting election involving three candidates. This useful tool was created by Donald G. Saari and uses an equilateral triangle with vertices corresponding to candidates A, B, and C. The triangle is subdivided into six regions by drawing the altitudes from each vertex to the opposite side. Each region has an integer assigned to it equal to the number of ballots for a particular ranking, as shown in **Figure 6.3.3**. For example, the number of ballots ranked ABC is placed in the region closest to A, next closest to B, and farthest from C. Likewise, the number of ballots with rank ACB is placed in the region closest to A, next closest to C, and farthest from B. The other regions are similarly labeled.

Displaying the ballot information in this way allows us to quickly determine both the winners according to the plurality, approval (vote-for-two), and Borda methods and any Condorcet candidate. For ease of computation, the Borda count will use an allotment of 2 points for each first-place vote, 1 point for a second-place vote, and 0 points for a third-place vote. (This distribution will produce the same Borda winner as with 3-2-1.) The following example demonstrates the procedure.

Example 4

Form the Saari triangle corresponding to the following preference table. Use it to find the election winners according to the plurality, approval (vote-for-two), and Borda methods and also to determine whether the table contains a Condorcet winner.

	11	3	6	5	4	8
First	A	A	B	B	C	C
Second	B	C	A	C	A	B
Third	C	B	C	A	B	A

⚙ Solution

The total number of first-place votes for each candidate is found from the Saari triangle by summing the numbers in the two regions closest to each vertex. For instance, the red-shaded regions in **Figure 6.3.4** indicate that candidate A received 3 + 11 = 14 first-place votes. The approval (vote-for-two) total for A must then include the two next closest regions to that vertex because they give the number of ballots that ranked A second. Therefore, we see A received 10 second-place votes by adding together the 4 and the 6 seen in green in Figure 6.3.4. Because we are assuming that approval votes are given to each ballot's top two choices, we add 14 to 10 to get 24 approval votes for A. Finally, the sum of the numbers in the first two rows

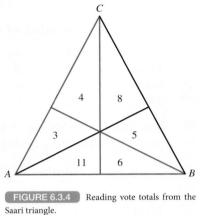

FIGURE 6.3.4 Reading vote totals from the Saari triangle.

must necessarily be the Borda count for each candidate because then each first-place vote gets counted twice and so produces the correct sum using a 2-1-0 point distribution. The vote count for each candidate by each method is given in the following table. We see that A was the plurality winner but that B was the approval and Borda winner.

	A	B	C
Plurality	14	11	12
Approval	24	30	20
Borda	38	41	32

To determine the Condorcet winner, we must find the dominant candidate in each one-on-one comparison. This is done for A and B by dividing the triangle in half by the altitude to the side joining A and B. The sum 4 + 3 + 11 = 18 on the half nearer to A (seen in red in **Figure 6.3.5**) is the number of voters who preferred A to B, and the sum 8 + 5 + 6 = 19 (seen in green in Figure 6.3.5) is the number of voters who preferred B to A. Therefore we have B > A. You can check in a similar fashion that A > C by a count of 20 to 17 and B > C by a count of 22 to 15. Thus B is a Condorcet winner. ◆

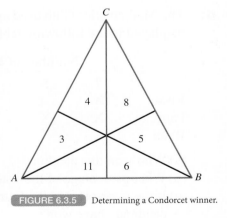

FIGURE 6.3.5 Determining a Condorcet winner.

The Saari triangle makes clear how the election outcomes would change due to modifications of the values of the various regions in the triangle. Shifts in voting patterns could be easily implemented, and the new results observed with little effort. It is a wonderful example of using the power of mathematical objects such as a simple triangle to help categorize and display information in a visual manner that is quite conducive to analysis.

Name _____

Exercise Set **6.3** ◯◯◯←──────────────────────

1. Suppose the voting method VM has declared A to be the winner of an election with the following preference table. Is VM a perfectly fair voting system? What criterion does it violate?

	Number of Ballots		
	3	2	2
First	A	B	C
Second	B	A	B
Third	C	C	A

2. Can you assign values to x, y, and z in the table shown here so that the winner is chosen by a perfectly fair voting system?

	Number of Ballots		
	x	y	z
First	A	B	C
Second	B	C	A
Third	C	A	B

3. Suppose the voting system VS has declared B to be the winner of an election with the following preference table. Is VS a perfectly fair voting system? What criterion does it violate?

	Number of Ballots			
	3	4	2	2
First	B	C	B	C
Second	A	B	C	A
Third	C	A	A	B

4. The Mathematics Club used approval voting to elect its president, and the voting results are displayed in the following table.

	Number of Ballots			
	7	5	3	2
Elsa	X		X	X
Patrick	X	X		X
Ryan			X	X

(a) Who won this election?
(b) What percentage of the voters approved of each candidate?
(c) If the two voters in the fourth column had not approved of any candidates, would the same candidate have won?

5. The Oregon Ecology Society used approval voting to elect its chair, and the voting results are displayed in the following table.

	Number of Ballots				
	10	7	6	4	1
Harrison	X		X	X	
Ajuin	X	X			
Elizabeth		X		X	

(a) Who won this election?

(b) What percentage of the voters approved of each candidate?

(c) If the voter in the rightmost column had approved of all the candidates, would the same candidate have won?

6. The option of a voter to cast no approval votes in a given election is called the *abstention strategy*. Why is the abstention strategy equivalent to the strategy of casting an approval vote for every candidate?

7. In an election with two candidates, A and B, there are three possible preferred subsets—$\{A, B\}$, $\{A\}$, and $\{B\}$—that represent approval for both A and B, approval for A but not for B, and approval for B but not for A, respectively. (No approval for either A or B is equivalent to approval for both. See the previous exercise.) List all possible nonequivalent preferred subsets that exist for an election with 3 candidates, 4 candidates, and n candidates.

8. Although unlikely, it is possible that the winner of an election using approval voting may not be acceptable to the majority of voters. Construct an approval table that illustrates this fact.

9. Three people are running for mayor of a small town. Of the three, candidate A is considered to be liberal; B, centrist (middle-of-the-road); and C, conservative. This is a progressive town that uses approval voting in all its elections. A poll taken 2 months before the election determined that the voters all had dichotomous preferences, with the following percentages of preferred subsets.

$\{A, B\}$	29%	$\{A\}$	18%
$\{B, C\}$	20%	$\{B\}$	12%
$\{A, C\}$	4%	$\{C\}$	17%

Candidate B is currently leading, and A is in second place. To which of the above six blocks of voters do you think A should try to appeal in order to swing enough voters to win the election? Explain your reasoning.

10. In negative approval voting, each person casts a ballot by indicating only those candidates of whom she or he disapproves. The candidate with the fewest negative votes wins. Do you think this method would differ from approval voting in any significant way?

11. In the preference table shown here, who would win by using the plurality method? Create an approval table by marking an approval vote for a first-place or second-place rank. Who wins according to this table?

	Number of Ballots			
	6	**3**	**2**	**2**
First	B	A	B	C
Second	D	D	C	D
Third	C	B	D	A
Fourth	A	C	A	B

12. In the preference table shown here, who would win by using the plurality method? Create an approval table by marking an approval vote for a first-place or second-place rank. Who wins according to this table?

	Number of Ballots			
	7	11	5	4
First	D	A	C	C
Second	B	D	A	A
Third	A	B	B	D
Fourth	C	C	D	B

13. Fill in the blanks in this example of the validity of Arrow's impossibility theorem.

Assume that a *V* is a voting scheme that satisfies the four fairness criteria we have examined and will produce at least one winner in an election with three candidates, *A*, *B*, and *C*. It is possible that this election could produce a preference table *T* in which *V* yields a winner, say *A*, that is a Condorcet candidate even though *A* did not receive a majority of first-place votes. Because *V* has to be _____ of irrelevant alternatives, we know that if we exchange the rankings of _____ with those of *C*, then _____ cannot win even if *C* gains a majority. However, if *C* does have a majority of first-place votes after a rank exchange with *B* on any ballot, then it would have to be true that *C* was preferred to *A* in a(n) _____ comparison in the original preference table *T*. This contradicts the assumption that *A* is a(n) _____ candidate. So this is a preference table for *V* in which it is impossible for all the fairness criteria to hold at the same time. Therefore, *V* cannot be _____.

Draw the Saari triangle that represents each of the following preference tables.

14.

	Ballots					
	12	9	8	4	3	1
First	A	C	B	C	B	A
Second	B	A	C	B	A	C
Third	C	B	A	A	C	B

15.

	Ballots				
	17	12	9	7	4
First	B	C	A	A	C
Second	A	A	B	C	B
Third	C	B	C	B	A

16.

	Ballots					
	28	25	20	11	9	7
First	C	A	B	B	C	A
Second	B	C	A	C	A	B
Third	A	B	C	A	B	C

Determine the plurality, approval (vote-for-two), and Borda winners for each of the following Saari triangles. Also, determine whether the triangle has a Condorcet winner.

17.

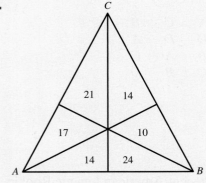

18.

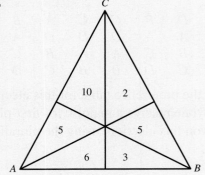

19.

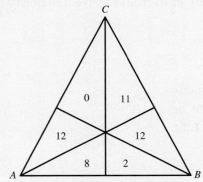

Name _____

use the extra space
to show your work

— Chapter Review Test **6** ☐☐☐ ←———————————————————————

1. The 12 members of the Students for Democracy club used preference ballots to vote for their president from the choices Allen (*A*), Boris (*B*), Catherine (*C*), and Dali (*D*). The individual ballots are given here.

First	*D*	*B*	*C*	*C*	*A*	*B*	*D*	*C*	*A*	*A*	*C*	*B*
Second	*A*	*A*	*D*	*D*	*D*	*A*	*A*	*D*	*D*	*D*	*D*	*A*
Third	*B*	*C*	*A*	*A*	*B*	*C*	*B*	*A*	*B*	*B*	*A*	*C*
Fourth	*C*	*D*	*B*	*B*	*C*	*D*	*C*	*B*	*C*	*C*	*B*	*D*

(a) Create the preference table for this election.
(b) Did anyone receive a majority of first-place votes?
(c) Who won this election, using the plurality method?

2. The article shown here displays the results of the voting for a recent American League Rookie of the Year. What voting method was used? Did the winner (Angel Berroa) receive a majority of first-place votes?

AMERICAN LEAGUE

Close Call

Royals shortstop Angel Barroa narrowly edged out Yankee Hideki Matsui to be the American League Rookie of the Year.

Average	.287
HR	17
RBI	73

Voting

First-, second-, and third-place votes and total points on a 5-3-1 basis.

Player	1st	2nd	3rd	Total
Angel Berroa	12	7	7	88
Hideki Matsui	10	9	7	84
Rocco Baldelli	5	5	11	51
Jody Geru	0	6	2	20
Mark Teixeira	1	1	1	9

3. In the table in the article shown above, suppose that 5 of the voters who ranked Berroa first also ranked Baldelli second and Matsui third. Show the new preference table that occurs if those 5 voters switch their placements of Baldelli and Matsui. Who wins with this new table? What fairness criterion does this violate?

4. The voting for the captain of the swim team produced the following preference table. Use the plurality, Borda, and Hare methods to compute the winner.

	Number of Ballots			
	10	8	6	3
First	Y	Z	X	Z
Second	X	X	Y	Y
Third	Z	Y	Z	X

5. The results of the voting for the 2008 NBA Sixth-Man-of-the-Year Award are given here. The winner was determined by using a Borda count. What point allotments were used for first place, second place, and third place?

Player	First Place	Second Place	Third Place	Total
Manu Ginobili, San Antonio	123	0	0	615
Leandro Barbosa, Phoenix	1	84	26	283
Jason Terry, Dallas	0	9	17	44
Kyle Korver, Utah	0	7	13	34
Ben Gordon, Chicago	0	6	9	27

6. Name the four fairness criteria. Which of these does the Borda count method satisfy?

7. If all voters have dichotomous preferences and they all vote for their preferred candidates, then which voting method must elect the Condorcet candidate?

8. In a recent polling of sportswriters to name the top team in college basketball, the following percentages of ballots resulted in the ranking of Connecticut (*C*), Illinois (*I*), Purdue (*P*), and Xavier (*X*).

Percentage of Ballots			
31	**27**	**25**	**17**
P	*I*	*X*	*C*
C	*X*	*P*	*P*
I	*P*	*C*	*X*
X	*C*	*I*	*I*

(First, Second, Third, Fourth)

(a) Does this table contain a 4-cycle? If so, what is it?
(b) Does this table contain a 3-cycle? If so, what is it?
(c) Does this table have a Condorcet candidate?

9. Complete this preference table of candidates *A*, *B*, and *C* so that it demonstrates a violation of the majority criterion by the Borda count method.

Number of Ballots		
23	**15**	**7**
A	*B*	*C*

(First, Second, Third)

10. The members of the Pinecrest University Ski Club voted on their choice of mountain for the spring ski trip. The choices were Killington, Greek Peak, Elk Mountain, and Hunter Mountain. They used approval voting, and the results are shown below.

	Number of Ballots						
	17	14	11	9	6	2	1
Killington	X		X		X	X	X
Greek Peak	X	X			X		X
Elk Mountain		X	X	X			X
Hunter Mountain	X		X	X		X	X

(a) Where are they headed for the trip?

(b) If the voter in the rightmost column had not approved of any of the candidates, would the results have been different?

11. Fill in the blanks in this example of the validity of Arrow's impossibility theorem.

Assume that V is a voting scheme that satisfies the four fairness criteria we have examined and will produce at least one winner in an election with three candidates, A, B, and C. It is possible that this election could produce a preference table T in which V yields a winner, say A, that is a Condorcet candidate even though A did not receive a majority of first-place votes. Because V has to be independent of _____ alternatives, we know that if we exchange the rankings of B with those of _____, then _____ cannot win even if C gains a majority. However, if C does have a majority of first-place votes after a rank exchange with B on any ballot, then it would have to be true that C was preferred to A in a(n) _____ comparison in the original preference table T. This contradicts the assumption that A is a(n) _____ candidate. So this is a preference table for V in which it is _____ for all the fairness criteria to hold at the same time. Therefore, V cannot be perfectly fair.

12. Determine the plurality, approval (vote-for-two), and Borda winners of the election represented by this Saari triangle. Find a Condorcet winner if it exists.

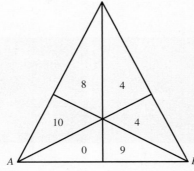

Chapter Objectives

check off when you've completed an objective

- [ ] Construct a sample space.
- [ ] Identify an event as a subset of a sample space.
- [ ] Apply operations of union, intersection, and complement to sets.
- [ ] Construct Venn diagrams.
- [ ] Know and use DeMorgan's laws for sets.
- [ ] Know and use the addition principle.
- [ ] Know the definition of probability.
- [ ] Determine the probability of an event.
- [ ] Know and use the properties of probability.
- [ ] Compute the expected value of an experiment.

Navigate Companion Website

go.jblearning.com/johnson

Visit go.jblearning.com/Johnson to access a Laboratory Manual, Student Solutions Manual, Interactive Glossary, and Interactive Flashcards.

chapter

7

Probability

7.1 Sample Spaces and Events

7.2 Introduction to Sets

7.3 Basic Probability

> They sit with papers before them scrawled over in pencil, note the strokes, reckon, deduce the chances, calculate, finally stake and—lose exactly as we simple mortals who play without calculations. On the other hand, I drew one conclusion which I believe to be correct: that is, though there is no system, there really is a sort of order in the sequence of casual chances—and that, of course, is very strange.
>
> —Fyodor Dostoyevsky, *The Gambler*

The "order in the sequence of casual chances," as spoken by the protagonist of Dostoyevsky's novel, has been noticed in games of chance and everyday life for centuries. Of course, gamblers are not the only ones interested in being able to effectively make predictions about the future. The intent of this chapter is to provide a brief exposure to how chance and probability are systematically studied, and we begin by introducing the basic concept of a *sample space* of outcomes that can result from some well-defined action or experiment.

7.1 Sample Spaces and Events

Suppose you decide to buy 10 shares of your favorite stock tomorrow. Is the price determined? Clearly the answer is no. You could look up today's opening price in the newspaper, but it certainly could change before you buy the stock tomorrow. It could even change while you are on the telephone placing your order. Similarly, do you know what the high temperature will be tomorrow or even whether it will rain? Your answer may depend on where you live and the time of year, but nevertheless, no one can be positive about the future state of the weather at any given time. What about your future career, the interest rate at the end of the year, your grade in your next mathematics course, or the roll of two dice? Although you may be able to make a good guess at what might happen, you cannot state with certainty the outcome in any of these cases. They cannot be predicted absolutely in the manner of *deterministic phenomena*, such as the effects of gravity or the result of mixing two specific chemicals in the laboratory. Throughout this chapter, we will be studying random **experiments**. These are actions or processes for which the outcome cannot be predicted with certainty. Probability is the study of patterns found in repeated occurrences of such experiments. Some examples of experiments are listed below.

experiments An action or a process for which the outcome cannot be predicted with certainty.

- Record the price of your favorite stock at the end of the next Tuesday.
- Record the high temperature for the day in your town tomorrow.
- Flip a coin and record the outcome.
- Pick a card from a standard (52-card) deck, and record the suit.
- Select five numbered balls from a barrel, and record the values.
- Roll two six-sided dice, and record the sum of the number of pips showing.
- Pick an integer between 1 and 10, inclusive.
- Do a Google search for a particular topic, and write down the number of results.

These are all examples of random experiments, because in each case, the outcomes cannot be predicted with certainty. Notice that you can list the possible outcomes (however unlikely) that may occur. For the coin toss, we know the coin must come up heads (*H*) or tails (*T*), so we could say the set {*H, T*} lists all possible outcomes from the experiment.

A set that lists all the distinct possible outcomes from a random experiment is called a **sample space**, usually denoted by *S*. *Sets*, collections of objects, are a useful tool in the study of probability, and we will study them in greater detail in the next section. For now, we examine some sample spaces that correspond to a few different experiments.

sample space The set of all distinct possible outcomes of an experiment.

? Example 1

■ Roll two six-sided dice, and record the sum of the number of pips showing. Then the sample space *S* can be given by $S = \{2, 3, 4, 5, 6, 7, 8, 9, 10, 11, 12\}$.

- Pick a card from a standard deck, and record the suit. Then the sample space can be given by $S = \{$heart, spade, club, diamond$\}$.
- Pick a card from a standard deck, and record the value (denomination) on its face. Then the sample space can be given by $S = \{2, 3, 4, 5, 6, 7, 8, 9, 10,$ jack, queen, king, ace$\}$. ♦

These examples illustrate an important point that you must consider in designing a sample space. The experiment involves not only completing an action (drawing a card) but also recording an outcome. In the second experiment, we were interested in recording the suit after selecting a card, and so we listed suits in our sample space. In the next instance, however, we were interested in recording the denomination on the face, and so our sample space consisted of a list of denominations. The next example also illustrates this distinction.

? Example 2

- Toss a fair coin twice, and record the outcomes. The sample space can be given by $S = \{HH, HT, TH, TT\}$, where H represents heads and T represents tails.
- Toss a fair coin twice, and record the number of heads. The sample space can be given by $S = \{0$ heads, 1 head, 2 heads$\}$. ♦

Different methods exist for listing the members of the sample's space, as seen in Example 3.

? Example 3

- Pick an integer between 1 and 10, inclusive. The sample space can be given by $S = \{1, 2, 3, 4, 5, 6, 7, 8, 9, 10\}$, which is the list of integers between 1 and 10, including both 1 and 10.
- Pick a four-digit number. Although it is large, we can list this sample space by establishing a pattern. The sample space is $S = \{1,000, 1,001, 1,002, \ldots, 9,998, 9,999\}$. Note that we did not list every outcome from this experiment, but we have developed enough of a pattern to fully describe the set.
- Toss a coin, and record the number of tosses required until a head is observed. The sample space can be listed as $S = \{1, 2, 3, 4, 5, 6, \ldots\}$. Notice that this is an infinite set, which is indicated by three dots trailing the list that defines the pattern. ♦

Using a listed set to describe outcomes is not always the most convenient way to write a sample space. Suppose you have two dice, one red and one green. You roll both dice and record the outcome. In this case, the most convenient form for a sample space might be a chart like the one in **Figure 7.1.1**.

$$
\begin{array}{c|cccccc}
6 & (1,6) & (2,6) & (3,6) & (4,6) & (5,6) & (6,6) \\
5 & (1,5) & (2,5) & (3,5) & (4,5) & (5,5) & (6,5) \\
4 & (1,4) & (2,4) & (3,4) & (4,4) & (5,4) & (6,4) \\
3 & (1,3) & (2,3) & (3,3) & (4,3) & (5,3) & (6,3) \\
2 & (1,2) & (2,2) & (3,2) & (4,2) & (5,2) & (6,2) \\
1 & (1,1) & (2,1) & (3,1) & (4,1) & (5,1) & (6,1) \\
\hline
 & 1 & 2 & 3 & 4 & 5 & 6
\end{array}
$$

Green Die (vertical axis), Red Die (horizontal axis)

FIGURE 7.1.1   Two-dice chart.

We now turn our attention to denoting a specific set of outcomes from an experiment. Suppose we toss a coin three times and record each outcome. The sample space could be written as $S = \{HHH, HHT, HTH, HTT, THH, THT, TTH, TTT\}$. If we want to know the chances of obtaining two heads out of three tosses, we are now interested in a particular **event**. Because an event is the occurrence of some combination of possible outcomes (including *no* outcomes), it is formally defined as any subset of a sample space.

> **event** A subset of outcomes from a sample space.

For our current space, suppose we describe event A as obtaining *exactly* two heads. According to our definition, we would list $A = \{HHT, HTH, THH\}$. Other events of interest might be obtaining *at least* two heads, which we list as $B = \{HHH, HHT, HTH, THH\}$; obtaining at least one tail, which we list as $C = \{HHT, HTH, HTT, THH, THT, TTH, TTT\}$; or obtaining a head on the first toss and a tail on the second toss, which we list as $D = \{HTH, HTT\}$. We return to our previous examples and describe some typical events.

? Example 4

- $S = \{2, 3, 4, 5, 6, 7, 8, 9, 10, 11, 12\}$ for the experiment of rolling two dice and adding the number of pips. Let A be the event that the sum is between 7 and 9, inclusive. We can list $A = \{7, 8, 9\}$. Let B be the event that the sum is odd. We can list $B = \{3, 5, 7, 9, 11\}$.

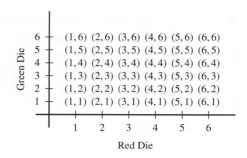

- $S = \{$heart, spade, club, diamond$\}$ for the experiment of drawing a card from a deck and recording the suit. Let R be the event that the card is red. We can list $R = \{$heart, diamond$\}$.
- $S = \{2, 3, 4, 5, 6, 7, 8, 9, 10,$ jack, queen, king, ace$\}$ for the experiment of drawing a card from a deck and recording the denomination. Let F be the event that the card is a face card. We can list $F = \{$jack, queen, king$\}$. ◆

? Example 5

- $S = \{HH, HT, TH, TT\}$ for the experiment of tossing two coins. Let E be the event that more than two heads occur. Then $E = \varnothing$ (the empty set) because none of the outcomes in S contain more than two heads.

© Zelenskaya/ShutterStock, Inc.

■ $S = \{0 \text{ heads, } 1 \text{ head, } 2 \text{ heads}\}$ for the experiment of tossing two coins and recording the number of heads. Let T be the event that we obtain exactly one tail. Because we must also have one head in that case, then $T = \{1 \text{ head}\}$. ◆

? Example 6

■ $S = \{1, 2, 3, 4, 5, 6, 7, 8, 9, 10\}$ for the experiment of picking an integer between 1 and 10, inclusive. Let P be the event that the integer is a prime number. We can list $P = \{2, 3, 5, 7\}$. Recall that a number is *prime* if it is larger than 1 and is divisible by only itself and 1.

■ $S = \{1{,}000, 1{,}001, 1{,}002, \ldots, 9{,}998, 9{,}999\}$ for the experiment of picking a four-digit number. Let F be the event that the first digit is 5. We can list $F = \{5{,}000, 5{,}001, 5{,}002, \ldots, 5{,}998, 5{,}999\}$. Note that we do not list every outcome in this event, but the pattern is clearly established.

■ $S = \{1, 2, 3, 4, 5, 6, \ldots\}$ for the experiment of tossing a coin and recording the number of tosses required until a head is observed. Let Q be the event that fewer than four tosses are required. We can list $Q = \{1, 2, 3\}$. ◆

Probability is concerned with measuring the likelihood that an event happens. As such, most probability problems begin with the establishment of a sample space and an event to measure. It is interesting to note a relationship that events have with sample spaces. In Example 4, we listed the event A that corresponded to the sum of pips on two dice as $A = \{7, 8, 9\}$. This was based on the sample space $S = \{2, 3, 4, 5, 6, 7, 8, 9, 10, 11, 12\}$. Yet we also established a separate sample space—let's call it S'—using a chart (Figure 7.1.1) that listed the outcomes of rolling two dice. Could S' be used to help define the event A? In **Figure 7.1.2**, we have indicated the outcomes that correspond to the sums listed in A.

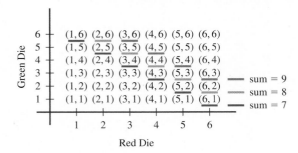

FIGURE 7.1.2 Two-dice chart.

So the event $A' = \{(6, 1), (5, 2), (4, 3), (3, 4), (2, 5), (1, 6), (6, 2), (5, 3), (4, 4), (3, 5), (2, 6), (6, 3), (5, 4), (4, 5), (3, 6)\}$ in the sample space S' corresponds to the possible dice combinations resulting in a sum of 7, 8, or 9. Which of these spaces is optimal for use in the area of probability? Typically, the sample space best suited to clarifying the situation when assigning probabilities is the one in which every outcome is equally likely to occur. This is the case here for S'; moreover, S' contains sufficient information to describe a greater number of events than S. For example, we can list the event of obtaining doubles as $B = \{(1, 1), (2, 2), (3, 3), (4, 4), (5, 5), (6, 6)\}$, whereas S does not contain such information. It would not be an appropriate sample space if obtaining doubles were of interest. This relationship between events and sample spaces will be revisited in Section 3, when we study the probability of an event occurring.

Name _____

Exercise Set **7.1** ☐☐☐←────────────────

1. Two nine-sided dice are rolled (numbered 1 through 9), and the sum of the pips is recorded.

   (a) List the sample space S.
   (b) List the event E that the sum is even.
   (c) List the event A that the sum is at least 10.
   (d) List the event L that the sum is less than 6.
   (e) List the event P that the sum is a prime number.

2. One red die and one green die (six-sided) are rolled, and the outcomes are recorded.

   (a) List the sample space S.
   (b) Let F be the event that both dice show an odd number of pips. List the event F. Be sure your description is consistent with your sample space from (a).
   (c) Let B be the event that the sum of pips is more than 8. List the event B.
   (d) Let N be the event that the number of pips showing on the red die is larger than on the green die. List the event N.
   (e) Let T be the event that the number of pips showing on the green die is exactly 2 more than on the red die. List the event T.
   (f) Let A be the event that the number of pips showing on the green die is at least 2 more than on the red die. List the event A.

3. Repeat Exercise 2, assuming the red die is four-sided and the green die is six-sided.

4. A student is asked to list a sample space for tossing two coins and lists $S = \{0$ heads, 1 head, 0 tails, 1 tail$\}$.

   (a) Explain why this is not a valid sample space for the experiment.
   (b) List a correct sample space.

5. Give the sample space for each of the following experiments.

   (a) A whole number is picked at random between 0 and 100, inclusive.
   (b) A five-digit number is picked at random.
   (c) A coin is tossed three times, and the number of tails is observed.
   (d) A coin is tossed three times, and the sequence of heads and tails is observed.

6. A liberal arts mathematics course contains freshmen, sophomores, juniors, and seniors. A student is picked at random, and her or his class and sex are recorded.

   (a) List the sample space S. (*Hint*: Use ordered pairs.)
   (b) Let F be the event that the student is a female. List the event F.
   (c) Let U be the event that the student is a junior or senior. List the event U.
   (d) Let M be the event that the student is a freshman or a male. List the event M.

7. Three fair coins are tossed, and the coin faces are recorded.

(a) List the sample space.
(b) List the event A that at least one head occurs.
(c) List the event B that exactly one tail occurs.
(d) List the event C that exactly two tails occur.
(e) List the event D that exactly no heads occur.

8. A fair coin is tossed four times, and the sequence of heads and tails is recorded.

(a) List the sample space S.
(b) Let T be the event that more than two heads are recorded. List the event T. Be sure your description is consistent with your sample space listed in (a).
(c) Let N be the event that either no heads or no tails are recorded. List the event N.
(d) Let F be the event that the first tail that appears must be directly preceded by a head. List the event F.
(e) Let E be the event that exactly three coins of the same face (e.g., three heads or three tails) appear. List the event E.
(f) Let A be the event that no two consecutive tosses result in the same side. List the event A.

9. The American roulette wheel shown here that is usually used in a casino has 38 slots numbered 1, 2, 3, . . ., 36 (colored either red or black) as well as the so-called house slots 0 and 00 (colored green). A ball is rolled around the wheel and randomly drops into one of the slots.

(a) List the sample space S.
(b) List the event R that the ball drops into a red slot.
(c) List the event B that the ball drops into a black slot.

© Zlatko Guzmic/ShutterStock, Inc.

(d) List the event M that the ball drops into a red slot numbered between 19 and 34.

(e) List the event G that the ball drops into a house slot.

(f) List the event E that the ball drops into an even slot or into a green slot.

10. In the game of blackjack, your objective is to have more points than the dealer without exceeding 21. We can use ordered pairs to represent outcomes. For example, (19, 17) could represent the result that you obtained 19 while the dealer obtained 17 (so you win!). Similarly, (23, 19) would represent the result that you obtained 23 while the dealer obtained 19 (you lose because you exceeded 21). Because of certain house rules for dealers, we'll assume the dealer's points can range between 17 and 26. Following accepted practices for players, we'll assume your points can range between 12 and 27.

(a) Using ordered pairs as described in this exercise, list the sample space S.

(b) You bust if your points exceed 21 (regardless of the dealer's points). List the event corresponding to "you bust."

(c) You win if your points exceed the dealer's points (and you don't bust) *or* if the dealer busts while your points remain 21 or lower. List the event corresponding to "you win."

(d) You push if your points are identical to the dealer's and neither one of you busts. List the event corresponding to "you push." Blackjack players note: For simplicity, we are representing a blackjack result by 21 without any special considerations.

(e) You lose if you bust *or* if your points are less than the dealer's and the dealer doesn't bust. List the event corresponding to "you lose."

7.2 Introduction to Sets

As Section 7.1 illustrated, the use of sets is essential in studying probability. Recall that the sample space for an experiment is the *set* of all possible outcomes and that an event is any *subset* of the sample space. For us to combine events, therefore, we must first be able to combine sets. Implicit in our discussion is an assumption of the existence of an all-encompassing set for a particular situation. A **universal set** is the collection of all elements for a situation; thus, the sample space serves as a universal set in an experiment, whose elements are the possible outcomes of the experiment. In the study of probability, we define a **set** as a collection of elements within a given universal set. For example, let $U = \{$all regularly enrolled college students$\}$ be the universal set, and let $Y = \{$students enrolled at Yale University$\}$. Then Y is a set, because it is a collection of elements (students) within U. It is important that a universal set be specified for any problem you encounter, just as a sample space is essential for any experiment. A universal set is situation-specific—that is, it may change from problem to problem.

> **universal set** A set of all possible elements associated with a certain phenomenon or experiment.
>
> **set** A collection of objects.

Let U be a universal set, and let A and B be sets. We say A is a **subset** of B, denoted by $A \subseteq B$, if every element of A is also an element of B. Using the symbol $\in$ for "is an element of," we can say that $x \in A$ implies that $x \in B$ if and only if $A \subseteq B$. The set containing no elements is called the **empty set** and is represented by the symbol $\varnothing$. It must necessarily be the case that $\varnothing$ and the universal set U are always subsets of U.

> **subset** Set A is a subset of set B if every element of A is also an element of B.
>
> **empty set** The set containing no elements. It is denoted by the symbol $\varnothing$.

Building on our previous example, consider these sets:

$$U = \{\text{all regularly enrolled college students}\},$$

$$Y = \{\text{students enrolled at Yale University}\},$$

$$E = \{\text{students majoring in English at Yale}\}.$$

Is $Y \subseteq E$, $E \subseteq Y$, neither, or both? If $Y \subseteq E$, then every student enrolled at Yale would have to be majoring in English. But this is not true because there are students enrolled at Yale who have majors other than English. So Y is not a subset of E, which we can denote $Y \nsubseteq E$. Our second question was whether $E \subseteq Y$. If $E \subseteq Y$, then every student majoring in English at Yale must be enrolled at Yale. Because this is true, we can say E is a subset of Y, denoted $E \subseteq Y$. The reasoning used here to show $Y \nsubseteq E$ is important enough to emphasize the following rule.

To show $A \nsubseteq B$, it suffices to find just one element of A that is not an element of B.

? Example 1

Let $U = \{$major league baseball players$\}$, $L = \{$left-handed hitters$\}$, and $R = \{$right-handed hitters$\}$. If $L \subseteq R$, then every left-handed hitter would have to be a right-handed hitter also. But this is not true, insofar as there are left-handed hitters who never bat right-handed. So $L \nsubseteq R$. Similar reasoning shows that $R \nsubseteq L$. ◆

❓ Example 2

Let U = {dice combinations from two six-sided die rolls}, D = {doubles are obtained}, E = {sum of dice is even}, and F = {both dice are odd}. Note that $D \not\subseteq F$, because there is a double—for example, $(4, 4)$—in which both dice are not odd. Also, $F \not\subseteq D$, because there are elements of F—for example, $(1, 5)$—that are not doubles. However, $D \subseteq E$, because adding the rolls of doubles always produces an even sum. You should check that $E \not\subseteq D$, $F \subseteq E$, and $E \not\subseteq F$. ◆

complement The complement $\overline{A}$ of a set A consists of the elements in the universal set that are not elements of A.

We now turn our attention to set operations. These are operations that create new sets out of existing sets. The **complement** of set A, denoted $\overline{A}$, is the set of all objects in the universal set that are not in A. Equivalently, we can write, $\overline{A} = \{x \mid x \in U \text{ and } x \notin A\}$. (Note the use of the symbol $\notin$ to represent "is not an element of.") From Example 2, the following complements should be evident:

$$\overline{D} = \{\text{doubles are not obtained}\} = \{\text{the two dice are not equal}\},$$

$$\overline{E} = \{\text{sum of dice is not even}\} = \{\text{sum of dice is odd}\},$$

$$\overline{F} = \{\text{it is not the case that both dice are odd}\} = \{\text{at least one die is even}\}.$$

It should also be evident that the complement of a complement is just the original set. Using complement notation, we can see that $\overline{\overline{A}} = A$, for any set A. Note that it must always be true that $\overline{U} = \emptyset$ and $\overline{\emptyset} = U$.

union of two sets The union $A \cup B$ of two sets A and B is the set of objects belonging to either A or B.

intersection of two sets The intersection $A \cap B$ of two sets A and B is the set of elements belonging to both A and B.

Two additional set operations that may be familiar to you are union and intersection. The **union of two sets** A and B, denoted $A \cup B$, is the set of objects belonging to A *or* B (or both). Equivalently, $A \cup B = \{x \mid x \in A \text{ or } x \in B\}$. The **intersection of two sets** A and B, denoted $A \cap B$, is the set of objects belonging to *both A and B*, or symbolically, $A \cap B = \{x \mid x \in A \text{ and } x \in B\}$.

From Example 1, the union of L and R is the entire universal set because all hitters must bat from one or the other side of the plate. So $L \cup R = U$, the universal set. The intersection of L and R is the set of hitters who bat from both left and right sides (called switch-hitters). So $L \cap R = \{\text{switch-hitters}\}$. Note that because $A \cup B$ contains elements in either A or B (or both), then it is always true that $A \cap B \subseteq A \cup B$.

❓ Example 3

Let U = {1, 2, 3, 4, 5, 6, 7, 8, 9, 10}, P = {2, 3, 5, 7}, D = {1, 3, 5, 7, 9}, and E = {2, 4, 6, 8, 10}. Find $\overline{P}$, $\overline{D}$, $P \cap D$, $\overline{P} \cap D$, $P \cup D$, $P \cap E$, $\overline{P} \cap E$, $P \cup E$, $D \cap E$, and $D \cup E$.

⚙ Solution

$$\overline{P} = \{1, 4, 6, 8, 9, 10\} \qquad \overline{D} = \{2, 4, 6, 8, 10\} = E$$

$$P \cap D = \{3, 5, 7\} \qquad \overline{P} \cap D = \{1, 9\}$$

$$P \cup D = \{1, 2, 3, 5, 7, 9\} \qquad P \cap E = \{2\}$$

$$\overline{P} \cap E = \{4, 6, 8, 10\} \qquad P \cup E = \{2, 3, 4, 5, 6, 7, 8, 10\}$$

$$D \cap E = \emptyset \text{ (the empty set)} \quad D \cup E = \{1, 2, 3, 4, 5, 6, 7, 8, 9, 10\} = U$$

Note that we list elements only once when performing unions even if the element appears in both sets. For example, in $P \cup D$ the element 5 is only listed once even though $5 \in P$ and $5 \in D$. ◆

---

disjoint sets Two sets whose intersection is empty; i.e., two sets A and B for which $A \cap B = \emptyset$.

mutually exclusive events Events that cannot occur at the same time.

partition For a set S, a pair of subsets A and B such that $A \cap B = \emptyset$ and $A \cup B = S$.

Example 3 brings to light several observations. For instance, an intersection can be empty. It is not necessary for two sets to have a common intersection point. In general, two sets that have an empty intersection are called **disjoint sets**. If these sets represent events in an experiment, then they are called **mutually exclusive events**, and such events cannot occur at the same time. In two coin tosses, the events $A = \{$obtain two heads$\}$ and $B = \{$obtain two tails$\}$ are mutually exclusive events because $A \cap B = \emptyset$. If instead we let $A = \{$obtain at least one head$\}$ and $B = \{$obtain at least one tail$\}$, note that $A \cap B = \{HT, TH\}$, and so now A and B are not mutually exclusive events.

We also note in Example 3 that although D and E have no elements in common ($D \cap E = \emptyset$), together they comprise the universal set ($D \cup E = U$). Here the result is not surprising, because $U = \{$whole numbers from 1 through 10$\}$, $D = \{$odd numbers from 1 through 10$\}$, and $E = \{$even numbers from 1 through 10$\}$. We call D and E a *partition* of U. In general, a collection of subsets of a set E form a **partition** of E if they are pairwise disjoint and their union is E. Notice that any set A and its complement must form a partition of the universal set because it is always true that

$$A \cap \overline{A} = \emptyset \quad \text{and} \quad A \cup \overline{A} = U.$$

We will utilize this idea later in the section.

❓ Example 4

Let U be a standard (52-card) deck of cards. Each of the following pairs of sets forms a partition of U.

(a) $R = \{$red cards$\}$ and $B = \{$black cards$\}$. Note that $\overline{R} = B$ and likewise $\overline{B} = R$.
(b) $N = \{2, 3, 4, 5, 6, 7, 8, 9, 10\}$ and $P = \{$jack, queen, king, ace$\}$.
(c) $E = \{2, 4, 6, 8, 10, \text{queen, ace}\}$ and $F = \{3, 5, 7, 9, \text{jack, king}\}$. ◆

Venn diagrams Picture of sets that displays a certain relationship among them.

Diagrams are also useful to describe set operations. In particular, pictures known as **Venn diagrams** are particularly useful to illustrate the ideas of complement, union, and intersection. They are named in honor of **John Venn** (1834–1923) (see **Figure 7.2.1**), who was born in Hull, England, and attended Gonville and Caius College Cambridge on a mathematics scholarship. Upon graduation, he become a Fellow of the College and, 2 years later, became an ordained priest as well. By 1862, he was a lecturer at Cambridge in moral science, philosophy, and probability and began to find himself drifting from the "orthodox clerical outlook." In the same year in which he was elected to the Royal Society, he left the priesthood but remained "throughout his life a man of sincere religious conviction."

FIGURE 7.2.1   John Venn. Taken from a Biographical history of Gonville and Caius college, 1349–1897.

In a typical Venn diagram, a rectangle is used to represent the universal set, circles are used to represent sets, and shading is used to represent the portion we are describing. Consider the Venn diagrams in **Figure 7.2.2**, which represent $\overline{A}$, $A \cup B$, and $A \cap B$.

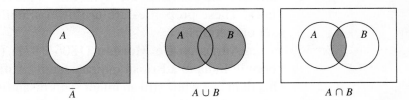

FIGURE 7.2.2   Venn diagrams for the basic set operations.

Venn diagrams are particularly useful for describing more complicated set operations. Consider the set $\overline{A \cup B}$. The union of the sets A and B is formed first, followed by taking the complement. We can use a sequence of Venn diagrams to illustrate these two operations. (See **Figure 7.2.3**.)

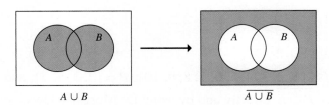

FIGURE 7.2.3   Venn diagram for $\overline{A \cup B}$.

A representation for $\overline{A} \cup \overline{B}$ can be developed by the sequence of Venn diagrams shown in **Figure 7.2.4**.

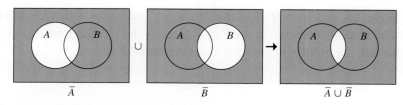

FIGURE 7.2.4   Venn diagram for $\bar{A} \cup \bar{B}$.

Comparing the diagrams for $\overline{A \cup B}$ and $\bar{A} \cup \bar{B}$ clearly shows, in general, that the two sets need not be equal. Consider the Venn diagram sequence for $\bar{A} \cap \bar{B}$ in **Figure 7.2.5**.

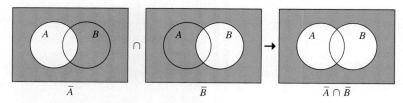

FIGURE 7.2.5   Venn diagram for $\bar{A} \cap \bar{B}$.

De Morgan's laws
For any sets A and B,
$\overline{A \cup B} = \bar{A} \cap \bar{B}$ and
$\overline{A \cap B} = \bar{A} \cup \bar{B}$.

Notice that this diagram does agree with that for $\overline{A \cup B}$. In fact, this equality is part of a famous pair of set equalities known as **De Morgan's laws**. For any sets A and B,

$$\overline{A \cup B} = \bar{A} \cap \bar{B} \quad \text{and} \quad \overline{A \cap B} = \bar{A} \cup \bar{B}.$$

Verification of the second law is left as an exercise. You may have already encountered the symbolic logic version of De Morgan's laws. These properties are useful in simplifying more complicated set operations.

Augustus De Morgan (1806–1871) (**Figure 7.2.6**), who was born in India to a British military family, suffered the loss of vision in his right eye while still an infant. This disability motivated teasing from his childhood classmates and perhaps influenced his development into a rather gruff and strongly opinionated individual. He entered Trinity College in Cambridge at the age of 16 and later became the first professor of mathematics at University College in London in 1828. He wrote several mathematics textbooks and numerous articles over his lifetime, and he was known as a great reformer of mathematical logic.

FIGURE 7.2.6   Augustus De Morgan.

? Example 5

Let $U = \{1, 2, 3, 4, 5, 6, 7, 8, 9, 10\}$, $P = \{2, 3, 5, 7\}$, and $E = \{2, 4, 6, 8, 10\}$. Find the set $\overline{P \cup \bar{E}}$ directly and by using De Morgan's laws.

⚙ Solution

Directly, $P = \{2, 3, 5, 7\}$ and $\bar{E} = \{1, 3, 5, 7, 9\}$, so $P \cup \bar{E} = \{1, 2, 3, 5, 7, 9\}$, and thus $\overline{P \cup \bar{E}} = \{4, 6, 8, 10\}$. By De Morgan's laws,

$$\overline{P \cup \overline{E}} = \overline{P} \cap E \left(\text{because } \overline{\overline{E}} = E\right)$$
$$= \{1, 4, 6, 8, 9, 10\} \cap \{2, 4, 6, 8, 10\}$$
$$= \{4, 6, 8, 10\}. \quad \blacklozenge$$

cardinality The cardinality $n(S)$ of a finite set S is the number of elements contained in S.

Notice that using De Morgan's laws in Example 5 simplifies the amount of set operations needed to solve such problems. Venn diagrams are also useful when counting is involved with set operations. The **cardinality** of a finite set A is the number of objects in the set and is denoted by $n(A)$. The concept of cardinality is used primarily in the study of infinite sets, which we will not consider in this text. So, for our purposes, cardinality is simply a *function* whose domain consists of the subsets of a universal set and whose range is the set of nonnegative integers (representing the number of elements in the set). Clearly, we see that $n(\varnothing) = 0$ and that if A and B are sets such that $A \subseteq B$, then $n(A) \leq n(B)$. We wish to develop rules for finding the cardinality of unions and intersections of sets whose cardinalities are known.

? Example 6

If $A = \{\text{letters of the English alphabet}\}$ and $B = \{x \mid 1 \leq x \leq 100 \text{ and } x \text{ is an even integer}\}$, then $n(A) = 26$ and $n(B) = 50$. Because $A \cap B = \varnothing$, we see that $n(A \cap B) = 0$ and $n(A \cup B) = 76$. $\quad \blacklozenge$

Suppose we survey 100 students and find that 48 are currently taking an English course, 56 are currently taking a mathematics course, and 22 are taking both English and mathematics. How can we find out how many students are taking neither English nor mathematics? Let E represent the event (set) of taking an English course and M the event of taking a mathematics course. The Venn diagram in **Figure 7.2.7** illustrates a partition of U (the universal set of 100 students surveyed) into *four* distinct sets. This is an example of the definition of a partition of U in which the union of several sets equals U and no pair of sets intersects.

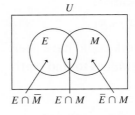

FIGURE 7.2.7 Partition of U into four sets.

$E \cap M = \{\text{students taking both English and mathematics}\}$

$E \cap \overline{M} = \{\text{students taking English but not mathematics}\}$

$\overline{E} \cap M = \{\text{students taking mathematics but not English}\}$

$\overline{E} \cap \overline{M} = \{\text{students taking neither English nor mathematics}\}$

We know there are 22 students taking both English and mathematics, so $n(E \cap M) = 22$. Now notice that $E \cap M$ and $\overline{E} \cap M$ form a partition of event M. Therefore,

$$n(E \cap M) + n(\overline{E} \cap M) = n(M).$$

We also know there are 56 students taking mathematics, so $n(M) = 56$. Substituting these values into the above equation, we get

$$22 + n\left(\overline{E} \cap M\right) = 56$$

$$n\left(\overline{E} \cap M\right) = 34.$$

Similarly, we partition event E into the two sets $E \cap M$ and $E \cap \overline{M}$ to obtain

$$n(E \cap M) + n\left(E \cap \overline{M}\right) = n(E)$$

$$22 + n\left(E \cap \overline{M}\right) = 48$$

$$n\left(E \cap \overline{M}\right) = 26.$$

Now we redraw the Venn diagram with these computed values inserted. (See **Figure 7.2.8**.) So we have

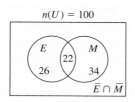

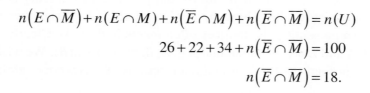

$$n\left(E \cap \overline{M}\right) + n(E \cap M) + n\left(\overline{E} \cap M\right) + n\left(\overline{E} \cap \overline{M}\right) = n(U)$$

$$26 + 22 + 34 + n\left(\overline{E} \cap \overline{M}\right) = 100$$

$$n\left(\overline{E} \cap \overline{M}\right) = 18.$$

FIGURE 7.2.8 Partition of U into four sets.

We see that 18 students surveyed were taking neither English nor mathematics. Note, therefore, that the number of students taking *either* English or mathematics must be

$$n(E \cup M) = 100 - 18 = 82.$$

Of course, this could also have been obtained by observing in Figure 7.2.8 that

$$n(E \cup M) = 26 + 22 + 34 = 82.$$

We emphasize that this is *not* equal to $n(E) + n(M) = 48 + 56 = 104$, because that would be counting the number of students in both sets, $n(E \cap M)$, twice. Note that if we subtract this value, 22, from 104, then of course, we would again arrive at the correct value of 82. This is shown by the equation

$$n(E \cup M) = n(E) + n(M) - n(E \cap M) = 48 + 56 - 22 = 82.$$

This a specific application of the property known as the **addition principle**:

> If A and B are sets, then $n(A \cup B) = n(A) + n(B) - n(A \cap B)$.

addition principle
For any sets A and B,
$n(A \cup B) = n(A) + n(B) - n(A \cap B)$.

Reviewing this last example, we see that most of the work could have been done by drawing a single Venn diagram and then subtracting values to find partition elements. (See **Figure 7.2.9**.)

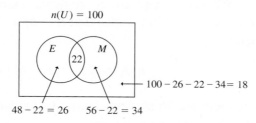

FIGURE 7.2.9 Simplified partition diagram for U.

So we see that partitions, Venn diagrams, and the addition principle are quite useful for working with counting problems of this type.

? Example 7

Let $A = \{1, 3, 4, 5, 7, 8, 10\}$ and $B = \{3, 4, 7, 9, 10\}$. Verify the addition principle by first computing $A \cup B$ and $A \cap B$ directly.

⚙ Solution

$$A \cup B = \{1, 3, 4, 5, 7, 8, 9, 10\} \quad \text{so} \quad n(A \cup B) = 8.$$
$$A \cap B = \{3, 4, 7, 10\} \qquad\qquad \text{so} \quad n(A \cap B) = 4.$$

Because $n(A) = 7$ and $n(B) = 5$, the formula $n(A \cup B) = n(A) + n(B) - n(A \cap B)$ is easily verified. ◆

? Example 8

A universal set U contains 200 elements. If $n(A) = 85$, $n(\bar{B}) = 138$, and $n(A \cup B) = 107$, for sets A and B, then find (a) $n(A \cap B)$, (b) $n(A \cap \bar{B})$, and (c) $n(\bar{A} \cap B)$.

⚙ Solution

(a) First, we must have $n(B) = n(U) - n(\bar{B}) = 200 - 138 = 62$. Then, because the addition principle implies that $n(A \cap B) = n(A) + n(B) - n(A \cup B)$, it must be true that

$$n(A \cap B) = 85 + 62 - 107 = 40.$$

(b) Because $A \cap B$ and $A \cap \bar{B}$ form a partition of A,

$$n(A \cap \bar{B}) = n(A) - n(A \cap B) = 85 - 40 = 45.$$

(c) Because $A \cap B$ and $\bar{A} \cap B$ form a partition of B,

$$n(\bar{A} \cap B) = n(B) - n(A \cap B) = 62 - 40 = 22. \quad ◆$$

Name _____

Exercise Set **7.2** ⬜⬜⬜ ←————————————————————————————

1. True or false? Let A and B be any subsets of the universal set U.

   (a) $A \subseteq A \cap B$
   (b) $A \subseteq A \cup B$
   (c) $n(A) + n(\overline{A}) = n(U)$
   (d) $A \cap \varnothing = \varnothing$
   (e) $\overline{A \cap B} = \overline{A} \cap \overline{B}$
   (f) If $A \subseteq B$, then $n(A) \leq n(B)$.
   (g) If A and B are disjoint subsets, then $\overline{A} \cap \overline{B} = \varnothing$.
   (h) If A and B are disjoint subsets, then $n(A \cup B) = n(A) + n(B)$.

2. Let $U = \{$all college students$\}$, $M = \{$students currently taking mathematics$\}$, $H = \{$students currently taking history$\}$, and $B = \{$students currently taking biology$\}$. Describe (in words) each of the following sets.

   (a) $M \cap B$　　　　(b) $\overline{H}$　　　　(c) $\overline{H} \cap \overline{B}$　　　　(d) $M \cup H$

3. Let $U = \{$people who like sports$\}$, $B = \{$people who like basketball$\}$, $F = \{$people who like football$\}$, and $H = \{$people who like hockey$\}$. Describe (in words) each of the following sets.

   (a) $B \cup H$　　　　(b) $\overline{F}$　　　　(c) $\overline{B} \cap F$　　　　(d) $\overline{H \cup F}$

4. Let $U = \{5, 6, 7, 8, 9, 10, 11\}$, $A = \{6, 7, 8\}$, and $B = \{7, 9, 11\}$. List the elements of each of the following sets.

   (a) $\overline{A}$　　　　(b) $A \cup B$　　　　(c) $\overline{A} \cap B$

   (d) $\overline{A \cap B}$　　　(e) $\overline{A} \cup \overline{B}$　　　(f) Do A and B form a partition of U?

5. Let $U = \{-3, -2, -1, 0, 1, 2, 3\}$, $A = \{-2, -1, 0, 1\}$, and $B = \{-1, 1, 2\}$. List the elements of each of the following sets.

   (a) $\overline{B}$　　　(b) $A \cap B$　　　(c) $\overline{A} \cup B$　　　(d) $\overline{A \cup B}$　　　(e) $\overline{A \cup \overline{B}}$

6. Draw Venn diagrams illustrating each of the following sets.

   (a) $A \cap \overline{B}$　　　　(b) $\overline{A} \cup B$　　　　(c) $\overline{A \cup B}$　　　　(d) $A \cap \overline{\overline{B}}$

7. Draw Venn diagrams illustrating each of the following sets.

   (a) $\overline{A} \cap B$　　　　(b) $A \cup \overline{B}$　　　　(c) $\overline{\overline{A} \cap B}$　　　　(d) $\overline{\overline{A} \cup B}$

8. De Morgan's second law states that $\overline{A \cap B} = \overline{A} \cup \overline{B}$. Draw Venn diagrams for $\overline{A \cap B}$ and $\overline{A} \cup \overline{B}$ to verify that these two sets are equal.

9. Let $U = \{1, 2, 3, 4, 5, 6, 7, 8\}$, $F = \{2, 4, 5, 7, 8\}$, and $S = \{3, 4, 8\}$.

   (a) Find the set $\overline{F \cap S}$ directly and by using De Morgan's laws.
   (b) Find the set $\overline{F} \cup \overline{S}$ directly and by using De Morgan's laws.
   (c) Find the sets $F \cap S$ and $F \cap \overline{S}$; then verify that they form a partition of F.
   (d) Find the sets $S \cap F$ and $S \cap \overline{F}$; then verify that they form a partition of S.

10. Let $U = \{a, b, c, d, e, f, g\}$, $P = \{a, d, f, g\}$, and $Q = \{b, c, d, g\}$.

   (a) Find the set $\overline{P \cup Q}$ directly and by using De Morgan's laws.
   (b) Find the set $\overline{P} \cap \overline{Q}$ directly and by using De Morgan's laws.
   (c) Find the sets $P \cap Q$ and $P \cap \overline{Q}$; then verify that they form a partition of P.
   (d) Find the sets $Q \cap P$ and $Q \cap \overline{P}$; then verify that they form a partition of Q.

11. Any group of people can be used as a universal set that can be partitioned by natural characteristics, such as sex and age. Describe some other characteristics that could be used to partition the students in your mathematics class.

12. Describe a characteristic that could be used to partition the members of the U.S. Congress.

13. For any set A, explain why $\varnothing \subseteq A$ and $A \subseteq A$.

14. For any sets A and B, explain why $A \cap B \subseteq A$ and $A \cap B \subseteq B$.

15. The **set difference** $A - B$ is defined by the equation $A - B = A \cap \overline{B}$. Draw Venn diagrams illustrating each of the following sets.

   (a) $A - B$
   (b) $B - A$
   (c) $A - \overline{B}$
   (d) $\overline{B - A}$

   > **set difference** The set $A - B$ consisting of those elements of A that are not also in B. It must necessarily be equal to the set $A \cap \overline{B}$.

16. The set $A - B$ was defined in Exercise 15. Use a Venn diagram to illustrate that
$$n(A \cup B) = n(A - B) + n(B - A) + n(A \cap B).$$

17. If $n(A \cup B) = 50$, $n(A - B) = 21$, and $n(B - A) = 10$, then find $n(A \cap B)$.

18. If $n(A \cup B) = 100$, $n(A - B) = 63$, and $n(B - A) = 15$, then find $n(A)$.

19. If $n(U) = 125$, $n(A \cup B) = 100$, $n(A - B) = 40$, and $n(B - A) = 53$, then find $n(A \cap B)$, $n(B)$, and $n(\overline{B})$.

20. Let g be a function that assigns to each set A the number of subsets that can be formed from A. For example, $g(\{0, 1\}) = 4$, because the subsets of $\{0, 1\}$ are $\varnothing$, $\{0\}$, $\{1\}$, and $\{0, 1\}$. (Recall that $\varnothing$ and A are always subsets of A.) Find the following.

(a) $g(\{1, 2, 3\})$ (b) $g(\{3\})$ (c) $g(\varnothing)$ (d) $g(\{a, b, c, d\})$

21. Based on your answers to Exercise 20, create a formula for $g(A)$ for any set A in terms of $n(A)$. How many subsets of cards exist for a deck of 52 cards?

22. Let $U = \{1, 2, 3, \ldots, 9, 10\}$. Define the set function f by the rule $f(A) = \overline{A}$ for any set A. Find the following.

(a) $f(\{1, 3, 5, 7, 9\})$ (b) $f(\{1, 4, 5, 9\})$ (c) $f(\varnothing)$ (d) $f(U)$

23. Let $U = \{\alpha, \beta, \chi, \delta, \varepsilon, \phi, \gamma\}$, $L = \{\alpha, \chi, \varepsilon, \phi\}$, and $M = \{\beta, \lambda\}$. Define the set function f by the rule $f(A) = \overline{A}$ for any set A. Find the following.

(a) $f(L)$ (b) $f(L \cap M)$ (c) $n(L)$
(d) $n(f(L))$ (e) $n(f(L \cap M))$

24. Let $U, L, M,$ and f be defined as in Exercise 23. Find the following.

(a) $f(M)$ (b) $n(M)$ (c) $n(f(M))$
(d) $f(U)$ (e) $n(f(U))$

25. In a survey of 1,000 college students, 430 are currently taking chemistry, 640 are currently taking history, and 225 are currently taking both chemistry and history.

(a) How many students are currently taking either chemistry or history?
(b) How many students are taking chemistry but not history?
(c) How many students are taking neither chemistry nor history?

26. A biologist examines 100 trees and finds that 80 of them are either pine trees or at least 20 feet (ft) high. She also finds that 60 of the trees are pine trees, and that 55 of the trees are at least 20 ft high.

(a) How many trees are pine trees that are at least 20 ft high?
(b) How many trees are not pine trees and are less than 20 ft high?
(c) How many trees are at least 20 ft high but are not pine trees?

27. A sample of 600 vegetables was exposed to a certain level of pesticide PCH-2 for several months. They were then tested, and 255 had become vitamin-depleted, 338 had become mineral-depleted, and 68 had lost both their vitamin and mineral content.

(a) How many vegetables were depleted in vitamins or minerals?
(b) How many vegetables were depleted in minerals but not in vitamins?
(c) How many vegetables were unaffected by PCH-2?

28. Human blood can contain no antigens, the A antigen, the B antigen, or both the A and B antigens. Blood types fall into one of the following four categories.

  Type A contains the A antigen but not the B antigen.
  Type B contains the B antigen but not the A antigen.
  Type AB contains both A and B antigens.
  Type O contains neither A nor B antigens.

In 100 samples of blood, 40 are found to contain the A antigen, 30 are found to contain the B antigen, and 8 are found to contain both antigens.

(a) How many of the samples are type A blood?
(b) How many of the samples are type B blood?
(c) How many of the samples are type O blood?

29. Antigens were defined in Exercise 28. A third antigen, called the *Rh antigen*, may or may not be present in human blood. Blood that contains the Rh antigen is called *positive*—thus, blood containing all three antigens is called type AB-positive—and blood that does not contain the Rh antigen is called *negative*. Thus, blood containing no antigens is called type O-negative. The Venn diagram in **Figure 7.2.10** summarizes samples of blood.

(a) How many samples are type O-negative?
(b) How many samples are type B-positive?
(c) How many samples are characterized as positive?
(d) How many samples contain exactly two antigens?
(e) How many samples contain exactly one antigen?

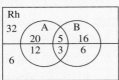

FIGURE 7.2.10 Blood antigens Venn diagram.

30. Use DeMorgan's laws and the addition principle to show that

$$n\left(\overline{A \cap B}\right) = n\left(\overline{A}\right) + n\left(\overline{B}\right) - n\left(\overline{A \cup B}\right).$$

31. Venn diagrams can be used to display situations involving more than two sets. The stained glass window shown here was created by Maria McClafferty in honor of John Venn and is located in the dining hall of Gonville and Caius College in Cambridge, UK. Let *A* be the interior of the top circle, let *B* be the interior of the bottom leftmost circle, and let *C* be the interior of the bottom rightmost circle.

Use set operations to denote each of the following regions.

(a) The three-sided figure in the middle.
(b) The part of *A* not contained in *B* or *C*.
(c) The part of *B* not contained in *A* or *C*.
(d) The part of *A* or *B* not contained in *C*.
(e) The part of *B* and *C* not contained in *A*.

7.3 Basic Probability

The first systematic study of probability was likely done by a gambler. Observation of patterns is the essence of inspiration for mathematics, and any successful gambler is certainly sure to be adept at identifying patterns that occur in card and dice games and using this knowledge to his or her advantage. An exceptional example of this is the renowned Italian scientist **Girolamo Cardano** (1501–1576) (**Figure 7.3.1**). In spite of suffering the social obstacles imposed by an illegitimate birth, Cardano (or the Latin *Cardan*) rose to fame and prominence as one of the most brilliant physicians and mathematicians of sixteenth-century Italy. At different times in his life, he was a university professor of mathematics and of medicine and even served as rector of the prestigious College of Physicians in Milan.

Unfortunately, his addiction to gambling not only led to much unhappiness in his personal life, but also robbed him of time that he could have used to produce even more contributions than the considerable amount he left us. One positive result of this affliction, however, was that he utilized his keen mathematical abilities to create one of the first formal analyses of the rules of chance and probability.

In examining random phenomena, we are interested in not only what can occur (the sample space), but also measuring the chances that a particular event occurs. For instance, if you toss a fair coin, what are the chances that the face shows a head? Intuitively we might say that the chances that a head appears are $\frac{1}{2}$, or 50%. What does this mean? Put another way, if we toss the coin a large number of times, we might expect that heads would occur in about one-half the number of tosses. More explicitly, we are using the *relative frequency approach* to defining *probability*. The **relative frequency** of an event A occurring is given by the fraction

$$\frac{\text{number of times } A \text{ occurs}}{\text{number of times experiment is repeated}}.$$

FIGURE 7.3.1   Girolamo Cardano (1501–1576).

relative frequency For an event A, it is defined by

$$\frac{\text{number of times } A \text{ occurs}}{\text{times experiment repeated}}.$$

Coin Tosses	No. of Heads	Relative Frequency
50	31	$31/50 = 0.62$
100	54	$54/100 = 0.54$
250	118	$118/250 = 0.472$
500	263	$263/500 = 0.526$
1,000	489	$489/1,000 = 0.489$
5,000	2,531	$2,531/5,000 = 0.506$

FIGURE 7.3.2   Relative frequencies of tossing heads.

probability The value approached by the relative frequency of the event as the number of repetitions of the experiment increases.

law of large numbers If p represents the probability that an event A occurs and r represents the relative frequency of A, then r approaches p as the number of repetitions increases.

As seen in **Figure 7.3.2**, in the experiment of tossing a fair coin, the fraction of heads gets increasingly closer to $\frac{1}{2}$ as you keep repeating the experiment. It seems reasonable to call this limiting value the **probability** of tossing a head, and this method for assigning probabilities is referred to formally as the **law of large numbers**.

If p represents the probability that an event A occurs and r represents the relative frequency of A, then r approaches p as the number of repetitions increases.

Therefore, we assign probabilities to events (or outcomes) in such a way that the relative frequency agrees with our assignments *in the long run*.

> **uniform sample space** A finite sample space in which each outcome is equally likely to occur.
>
> **probability of event A** (in a uniform sample space S) A function p with domain consisting of the subsets (events) of S and range consisting of the unit interval [0, 1]. For any event A,
>
> $$p(A) = \frac{n(A)}{n(S)}.$$

Next, we need to develop a reasonable way to assign probabilities that satisfy the law of large numbers. First, we define a **uniform sample space** to be a sample space for an experiment that has a finite number of outcomes, each of which is equally likely to occur. For example, $S_1 = \{H, T\}$ is a uniform sample space for tossing a fair coin once. Similarly, $S_2 = \{HH, HT, TH, TT\}$ is a uniform sample space for tossing a fair coin twice (or tossing two fair coins). It is important to distinguish S_2 from $S_3 = \{0 \text{ heads}, 1 \text{ head}, 2 \text{ heads}\}$, which is also a sample space for tossing a fair coin twice, because S_3 is not a *uniform* sample space. As you can see by comparing S_2 and S_3, the outcome 1 head will occur twice as often as the outcomes 0 heads or 2 heads—thus, S_3 does not contain equally likely outcomes.

We can now offer the following definition. If S is a uniform sample space and A is an event of S, then the **probability of event A**, denoted $p(A)$, is the *function* defined by

$$p(A) = \frac{n(A)}{n(S)}.$$

It is important to point out that probability is a function whose domain is all events (subsets) of a sample space and whose range is the closed interval of real numbers [0, 1]. Note that $p(A) = 0$ must mean that A contains no elements (meaning that event A *cannot* occur), and so A must be the empty set $\emptyset$. Furthermore, $p(A) = 1$ must mean that $n(A) = n(S)$, so A contains all elements of the sample space (meaning that A *must be S*).

? ## Example 1

A fair coin is tossed twice. Then $S = \{HH, HT, TH, TT\}$ can be used to represent a uniform sample space for this experiment. Events A, B, and C can be listed as follows.

   A: at least 1 head occurs $= \{HH, HT, TH\}$
   B: exactly 1 head occurs $= \{HT, TH\}$
   C: more than 2 heads occur $= \emptyset$

Then the probabilities of A, B, and C are given as follows.

$$p(A) = \frac{n(A)}{n(S)} = \frac{3}{4}$$

$$p(B) = \frac{n(B)}{n(S)} = \frac{2}{4} = \frac{1}{2}$$

$$p(C) = \frac{n(C)}{n(S)} = \frac{0}{4} = 0. \quad \blacklozenge$$

Example 2 may be of interest to those with an interest in the gambling game of craps played at all casinos.

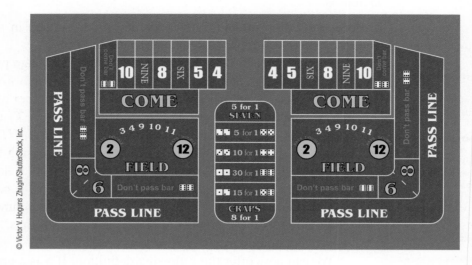

A typical craps gaming table.

? Example 2

Two fair dice are rolled, and their outcomes are recorded. Then S can be listed as in **Figure 7.3.3**.

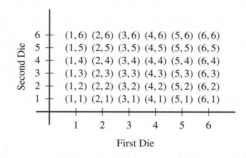

FIGURE 7.3.3 Two-dice chart.

Note that S is a uniform sample space. Find the probabilities of the following events.
(a) Doubles are rolled.
(b) The sum of the dice is greater than 8.
(c) The sum of the dice is even, and the first die is odd.

⚙ Solution

(a) Let D = doubles = $\{(1, 1), (2, 2), (3, 3), (4, 4), (5, 5), (6, 6)\}$.
Then $n(D) = 6$, whereas $n(S) = 36$. So,

$$p(D) = \frac{n(D)}{n(S)} = \frac{6}{36} = \frac{1}{6}.$$

Note that we can also list D graphically, where doubles are in color. (See **Figure 7.3.4**.)

Second Die
6 (1,6) (2,6) (3,6) (4,6) (5,6) (6,6)
5 (1,5) (2,5) (3,5) (4,5) (5,5) (6,5)
4 (1,4) (2,4) (3,4) (4,4) (5,4) (6,4)
3 (1,3) (2,3) (3,3) (4,3) (5,3) (6,3)
2 (1,2) (2,2) (3,2) (4,2) (5,2) (6,2)
1 (1,1) (2,1) (3,1) (4,1) (5,1) (6,1)
1 2 3 4 5 6
First Die

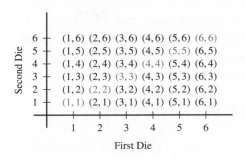

FIGURE 7.3.4 Two-dice chart for doubles.

(b) Let G = "The sum of the dice is greater than 8." The sum must therefore be 9, 10, 11, or 12. These pairs are colorized in **Figure 7.3.5**.

Second Die
6 (1,6) (2,6) (3,6) (4,6) (5,6) (6,6)
5 (1,5) (2,5) (3,5) (4,5) (5,5) (6,5)
4 (1,4) (2,4) (3,4) (4,4) (5,4) (6,4)
3 (1,3) (2,3) (3,3) (4,3) (5,3) (6,3)
2 (1,2) (2,2) (3,2) (4,2) (5,2) (6,2)
1 (1,1) (2,1) (3,1) (4,1) (5,1) (6,1)
1 2 3 4 5 6
First Die

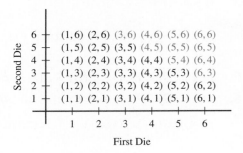

FIGURE 7.3.5 Two-dice chart for sum greater than 8.

Then $n(G) = 10$, whereas $n(S) = 36$. So,

$$p(G) = \frac{n(G)}{n(S)} = \frac{10}{36} = \frac{5}{18}.$$

(c) Let B = "The sum of the dice is even, and the first die is odd." The possible dice combinations are underlined in **Figure 7.3.6**.

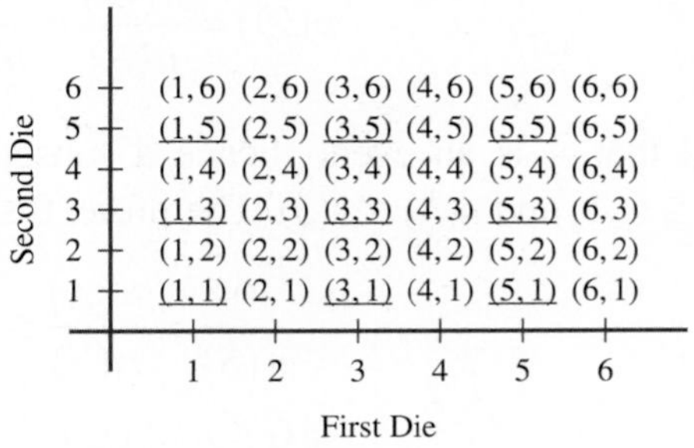

Second Die
6 (1,6) (2,6) (3,6) (4,6) (5,6) (6,6)
5 (1,5) (2,5) (3,5) (4,5) (5,5) (6,5)
4 (1,4) (2,4) (3,4) (4,4) (5,4) (6,4)
3 (1,3) (2,3) (3,3) (4,3) (5,3) (6,3)
2 (1,2) (2,2) (3,2) (4,2) (5,2) (6,2)
1 (1,1) (2,1) (3,1) (4,1) (5,1) (6,1)
1 2 3 4 5 6
First Die

FIGURE 7.3.6 Two-dice chart for part (c).

Then $n(B) = 9$, whereas $n(S) = 36$, so

$$p(B) = \frac{n(B)}{n(S)} = \frac{9}{36} = \frac{1}{4}. \quad \blacklozenge$$

It is important to note that we are not considering the probability of events in which the sample space is not uniform.

In summary, we have defined probability as a function whose domain is all subsets A of a uniform sample space and whose range is the interval $[0, 1]$. We have seen that the functional assignment to each subset is a matter of counting outcomes in the subset and in our sample space (assuming it is finite). Some very useful properties of the probability function are consequences of this definition.

Properties of Probability

If S is a uniform sample space that is finite and nonempty and if A and B are events, the following properties hold.

1. $p(S) = 1$
2. $p(\varnothing) = 0$
3. $0 \le p(A) \le 1$
4. $p(x) = \dfrac{1}{n(S)}$ for any outcome $x \in S$
5. $p(A \cup B) = P(A) + P(B) - P(A \cap B)$
6. $p(A \cup B) = P(A) + P(B)$ if A and B are mutually exclusive
7. $p(\overline{A}) = 1 - p(A)$ where $\overline{A}$ is the complement of A

We will verify some of these properties.

1. $p(S) = 1$ directly from the definition $p(A) = \dfrac{n(A)}{n(S)}$.

$$p(S) = \frac{n(S)}{n(S)} = 1.$$

Recall that $p(A)$ is a *function*, so $p(S)$ simply means to evaluate that function at $A = S$.

2. $p(\varnothing) = 0$ directly from the definition.

$$p(\varnothing) = \frac{n(\varnothing)}{n(S)} = \frac{0}{n(S)} = 0.$$

3. Recall that A is an event (hence a subset of S), so $\varnothing \subseteq A$ and $A \subseteq S$. Then $n(\varnothing) \le n(A)$ and $n(A) \le n(S)$. Therefore, $0 \le n(A) \le n(S)$. Dividing by $n(S)$, we get

$$\frac{0}{n(S)} \le \frac{n(A)}{n(S)} \le \frac{n(S)}{n(S)}$$

$$0 \le p(A) \le 1.$$

4. This proof is left as an exercise.
5. This property should look similar to the addition principle:

$$n(A \cup B) = n(A) + n(B) - n(A \cap B).$$

Dividing each term by $n(S)$, we get

$$\frac{n(A \cup B)}{n(S)} = \frac{n(A)}{n(S)} + \frac{n(B)}{n(S)} - \frac{n(A \cap B)}{n(S)}$$

$$p(A \cup B) = p(A) + p(B) - p(A \cap B).$$

The proofs of properties 6 and 7 are left as exercises. Example 3 illustrates some of these properties.

❓ Example 3

Suppose A and B are events such that $p(A) = \dfrac{1}{2}, p(B) = \dfrac{1}{3}$, and $p(A \cap B) = \dfrac{1}{6}$. Find the following probabilities.

(a) $p(\bar{A})$ (b) $p(\bar{B})$ (c) $p(A \cup B)$ (d) $p(\bar{A} \cup \bar{B})$ (e) $p(\bar{A} \cap \bar{B})$

⚙ Solution

(a) Using property 7, we have

$$p(\bar{A}) = 1 - p(A) = 1 - \frac{1}{2} = \frac{1}{2}.$$

(b) Using property 7, we have

$$p(\bar{B}) = 1 - p(B) = 1 - \frac{1}{3} = \frac{2}{3}.$$

(c) Using property 5, we have

$$p(A \cup B) = p(A) + p(B) - p(A \cap B) = \frac{1}{2} + \frac{1}{3} - \frac{1}{6} = \frac{2}{3}.$$

(d) Applying De Morgan's law, $\bar{A} \cup \bar{B} = \overline{A \cap B}$, we use property 7.

$$p(\bar{A} \cup \bar{B}) = p(\overline{A \cap B}) = 1 - p(A \cap B) = 1 - \frac{1}{6} = \frac{5}{6}.$$

(e) Applying De Morgan's law, $\bar{A} \cap \bar{B} = \overline{A \cup B}$, we use property 7 and our result from part (c).

$$p(\bar{A} \cap \bar{B}) = p(\overline{A \cup B}) = 1 - p(A \cup B) = 1 - \frac{2}{3} = \frac{1}{3}. \quad \blacklozenge$$

❓ Example 4

Let $S = \{1, 2, 3, 4, 5, 6, 7, 8\}, A = \{1, 3, 4, 5\}, B = \{2, 4, 6\}$, and $C = \{1, 3, 5, 8\}$. Find each of the following probabilities.

(a) $p(A)$ and $p(B)$ (b) $p(A \cap C)$

(c) $p(A \cup C)$ (d) $p(\bar{B})$

(e) $p(\bar{A} \cap \bar{C})$ (f) $p(B \cup C)$

⚙ Solution

(a) Because $n(A) = 4$, $n(B) = 3$, and $n(S) = 8$, we have

$$p(A) = \frac{n(A)}{n(S)} = \frac{4}{8} = \frac{1}{2} \quad \text{and} \quad p(B) = \frac{n(B)}{n(S)} = \frac{3}{8}.$$

(b) First, we find $A \cap C = \{1, 3, 5\}$. So $n(A \cap C) = 3$ and

$$p(A \cap C) = \frac{n(A \cap C)}{n(S)} = \frac{3}{8}.$$

(c) Because $p(A \cap C) = \frac{3}{8}$, $p(A) = \frac{1}{2}$, and $p(C) = \frac{n(C)}{n(S)} = \frac{4}{8} = \frac{1}{2}$, we can use property 5.

$$p(A \cup C) = p(A) + p(C) - p(A \cap C) = \frac{1}{2} + \frac{1}{2} - \frac{3}{8} = \frac{5}{8}.$$

(d) Because $\bar{B}$ is the complement of B and because $p(B) = \frac{3}{8}$, we can use property 7.

$$p(\bar{B}) = 1 - p(B) = 1 - \frac{3}{8} = \frac{5}{8}.$$

(e) There are a couple of approaches we can use for this problem. We can do the problem directly by finding the required sets.

$$\bar{A} = \{2, 6, 7, 8\} \quad \bar{C} = \{2, 4, 6, 7\} \quad \bar{A} \cap \bar{C} = \{2, 6, 7\}$$

So the required probability is given by the following equations.

$$p(\bar{A} \cap \bar{C}) = \frac{n(\bar{A} \cap \bar{C})}{n(S)} = \frac{3}{8}.$$

A quicker approach is to use De Morgan's laws, which state that

$$\bar{A} \cap \bar{C} = \overline{A \cup C}.$$

Because $p(A \cup C) = \frac{5}{8}$, then using property 7, we get

$$p(\overline{A \cup C}) = 1 - p(A \cup C) = 1 - \frac{5}{8} = \frac{3}{8}.$$

(f) Because $B \cap C = \varnothing$, then B and C are mutually exclusive; by property 6, we have

$$p(B \cup C) = p(B) + p(C) = \frac{3}{8} + \frac{1}{2} = \frac{7}{8}. \quad \blacklozenge$$

One application of probability that is useful in making decisions involving financial risk, such as buying a lottery ticket or gambling in a casino, concerns making predictions about your average return. More generally, if each outcome of an experiment is a

> **expected value** The average value per repetition of a numerically valued experiment after a large number of repetitions. Signified as $E(X)$.

numerical quantity (such as the payoff from a wager), then the *expected value* of the experiment is the average value per occurrence of the experiment after a large number of repetitions. If X represents this numerical quantity, then the expected value of X is usually denoted by $E(X)$. For instance, suppose your friend offers to pay you $1 every time you randomly draw a heart from a deck of bridge cards but to collect 50 cents from you if you draw a club, spade, or diamond. If this is repeated indefinitely, you would win $1 one-fourth of the time but lose 50 cents three-fourths of the time. In the long run, you would expect a return of $E(X) = (+1)\left(\dfrac{1}{4}\right) + (-0.50)\left(\dfrac{3}{4}\right) = -0.125$ dollars per repetition of this game. In general, if X may be any one of a finite number of numerical values $x_1, x_2, \ldots, x_n$ associated with n outcomes of an experiment, then the **expected value** is defined to be

$$E(X) = x_1 p(x_1) + x_2 p(x_2) + \ldots x_n p(x_n).$$

 ## Example 5

Your school sponsors a raffle in which 1,000 tickets are sold for $1 each. Three tickets will be randomly drawn. One ticket number will win $100, the second ticket will win $50, and the third will win $25. If you buy one ticket, what is the expected value of your investment?

Solution

The probability that you win any single one of the prizes is 1/1,000. Therefore, the expected value is

$$E(X) = (+99)\left(\dfrac{1}{1,000}\right) + (+49)\left(\dfrac{1}{1,000}\right) + (+24)\left(\dfrac{1}{1,000}\right) + (-1)\left(\dfrac{997}{1,000}\right) = -0.825.$$

This means that if this raffle were held many times and you bought one ticket each time, the average amount you would win is a negative—that is, −$0.825 per purchase. In other words, you could expect to lose an average amount of 82.5 cents per raffle. Of course, most people enter raffles for the fun of it and to support a good cause. Just don't expect to get rich! ◆

We will conclude this brief look at probability by looking at an application in genetics discovered by **Gregor Mendel** (1822–1884) while he was cross-pollinating garden pea plants. He found that the flower color of a plant could be reasonably predicted by the flower color of its parents if he knew the genes responsible for the flower color. He discovered that each gene consists of a pair of *alleles*, one of which is dominant and one of which is recessive. For the gene-determining color, we let R represent the dominant red allele and let w represent the recessive white allele. Then the possible combinations of alleles (called *genotypes*) are as follows.

> RR represents red flower.
> Rw represents red flower (because R is dominant and w is recessive).

© Filip Fuxa/ShutterStock, Inc.

wR represents red flower (because *R* is dominant and *w* is recessive).
ww represents white flower.

In one experiment, Mendel crossed pure reds (*RR*) with pure whites (*ww*), and he obtained all red-flowered peas. He then crossed these offspring, and he obtained 705 red-flowered and 224 white-flowered offspring. If *R* is dominant and *w* is recessive, how can we explain these results? We can describe the allele mixing by using a chart, called a *Punnett square*. If we crossed pure reds (*RR*) with pure whites (*ww*), the Punnett square would look like the one that follows.

	w	*w*
R	*Rw*	*Rw*
R	*Rw*	*Rw*

Note that all offspring are *Rw*, which are red-flowered (remember *R* is dominant and *w* is recessive). This is consistent with the first stage of Mendel's experiment. If we now crossed these offspring, the Punnett square would look like the one that follows.

	R	*w*
R	*RR*	*Rw*
w	*wR*	*ww*

Considering this last set of genotypes as a sample space, we have $S = \{RR, Rw, wR, ww\}$. Because *RR*, *Rw*, and *wR* all produce red-flowered peas whereas *ww* produces white-flowered peas, we have the following probabilities.

$$p(\text{red-flowered}) = \frac{3}{4} = 0.75$$

$$p(\text{white-flowered}) = \frac{1}{4} = 0.25$$

To compare with Mendel's results on the $705 + 224 = 929$ plants, let's compute the relative frequencies of each color.

$$r(\text{red-flowered}) = \frac{705}{929} \approx 0.76$$

$$r(\text{white-flowered}) = \frac{224}{929} \approx 0.24$$

Our probabilities are consistent with the relative frequencies obtained by Mendel. This approach used in genetics extends to more complicated genetic traits and can be used by genetic counselors when advising prospective parents who may carry genes for some human diseases.

Name _____

Exercise Set **7.3** ◯◯◯←

1. A fair coin is tossed twice, and the coin faces are recorded.

    (a) List the sample space S.
    (b) Find the probability that at most one head occurs.
    (c) Find the probability that more than one tail occurs.
    (d) Find the probability that the two coin faces are the same.

2. A fair coin is tossed three times, and the coin faces are recorded.

    (a) List the sample space S.
    (b) Find the probability that at least two heads occur.
    (c) Find the probability that no more than one head occurs.
    (d) Find the probability that more than three tails occur.

3. Two fair dice (six-sided) are rolled, and their outcomes are recorded.

    (a) List the sample space S.
    (b) Find the probability that the sum of the pips is at most 6.
    (c) Find the probability that the sum of the pips is odd and the first die is at least 5.
    (d) Find the probability that the first die is 2 less than the second.
    (e) Find the probability that the first die is at least 2 less than the second.

4. Explain why property 4 (under Properties of Probability) is true. (*Hint:* How many outcomes are there in S?)

5. Prove property 6 (under Properties of Probability).

6. Prove property 7 (under Properties of Probability).

7. A positive integer is picked at random between the numbers 10 and 20, inclusive, and its value is recorded.

    (a) List the sample space S.
    (b) Find the probability that the number is even.
    (c) Without listing the event, find the probability that the number is odd.
    (d) Find the probability that the number is prime. (*Hint:* Recall that a prime number is evenly divisible by only 1 and itself.)
    (e) Find the probability that the number is a solution to the inequality $x^2 < 300$.

8. In Exercise 7, let $A = \{10, 12, 13, 15, 16\}$ and $B = \{12, 13, 14, 17, 18, 20\}$. Find each of the following probabilities.

    (a) $p(A)$ and $p(B)$ (b) $p(\bar{A})$ and $p(\bar{B})$
    (c) $p(A \cap B)$ and $p(A \cup B)$ (d) $p(\bar{A} \cap \bar{B})$ and $p(\bar{A} \cup \bar{B})$
    (e) $p(\bar{A} \cap B)$ and $p(\bar{A} \cup B)$ (f) $p(A \cap \bar{B})$ and $p(A \cup \bar{B})$

9. Let A and B be events such that $p(A) = \frac{1}{2}$, $p(B) = \frac{3}{4}$, and $p(A \cap B) = \frac{3}{8}$. Find each of the following probabilities.

(a) $p(\overline{A})$ (b) $p(\overline{B})$ (c) $p(A \cup B)$ (d) $p(\overline{A} \cap \overline{B})$ (e) $p(\overline{A} \cup \overline{B})$

10. Let A and B be events such that $p(A) = \frac{2}{3}$, $p(B) = \frac{1}{4}$, and $p(A \cup B) = \frac{5}{6}$. Find each of the following probabilities.

(a) $p(\overline{A})$ (b) $p(\overline{B})$ (c) $p(A \cap B)$ (d) $p(\overline{A} \cap \overline{B})$ (e) $p(\overline{A} \cup \overline{B})$

11. Sets A, B, and C form a partition of U if $U = A \cup B \cup C$, $A \cap B = \varnothing$, $A \cap C = \varnothing$, and $B \cap C = \varnothing$. In this case, show that $p(A) + p(B) + p(C) = 1$.

12. A single card is picked at random from a deck of 52 cards. Find the probability that the card is (a) a king, (b) red, (c) a spade, (d) a king or a spade, (e) a black ace, and (f) a face card. (*Note*: Face cards are jacks, queens, and kings.)

13. A four-digit number is picked at random. (The first digit must be nonzero.) Find each of the following probabilities.

(a) The first digit is a 5.
(b) The number is a multiple of 100.
(c) The number is even.
(d) The number is not a multiple of 10. (*Hint*: Find the probability that the number *is* a multiple of 10; then apply a property of probability.)

14. A fair coin is tossed four times, and the coin faces are recorded. Find each of the following probabilities.

(a) Two heads occur. (*Note*: Recall that this means *exactly* two heads occur.)
(b) More than two heads occur.
(c) At most two heads occur.
(d) At least one head occurs.

15. One red die and one green die are rolled, and the outcomes are recorded. Both dice are four-sided with one, two, three, and four pips on each side. Find the probabilities of the following events.

(a) Both dice show an odd number of pips.
(b) At least one die is odd.
(c) The sum of the pips is less than 7.
(d) The number of pips showing on the red die is larger than the number on the green die.

16. A liberal arts mathematics class has the following composition of students.

	Male	Female
Freshmen	2	3
Sophomores	3	8
Juniors	1	4
Seniors	5	6

If a student is picked at random, find each of the following probabilities.

(a) The student is male.
(b) The student is a junior.
(c) The student is not a senior.
(d) The student is a female upperclassman (i.e., junior or senior).

17. In Exercise 16, find the following probabilities.

(a) The student is female.
(b) The student is not a junior.
(c) The student is a male freshman or male senior.
(d) The student is a male freshman or female junior.

18. Let $U = \{-3, -2, -1, 0, 1, 2, 3\}$, $A = \{-2, -1, 0, 1\}$, and $B = \{-1, 1, 2\}$. Find the probability of each of the following events, assuming a number is picked at random from U.

(a) $A \cap \bar{B}$ (b) $\bar{A} \cup B$ (c) $\overline{\bar{A} \cap B}$ (d) $\overline{A \cup \bar{B}}$

19. In a survey of 1,000 college students, 430 are currently taking chemistry, 640 are currently taking history, and 225 are currently taking both chemistry and history. (See Exercise 25 from Section 7.2.) What is the probability that a student chosen from this survey

(a) is taking either chemistry or history?
(b) is taking chemistry, but not history?
(c) is taking neither chemistry nor history?

20. A biologist examines 100 trees and finds that 80 of them either are pine trees or are at least 20 ft high. She also finds that 60 of the trees are pine trees and 55 of the trees are at least 20 ft high. (See Exercise 26 from Section 7.2.) What is the probability that a tree in this study

(a) is at least 20 ft high?
(b) is less than 20 ft high?
(c) is at least 20 ft high but not a pine?

21. A roulette wheel used in a casino has 38 slots numbered 1, 2, 3, . . ., 36, as well as 0 and 00. The odd numbers are colored black, the even numbers are colored red, and the numbers 0 and 00 are colored green. A ball is rolled around the wheel and randomly drops into one of the slots. Find each of the following probabilities.

(a) The ball lands in a red slot.
(b) The ball lands in a green slot.
(c) The ball lands on a prime number.
(d) The ball lands in a slot numbered 18–25.
(e) The ball lands in an even slot or in a green slot.

22. In Exercise 21, suppose a $1 bet on a red slot returns $1 (plus the $1 you bet) if the ball lands on red. What is the expected value of such a bet?

23. Each week, the state lottery pays out $100,000 to the person who has chosen the same six-digit number that is randomly drawn from a barrel with 1 million such numbers in it. If each lottery ticket (with one number on it) costs $2, what is your expected value if you purchase one ticket?

24. Suppose you work in an office where there is a weekly football pool. For $5, you get to pick one digit. If your pick is the last digit in the aggregate total of points scored in professional football games that week, then you win $45. What is the expected value of this bet?

25. Suppose we cross-pollinate mixed-red peas (*Rw*) with white peas (*ww*). Construct a Punnett square and compute the probability of obtaining red-flowered peas.

26. If we then cross-pollinate the red-flowered offspring from Exercise 25, what percentage of *their* offspring would you expect to be white?

27. Human hair color (brown and blonde) can be examined with Punnett squares, where brown (*B*) is dominant and blonde (*b*) is recessive. If two brown-haired people have a child, is it possible for that child to be blonde? In this case, what probability would you associate with the child having blonde hair?

28. In humans, left-handedness appears to be recessive to the dominant right-handedness. Suppose a right-handed person *A* has a parent who is left-handed.

(a) If *A* mates with a left-handed person, what is the probability that the couple produces a left-handed child?
(b) If *A* mates with a right-handed person who has a left-handed parent, what is the probability that the couple produces a left-handed child?
(c) If *A* mates with a right-handed person, both of whose parents are right-handed, is it possible for the couple to have a left-handed child?

29. A type of dwarfism (characterized by short limbs) is controlled by a dominant allele. Is it possible for two dwarf parents to produce a normally proportioned child? In this case, what probability would you associate with the child being normally proportioned?

30. In the game of backgammon, you can "bump" your opponent based on the roll of two dice. If your opponent leaves a piece unprotected 8 spots from you and your dice total 8 (say, 5 and 3), then you can move to his position and bump him to the beginning of the board. If his unprotected piece is 5 spots from you, your dice can either total 5 (say, 3 and 2), or either of the two dice can be a 5 (say, 4 and 5), in order to bump him. Find the probability that you can bump your opponent based on the following situations.

(a) His unprotected piece is 7 spots from you.
(b) His unprotected piece is 5 spots from you.
(c) He has two unprotected pieces 4 and 8 spots from you. (*Note*: You are finding the probability that either (or both) can be bumped.)

31. (Continuation of Exercise 30.) If your opponent leaves one piece unprotected, how many spots from you will have the highest probability that you can bump him or her? (*Hint*: Bumps can occur from positions 1–12, inclusive. You will need to compute the probability for each position to find the highest value.)

32. Related to the idea of probability is that of odds. Whereas the probability of an event is

$$p(E) = \frac{n(E)}{n(S)} = \frac{\text{number of favorable outcomes}}{\text{number of total outcomes}},$$

we define the **odds for an event** as the ratio $p(E)/p(\bar{E})$. This, in turn, is then equal to

> **odds for an event** The ratio of the probability that the event does occur to the probability that it does not occur.

$$\frac{p(E)}{1-p(E)} = \frac{n(E)/n(S)}{1-n(E)/n(S)}$$

$$= \frac{n(E)}{n(S)-n(E)}$$

$$= \frac{\text{number of favorable outcomes}}{\text{number of unfavorable outcomes}}.$$

For example, if a fair coin is tossed twice, we can list the outcomes as $\{HH, HT, TH, TT\}$, so the odds of obtaining at least one head are $\dfrac{3\,\text{favorable}}{1\,\text{unfavorable}}$, or 3 to 1. This is usually denoted $3:1$, and it can be interpreted that three times as many outcomes have at least one head rather than no heads. Find the odds for the following events and experiments.

(a) Experiment: Toss a fair coin three times. Event: Obtain at least two heads.

(b) Experiment: Roll two fair dice. Event: Sum of dice is at most 5.

(c) Experiment: Pick a person at random from a group of four, including Hillary. Event: Hillary is picked.

(d) Experiment: Pick a card at random from a standard 52-card deck. Event: The card is red. (*Note*: This situation is called "even odds.")

33. In horse racing, odds are usually posted *against* a horse. The **odds against an event** are defined as the reciprocal of the odds in favor, that is, $p(\overline{E})/p(E)$ for an event E. This is then the reciprocal of all the ratios listed in Exercise 32. In horse racing, odds of $5:2$ against the horse winning would mean that for every \$5 bet against the horse, \$2 is bet in favor of the horse. List the odds (reduced form) against the horse winning for the following betting amounts.

> **odds against an event** The ratio of the probability that the event does not occur to the probability that it does occur.

(a) \$20,000 against, \$6,000 in favor

(b) \$5,000 against, \$5,000 in favor

(c) \$2,000 against, \$7,000 in favor

Name _____

Chapter Review Test **7** ◯◯◯ ←

1. True or false? Let A and B be subsets.

(a) $A \cap B \subseteq A \cup B$
(b) If A is a subset of the universal set U, then $A \cup \bar{A} = U$.
(c) $A \subseteq A \cap B$
(d) $\overline{A \cap B} = \bar{A} \cap \bar{B}$
(e) If A and B are mutually exclusive events, then $p(A \cap B) = 0$.
(f) If U has 5 elements in it, then the total number of subsets of U is 10.
(g) $A = (A - B) \cup (A \cap B)$
(h) If $A \subseteq B$, then $n(B) \leq n(A)$.

2. Consider the experiment of throwing a six-sided die twice and recording the sum of the numbers that occur. Is $S = \{0, 1, 2, 3, 4, 5, 6\}$ a valid sample space? Why or why not?

3. Let $U = \{s, t, u, v, w, x, y, z\}$, $A = \{s, u, v, y\}$, and $B = \{t, u, w, x, y\}$. Determine each of the following.

(a) $\bar{A}$ (b) $A \cap B$ (c) $A \cap \bar{B}$ (d) $\bar{A} \cup B$

(e) $n(B)$ (f) $n(\bar{A} \cap B)$ (g) $n(\overline{A \cup B})$

4. Draw Venn diagrams illustrating each of the following sets.

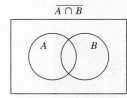

$\overline{A \cap B}$

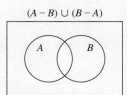

$(A - B) \cup (B - A)$

5. If $n(A \cup B) = 205$, $n(A) = 143$, and $n(B) = 87$, then find $n(A \cap B)$.

←

6. If $n(A \cup B) = 400$, $n(A - B) = 229$, and $n(B - A) = 150$, then find $n(A)$.

7. Let A and B be events such that $p(A) = \dfrac{1}{2}$, $p(B) = \dfrac{3}{4}$, and $p(A \cap B) = \dfrac{3}{8}$. Find the following probabilities.

(a) $p(\overline{A})$ (b) $p(\overline{B})$ (c) $p(A \cup B)$ (d) $p(\overline{A} \cap \overline{B})$ (e) $p(\overline{A} \cup \overline{B})$

8. In a survey of 1,000 marathon runners, 425 drink Pepsi every day, 703 drink Gatorade every day, and 288 drink both of these every day

(a) How many runners drink either Pepsi or Gatorade (or both)?
(b) How many runners drink Pepsi but not Gatorade?
(c) How many runners drink neither Pepsi nor Gatorade?

9. The probability that a southern pine beetle infects a Virginia pine tree is 0.17. The probability that an engraver beetle infects a Virginia pine tree is 0.11. The probability that both of these beetles infect the same tree is 0.05. What is the probability of a Virginia pine being infected by neither the southern pine beetle nor the engraver beetle?

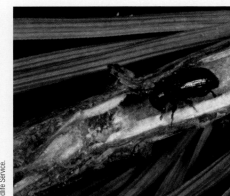

10. You are paying roulette at a fair wheel with 38 slots numbered 0, 00, and 1–36. The positive even-numbered slots are red, the odd-numbered slots are black, and 0 and 00 are green. What are the probabilities of the following events associated with a spin of the wheel?

   (a) The ball lands on an odd number.
   (b) The ball lands on a red slot or a green slot.
   (c) The ball lands on numbers 1–12 or a red slot.

11. Draw a Punnett square to help answer each of the following questions. In humans, left-handedness appears to be a recessive trait compared to the dominant right-handedness. Suppose Lars, a right-handed person, has a parent who is left-handed (i.e., his genotype has one of each allele).

   (a) If Lars mates with a left-handed person, what is the probability that the couple produces a left-handed child?
   (b) If Lars mates with a right-handed person who has a left-handed parent, what is the probability that the couple produces a left-handed child?

12. Each week, the state lottery pays out $500,000 to the person who has chosen the same seven-digit number that is randomly drawn from a barrel with 2 million such numbers in it. If each lottery ticket (with one number on it) costs $2, what is your expected value if you purchase one ticket?

Chapter Objectives ☑◻◻ ← *check off when you've completed an objective*

- ◻ Construct a histogram.
- ◻ Compute the mean and median of a small data set.
- ◻ Estimate the mean and median from a histogram or grouped data.
- ◻ Know the difference between a skewed and a symmetric distribution.
- ◻ Compute the standard deviation of a small data set.
- ◻ Estimate the standard deviation from a histogram or grouped data.
- ◻ Use a density function to find probabilities.
- ◻ Know the properties of a normal distribution.
- ◻ Use the standard normal distribution table to find percentages and probabilities of intervals in a population.
- ◻ Know the central limit theorem.
- ◻ Learn about Carl Friedrich Gauss.
- ◻ Know the properties of a sampling distribution.
- ◻ Construct a confidence interval with a given confidence level.
- ◻ Know how to interpret the results of an opinion poll.

Navigate Companion Website

go.jblearning.com/johnson

Visit go.jblearning.com/Johnson to access a Laboratory Manual, Student Solutions Manual, Interactive Glossary, and Interactive Flashcards.

8.1 Representation of Data

8.2 Dispersion of Data

8.3 The Normal Distribution

8.4 Confidence Intervals

chapter 8 Statistics

"The great body of physical science, a great deal of the essential fact of financial science, and endless social and political problems are only accessible and only thinkable to those who have had a sound training in mathematical analysis, and the time may not be very remote when it will be understood that for complete initiation as an efficient citizen of one of the new great complex worldwide States that are now developing, it is as necessary to be able to compute, to think in averages and maxima and minima, as it is now to be able to read and write."

—H. G. Wells

Numeracy, a term coined by the American mathematician John Paulos, describes the ability of a person to use numbers to enhance understanding. This term, you'll note, is an echo of the word *literacy*, the ability of someone to use letters to read and write. The advance of society in recent years has been accompanied by an explosion in new ways to exploit and utilize numbers. To be prepared to meet new challenges in a chosen career as well as make informed and responsible decisions in daily life, today's citizen simply cannot afford to be innumerate any more than he or she can afford to be illiterate. **Statistics** can be defined as the science of collecting, organizing, and analyzing large amounts of data in a cohesive manner for the purpose of reaching meaningful conclusions.

> **statistics** The science of collecting, organizing, and analyzing large collections of data in a manner conducive to reaching meaningful conclusions.

Statistics are everywhere. Newspapers, magazines, and TV news shows are full of statistical information on everything from weather forecasts, to the performance of sports stars, to the health of the economy. Control of the quality of mass-produced items as dissimilar as walnuts and automobiles is dependent on the statistical analysis of small samples of the product. Huge industries make major decisions based on marketing surveys of public opinion. Nielson ratings, for example, determine the fate of every television program. Insurance companies decide on the payments and premiums of millions of people based on the conclusions and decisions reached by their statisticians (known as *actuaries*). The winners in many areas of today's job market will be those best equipped to tackle problems such as these. An understanding of some of the basic principles of statistical methods gives you a snowplow with which to penetrate and organize what sometimes seems to be a blizzard of numbers produced by the modern world. Your future career may depend on your possession of such a tool.

The table below was posted on the U.S. Bureau of Labor Statistics website in April 2010. By the end of this chapter, we will be able to understand the rightmost column regarding the test of significance based on a 90% confidence interval.

Estimated Employment Change March to April 2010

	Not Seasonally Adjusted		Seasonally Adjusted		
	Normal Seasonal Movement	Estimated Monthly Change	Estimated Monthly Change	Minimum Significant Change*	Pass Significance Test
Total nonfarm	868.0	1,158.0	290.0	100.8	Yes
Total private	853.0	1,084.0	231.0	97.1	Yes
Goods-producing	218.0	283.0	65.0	34.8	Yes
Service-providing	650.0	875.0	225.0	94.3	Yes
Private service-providing	635.0	801.0	166.0	90.3	Yes
Mining and logging	4.0	11.0	7.0	7.3	
Logging	−1.4	−0.8	0.6	1.3	
Mining	4.8	11.6	6.8	7.1	
Oil and gas extraction	−0.5	0.6	1.1	2.7	
Mining, except oil and gas	5.8	6.1	0.3	2.5	
Coal mining	−0.3	−0.1	0.2	1.4	

	Not Seasonally Adjusted		Seasonally Adjusted		
	Normal Seasonal Movement	**Estimated Monthly Change**	**Estimated Monthly Change**	**Minimum Significant Change***	**Pass Significance Test**
Support activities, mining	−0.5	4.9	5.4	5.3	Yes
Construction	195.0	209.0	14.0	22.9	
Construction of buildings	21.2	26.4	5.2	10.1	
Residential building	9.7	5.7	−4.0	5.9	
Nonresidential building	11.5	20.7	9.2	7.9	Yes
Civil engineering construction	59.5	68.7	9.2	9.3	

*Note: Significant over-the-month changes are calculated at a 90% confidence level. The standard error is used for a 1-month change.

Source: U.S. Bureau of Labor Statistics.

Normal seasonal movements, estimated employment change between March and April 2010, in thousands, and test of significance. (*Bureau of Labor Statistics.*)

8.1   Representation of Data

Suppose you were asked to clean up a messy room strewn with shoes, socks, shirts, books, papers, CDs, and candy bars. What would you do first? Making sense of a large collection of objects usually involves organization into categories. You would probably put your shoes in the closet, your socks in a dresser drawer, shirts on hangers, books on a shelf, and so on. The same type of strategy applies to organizing a large collection of numbers. Debra is a certified public accountant (CPA) and manager for a large accounting firm who wishes to present an overview to a prospective client of the amounts of federal income tax paid out by a particular group of 40 people. The dollar amounts are shown in the following table.

8,030	35,362	4,215	25,339	12,325	17,112	12,607	9,288
9,505	7,450	14,310	11,608	44,501	25,815	10,204	7,700
15,108	13,004	9,883	39,440	29,210	24,225	8,024	14,811
48,204	6,707	31,374	11,258	18,200	31,290	22,104	27,883
35,692	21,450	8,404	6,730	30,050	19,455	18,310	26,005

interval classes
Adjacent intervals of real numbers used for sorting a large collection of data values.

boundary markers
The numbers that form the left and right endpoints of the interval classes.

range The difference between the largest and smallest values in a collection.

A simple list of these numbers would not reveal much information at a glance. Therefore, as we do when cleaning a messy room by sorting items, we shall place these numbers into separate—and adjacent—categories known as **interval classes** that are separated by divisional **boundary markers**. To form the proper intervals, we first survey the set shown here and note that the least and greatest tax amounts paid are $4,215 and $48,204, respectively. The difference between the high and low values is called the **range**, and in this case, it is equal to $48,204 − $4,215 = $43,989, which provides a rough idea of the total spread of the data. To accommodate every number in the list, we must ensure that the distance from the left boundary of the first interval class to the right boundary of the last interval class is larger than the range. If we wish to sort our numbers into classes of width 10,000 each, then the number of classes we will need must be greater than $\dfrac{43{,}989}{10{,}000} \approx 4.4$. For instance, five interval classes separated by the boundary markers 0, 10,000, 20,000, 30,000, 40,000, and 50,000 would include every value in the collection. We then count the number of values in each interval, known as the **frequency**. There are 11 tax amounts between 0 and 10,000; 13 amounts between 10,000 and 20,000; and so on. For consistency, we agree to count a data value that falls right on a boundary as belonging to the interval class in which it is the *smallest* member. In the present example, for instance, we would count 20,000 as belonging to the interval 20,000–30,000. The **relative frequency** of each class, denoted by the Greek letter rho ρ (pronounced "row"), is the fractional portion of the entire collection that lies in that class. In other words,

frequency The number of data values in an interval class.

relative frequency The fractional portion of a collection of data values that lie in a particular class. It is obtained by dividing the frequency of the class by the total number of values in the collection.

$$\rho = \frac{\text{frequency of class}}{\text{total number of data values}}.$$

This information is summarized in the table in **Figure 8.1.1**. Such a table is known as **relative frequency distribution**.

relative frequency distribution A table displaying interval classes and their corresponding relative frequencies.

Class	Frequency	Relative Frequency (ρ)
0–10,000	11	$11/40 = 0.275$
10,000–20,000	13	$13/40 = 0.325$
20,000–30,000	8	$8/40 = 0.2$
30,000–40,000	6	$6/40 = 0.15$
40,000–50,000	2	$2/40 = 0.05$

FIGURE 8.1.1   Relative frequency distribution.

Note that the sum of the relative frequencies in the rightmost column must be equal to 1. We summarize this fact by using the notation $\Sigma\rho = 1$. (Σ is the capital Greek letter sigma, usually standing for "sum.")

A large set of values arranged in groups, such as in the table in Figure 8.1.1, is also commonly called *grouped data*. Many textbooks and journals favor displaying grouped data in a form that is easily assessed by the human eye. By using blocks whose heights correspond to relative frequency in a coordinate system featuring the interval classes on the horizontal axis, we get a picture known as a **histogram**, as shown in **Figure 8.1.2**.

histogram A pictorial display of a relative frequency distribution with the interval classes represented on a horizontal axis and the relative frequencies represented by areas of rectangles situated over the intervals.

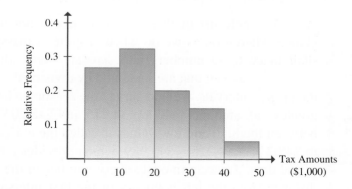

FIGURE 8.1.2   Histogram of federal income tax amounts.

We readily see that the interval containing the greatest portion of the original list is the 10,000–20,000 class, because the biggest block rests on that interval. Note that it is the size or *area* of that block that attracts our attention. Therefore, we want the area of each block to be proportional only to the relative frequency of the interval class upon which it rests. Because each ρ corresponds to the height of the block, we want all the blocks in our histogram to have the *same* width. By standardizing the widths (making each equal to, say, 1 unit), we then, for convenience, actually consider the *areas* of the blocks to be equal to the relative frequency of the interval class. In Figure 8.1.2, for example, we think of the first block as having area 0.275, the second block as having area 0.325, and so on.

The choice of the common interval width is somewhat subjective. Overly wide intervals result in a loss of accuracy in the data representation, yet a narrow width results in perhaps too many classes and therefore a rather raggedy picture. In practice, one rule of thumb is to choose a width leading to a total of 5 to 8 intervals, with the actual number depending on the purpose of the histogram. We saw in our example of grouping tax amounts that this number must be at least as large as $\left(\dfrac{\text{range}}{\text{width}}\right)$. A convenient first left boundary marker is chosen. (Clearly, this must be less than the smallest data value.) You add the width to obtain the right boundary of the first class, which serves also as the left boundary of the second class. This process is repeated until you have formed enough classes to ensure that the largest data value will be in the last class.

We noted that a true histogram must be faithful to the premise of equal interval widths, or else a false picture of the data can result. Observe what occurs when we change the boundaries of the last interval in Figure 8.1.1 to 30,000–50,000. (See **Figure 8.1.3**.)

Class	Frequency	Relative Frequency (ρ)
0–10,000	11	$11/40 = 0.275$
10,000–20,000	13	$13/40 = 0.325$
20,000–30,000	8	$8/40 = 0.2$
30,000–50,000	8	$8/40 = 0.2$

FIGURE 8.1.3   New frequency distribution with amended last class.

This, in turn, changes the corresponding histogram by increasing the total area over the new interval. The amount of increase is indicated by the light-shaded region in **Figure 8.1.4**.

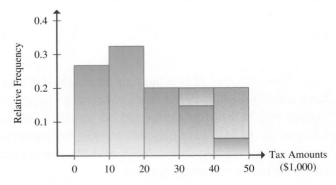

FIGURE 8.1.4   The changed picture as a result of unequal classes.

The misleading picture in Figure 8.1.4 indicates why equal-width intervals must be maintained in creating a histogram. It is clear that someone wishing to stress a certain aspect of the data could structure the interval classes in a favorable way, although that person would be running the risk of being charged with a misuse of statistics. **Figure 8.1.5** illustrates a type of picture graph other than a histogram that uses area to intentionally misrepresent information. Ironwheel Tire Company is showing how its share of the tire market has doubled from 15% to 30% between 2005 and 2013 by doubling the height of the second tire. However, because the width must also be doubled to keep the proper shape, the *area* of the second tire has actually increased by a factor of $2^2 = 4$. Someone rapidly scanning an advertisement such as this might very well receive the false impression of Ironwheel's market share increasing by a factor greater than 2.

FIGURE 8.1.5   A misleading picture graph. Although the percentage only doubled, the tire on the right is four times larger than the tire on the left.

Once Debra shows a prospective client her histogram from Figure 8.1.2, one question usually asked concerns the typical or average value of her data. This is a number that is supposed to be indicative of the center of the collection or, as statisticians like to call it, a **measure of central tendency**. We now define two such measures.

The **mean** $\overline{X}$ of a set of data values is obtained by dividing the sum of the values by the total number n of values in the set. Symbolically, we write

$$\overline{X} = \frac{\sum x}{n}.$$

The **median** M is a number in the middle of a collection whose members have already been ordered from smallest to largest value. The median will then exceed one-half of the values and be less than one-half of the values. The proximity of $\overline{X}$ to M always depends on how the numbers are distributed or scattered around them. We first determine $\overline{X}$ and M for some small data sets, using the same rule adopted in previous chapters:

> Use the full accuracy of your calculator for all intermediate calculations, but round off the final answer to one more significant digit than that used in the given data.

measure of central tendency The typical or average member of a collection of data.

mean The sum of the values in a collection of data, divided by the number of values. Signified by $\overline{X}$.

median A number that is greater than 50% of the values in a collection and less than 50% of the values. Signified by M.

? Example 1

Bernhard golfed every day last week and recorded scores of 96, 92, 94, 80, 88, 94, and 91. His mean score is

$$\bar{X} = \frac{96+92+94+80+88+94+91}{7} = \frac{635}{7} = 90.7$$

To find the median, you must first place the data in order from smallest to largest.

$$80 \quad 88 \quad 91 \quad \boxed{92} \quad 94 \quad 94 \quad 96$$

The median must be the value that is in the middle: $M = 92$. ♦

Note in Example 1 that the mean and median turned out to be different numbers. This is a typical occurrence for small sets. You should also observe that the presence of a rather unusually low score (compared to the rest), that is, 80, unduly affected $\bar{X}$. If we were to change 80 to 77, the new mean would be decreased even more—to 90.3—but the median, in contrast, would remain unchanged. This demonstrates both the sensitivity of $\bar{X}$ and the resistance of M to changes in extreme values, thus suggesting why the median is sometimes thought to be a better indicator of a central value.

? Example 2

The Sheffields want to create a family monthly budget. By examining the checkbook, they totaled their expenditures for rent, utilities, food, clothing, and insurance for each month of the previous year.

Jan	Feb	Mar	Apr	May	Jun	Jul	Aug	Sep	Oct	Nov	Dec
2,472	2,565	2,410	2,875	2,184	2,113	2,076	3,318	2,204	2,826	2,340	3,780

The mean is

$$\bar{X} = \frac{2,472+2,565+\cdots+3,780}{12} = \frac{31,163}{12} = 2,596.9.$$

To compute the median, we order the list.

$$2,076 \quad 2,113 \quad 2,184 \quad 2,204 \quad 2,340 \quad 2,410 \quad 2,472 \quad 2,565 \quad 2,826 \quad 2,875 \quad 3,318 \quad 3,780$$

Because there are an even number (12) of values here, the median M must be greater than the first six values and less than the last six. So it must be between 2,410 and 2,472. The convention is to use the number midway between them, giving us $M = \frac{2,410+2,472}{2} = 2,441$. The Sheffields noticed that the especially expensive months were April and October (insurance premiums due), August (vacation time), and especially December (Christmas). In fact, $\bar{X}$ is greater than M in large part because 3,318 and 3,780 are much farther above the median than the two smallest values, 2,076 and 2,113, are below it. ♦

These computations are fine for small sets of numbers, but what of larger collections? Is there a way to get a fairly good estimate of the mean and median without having to figure in every single value? Returning to Debra and her 40 tax amounts, we find that the mean turns out to be $\bar{X} = 19{,}279.5$. But suppose we only had the relative frequency distribution, repeated here, available to us.

Class	Frequency	Relative Frequency
0–10,000	11	11/40 = 0.275
10,000–20,000	13	13/40 = 0.325
20,000–30,000	8	8/40 = 0.2
30,000–40,000	6	6/40 = 0.15
40,000–50,000	2	2/40 = 0.05

group estimate of the mean An estimate of the mean made by summing the products of the midpoints and the corresponding relative frequencies of the interval classes of a relative frequency distribution.

The way to proceed to find the mean amount is similar to finding the mean mass per fruit in a truckload of fruit in which you have not weighed every piece but you do know that the load consists of 20% bananas, 35% oranges, and 45% apples. If you also know the average mass of a banana in this truck to be 225 grams (g), an orange to be 190 g, and an apple to be 210 g, it makes sense to *estimate* $\bar{X}$ by totaling the relative contributions made by each fruit type. For instance, the contribution of the bananas to $\bar{X}$ is 20% of 225, or $(0.20)(225) = 45$ g. We call this the **group estimate of the mean** (see **Figure 8.1.6**), and it is denoted as

$$\bar{X}_g = (0.20)(225) + (0.35)(190) + (0.45)(210) = 206.$$

$$(0.20)(225) = 45$$
$$(0.35)(190) = 66.5$$
$$(0.45)(210) = 94.5$$
$$\overline{}$$
$$206$$

FIGURE 8.1.6   Group estimate of a mean.

Similarly, we can compute a group estimate of the mean for the tax amounts by using the center or midpoint of each interval class as the average value from that interval. The midpoint of the first class in Figure 8.1.1 is 5,000, that of the second class is 15,000, and so on. We then measure the contribution made by the values from each class by multiplying the midpoint by the relative frequency. Adding the contributions gives us

$$\bar{X}_g = (0.275)(5{,}000) + (0.325)(15{,}000) + (0.20)(25{,}000) + (0.15)(35{,}000) + (0.05)(45{,}000)$$
$$= 18{,}750.$$

Because the actual mean $\bar{X} = 19{,}279.5$, this estimate is in error by 529.5, the price we pay for seeking a quicker and less tedious method. A general formula would be

$$\bar{X}_g = \sum \rho c,$$

distribution The manner in which a set of numbers is apportioned or arranged between the extremes.

which indicates that you sum all the products ρc, where ρ represents the relative frequency of each class and c (standing for center) is the corresponding midpoint.

The manner in which a set of numbers is apportioned or arranged between the extremes is known as the **distribution** of the data. This often takes the form of a summary presentation of correspondences between intervals of data and the percentages that those intervals constitute of the entire set. The histogram has a distinct advantage over other types of bar graph that give a picture of a distribution. First, we recall that the area of each block of the histogram of a data set represents the relative frequency of the associated interval or, equivalently, the percentage of the set that lies between the left and right endpoints of that interval. This would imply, in turn, that the area of *any* portion of the histogram between two vertical lines crossing the horizontal axis at values a and b represents the percentage of values that lie between a and b.

Suppose, for example, that we wish to estimate the percentage of values between 2.5 and 5 in the collection represented by the histogram in **Figure 8.1.7**. This corresponds to the first shaded portion. Because 2.5 is the midpoint of the interval [0, 5], the area of this portion is equal to one-half of the area of the first block. Therefore, about $\frac{1}{2}(0.10) = 0.05 = 5\%$ of this collection of numbers lies between 2.5 and 5. Similarly, the next shaded portion indicates that about $\frac{1}{2}(0.16) + \frac{1}{2}(0.25) = 0.205 = 20.5\%$ of the collection lies between 7.5 and 12.5.

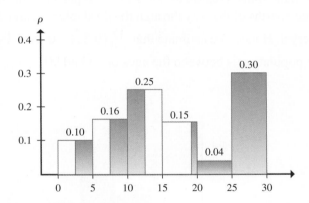

FIGURE 8.1.7   Using a histogram to estimate percentages.

Finally, if we wish to estimate the percentage between 19 and 30, then we do so by computing $\frac{1}{5}(0.15) + 0.04 + 0.30 = 0.37 = 37\%$, because 19 is $\frac{4}{5}$ of the way between 15 and 20.

? Example 3

The histogram in **Figure 8.1.8** displays the relative frequencies of age groups for the inhabitants of Kingsbridge County. Estimate the percentage of the population that is

(a)   between 20 and 40 years old.
(b)   between 10 and 30 years old.
(c)   between 55 and 90 years old.
(d)   less than 65 years old.
(e)   more than 65 years old.

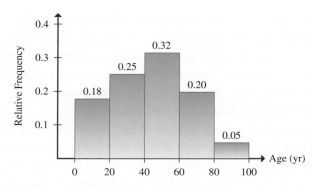

FIGURE 8.1.8   Distribution of ages in Kingsbridge County.

⚙ Solution

(a)   The second block tells us that 25% of the people of the county are between 20 and 40 years old.

(b)   Because 10 lies halfway through the first interval and 30 lies halfway through the second interval, we estimate that $\frac{1}{2}(0.18) + \frac{1}{2}(0.25) = 0.215 = 21.5\%$ of the population is between 10 and 30 years old.

(c)   The dark-shaded region in **Figure 8.1.9** represents the desired percentage. Age 55 is three-fourths of the way through the third interval, and 90 is in the middle of the last interval. Hence, we estimate that $\frac{1}{4}(0.32) + 0.20 + \frac{1}{2}(0.05) = 0.305 = 30.5\%$ of the population is between the ages of 55 and 90.

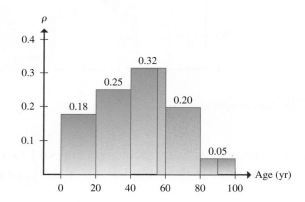

FIGURE 8.1.9   The percentage of the people of Kingsbridge County between 55 and 90 years old.

(d)   The percentage of people younger than 65 years old is $0.18 + 0.25 + 0.32 + \frac{1}{4}(0.20) = 0.80 = 80\%$.

(e)   The percentage of people older than 65 years old is $\frac{3}{4}(0.20) + 0.05 = 0.20 = 20\%$. Because the entire area of any histogram must be equal to 1, note that we also could have computed this percentage by subtracting the answer to part (d) from 1 to get $1 - 0.80 = 0.20 = 20\%$. ◆

We now wish to develop a method for estimating the median of a collection of grouped data given by a histogram or relative frequency distribution. The basic idea is to find the vertical line that divides the total area of the histogram into two equal halves. The interval class that the line intersects must contain the median and hence is called the **median interval**. So in Example 3, we first observe that the first two interval classes contain $0.18 + 0.25 = 0.43 = 43\%$ of the total collection. Because the median must exceed exactly 50% of the data, it must be a member of the third interval class. (See **Figure 8.1.10**.) This is the median interval, and the place where the line cuts the axis is the **group estimate of the median**, denoted by M_g. Note in Figure 8.1.10 that the light-shaded region corresponds to the lower half of the distribution and the dark-shaded region to the upper half. By assuming that the values in each interval increase uniformly from the left to the right boundary, we can compute M_g by adding some amount to 40, the left endpoint of the median interval. This amount is the difference $M_g - 40$. The ratio of $M_g - 40$ to the common width 20 of the interval must be equal to the ratio of the area of the green-shaded rectangle over the interval $[40, M_g]$ to the area of the entire block over $[40, 60]$. The area of the green- shaded rectangle must be $0.50 - 0.43 = 0.07$, and so in this case, we have

> **median interval** The interval class in a histogram or relative frequency distribution that contains the median.

> **group estimate of the median** An estimate of the median made by locating the median interval and approximating its location in the interval. Denoted by M_g.

$$\frac{M_g - 40}{20} = \frac{0.07}{0.32}.$$

Solving for M_g, we get

$$M_g - 40 = \left(\frac{7}{32}\right)(20)$$

$$= 4.375$$

$$M_g = 44.375$$

$$\approx 44.4.$$

In summary, the group estimate of the median is found by solving for M_g in the equation

$$\frac{M_g - L}{w} = \frac{\alpha}{A},$$

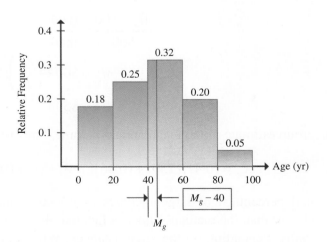

FIGURE 8.1.10 Estimating the median.

where

$$L = \text{left boundary of median interval}$$
$$w = \text{interval width}$$
$$\alpha = 0.50 - \text{total relative frequency to left of } L$$
$$A = \text{relative frequency of median interval.}$$

We have seen that the shape of a histogram tells us much about the distribution of any large collection of data. We have also observed that, of our two measures of central tendency, the mean is more sensitive than the median to the presence of extreme values. The mean tends to be pulled to the upper end when there are very large values present or to the lower end for a collection comprising quite small numbers. How does this fact affect the appearance of a histogram? A classic example is the distribution of wealth in developing countries that possess a small but wealthy ruling class. The histogram in **Figure 8.1.11** shows that 87% of the people of Zetonia earn less than $1,000 per year. The group estimate of the median income is determined by solving the following equation:

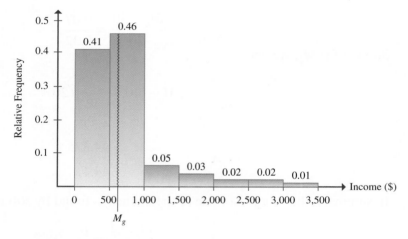

FIGURE 8.1.11   Distribution of the annual income of the people of Zetonia.

$$\frac{M_g - 500}{500} = \frac{0.09}{0.46}$$

$$M_g = 500 + \left(\frac{9}{46}\right)(500)$$

$$\approx \$598.$$

The group estimate of the mean annual income, however, is

$$\bar{X}_g = (0.41)(250) + (0.46)(750) + \cdots + (0.01)(3,250) \approx \$695.$$

A small percentage of large incomes produced a mean annual income that is almost $100 larger than the median. This is a fact that the king of Zetonia would certainly exploit differently, depending on the circumstances. When appealing for financial aid, he might give the (lower) median value in an effort to show a more pitiful situation. If he is trying make a good impression of the general welfare of his people, however, the mean would probably be his measure of choice, albeit a misleading one.

skewed distribution
A distribution in which the main body of the values is clustered at one end. The mean is not equal to the median in a skewed distribution..

symmetric distribution A distribution where a vertical line through the median divides the histogram into two mirror images. The mean is equal to the median in a symmetric distribution.

A distribution such as the one in Figure 8.1.11 is known as a **skewed distribution** and is characterized by the fact that the main body of values is clustered at one end of the scale. In a skewed distribution, the mean and median are significantly different. In contrast, a **symmetric distribution** is one in which the lower half is spread out exactly like the upper half, resulting in a histogram whose region to the left of a vertical line through the median is a mirror image of the region to the right. In such a distribution, the mean and the median are equal. The distribution in **Figure 8.1.12** is approximately symmetric.

The mean and median are two very important indicators of the center of any large collection of data. An equally important aspect is the manner in which the data values are distributed with respect to the center. What portions of the data are close to the center, slightly farther away, or quite distant? Questions like these refer to the *dispersion* of the data, the topic of the next section.

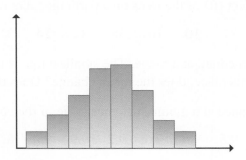

FIGURE 8.1.12   An approximately symmetric distribution.

Name _____

Exercise Set **8.1** ☐☐☐←———————

Compute all answers to one more significant digit than that of the given data. Data sets are given in the Appendix.

1. The ages of the people working in Rebecca's law office are as follows:

   53   42   23   26   34   37   62   55   43   18   29.

Find the mean, median, and range. If the oldest person in the office retires and is replaced by someone 56 years old, does the mean change? Does the median?

2. The heights in feet (ft) of the trees on a particular acre of Sven's property are as follows:

   21   10   16   38   32   24   31   27   19   22.

Find the mean, median, and range. The smallest tree is uprooted in a windstorm. If Sven plants a new 6-ft tree in its place, does the mean change? Does the median?

3. After a performance at a gymnastics meet, one of the contestants received these scores:

   9.5   9.6   9.6   9.2   9.9   9.0   8.2   9.3   9.5   9.4.

Find the mean and median. A standard practice is to throw out the high and low scores. After this is done, what are the new mean and median? Do you think this is a good practice to follow? Why or why not?

4. Every cereal lists the quantity of sodium contained in a serving of 1 ounce (oz). The following values are the amounts of sodium, in milligrams (mg), found in single servings of 20 different cereals. Construct a relative frequency distribution and a corresponding histogram that uses an interval width of 12 mg and has a first left boundary marker of 160. Compute the mean and median and the group estimates of the mean and median.

| 170 | 210 | 160 | 200 | 165 | 200 | 220 | 215 | 190 | 240 |
| 185 | 205 | 170 | 235 | 210 | 195 | 190 | 225 | 230 | 230 |

5. Data Set A in Appendix II lists the number of games won during a season by the Chicago Cubs baseball team from the years 1901 to 1940. Construct a relative frequency distribution and a corresponding histogram that uses an interval width of 10 wins and has a first left boundary marker of 50. Compute the mean and median and the group estimates of the mean and median. Would you label this an approximately symmetric distribution?

6. The absolute magnitude of a star is a measure of its intrinsic brightness. (Lower magnitude numbers indicate brighter stars.) Data Set B in Appendix II lists the absolute magnitudes of 25 of the nearest stars to Earth. Construct a relative frequency distribution and a corresponding histogram that uses an interval width of 3.0 magnitudes and has a first left boundary marker of 1.0. Compute the mean and median and the group estimates of the mean and median. Would you call this an approximately symmetric distribution?

7. The following table contains the top 25 football scorers in the Spanish Primera Division of the La Liga Football League in 2008/2009. Construct a relative frequency distribution and a histogram of these data, using 4 as the first left boundary marker and an interval width of 5. Compute the group estimates of the mean and median. (Note that more than 25 players are listed because of ties.)

Rank	Player	Team	Goals
1	Diego Forlan	Atlético Madrid	32
2	Samuel Eto'o	Barcelona	30
3	David Villa	Valencia	28
4	Lionel Messi	Barcelona	23
5	Gonzalo Higuaín	Real Madrid	22
6	Thierry Henry	Barcelona	19
	Álvaro Negredo Frederic	Almeria	19
8	Kanouté	Sevilla FC	18
	Raúl	Real Madrid	18
10	Sergio Aguero	Atlético Madrid	17
11	Joseba Llorente	Villarreal	15
12	Roberto Soldado	Getafe	13
	Fernando Llorente	Athletic Bilbao	13
	Nikola Zigic	Racing Santander	13
15	Mate Bilic	Sporting Gijon	12
	Jose Barkero	Numancia	12
	Giuseppe Rossi	Villarreal	12
18	Walter Pandiani	Osasuna	11
	Aritz Aduriz	Mallorca	11
	Achille Emana	Real Betis	11
	Mata	Valencia	11
22	David Barral	Sporting Gijon	10
	Henok Goitom	Valladolid Recreativo	10
	Javier Camuñas	Huelva	10
25	Sergio García	Real Betis	9
	Nabil Baha	Málaga	9
	Jose Jurado	Mallorca	9
	Apono	Málaga Recreativo	9
	Adrián Colunga	Huelva	9

8. Is the mean or median larger in Exercise 7? Explain the reason for this difference.

9. The following is a list of the ages of 42 U.S. presidents at the times of their inaugurations.

57	61	57	57	58	57	61	54	68	51	49	64	50	48
65	52	56	46	54	49	50	47	55	55	54	42	51	56
55	51	54	51	60	62	43	55	56	61	52	49	65	46

Construct a relative frequency distribution and a histogram of these data, using 40 as the first left boundary marker and an interval width of 3. Compute the group mean and median.

10. The following table shows the frequencies of birth weights in pounds (lb) for a collection of 35 infants. Compute the relative frequencies for each interval class, and then determine the group estimates of the mean and the median.

Birth Weight (lb)	Frequency	Relative Frequency
1.0–3.0	1	
3.0–5.0	8	
5.0–7.0	14	
7.0–9.0	9	
9.0–11.0	2	
11.0–13.0	1	

11. The Hartsville State College newspaper wanted to print an article describing the average student at the college. To arrive at a height, in inches (in.), for this person, the newspaper conducted a random survey of students and arrived at the following relative frequency distribution.

Height (in.)	ρ
58–61	0.04
61–64	0.19
64–67	0.23
67–70	0.28
70–73	0.19
73–76	0.05
76–79	0.02

Draw a histogram that displays this distribution. Is it approximately symmetric? What is the group estimate of the mean and median?

12. Data Set D in Appendix II lists the number of earthquakes in the United States for the years 2000–2009, using interval classes of Richter-scale magnitudes.

(a) Construct a relative frequency distribution and histogram for the year 2009, starting at magnitude 1.0 and ending at 6.0. Use an interval width of 1.0.

(b) Construct a relative frequency distribution and histogram for the year 2008, starting at magnitude 1.0 and ending at 7.0. Use an interval width of 1.0.

(c) Compute the group estimates for the mean and median for the years 2008 and 2009.

13. The results of 485 students at Roosevelt High School who took the SAT verbal achievement test one year are listed here.

Scores	ρ
200–300	0.02
300–400	0.15
400–500	0.36
500–600	0.33
600–700	0.12
700–800	0.02

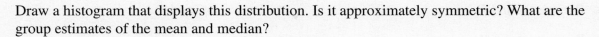

Draw a histogram that displays this distribution. Is it approximately symmetric? What are the group estimates of the mean and median?

14. For the distribution given in Exercise 13, estimate the percentages of students who scored in each of the following categories.

(a) Between 300 and 400
(b) Between 250 and 400
(c) Between 350 and 475
(d) Greater than 650
(e) Less than 650

15. The average 6-month auto insurance premiums per state for each of the 50 states plus the District of Columbia are distributed according to the following table.

Average Premium ($)	Number of States	ρ
250–350	5	0.10
350–450	17	0.33
450–550	14	0.27
550–650	11	0.22
650–750	3	0.06
750–850	1	0.02

The actual mean of the 51 numbers is $517.71. Compute the group estimate of the mean and compare. Also estimate the percentage of states whose average premium is

(a) less than 400.
(b) between 500 and 750.
(c) between 375 and 600.
(d) greater than 625.

The capital letters in the histogram shown here represent the boundary markers of their respective interval classes. The lowercase letters represent the midpoints. For Exercises 16–23, estimate the percentage of the distribution that is represented by the indicated interval(s).

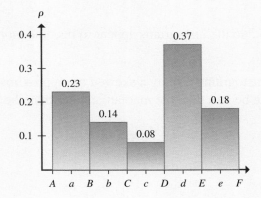

16. Between C and D

17. Between C and F

18. Between a and B

19. Between A and c

20. Greater than C

21. Between b and e

22. Less than d

23. Between A and F

The following histograms give the distribution of annual individual income for a particular city in 2013 for three different age brackets. The units on the horizontal axis are thousands of dollars. In each case, the last block actually represents the portion of household incomes over $60,000.

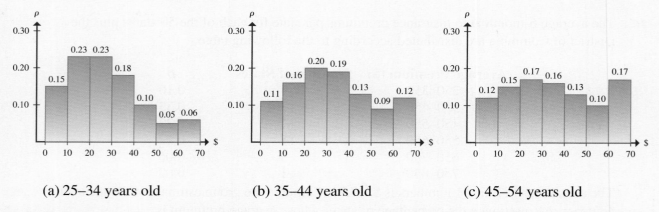

(a) 25–34 years old   (b) 35–44 years old   (c) 45–54 years old

24. Estimate the percentage of individuals who earn $10,000–$30,000 annually for each of the three age brackets.

25. Estimate the percentage of individuals who earn $15,000–$35,000 annually for each of the three age brackets.

26. Estimate the percentage of individuals who earn more than $45,000 annually for each of the three age brackets.

27. Is the percentage of individuals who earn more than $30,000 annually an increasing or decreasing function of age bracket? Is the percentage who earn less than $30,000 an increasing or decreasing function of age bracket?

28. If you ignore the last block, do the histograms appear to become more or less symmetric as the age bracket increases?

29. Do each of the following histograms display a skewed or (approximately) symmetric distribution? Would the mean of the data be less than the median, greater than the median, or about the same?

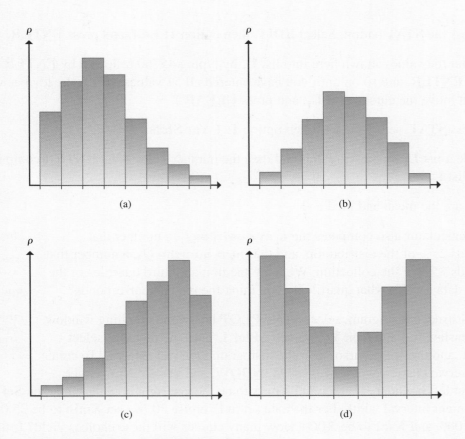

(a)   (b)

(c)   (d)

Using a Calculator with a Statistics Capability

Exercises 30–34 are based on the following data set.

The following list is composed of the average secondary school teacher (except special and vocational education) annual salaries (in dollars) for each state and the District of Columbia in May 2008.

45,760 58,950 42,690 45,050 63,960 48,630 65,280 57,990 53,210 52,520 48,030 51,760
69,830 48,310 39,150 41,300 48,340 43,910 44,360 58,100 59,310 54,970 48,540 42,470
42,770 39,590 43,720 47,390 50,390 63,720 53,530 66,860 42,970 41,490 55,050 41,110
49,610 53,380 61,830 46,660 37,350 44,610 47,730 50,760 50,950 44,450 59,050 56,460
41,840 48,970 54,230

Most modern calculators have the capability to compute the mean, median, and other statistics of any list of numbers that you input to it. To use one of the typical models from Texas Instruments, you perform the following steps. Other calculators have similar commands.

Options that appear in the window are indicated in bold. **Buttons** are indicated by underlined words in bold.

1. Press the **STAT** button. Select **EDIT**, then option **1: Edit**, and press **ENTER**.

2. Input the values shown here into list L_1 by typing 45760 followed by **ENTER**, 58950 followed by **ENTER**, and so on until you have entered all 51 values. (If L_1 already has values in it, then first move the cursor over L_1, and press **CLEAR**.)

3. Press **STAT**, select **CALC**, then option **1: 1-Var Stats**.

4. Select list L_1 by pressing **2nd** and then the number **1**. Pressing **ENTER** computes the statistics of list L_1.

30. What are the mean and median?

31. Your calculator also computes the **first quartile** Q_1, a number that exceeds 25% of the distribution, and the **third quartile** Q_3, a number that exceeds 75% of the collection. What are the first and third quartiles of the given data set? To what quartile do you think the median corresponds?

> **first quartile** A number that is greater than 25% of the values in a collection and less than 75% of the values.

> **third quartile** A number that exceeds 75% of the collection.

32. To construct a histogram, select **STAT PLOT** below the graphing window by pressing **2nd** and then **Y=**. Select **1:Plot 1**. Turn the plot **On**, select **Type:**, and move the cursor over the histogram icon and select it. To create the interval classes, you now press **WINDOW**, and you set **Xmin** to be the first left boundary marker and **Xmax** to be the last right boundary marker. Set **Xsc1** to be a convenient interval width. For the data set in Exercise 30, we set **Xmin** to be 35,000, **Xmax** to be 70,000, and **Xsc1** to be 5,000. How many classes will these choices yield? Instead of relative frequency, the vertical axis on a histogram created by a calculator usually represents the actual frequencies of the classes. Here we set **Ymin** to be 0, **Ymax** to be 20, and **Ysc1** to be 5. Now press **GRAPH**. Does this histogram indicate a skewed or symmetric distribution?

33. While viewing a histogram, pressing the **TRACE** button will initiate a flashing point on the top of the first interval class and display the frequency of the interval as well as the left and right boundary markers. Moving the cursor moves the point from class to class. What is the frequency of each class in the histogram created in Exercise 32?

34. Altering the class width changes the number of classes and the look of the histogram. Change the class width by pressing **WINDOW** and resetting **Xsc1** to 4,000. How many classes are created? Now change **Xsc1** again to 3,000. How many classes are there now? What is the frequency of each class?

35. Katie kept a record of her cross-country times for 4 years. Use **STAT PLOT** to create a histogram with a left boundary marker of 20.5, using the following interval widths.

(a) If the interval width is 1.0, what is the frequency of each class?
(b) Change the interval width to 0.8. Now what is the frequency of each class?
(c) Change the interval width to 0.5. Now what is the frequency of each class?

(Timed events are typically measured in minutes and seconds, but all values below have been converted to minutes for convenience. For example, 21:30 = 21.50 minutes.)

20.93	20.95	21.05	21.07	21.07	21.90
21.92	22.07	22.12	22.13	22.23	22.23
22.33	22.33	22.40	22.50	22.83	22.83
22.97	22.98	23.03	23.08	23.08	23.08
23.12	23.17	23.18	23.18	23.23	23.50
23.53	23.57	23.83	23.87	23.90	23.92
23.97	24.00	24.05	24.08		

36. Find the mean, first quartile, median, and third quartile of the data set in Exercise 35.

37. Data Set C in Appendix II lists the percentage of high school students in each state who graduated from high school in 2006. Use your calculator to compute the mean and median of the graduation rates. (You may want to save this list in your calculator for use in the exercise set in the next section.)

38. Use your calculator to construct a histogram for Data Set C that uses an interval width of 10 percentage points and has a first left boundary marker of 45%.

39. Use your calculator to compute the mean, median, and first and third quartiles of the 2005–2006 annual healthcare costs in the major metropolitan areas listed in Data Set E in Appendix II. (You may want to save this list in your calculator for use in the exercise set in the next section.)

40. Use your calculator to compute the mean, median, and first and third quartiles of the fuel costs of the entire collection of cars in Data Set F in Appendix II. (You may want to save this list in your calculator for use in the exercise set in the next section.)

41. Data Set G in Appendix II contains the runs, batting averages, and home runs by each National League baseball team during the 2008 season. Use your calculator to compute the mean, median, and first and third quartiles of the team runs. (You may want to save these lists in your calculator for use in the exercise set in the next section.)

42. Use your calculator to compute the mean, median, and first and third quartiles of the team home runs hit by the National League baseball teams during the 2008 season. They are listed in Data Set G in Appendix II.

43. Data Set H in Appendix II contains some passing statistics for 20 quarterbacks in the National Football League for the 2008 season. Use your calculator to compute the mean, median, and first and third quartiles for passing yards. Pick a quarterback and see if he is above or below the median.

44. The website stats.org written and maintained by George Mason University is an excellent source for showcasing examples of the misuse of statistics in the public media. One such error occurs in the practice of quoting the average number of daily hits as a means for gauging the popularity of a website. Suppose the average hit count for a site is 1,000 per day. Do you think this is indicative of the actual number of different people who visit the site daily? What factors might be taken into account? Would this number imply that 7,000 different people visit the site per week and more than 30,000 per month?

Web-Generated Histograms

45. The Illuminations website of the NCTM (National Council of the Teachers of Mathematics) provides an online histogram generator (illuminations.nctm.org/ActivityDetail.aspx?id=78). Use this program to create a histogram for the high school graduation rates from Data Set C in Appendix II, using an interval width of 10%. Allow the interval width to vary among 5%, 10%, 15%, and 20%. How does the shape of the histogram change?

46. Use the website described in Exercise 45 to create a histogram for the 2008 season passing yards in the U.S. National Football League given in Data Set H in Appendix II. Allow the interval width to vary among 200, 400, 600, 800, and 1,000 yards (yd). How does the shape of the histogram change?

8.2 Dispersion of Data

dispersion The spread of any collection of data values.

We have seen that measures of central tendency give us a feel for the average value of a set of numbers. But what about the **dispersion** or spread of all the numbers from that average value? We can see that the two histograms in **Figure 8.2.1** have the same mean but vary significantly in how they are distributed about that mean. Suppose you received a score of 70 in a science achievement test given to several hundred students in your school, and you have just learned that the mean score on the exam was 65 (out of 100).

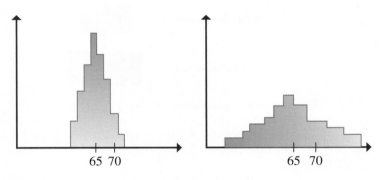

FIGURE 8.2.1   Two histograms with the same mean but distributed differently.

You know something important about the distribution of scores, but you still do not really know how your 70 compares to the rest of the scores. Okay, you know you scored above the average, but how *far* above? In which of the distributions in Figure 8.2.1 would you prefer your score to be a member? The answer to this question requires the definition of a specific unit designed to measure distance in a distribution. A unit such as a centimeter is not meaningful here because the numbers in our data collections could represent anything—heights, weights, times, test scores, inflation rates, etc.—and so a standard-sized unit is not flexible enough. We need a unit that will calibrate distance from the mean in *any* distribution, implying that the unit must be a *function* of the distribution we are analyzing.

deviation The difference between a specific data value and the mean of the collection of which the value is a member.

standard deviation The primary measure of dispersion of any collection of data. It is the square root of the mean of the squared deviations.

The difference 5 between your score 70 and the mean 65 is called the **deviation** of 70 from the mean. The unit of measure we are seeking is known as the **standard deviation** and represents a type of average deviation of all the values in a collection. It is denoted by the Greek letter sigma σ and defined by

$$\sigma = \sqrt{\frac{\sum\left(x - \bar{X}\right)^2}{n}},$$

where x varies over all the data values of a collection of size n and having mean $\bar{X}$. Note that we first compute the mean of the squared deviations and then take the square root of the result. This procedure is used to smooth out the data in an analytic sense. (The use of $\sum\left(x - \bar{X}\right)$ would result in an unreliable distance indicator because of the cancellation of positive and negative contributions.) Examining this formula, we use a flowchart to demonstrate the order of operations. (See **Figure 8.2.2**.)

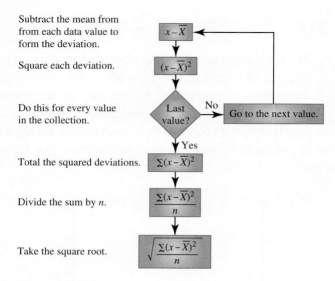

Subtract the mean from from each data value to form the deviation.

Square each deviation.

Do this for every value in the collection.

Total the squared deviations.

Divide the sum by n.

Take the square root.

FIGURE 8.2.2   Computing the standard deviation.

❓ Example 1

At a recent swim meet, the recorded times for the 8-year-old boys in the 25-yd breaststroke event are given in the following table. Find the mean and standard deviation of this set of times.

23.20	23.49	24.17	24.20	24.49	25.17	25.37	25.41
26.27	26.36	27.41	27.60	27.72	27.84	27.89	

⚙ Solution

First, we find the mean to be

$$\overline{X} = \frac{23.20 + 23.49 + \cdots + 27.89}{15} = 25.77 \,.$$

To compute the standard deviation, it is convenient to use a table.

x	$x - \overline{X}$	$\left(x - \overline{X}\right)^2$
23.20	−2.57	6.60
23.49	−2.28	5.20
24.17	−1.6	2.56
24.20	−1.57	2.46
24.49	−1.28	1.64
25.17	−0.6	0.36
25.37	−0.4	0.16
25.41	−0.36	0.13
26.27	0.5	0.25
26.36	0.59	0.35
27.41	1.64	2.69
27.60	1.83	3.35
27.72	1.95	3.80
27.84	2.07	4.28
27.89	2.12	4.49
		38.32

Summing the numbers in the last column, we get

$$\sum(x-\overline{X})^2 = 38.32$$

$$\frac{\sum(x-\overline{X})^2}{15} = \frac{38.32}{15} = 2.55$$

$$\sigma = \sqrt{\frac{\sum(x-\overline{X})^2}{15}} = \sqrt{2.55} = 1.60. \quad \blacklozenge$$

In Exercise Set 8.1, we learned how to use a calculator to input a list of numbers and compute a mean and median. You may have observed at that time that the standard deviation is also a value that is included with the output of the statistical calculations. Next, we examine how the standard deviation is used to mark the position of a specific data value in any collection with respect to the mean. Repeating our two distributions of test scores in **Figure 8.2.3**, we observe that the tall, narrow histogram on the left must represent a distribution possessing a smaller standard deviation than the wide histogram on the right because more of its values lie closer to the mean. The larger the portion of values that lie close to the mean, the smaller are the squared deviations and thus the smaller is the corresponding standard deviation. So, generally speaking, a large σ value is indicative of a spread out distribution.

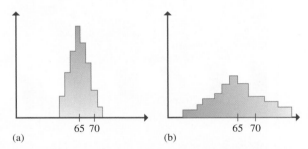

(a)   (b)

FIGURE 8.2.3 Two histograms of data collections with the same mean but distributed differently. The standard deviation of the distribution of (a) is less than that of (b), and so the z-score of 70 is greater in (a) than in (b).

Reasonable values for σ in the two distributions in Figure 8.2.3 would be $\sigma = 2$ for histogram (a) and $\sigma = 5$ for histogram (b). We now see how to use them as units of measurement. In both situations, 70 lies 5 points above the mean of 65. However, because $\sigma = 2$ in the left-hand distribution and $5 = (2.5)(2) = 2.5\sigma$, we say that 70 lies 2.5 standard deviations above the mean. On the other hand, $\sigma = 5$ in the right-hand histogram, from which it follows that 70 lies only 1 standard deviation above the mean in this case. This implies that the value 70 exceeds a greater percentage of values in distribution (a) than in (b). In the second distribution, your score of 70 would only be moderately above average, but in the first distribution you would rank almost at the top of your class!

In the language of statistics, the number of standard deviations that any value lies from the mean is called its **z-score**. The z-score of x in a collection of numbers with mean $\overline{X}$ and standard deviation σ is defined by

z-score A measure of distance of any data value from the mean. It is equal to the deviation of the value divided by the standard deviation.

$$z_x = \frac{x-\overline{X}}{\sigma}.$$

Note that the z-scores were $z_{70} = \dfrac{70-65}{2} = 2.50$ for the first of the distributions in Figure 8.2.3, but $z_{70} = \dfrac{70-65}{5} = 1.00$ for the second. So the z-score of a number x is a measure of the distance that x is from the mean in a *particular* distribution. The larger the z-score (in absolute value) of the x value in the collection, the farther away x is located from the mean. The sign of the z-score indicates the direction of the value. Any number greater than $\overline{X}$ must have a positive z-score, whereas those less than $\overline{X}$ must have a z-score that is negative. Only $\overline{X}$ itself has a z-score of 0.

? Example 2

Anthony, Brent, and Jesse are three of the 8-year-old boys who participated in the 25-yd breaststroke in Example 1. If their times were 24.17, 25.37, and 27.84, what are the corresponding z-scores? (Typically z-scores are rounded to the nearest hundredth.)

© Suzanne Tucker/Shutterstock, Inc.

⚙ Solution

Recall from Example 1 that $\overline{X} = 25.77$ and $\sigma = 1.60$.
The z-score of 24.17 (Anthony) is

$$z_{24.17} = \frac{24.17 - 25.77}{1.60} = -1.00.$$

The z-score of 25.37 (Brent) is

$$z_{25.37} = \frac{25.37 - 25.77}{1.60} = -0.25.$$

The z-score of 27.84 (Jesse) is

$$z_{27.84} = \frac{27.84 - 25.77}{1.60} = 1.29. \quad \blacklozenge$$

Suppose now that, given a particular collection of values, we add the constant number k to each of those values to generate a second collection. It seems reasonable that the mean of the new collection would also increase by k, but would the standard deviation? No, it would not, because each deviation $(x - \bar{X})$ used in the formula would remain unchanged. **Figure 8.2.4** presents an example of this situation ($k = 7$) in which we assume identical scales for both sets of axes. Both histograms have the same shape, with the second one translated 7 units to the right of the first one.

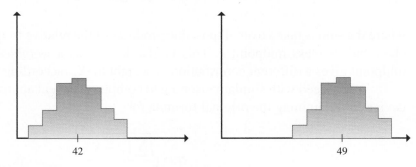

42   49

Two histograms with the same standard deviation but different means.

Two sets of data associated with the same phenomenon but different with respect to a single variable often have distributions that display this characteristic. For instance, the heights of two groups of the same species of tree may possess roughly the same standard deviation, but they could have quite different means if one group was growing at a substantially different altitude or in a different climate than the other group. Monthly food prices recorded in years from different decades would be another good example. The 12 monthly prices for a loaf of bread in 2015 would probably have a similar pattern of clustering as those in 2005, but they undoubtedly would have a higher mean.

Example 3

Suppose the same set of boys is swimming the same event at a meet one year later in the nine-year-old division. It is reasonable to assume that the mean time of the boys would be lower than the previous year, but that the standard deviation might change very little if each boy had improved his time by about the same amount. Suppose it turns out that $\bar{X} = 21.85$ and $\sigma = 1.45$ for this new set of times. If Anthony, Brent, and Jesse have the same z-scores in this distribution as they did last year, what are their times?

Solution

We again use the formula for the z-score, but we solve for x, in seconds(s), this time instead of z. Recall that the z-scores of the three boys were -1.00, -0.25, and 1.29. We wish to find the times—that is, the x values—that correspond to these scores.

Anthony	Brent	Jesse
$\dfrac{x - 21.85}{1.45} = -1.00$	$\dfrac{x - 21.85}{1.45} = -0.25$	$\dfrac{x - 21.85}{1.45} = 1.29$
$X - 21.85 = -1.45$	$x - 21.85 = -0.36$	$x - 21.85 = 1.87$
$x = 20.40\,\text{s}$	$x = 21.49\,\text{s}$	$x = 23.72\,\text{s}$ ◆

Meaningful statistical analysis usually requires large collections of data, and we have seen that most large collections are presented via the use of histograms and relative frequency distributions. Our method for estimating the standard deviation from such grouped data is analogous to that for estimating the mean. Recall the computation for $\bar{X}_g$ to be

$$\bar{X}_g = \sum \rho c,$$

where the sum is taken over all possible products of the relative frequency ρ of each interval class and its class midpoint c. This is also known as a *weighted average* because each midpoint gives a different contribution or weight to $\bar{X}_g$ according to its relative frequency.

We proceed with similar reasoning to compute a decent approximation to the standard deviation. Examining the original formula for

$$\sigma = \sqrt{\frac{\sum (x - \bar{X})^2}{n}},$$

group estimate of the standard deviation An estimate of the standard deviation from a relative frequency distribution made by taking the square root of a weighted average of the squared deviations.

we observe that the square root is taken of the mean of the squared deviations. Because we are dealing with grouped data, we simply replace the expression under the square root with a weighted average of the squared deviations. That gives us a **group estimate of the standard deviation**

$$\sigma_g = \sqrt{\sum \rho (c - \bar{X}_g)^2},$$

where ρ is the relative frequency of each interval class having midpoint c. Note that this formula uses the *group estimate* of the mean rather than the exact mean, which, of course, is all that is available to you when you are restricted to grouped data. The flowchart in **Figure 8.2.5** formats the steps in this calculation.

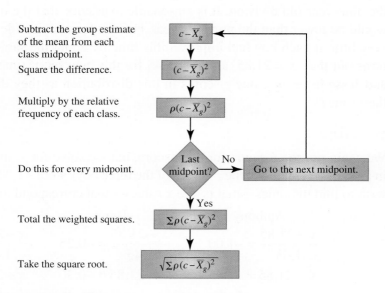

FIGURE 8.2.5   Estimating the standard deviation from a relative frequency distribution.

? Example 4

The following relative frequency distribution shows the daily amounts of time in minutes (min) spent commuting back and forth to work by people living in and around London, England. Find the group estimates of the mean and the standard deviation. Approximately what times would have z-scores of -2, -0.3, 1.2, and 2.5?

Commute Time (min)	Relative Frequency
0–15	0.07
15–30	0.16
30–45	0.28
45–60	0.23
60–75	0.17
75–90	0.09

⚙ Solution

$$\bar{X}_g = (7.5)(0.07) + (22.5)(0.16) + (37.5)(0.28) + (52.5)(0.23) + (67.5)(0.17) + (82.5)(0.09)$$
$$= 45.6$$

c	$c - \bar{X}_g$	$\left(c - \bar{X}_g\right)^2$	ρ	$\rho\left(c - \bar{X}_g\right)^2$
7.5	−38.1	1,451.61	0.07	101.61
22.5	−23.1	533.61	0.16	85.38
37.5	−8.1	65.61	0.28	18.37
52.5	6.9	47.61	0.23	10.95
67.5	21.9	479.61	0.17	81.53
82.5	36.9	1,361.61	0.09	122.54
				420.38

$$\sigma_g = \sqrt{\sum \rho\left(c - \bar{X}_g\right)^2} = \sqrt{420.38} = 20.5$$

Using these *estimates* for the mean and the standard deviation, we can now *estimate* the times having the desired z-scores.

$$\frac{x - 45.6}{20.5} = -2 \qquad\qquad \frac{x - 45.6}{20.5} = -0.3$$
$$x = 45.6 - 2(20.5) \qquad\qquad x = 45.6 - 0.3(20.5)$$
$$x = 4.6 \text{ min} \qquad\qquad\qquad x = 39.5 \text{ min}$$

$$\frac{x - 45.6}{20.5} = 1.2 \qquad\qquad \frac{x - 45.6}{20.5} = 2.5$$
$$x = 45.6 + 1.2(20.5) \qquad\qquad x = 45.6 + 2.5(20.5)$$
$$x = 70.2 \text{ min} \qquad\qquad\qquad x = 96.9 \text{ min} \quad \blacklozenge$$

Name _____

Exercise Set **8.2** ☐☐☐←———————————

Exercises 1–11 are based on data presented in Exercise Set 8.1. Use your means from that section to help answer these questions.

1. The ages of the people working in Rebecca's law office were as follows:

   53   42   23   26   34   37   62   55   43   18   29.

   Find the standard deviation. If the oldest person in the office retires and is replaced by someone 40 years old, would the standard deviation increase or decrease? (Do not compute it.)

2. The heights in feet (ft) of the trees on a particular acre of Brandon's property were as shown here:

   21   10   16   38   32   24   31   27   19   22.

   Find the standard deviation. If the smallest tree were uprooted in a windstorm, would the standard deviation get smaller or larger? (Do not compute it.)

3. After a performance at a gymnastics meet, one of the contestants received these scores:

   9.5   9.6   9.6   9.2   9.9   9.0   8.2   9.3   9.5   9.4.

   Find the standard deviation. A standard practice is to throw out the high and low scores. After this is done, will the new standard deviation be larger or smaller? Compute the new standard deviation to verify your answer.

4. The following are the amounts of sodium in milligrams (mg) found in single servings of 20 different cereals:

   | 170 | 210 | 160 | 200 | 165 | 200 | 220 | 215 | 190 | 240 |
   | 185 | 205 | 170 | 235 | 210 | 195 | 190 | 225 | 230 | 230. |

   (a) The mean found in Exercise Set 8.1 is $\bar{X} = 202.3$. Compute the standard deviation.
   (b) The group estimate of the mean found in Exercise Set 8.1 is $\bar{X}_g = 86.0$. The relative frequency distribution you created is repeated here. Compute the group estimate of the standard deviation. How do these two standard deviations compare?

Sodium (mg)	Frequency	p
160–172	4	0.20
172–184	0	0.00
184–196	4	0.20
196–208	3	0.15
208–220	3	0.15
220–232	4	0.20
232–244	2	0.10

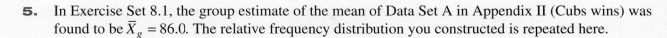

5. In Exercise Set 8.1, the group estimate of the mean of Data Set A in Appendix II (Cubs wins) was found to be $\overline{X}_g = 86.0$. The relative frequency distribution you constructed is repeated here.

Wins	Frequency	ρ
50–60	1	0.025
60–70	4	0.10
70–80	6	0.15
80–90	14	0.35
90–100	10	0.25
100–110	4	0.10
110–120	1	0.025

(a) Compute the group estimate of the standard deviation.
(b) The actual mean was computed to be $\overline{X} = 85.3$. Refer to Data Set A in Appendix II, and compute the actual standard deviation. (Use the statistical package on your calculator, if available.) How do these two standard deviations compare?

6. In Exercise Set 8.1, the group estimate of the mean of Data Set B in Appendix II (star magnitudes) was found to be $\overline{X}_g = 10.66$. The relative frequency distribution you constructed is repeated here.

Magnitude	Frequency	ρ
1.0–4.0	2	0.08
4.0–7.0	4	0.16
7.0–10.0	4	0.16
10.0–13.0	5	0.20
13.0–16.0	9	0.36
16.0–19.0	1	0.04

(a) Compute the group estimate of the standard deviation.
(b) The actual mean was computed to be $\overline{X} = 10.27$. Refer to Data Set B in Appendix II, and compute the actual standard deviation. How do these two standard deviations compare?

7. In Exercise Set 8.1, the group estimate of the mean of the top 25 football scorers in the Spanish La Liga in 2008/2009 was found to be $\overline{X}_g = 14.25$. The relative frequency distribution you constructed is repeated below. Compute the group estimate of the standard deviation.

Goals	Frequency	Relative Frequency
4–9	5	5/29 = 0.17
9–14	13	13/29 = 0.45
14–19	6	6/29 = 0.21
19–24	2	2/29 = 0.07
24–29	1	1/29 = 0.03
29–34	2	2/29 = 0.07

8. Use the estimates from Exercise 7 to find the z-scores of the goals scored by Sergio García (9), Joseba Liorente (15), Lionel Messi (23), and Diego Forlan (32).

9. In Exercise Set 8.1, the group estimate of the mean of the ages of 42 U.S presidents at the time of inauguration was found to be $\overline{X}_g = 54.8$ years. The relative frequency distribution and histogram you constructed are repeated here. Find the group estimate of the standard deviation.

Ages	Frequency	ρ
40–44	2	0.05
44–48	3	0.07
48–52	10	0.24
52–56	10	0.24
56–60	8	0.19
60–64	5	0.12
64–68	3	0.07
68–72	1	0.02

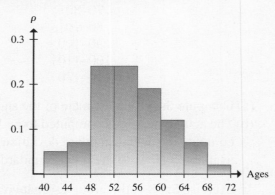

10. The birth weights for a collection of 35 infants were given in Exercise 10 in Exercise Set 8.1. The group estimate of the mean was computed to be $\overline{X}_g = 6.42$. Using the relative frequency distribution you constructed, estimate the standard deviation.

11. The survey results of the heights of Hartsville State College students were given in Exercise 11 in Exercise Set 8.1. The group estimate of the mean was computed to be $\overline{X}_g = 67.4$. Estimate the standard deviation.

12. The relative frequency distribution for the earthquakes that occurred one year in California is given here. The group estimate of the mean computed from this distribution is 3.38.

Magnitude	Frequency	ρ
2.0–2.5	3	0.06
2.5–3.0	8	0.17
3.0–3.5	18	0.38
3.5–4.0	13	0.27
4.0–4.5	4	0.08
4.5–5.0	2	0.04

(a) Estimate the standard deviation.
(b) Find the z-scores of magnitudes 3.1 and 4.3, using these estimates.
(c) Find the earthquake magnitudes having z-scores of 2.0 and −1.4.

13. A relative frequency distribution of the verbal SAT scores of the students at Roosevelt High School for one year were given in Exercise 13 in Exercise Set 8.1. It is repeated here. The group estimate of the mean was computed to be $\overline{X}_g = 494$. Estimate the standard deviation.

Scores	ρ
200–300	0.02
300–400	0.15
400–500	0.36
500–600	0.33
600–700	0.12
700–800	0.02

14. Use the estimates of the mean and the standard deviation that you found in Exercise 13 to find the z-scores of the values 330, 400, 530, and 650.

15. Use the estimates of the mean and the standard deviation that you found in Exercise 13 to estimate the SAT scores that correspond to z-scores of −2.5, −0.8, 1.2, and 2.4.

16. The following table lists age distributions (by percentage of total population) of U.S. residents in the years 1900 and 1986.

Age	1900 Population (%)	1986 Population (%)
0–5	12.1	7.5
5–15	22.3	14.0
15–25	19.7	16.3
25–35	16.0	17.8
35–45	12.2	13.7
45–55	8.5	9.4
55–65	5.3	9.2
65+	4.1	12.1

(a) Compute the group estimate of the mean for each year, using 70 as the midpoint for the 65+ class.

(b) Estimate the standard deviation for each year.

(c) What conclusions can you draw from your answers from parts (a) and (b)?

(d) Find the z-scores of 12, 32, and 54 in the 1900 data. Then find the ages in the 1986 data that correspond to those z-scores.

For the data in Exercises 17–22, round the desired z-scores to the nearest hundredth.

17. Length (inches) of male infants: $\overline{X} = 19.69$, $\sigma = 0.63$. Find the z-score of the length of a newborn boy who is 21.7 in. long.

18. Length in centimeters (cm) of adult male femurs: $\bar{X} = 45.01$, $\sigma = 2.08$. Find the z-score of the length of a male femur that is 40.96 cm long.

19. Body temperature (degrees Fahrenheit) of young girls: $\bar{X} = 99.13$, $\sigma = 0.40$. Find the z-score of a 99.3°F body temperature of a young girl.

20. Heat production, in calories per square inch (cal/in²), of adult males: $\bar{X} = 839.38$, $\sigma = 54.55$. Find the z-score of one man's heat production of 750 cal/in².

21. Duration of pregnancy (days) of German women: $\bar{X} = 287.13$, $\sigma = 14.77$. Find the z-score of a pregnancy that lasts 300 days.

22. Weight in kilograms (kg) of 6-month-old boys: $\bar{X} = 8.23$, $\sigma = 0.83$; weight (kg) of 6-month-old girls: $\bar{X} = 7.60$, $\sigma = 0.75$. Find the z-score of a 9.59-kg boy. What would be the weight of a girl with this z-score?

23. A doctor is concerned about a patient's blood platelet count and knows that a z-score of -3.2 or lower is a symptom of a severe blood disorder. If the mean platelet count is 293.97 and the standard deviation is 43.00, at what platelet count should the doctor be concerned?

24. The mean blood pressure, in millimeters (mm) mercury, of an adult male is 130.0 with a standard deviation of 13.4. If a blood pressure with a z-score greater than 2.5 is considered dangerous, what is the corresponding blood pressure?

25. The mean number of digits that can be correctly repeated without error by adults is 6.60, with a standard deviation of 1.13. If an adult with a z-score less than -2 could have a memory retention problem, what is the corresponding number of digits?

26. The mean level of blood sugar in male adults is 96.99 mg [per 100 cubic centimeters (cc)] with a standard deviation of 6.90. If Richard is concerned about his blood sugar level and must have his z-score less than 2.0, at what blood sugar level should he be concerned?

Using a Calculator with Statistics Capability

27. Data Set C in Appendix II lists the high school graduation rates in the United States in 2006. Use your calculator to find the mean and standard deviation.

28. Use Data Set C in Appendix II to find the z-scores of the graduation rates for California, Pennsylvania, and North Dakota.

29. Use Data Set C in Appendix II to find the graduation rates associated with z-scores of -1.0, 0, and 1.0.

30. Data Set E in Appendix II lists the annual healthcare costs in major metropolitan areas in 2005–2006.

(a) Compute the mean and the standard deviation.
(b) Find the z-scores of the urban healthcare costs of Baltimore, Dallas–Fort Worth, and Phoenix–Mesa.
(c) What healthcare costs are associated with z-scores of -2.0, -1.0, 0, 1.0, and 2.0?

31. Data Set D in Appendix II gives a frequency distribution for earthquake magnitudes in the United States during 2000–2009. Find the mean and standard deviation of the magnitudes for 2009. (Round up the right endpoint of each interval when you determine the center.)

32. Using the values determined in Exercise 31, determine the z-score of 3.6 in Data Set D.

33. Data Set F in Appendix II lists estimated fuel costs to drive automobiles for 15,000 miles (mi). Compute the mean and standard deviation of the entire collection. What is the z-score of a Saturn? Chevrolet Lumina? Dodge Grand Caravan?

34. Compute the mean of each vehicle class in Data Set F. Using the values from Exercise 33, find the z-scores of these class means within the entire collection.

35. The following list is composed of the average secondary school teacher (except special and vocational education) annual salaries (in dollars) for each state and the District of Columbia in May 2008.

45,760 58,950 42,690 45,050 63,960 48,630 65,280 57,990 53,210 52,520 48,030 51,760
69,830 48,310 39,150 41,300 48,340 43,910 44,360 58,100 59,310 54,970 48,540 42,470
42,770 39,590 43,720 47,390 50,390 63,720 53,530 66,860 42,970 41,490 55,050 41,110
49,610 53,380 61,830 46,660 37,350 44,610 47,730 50,760 50,950 44,450 59,050 56,460
41,840 48,970 54,230

Find the mean and standard deviation. What salary would have a z-score of -1.5? 0? 1.5?

36. In the list in Exercise 35, if the top two salaries were replaced by 70,000 and 75,000, respectively, would the mean change? Median? Standard deviation?

37. The World Wide Web is a great source for lists of numbers. Data Set G in Appendix II was obtained from ESPN's website and contains the National League team batting averages and home runs from the 2008 season. Find the mean and standard deviation for both sets of data.

38. Using your results from Exercise 37, find the z-scores of the team average and number of home runs for Arizona. Are they about the same? Do the same computation for another team.

39. Data Set H in Appendix II gives the 2008 season passing yards for 20 quarterbacks in the National Football League. Find the mean and standard deviation.

40. Using your results from Exercise 39, find the z-scores of the number of passing yards for Peyton Manning and Eli Manning.

8.3 The Normal Distribution

> Will this course be graded on the curve?
>
> —Joe R. Student

Certainly, you have heard this question asked on several occasions in your educational experience. Do you know what it means? What is "the curve"? Is it a good thing or a bad thing? Are grades the only objects that can be measured with it? We will be exploring the answers to these questions in this section.

We have seen how histograms assist us in displaying large collections of data and how the shape of a histogram tells us important information concerning the distribution of those data. We now wish to observe what changes take place in a histogram associated with a particular phenomenon as the available number of data values increases. A small number of values (say, less than 20) permits only a fairly crude picture. However, as more data are collected, we can keep creating more interval classes of smaller width until, eventually, the tops of the blocks begin to resemble a bumpy curve. Recall that we can use any histogram to *estimate* the percentage of the collection that lies between values a and b by determining the area of the histogram that lies above the interval $[a, b]$. If we imagine our data set becoming larger and larger. we can often find a smooth curve such that the area under that curve and directly above $[a, b]$ is *exactly* equal to the fractional portion of the collection that lies between a and b. If the area under the entire curve is exactly equal to 1 and the curve is the *graph* of an actual function f, then f is known as a **density function**. (See **Figure 8.3.1**.)

density function
A function whose graph specifies how a population is distributed. The area under the graph and over the domain of the function must equal 1.

population
An infinite (or very large) collection of data.

parameters
Measurements of central tendency and dispersion associated with a population.

We need some new terminology for dealing with extremely large collections of data or with infinite sets consisting of potential data values. Usually, such a collection is known as a **population**, and measures of central tendency and dispersion of a population are generally called **parameters**. Knowledge of the density function associated with a particular population is of tremendous use in determining how the data are distributed. We can use it to determine the precise portion of the population that lies in any interval or set of intervals. For example, the shaded area in **Figure 8.3.2** represents the fraction, 0.42, of this population that lies between 5 and 12.

Alternatively, we can think of any population as an infinite *sample space* and any subset A of data from that population as an *event*. The probability of a member of the population being chosen at random from A, denoted by $p(A)$, is defined to be the fraction of the population that the data values of A comprise. This fraction is then, in turn, given by

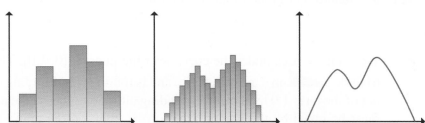

FIGURE 8.3.1   As more data values are added into a collection, the histogram can eventually be replaced by a smooth curve. This curve is the graph of the density function.

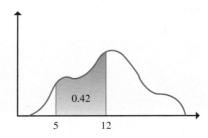

0.42

5   12

FIGURE 8.3.2   The fraction of the population between 5 and 12 is 0.42. We can also say that the probability of randomly choosing a member of this population between 5 and 12 is 0.42.

the appropriate area under the graph of the density function. In Figure 8.3.2, we say that the probability of randomly choosing a value between 5 and 12 is equal to 0.42, or that $p(A) = 0.42$ if A is the interval $[5, 12]$. Equivalently, if we let the variable x represent our selection before it is made, then we may write $p(5 \leq x \leq 12) = 0.42$.

Law of Large Numbers
The relative frequency of a set A approaches the probability of A occurring as the number of data values increases.

We can state the **Law of Large Numbers** to fit our current context.

If $p(A)$ represents the probability that a data value in set A is chosen at random from the population and ρ represents the relative frequency of A, then ρ approaches $p(A)$ as the number of data values in the population increases.

❓ Example 1

Aubrey enjoys hiking in the mountains at speeds that are distributed linearly up to 2.5 miles per hour (mi/h) but never exceed 2.5 mi/h. The function $f(x) = 0.32x$ is the density function for this distribution of hiking speeds. (See Figure 8.3.3.) We imagine that we have measured Aubrey's speed an infinite number of times so that our population theoretically consists of all possible real numbers between 0 and 2.5. Note that the right endpoint of the line segment is (2.5, 0.8), and so the area under the entire curve is equal to that of a right triangle with sides of length 0.8 and 2.5. Its area is, therefore, equal to $\frac{1}{2}(0.8)(2.5) = 1$. (This has to be the case for the area under *any* density function.) If S is the set of all possible speeds by Aubrey, then we write $p(S) = 1$. We may word the following questions in two different ways:

"What fraction of Aubrey's speeds are . . ."

or

"What is the probability that, at any random time, Aubrey is traveling at a speed . . ."

(a) at most 1.0 mi/h?   (b) at least 1.25 mi/h?
(c) exactly equal to 1.5 mi/h?   (d) between 0.5 and 1.75 mi/h?
(e) less than 0.5 or greater than 1.75 mi/h?

⚙ Solution

(a) We want the area under the curve over the interval $[0, 1.0]$. This is given by the triangle ◤ in **Figure 8.3.3(a)** and is found by multiplying one-half the product of the base 1.0 by the corresponding altitude. Because the altitude (or height) must be given by the y-coordinate of the point on the graph with x-coordinate 1.0, it is equal to $f(1.0) = 0.32(1.0) = 0.32$. Therefore,

$$p(x \leq 1.0) = \frac{1}{2}(1.0)(0.32) = 0.16.$$

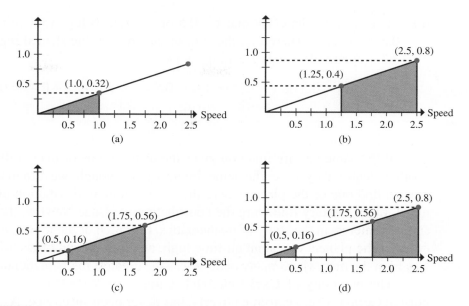

The graph of the density function $f(x) = 0.32x$. The areas of the shaded portions give the probability, at any given time, that the hiker is traveling at a speed of (a) at most 1.0 mi/h; (b) at least 1.25 mi/h; (c) between 0.5 and 1.75 mi/h; and (d) less than 0.5 or greater than 1.75 mi/h.

(b) We want $p(x \geq 1.25) = p(1.25 \leq x \leq 2.5)$, which is equal to the shaded area ▮ in **Figure 8.3.3(b)**. There are several ways to determine this area, one of which is to first find the area of the (unshaded) triangle. Its height is $f(1.25) = (0.32)(1.25) = 0.4$, giving its area as $\frac{1}{2}(1.25)(0.4) = 0.25$. Because the entire area under the graph equals 1, it must be true that

$$p(x \geq 1.25) = 1 - 0.25 = 0.75.$$

(c) Remember that we are not dealing with discrete probabilities associated with a finite sample space here. Aubrey can be hiking at a rate in miles per hour equal to *any* real number at all between 0 and 2.5 mi/h. The probability that she is traveling a certain speed at any particular instant is equal to that one number out of an infinite number of possibilities and so must be equal to 0. As with all infinite data sets, it makes sense to speak only of the probability of making a choice from an *interval* of data. *The probability of choosing a member of an interval will be the same, regardless of whether the endpoints are included.* In other words,

$$p(a < x < b) = p(a \leq x < b) = p(a < x \leq b) = p(a \leq x \leq b).$$

(d) One way to find $p(0.5 < x < 1.75)$ in this case is to subtract the area of the small (unshaded) right triangle in Figure 8.3.3(c) from the area of the large right triangle whose base is the interval $[0, 1.75]$. The heights of the two triangles are $f(0.5) = 0.16$ and $f(1.75) = 0.56$. This yields

$$p(0.5 < x < 1.75) = \frac{1}{2}(1.75)(0.56) - \frac{1}{2}(0.5)(0.16) = 0.49 - 0.04 = 0.45.$$

(e)   First, we note the event that $x < 0.5$ or $x > 1.75$ is the event complementary to $0.5 < x < 1.75$. Therefore, the sum of the areas of the shaded regions in Figure 8.3.3(c) and (d) equals 1, and we must have

$$p(x < 0.5 \text{ or } x > 1.75) = 1 - p(0.5 < x < 1.75) = 1 - 0.45 = 0.55. \quad \blacklozenge$$

If the same density function gives the distributions of many different phenomena, it is often identified by a specific name. Interestingly enough, we turn to the subject of astronomy to find one of the places where the most common distribution of all was first recognized. In the century following the contributions of Isaac Newton, there was a substantial surge of interest in the study of mathematics. Research of the movements and orbital properties of the planets was at an all-time high, and into this vigorous academic environment was born the man whom many believe was the greatest mathematician who ever lived.

The precocity of **Carl Friedrich Gauss** (1777–1855) (**Figure 8.3.4**) as a child, born in Germany into extreme poverty, has never been surpassed. As a tender 3-year-old, it is said that he discovered an error in a long list of computations in his father's bookkeeping which, upon rechecking, showed the toddler to be correct. As a 10-year-old, he was a student of a harsh and uninspiring teacher who routinely ordered his pupils to add up long lists of numbers as a form of discipline. Gauss turned over his slate almost immediately and then sat at his desk under the unrelenting glare of the schoolmaster as they waited almost an hour for the rest of the class to finish the tedious assignment. When everyone was done, it turned out that young Gauss had the only correct answer. He had intuitively created a clever algorithm to find the required sum. The fame of his incredible genius eventually spread to the Duke of Brunswick, who was so impressed by the amazing stories of this youth that he paid for the rest of Gauss's education.

In his early adult life, Gauss was possessed with such a torrent of ideas that he could barely get them all down on paper. He went on to incomparable greatness in mathematics, physics, and astronomy, advancing these disciplines with an incredible number of both theoretical and practical contributions. He combined a prodigious ability at computation with profound and penetrating insight. It is impossible to open a modern textbook on calculus, differential geometry, physics, number theory, linear algebra,

FIGURE 8.3.4   Carl Gauss.

probability, statistics, or any of the sciences empowered by these tools without the name of Gauss peppering its pages. He has been properly labeled by science historians as the Prince of Mathematicians.

Gauss appears now in our study of statistics primarily as a result of his observation of a pattern among the errors that occurred in his astronomical measurements of the asteroid Ceres. The Titius-Bode rule (discovered in 1766) was presented as a specific sequence of numbers that matched closely the mean distances of the planets from the sun. Because this rule predicts the presence of a planet between the orbits of Mars and Jupiter, many telescopic observers at that time searched tirelessly for the alleged object. Later, it was determined that this region is occupied by thousands of small, irregularly shaped bodies known as asteroids; but when Ceres, the largest of these bodies, was first discovered, it caused a great deal of excitement. In the first part of the nineteenth century, astronomers were eager to test the laws of celestial mechanics; and when Ceres disappeared behind the sun during its orbit, Gauss accepted the challenge of predicting the location of its reappearance. It seems that as he studied his measurements of Ceres, he began to think of the inherent errors associated with the measuring process, and this, in turn, led to what he termed the *distribution of errors* and what we today call the **Gaussian (normal) distribution**.

> **Gaussian (normal) distribution**
> A distribution discovered by Carl Gauss whose z-scores form a population having density function
> $$f(z) = \frac{1}{\sqrt{2\pi}} e^{-\left(z^2/2\right)}.$$

We can all appreciate the fact that every time a measurement is made (of distance, size, location, etc.), it is probably not exactly correct because it is dependent on both the measuring device and the human making the measurement—both of which are imperfect. Suppose, for example, you take a number of measurements of your pet alligator in an attempt to determine its length. One may imagine all types of problems associated with the physical measuring process! Over a period of several weeks, at great risk to your personal safety, you give your best efforts to obtain a number of different measurements. You recorded 175.0 centimeters (cm) on Monday, 178.2 cm on Tuesday, 173.5 cm on Wednesday, 176.4 cm on Thursday, and so forth (preferably with a different meter stick in each instance to ensure an averaging out of instrumentation error). At the end of this process, you could compose a histogram of your results and see the shape that it assumes. If you amass a huge collection of measurements, the resulting distribution you obtain will turn out to approximate the normal distribution of Gauss whose curve is the famous bell-shaped **normal curve** shown in **Figure 8.3.5**. The mean (176.0 cm) of this distribution will likely be very close to be the true length of the alligator and will be positioned in the middle of your population of measurements. This normal curve will display the same properties that Gauss first observed in his astronomical data:

> **normal curve**
> The bell-shaped graph of a Gaussian or normal distribution.

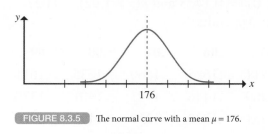

FIGURE 8.3.5   The normal curve with a mean $\mu = 176$.

1. One-half of the area under the curve lies to the left of the mean and one-half to the right. In other words, the *mean is equal to the median.*

2. The curve has a bell-like shape and is *symmetric* with respect to a vertical line drawn through the mean.

3. As for all distributions, the area under the entire curve is equal to 1.

4. The farther you recede to the left of the mean or advance to the right of the mean, the closer the curve comes to the horizontal axis.

The mean of a population is usually denoted by μ (the Greek letter mu), and we continue to use σ to stand for the standard deviation. The beauty of the normal curve is that its associated density function was determined in 1733 by the English mathematician **Abraham DeMoivre** (1667–1754). Knowing this function allows us to estimate areas under the curve over any interval [a, b] by first *standardizing* the population—i.e., by converting all the values to their corresponding z-scores. For any value x in a population, its z-score is again given by

$$z_x = \frac{x - \mu}{\sigma}.$$

Suppose the standard deviation of your alligator's-length measurements was $\sigma = 2.0$ cm. Some of your conversions would be

$$z_{175} = \frac{175 - 176}{2.0} = -0.5 \quad z_{178.2} = \frac{178.2 - 176}{2.0} = 1.1 \quad z_{173.5} = \frac{173.5 - 176}{2.0} = -1.25.$$

> **standard normal distribution**
> The distribution whose density function is given by $f(z) = \frac{1}{\sqrt{2\pi}} e^{-(z^2/2)}$. It has mean 0 and standard deviation 1.

If this is done to *every* value, the resulting distribution of z-scores will necessarily have a mean of 0 and a standard deviation of 1 and is called the **standard normal distribution**. DeMoivre proved that the density function of the standard normal distribution is

$$f(z) = \frac{1}{\sqrt{2\pi}} e^{-(z^2/2)},$$

where the base $e \approx 2.7183$ is that same magic number honored with its own special button on your calculator. Graphing this function of z gives us the normal curve, which we align with the corresponding value axis, as shown in **Figure 8.3.6**.

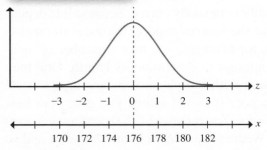

Even though this density function graph does not yield nice triangular areas for us to determine, we can still estimate any area we wish because of a branch of mathematics and computer science known as *numerical analysis*. The techniques, however, are beyond the scope of this text, and we shall content ourselves with using a table. To find the probability $p(z > \tau)$ for any z-score τ, we want the area under the bell curve shown here.

FIGURE 8.3.6 The standard normal curve is the graph of $f(z) = \frac{1}{\sqrt{2\pi}} e^{-(z^2/2)}$.

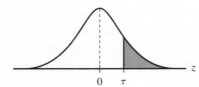

This area is located in the table shown in **Figure 8.3.7** at the intersection of the row with the first two digits of the z-score at the far left and the column headed by the third digit. For example, the boxed numbers below tell us that $p(z > 1.07) = 0.1423$ and $p(z > 1.92) = 0.0274$. The complete table is given in Appendix I.

π	.00	.01	.02	.03	.04	.05	.06	.07	.08	.09
0.9	.1841	.1814	.1788	.1762	.1736	.1711	.1685	.1660	.1635	.1611
1.0	.1587	.1562	.1539	.1515	.1492	.1469	.1446	.1423	.1401	.1379
1.1	.1357	.1335	.1314	.1292	.1271	.1251	.1230	.1210	.1190	.1170
. . . .										
1.8	.0359	.0352	.0344	.0336	.0329	.0322	.0314	.0307	.0301	.0294
1.9	.0287	.0281	.0274	.0268	.0262	.0256	.0250	.0244	.0239	.0233
2.0	.0228	.0222	.0217	.0212	.0207	.0202	.0197	.0192	.0188	.0183

FIGURE 8.3.7 Portion of the standardized normal distribution table.

? Example 2

The symmetry of the normal curve permits the computation of probabilities and percentages related to all types of intervals. The following questions refer to the normal distribution of length measurements of your pet alligator, as previously discussed. Recall that the mean $\mu = 176.0$ and the standard deviation $\sigma = 2.0$. (All values are in centimeters.)

(a) What percentage of the population is greater than 178?
(b) What percentage of the population is less than 178?
(c) What percentage of the population is less than 175.2?
(d) If you were to make another measurement tomorrow, what is the probability that it would lie between 177 and 180.3?
(e) If you were to make another measurement tomorrow, what is the probability that it would lie between 173.5 and 177.2?

⚙ Solution

(a) We can make use of the standardized normal distribution table by first converting 178 to a z-score:

$$z_{178} = \frac{178 - 176}{2.0} = 1.00.$$

The main concept is to realize that $p(x > 178) = p(z_x > z_{178}) = p(z_x > 1.00)$ and that this probability is given by the area in the top picture of Figure 8.3.8. Consulting the table in Figure 8.3.7, we see that this area is equal to 0.1587. Because this means that $p(x > 178) = 0.1587$, we conclude that the percentage of the population greater than 178 is 15.87%.

(b) First note that $p(x < 178) = p(x \le 178)$, because $p(x = 178) = 0$. Next, because the events $x > 178$ and $x < 178$ are complementary, $p(x < 178) = 1 - p(x > 178)$. Note that this correlates to the fact that the area under the entire curve is equal to 1, and the area we require is shown in the bottom graph of **Figure 8.3.8**. You can see that the sum of the areas of the two shaded regions is equivalent to the entire

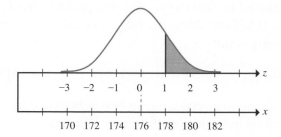

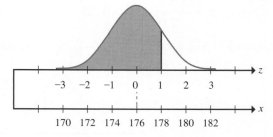

FIGURE 8.3.8   The top area corresponds to $p(x > 178)$ and the bottom to $p(x < 178)$.

region under the curve. Therefore, $p(x < 178) = 1 - 0.1587 = 0.8413$, and so the percentage of the population less than 178 is 84.13%.

(c) We now want the probability $p(x < 175.2) = p(z_x < -0.40)$, because $z_{175.2} = \dfrac{175.2 - 176}{2.0} = -0.40$. This is given by the area of the region in this next graph.

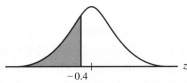

The table in Figure 8.3.7 gives us probabilities only associated with positive z-scores, but the symmetry of the curve implies that the area in the preceding graph is identical to the one shown in the next graph.

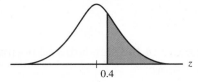

This implies that $p(z_x < -0.4) = p(z_x > 0.4) = 0.3446$, and so it is true that 34.46% of our population is less than 175.2.

(d) First, we find $z_{177} = \dfrac{177 - 176}{2.0} = 0.50$ and $z_{180.3} = \dfrac{180.3 - 176}{2.0} = 2.15$. The area giving us $p(177 < x < 180.3) = p(0.5 < z_x < 2.15)$ is shown here.

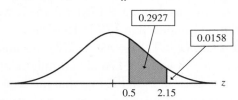

The standardized normal distribution table (Figure 8.3.7) gives only areas to the right of the listed z-scores, but the picture shown here indicates that it can be obtained by subtracting the area to the right of $z = 2.15$ from that to the right of $z = 0.5$. Consulting the table and subtracting these probabilities lead to the following result.

$$p(0.5 < z_x < 2.15) = p(z_x > 0.5) - p(z_x > 2.15) = 0.3085 - 0.0158 = 0.2927$$

(e) In this case, we need a region that is bounded by vertical lines through z-scores located on both sides of the mean. The required z-scores are

$$z_{173.5} = \frac{173.5 - 176}{2.0} = -1.25 \qquad z_{177.2} = \frac{177.2 - 176}{2.0} = 0.60.$$

We want $p(173.5 < x < 177.2) = p(-1.25 < z_x < 0.6)$.

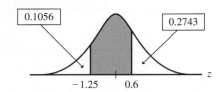

The sum of the shaded and the two unshaded regions must equal 1. We find the areas of the unshaded regions from Figure 8.3.7 to be $p(z_x > 0.6) = 0.2743$ and $p(z_x < -1.25) = p(z_x > 1.25) = 0.1056$. Therefore,

$$p(173.5 < x < 177.2) = p(-1.25 < z_x < 0.6) = 1 - (0.2743 + 0.1056) = 0.6201. \quad \blacklozenge$$

A large variety of data collections are normally (or approximately normally) distributed. When you stop to think about it, this makes sense—the same reasoning that explains a normal distribution of errors when repeatedly measuring the same object applies to the distribution of values that are associated with the same biological, physical, or social phenomenon. It stems from the fact that the numbers that are the members of these collections are essentially generated by summing over a group of independent factors. In the case of measurement errors, we saw that those factors consisted of such things as a faulty measuring stick and shaky hands. Similarly, the factors that contribute, say, to the growth of a particular type of tree in a certain geographic region are also basically independent of one another—rainfall, soil composition, competing species, and so on. A central governing principle in the science of statistics, called the **Central Limit Theorem**, is stated as follows:

> **Central Limit Theorem**
>
> If each value in a population is the result of summing over the same independent factors, then the distribution of that population approaches a normal distribution as the number of factors increases.

> If each value in a population is the result of summing over the same independent factors, then the distribution of that population approaches a normal distribution as the number of factors increases.

❓ Example 3

In recent years, Scholastic Aptitude Test (SAT) scores in mathematics were approximately normally distributed, with a mean of 476 and a standard deviation of 80. What percentage of students taking the test had mathematics SAT scores

(a)   between 400 and 500?
(b)   more than 550?
(c)   less than 425?
(d)   within 1 standard deviation of the mean?
(e)   more than 385?

⚙ Solution

(a)   We first need to compute the required z-scores:

$$z_{400} = \frac{400 - 476}{80} = -0.95 \quad z_{500} = \frac{500 - 476}{80} = 0.30.$$

We are asked to find $p(400 < x < 500) = p(-0.95 < z_x < 0.30)$, which is equivalent to finding the shaded area in the following figure.

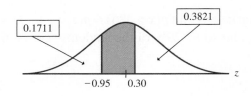

Using symmetry properties and the standard normal distribution table in Appendix I, we see that $p(-0.95 < z_x < 0.30) = 1 - (0.1711 + 0.3821) = 1 - 0.5532 = 0.4468$. Thus, 44.68% of the students scored between 400 and 500.

(b) We first compute the z-score:

$$z_{550} = \frac{550 - 476}{80} = 0.93.$$

We are asked to find $p(x > 550) = p(z_x > 0.93)$, which is found directly in the standard normal distribution table as 0.1762. So 17.62% of the students scored more than 550.

(c) We first compute the z-score:

$$z_{425} = \frac{425 - 476}{80} = -0.64.$$

We are asked to find $p(x < 425) = p(z_x < -0.64)$. Look at this next figure.

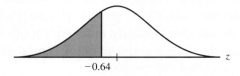

From this graph, we have $p(z_x < -0.64) = p(z_x > 0.64) = 0.2611$ from the standard normal distribution table. So 26.11% of the students scored less than 425.

(d) Because the standard normal distribution has a mean of $z = 0$ and a standard deviation of 1, we are asked to find $p(-1 < z_x < 1)$. This is equivalent to finding the shaded area in the next graph.

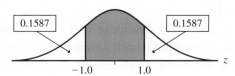

The symmetry of the normal curve implies that $p(z_x < -1) = p(z_x > 1) = 0.1587$, and so we see that

$$p(-1 < z_x < 1) = 1 - 2(0.1587) = 0.6826.$$

Thus 68.26% of the students scored within 1 standard deviation of the mean.

(e) We first compute the z-score:

$$z_{385} = \frac{385 - 476}{80} = -1.14.$$

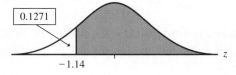

We are asked to find $p(x > 385) = p(z_x > -1.14) = 1 - p(z_x < -1.14) = 1 - 0.1271 = 0.8729$, and so 87.29% of the students scored more than 385. ◆

? Example 4

We are now in a position to discuss grading on the curve. Suppose your instructor in a particular class has given the exact same test a great number of times over the course of many years. Because each individual student score on the test is the result of many independent factors, such as native ability, work ethic, studying environment, and so on, the resulting distribution is surely a normal one. Knowing the mean of the current distribution of scores to be $\mu = 70$ (out of 100) and the standard deviation to be $\sigma = 11$, your instructor decides that grades will be determined by the z-scores of the test results according to the information shown in the following table.

z-Score	Grade
$z \geq 2.0$	A
$1.0 \leq z < 2.0$	B
$-1.0 \leq z < 1.0$	C
$-2.0 \leq z < -1.0$	D
$z < -2.0$	F

(a) What is the range of test scores that corresponds to each grade?
(b) What percentage of A's, B's, C's, D's, and F's can the instructor expect?

⚙ Solution

(a) Using the formula $z_x = \dfrac{x - \mu}{\sigma}$, we can find each score as follows.

$$\frac{x - 70}{11} = 2.0 \qquad \frac{x - 70}{11} = 1.0 \qquad \frac{x - 70}{11} = -1.0 \qquad \frac{x - 70}{11} = -2.0$$

$$x - 70 = 22 \qquad x - 70 = 11 \qquad x - 70 = -11 \qquad x - 70 = -22$$

$$x = 92 \qquad\qquad x = 81 \qquad\qquad x = 59 \qquad\qquad x = 48$$

The ranges of test scores are thus A(92–100), B(81–91), C(59–80), D(48–58), and F(0–47).

(b) Use the standard normal distribution for the following calculations:

$$p(z \geq 2.0) = 0.0228$$
$$p(1.0 \leq z < 2.0) = p(z \geq 1.0) - p(z \geq 2.0) = 0.1587 - 0.0228 = 0.1359$$
$$p(-1.0 \leq z < 1.0) = 1 - 2p(z \geq 1.0) = 1 - 2(0.1587) = 0.6826$$
$$p(-2.0 \leq z < -1.0) = p(1.0 \leq z < 2.0) = 0.1359$$
$$p(z < -2.0) = p(z > 2.0) = 0.0228.$$

The instructor can expect approximately 2% A's, 14% B's, 68% C's, 14% D's, and 2% F's. ◆

We have seen that the individual z-scores of the standard normal distribution are associated with specific areas under the normal curve. In particular, Example 4 demonstrates that the area under the curve between $z = -1.0$ and $z = 1.0$ is about 0.68. Equivalently, we could say that about 68% of all values in any normal distribution lie within 1 standard deviation of the mean. (See **Figure 8.3.9**.)

We can also refer to Example 4 to find that $p(-2.0 \leq z \leq 2.0) = 1 - 2p(z \geq 2.0) = 1 - 2(0.0228) = 0.9544$. This tells us that 95.44% of a normal distribution lies within 2 standard deviations of the mean. One could then reasonably predict that 95% of the distribution lies just a bit closer. Example 5 illustrates how to find the more accurate z-score that corresponds to this percentage.

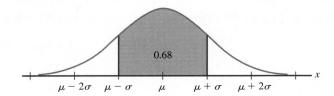

0.68

FIGURE 8.3.9   About 68% of all values in a normal distribution lie within 1 standard deviation of the mean.

? Example 5

Estimate a distance from either side of the mean that encompasses 95% of a normal distribution.

⚙ Solution

We wish to find z^* such that $p(-z^* \leq z \leq z^*) = 0.95$. As we have seen, the symmetry of the normal curve means we need to search the standard normal distribution table for a z^* such that $p(z \geq z^*) = \frac{1}{2}(1 - 0.95) = \frac{1}{2}(0.05) = 0.025$. Consulting the standardized normal distribution table, we find that $z^* = 1.96$ is the required z-score. So we have found that about 95% of the values in any normal distribution lie within 1.96 standard deviations of the mean. (See **Figure 8.3.10**.) This fact plays a central role in the construction of *confidence intervals*, which are the subject of the next section.   ◆

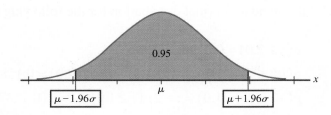

0.95

FIGURE 8.3.10   About 95% of all values in a normal distribution lie within 1.96 standard deviations of the mean.

Name _____

Exercise Set **8.3** ◯◯◯←

1. Let $f(x) = 0.5x$ be the density function for x, where $0 \leq x \leq 2$. Find the probability that x is in the following intervals.

(a) less than 1
(b) less than 1.8
(c) greater than 1.8
(d) between 0.5 and 1.5

2. Let $f(x) = \dfrac{2}{9}x$ be the density function for x, where $0 \leq x \leq 3$. Find the probability that x is in the following intervals.

(a) less than 1
(b) between 1 and 2
(c) greater than 2.4
(d) less than 3

3. Let $f(x) = 0.125x$ be the density function for x, where $0 \leq x \leq 4$. Find the probability that x is in the following intervals.

(a) between 0 and 2
(b) greater than 1
(c) at least 3.2
(d) between 1.7 and 2.9

4. Let $f(x) = \dfrac{8}{9}x$ be the density function for x, where $0 \leq x \leq 1.5$. Find the probability that x is in the following intervals.

(a) greater than 0.5
(b) less than 1.1
(c) at most 0.8
(d) more than 1.6

5. Let $f(x) = 0.08x$ be the density function for x, where $0 \leq x \leq 5$. Find the probability that x is in the following intervals.

(a) less than 2.3
(b) more than 4.2
(c) between 1.6 and 4.1
(d) within 1 unit of 2.5

6. Let $f(x) = 0.02x$ be the density function for x, where $0 \leq x \leq 10$. Find the probability that x is in the following intervals.

(a) between 6 and 7
(b) exactly 5
(c) between 4.5 and 5.5
(d) at most 6.2

7. Given a population with the following symmetric distribution curve, find the portions of the population that lie in the stated intervals.

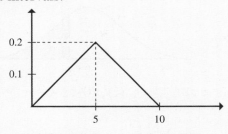

(a) $[0, 5]$ (b) $[5, 10]$ (c) $[0, 2.5]$ (d) $[7.5, 10]$ (e) $[2.5, 7.5]$

8. What is the median of the population with the distribution depicted in Exercise 7? What is the mean?

9. The *uniform distribution* is a density function that is constant for a particular interval of values. For example $f(x) = \dfrac{1}{8}$ is a uniform distribution where $0 \le x \le 8$. Find the following probabilities.

(a) $p(0 \le x \le 2)$ (b) $p(5 \le x \le 7)$ (c) $p(x \le 3)$ (d) $p(x \ge 3)$

10. The uniform distribution is defined in Exercise 9.

(a) If a uniform distribution exists for x on the interval $[0, b]$, what is the probability density function $f(x)$? (*Hint:* What must be the area under $f(x)$ on $[0, b]$?)
(b) If a uniform distribution exists for x on the interval $[a, b]$, what is the probability density function $f(x)$? (*Hint:* How wide is the interval $[a, b]$?)

11. Determine the median of a population having the density function $f(x) = 0.32x$ for $0 \le x \le 2.5$.

12. Determine the median of a population having the density function $f(x) = 0.08x$ for $0 \le x \le 5$.

13. If a population is distributed according to the linear density function $f(x) = mx$ over the interval $0 \le x \le r$, find an expression for the median M as a function of the slope m. (Assume $m > 0$.)

14. If a population is distributed according to the linear density function $f(x) = mx$ over the interval $0 \le x \le r$, then the mean μ must be given by $\mu = \dfrac{2r}{3}$, and the median M must be given by $M = \dfrac{r}{\sqrt{2}}$. Which is always the larger value, the mean or the median?

15. Use the functions in Exercise 14 to find the mean and median for the population given in Exercise 11. Does your value for the median match your previous computation?

16. Use the functions in Exercise 14 to find the mean and median for the population given in Exercise 12. Does your value for the median match your previous computation?

The capital letters in the graphs for Exercises 17–18 represent areas of the indicated regions under the standard normal curve. In each case, write an expression that gives the probability.

17.

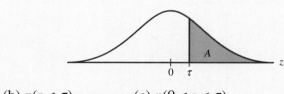

(a) $p(z > \tau)$ (b) $p(z < \tau)$ (c) $p(0 < z < \tau)$

18.

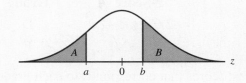

(a) $p(z > b)$ (b) $p(z < a)$ (c) $p(z > a)$ (d) $p(a < z < b)$

19. Determine the following probabilities for the standard normal distribution.

(a) $p(z > 2.11)$ (b) $p(0 \leq z < 1.30)$ (c) $p(z < 1.63)$
(d) $p(-1.23 < z < 1.84)$ (e) $p(0.32 < z \leq 1.47)$ (f) $p(z > -2.3)$

20. Determine the following probabilities for the standard normal distribution.

(a) $p(z \geq 0.64)$ (b) $p(0 \leq z < 1.31)$ (c) $p(z < -2.03)$
(d) $p(-1.02 < z \leq -0.23)$ (e) $p(-1.21 \leq z < 0.50)$ (f) $p(z < 1.0)$

21. Find the area under the standard normal curve.

(a) between 1 and 2   (b) to the right of 0.43
(c) to the left of −0.4   (d) between −0.5 and 0.8

22. Find the area under the standard normal curve.

(a) to the left of 0.45   (b) to the right of 0
(c) between −2.6 and −0.3   (d) between −0.71 and 0.18

23. A population is normally distributed with a mean of 40 and a standard deviation of 6. Find the portions of this population that are

(a) greater than 45.   (b) less than 36.1.   (c) between 37.5 and 46.3.
(d) greater than 38.   (e) exactly equal to 40.   (f) between 30 and 38.

24. A population is normally distributed with a mean of 28 and a standard deviation of 4.2. Find the portions of this population that are

(a) greater than 30.   (b) less than 25.5.   (c) between 26 and 31.
(d) between 22.3 and 27.5.   (e) greater than 25.3.   (f) exactly equal to 28.

25. Cynthia heard that the mean score in Easy Ed's statistics class last semester was 92. However, Cynthia did not hear that the standard deviation was 3.2 and that Professor Ed assigned grades on the curve according to the following scale.

z-Score	Grade
$z \geq 2.0$	A
$1.0 \leq z < 2.0$	B
$-1.0 \leq z < 1.0$	C

z-Score	Grade
$-2.0 \leq z < -1.0$	D
$z < -2.0$	F

(a) What is the range of test scores that corresponds to each grade?
(b) What percentage of A's, B's, C's, D's, and F's can the instructor expect?

26. The mean score in Easy Ed's statistics class this semester was 66 with a standard deviation of 7.3, but Professor Ed continued to grade on the curve according to the scale shown in Exercise 25.

(a) What is the range of test scores that corresponds to each grade?
(b) Do the percentages of A's, B's, C's, D's, and F's change from those in Exercise 25?
(c) What grade would a score of 82 be assigned in both this problem and the one in Exercise 25?

27. Mystery writers such as Raymond Chandler, Elmore Leonard, and Walter Mosley usually keep the lengths of their novels to 200 to 250 pages. The collected works of the little-known but critically acclaimed author Scott Nicholson have lengths that are normally distributed with a mean of 230 pages and a standard deviation of 15 pages.

(a) What percentage of his books is more than 250 pages in length?
(b) What percentage of his books is fewer than 200 pages in length?
(c) What percentage of his books is between 200 and 250 pages in length?

28. In Anywhere, USA, the household incomes are normally distributed with a mean of $54,000 and a standard deviation of $5,000. What percentage of the city households has incomes in the following ranges?

(a) more than $65,000
(b) less than $50,000
(c) between $52,000 and $60,000
(d) at least $48,000

29. The speeds of cars traveling on a freeway are normally distributed with a mean of 68 mi/h and a standard deviation of 4.5 mi/h. Find the percentage of cars with the following speeds.

(a) more than 75 mi/h
(b) less than 68 mi/h
(c) between 70 and 80 mi/h
(d) between 65 and 75 mi/h

30. The life spans of lightbulbs of 100 watts (W) produced by Wizard Electronics are normally distributed with a mean of 750 h and a standard deviation of 25 h. If you buy all your bulbs from this company, what portion of the 100-W bulbs will last

(a) more than 800 h?
(b) fewer than 740 h?
(c) between 700 and 760 h?
(d) between 710 and 810 h?

31. In one biology lab experiment, the students are asked to determine the mass of a certain type of insect. Historically, this experiment has yielded masses that are normally distributed with a mean of 0.26 g and a standard deviation of 0.08 g. The professor gives an A to those students whose results have a z-score between −1.0 and 1.0 and a C to those whose results have a z-score more than 2.0 or less than −2.0.

(a) What range of masses will earn a student an A?
(b) What percentage of students receives an A for this lab?
(c) What range of masses will earn a student a C?
(d) What percentage of students receives a C for this lab?

32. Estimate a distance (in units of standard deviations) from either side of the mean that encompasses 99% of a normal distribution.

33. Estimate a distance (in units of standard deviations) from either side of the mean that encompasses 90% of a normal distribution.

34. A machine produces bolts whose lengths are normally distributed with a mean of 6.4 millimeters (mm) and a standard deviation of 0.5 mm. What percentage of the bolt lengths is

(a) between 6.0 and 6.6 mm?   (b) more than 7.0 mm?

35. Cassandra is a forest service biologist who wants to determine the mean and standard deviation of the diameters of lodge pole pine trees in a certain section of Yellowstone National Park. She begins by recording the diameters in centimeters (cm) of a set of 50 trees and creating a relative frequency distribution, resulting in the following table.

Class (cm)	Frequency	Relative Frequency
0–10	4	4/50 = 0.08
10–20	9	9/50 = 0.18
20–30	15	15/50 = 0.30
30–40	13	13/50 = 0.26
40–50	7	7/50 = 0.14
50–60	2	2/50 = 0.04

Thinking that a bigger database would yield more-accurate information for her study, she measured an additional 150 trees, bringing the total to 200. This naturally affected her relative frequency distribution, as shown in the following table.

Class (cm)	Frequency	Relative Frequency
0–10	14	14/200 = 0.07
10–20	38	38/200 = 0.19
20–30	52	52/200 = 0.26
30–40	50	50/200 = 0.25
40–50	34	34/200 = 0.17
50–60	12	12/200 = 0.06

(a) Draw histograms for both of these relative frequency distributions.

(b) If Cassandra measures and records the diameters of a very large number of trees, she obtains a normal distribution with a mean of $\mu = 28.5$ and a standard deviation of $\sigma = 11.8$. Find $p(10 < x < 20)$ and $p(30 < x < 40)$. How do these probabilities compare with the relative frequencies of these intervals in both of the relative frequency distributions?

Use a calculator to find areas under the graphs in Exercises 36–39.

36. Consider the density function $f(x) = 0.375x^2$ for the domain $0 \le x \le 2$. We cannot use the triangle formula to evaluate areas under this curve because the function is not linear. Moreover, appropriate tables are not available. However, most graphing calculators have an option for determining such areas by using what is known in calculus as *numerical integration*. We demonstrate this capability on the TI by verifying that the area under this curve over [0, 2] is equal to 1. (Recall that this must be true for *any* density function.)

(a) Press **Y =** and input $Y_1 = 0.375$ X^2. Press **WINDOW** and size your graphing window by setting **Xmin** = 0, **Xmax** = 2, **Xscl** = 1, **Ymin** = 0, **Ymax** = 3, and **Yscl** = 1. Press **GRAPH**.

(b) Select **CALC** by pressing **2nd** and **TRACE**. Choose the option **labeled** $\int f(x)\, dx$. Your graph will reappear, and you will be asked for a lower limit—that is, the left endpoint of the interval. In the present case, we input 0 by pressing **0** and then **ENTER**. You will then be asked for an upper limit—that is, the right endpoint of the interval. In the present case, we input 2. You will observe that the appropriate region is shaded in the window and the area is given at the bottom.

(c) We can then find any other area by first clearing the shaded region and repeating the above steps in parts (a) and (b). To clear the region, select **DRAW** by pressing **2nd** and then **PRGM**. Select **1: ClrDraw** and press **ENTER**.

37. Consider the density function given in Exercise 36, namely $f(x) = 0.375x^2$ for the domain $0 \le x \le 2$. Find the probabilities (to three decimal places) that x is in each of the following intervals.

(a) $[0, 0.75]$ (b) $[0.75, 2]$ (c) $[0.5, 1.5]$ (d) $[1.5, 1.75]$

38. Consider the density function $f(x) = 0.1875x^{1/2}$ for the domain $0 \le x \le 4$. Verify that this is a density function by finding the area under the curve over the given domain. Set the graphing window according to **Xmin** = 0, **Xmax** = 5, **Xscl** = 1, **Ymin** = 0, **Ymax** = 0.5, and **Yscl** = 1.

39. Consider the density function $f(x) = 0.1875x^{1/2}$ for the domain $0 \le x \le 4$. Find the probabilities (to three decimal places) that x is in each of the following intervals.

(a) $[0, 1]$ (b) $[1, 2]$ (c) $[2, 3]$ (d) $[3, 4]$

8.4 Confidence Intervals

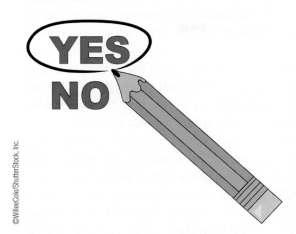

©WilleeCole/ShutterStock, Inc.

margin of error
A number that is added and substracted from an estimate ($\bar{x}$ or $\hat{p}$) to create a confidence interval for a parameter (μ or p).

sample A subset of a larger collection or population.

Polls and surveys of many kinds are a staple of weekly news magazines, particularly around election time, and so the accuracy of such findings is an important issue. Typically, the **margin of error** and the number of people polled are given in a footnote at the bottom of the page where the poll appears. What do these numbers mean? You probably associate them with a measure of the reliability of the poll results, but in what manner? Suppose you conducted your own poll. If you asked each student in your current mathematics course whether she or he were enjoying the class, would you interpret the percentage of yes responses to be indicative of the entire contingent of students everywhere who have also taken this course? In other words, is your class a representative subset or **sample** of a much larger collection or population? This section discusses the answers to questions such as these and, in so doing, offers some insight into the underpinnings of how statistics may be properly used in a deductive manner to reach reasonable conclusions from polls and sampling, a process known as *inference*.

The normal distribution has shown us how most large populations are scattered about the mean, but in practical applications, we need a trustworthy method for *estimating* the mean itself. This is a tricky business because we may not know anything at all about the population of values that we are investigating. For example, what would you guess to be the mean value of each of the following populations?

- Square meters of floor space in houses in Barcelona, Spain
- Diameters of headlights on Ford trucks
- Weights of athletic socks used by 10-year-old Little League players
- Bushels of corn per acre on farms in Germany
- Percentages of acetaminophen in pain relief capsules of a particular brand
- Amounts of selenium per gallon of water in the Danube River
- Shares of stock traded per day on the New York Stock Exchange over the last 25 years
- Lifetimes of stars in the Milky Way galaxy

Even though the data for some of these examples may actually be available, the computation of a mean by adding all the possible numbers and dividing would be a brutishly difficult task at best. Surely, you have better ways to spend your time! A more time-efficient procedure—and one that is effective even in the absence of many of the actual data—involves first choosing a small subset or *sample* of the population. We then compute a *mean of the sample*, call it $\bar{x}$, and use this as an *estimate* of the mean μ of the population. A number computed from a sample is referred to as a *statistic* to distinguish it from any value associated with the parent population, which is a *parameter*. So a sample mean $\bar{x}$ is a statistic, and the population mean μ is a parameter. This section addresses the answers to the following three questions.

- Is our sample representative of the population as a whole?
- How do we express our estimation in a meaningful way?
- How is the size of the sample related to the accuracy of the related statistics?

random sample
A sample gathered from a population in such a way that every member of the population has an equal chance of being selected.

The first question is concerned with the process of gathering your sample. The essential requirement is that *every member of the population have an equal chance of being selected.* If this is ensured, the handful of numbers that have been chosen is known as a **random sample**. How do you ensure randomness in your selection process?

Suppose, for instance, that you wished to estimate the mean weight of the 8,000 students who attend your school. Would you station yourself with a scale outside the locker room for the football team and weigh the first 30 people who came out? Of course not. The $\bar{x}$ of this sample would almost certainly be too high. This, in fact, is a clear case of a **biased sample**, or one that is produced by a technique in which the computed sample mean is consistently too high or too low. (See **Figure 8.4.1**.) Instead, you need a selection methodology that ensures

biased sample
A sample gathered by a process in which the computed statistic is consistently too high or too low.

that all students at the school have the same chance of being chosen. Let's assume that each student has a unique four-digit ID number and you wish to pick a sample of about 30. You cannot just look at a student roster and pick numbers "at will," for it is quite possible you may have a personal inclination toward choosing particular numbers with, say, a 3 or a 7 in them. One common method for ensuring a truly random draw is to use an official random digit table produced by a computer algorithm or other means. Generation of a random digit table is not as easy as it sounds because the author must be able to prove its authenticity. Using the so-called official table given in **Figure 8.4.2**, we make our selections by simply moving across each row from left to right, picking four consecutive digits, and including the weight of that particular student in our sample if there is, in fact, a student with that ID number. If not, we simply discard those digits, move on to the next four, and repeat the process. In this case, if there is a student with ID number 5028, then we use the weight of that student (presumably readily avail-

FIGURE 8.4.1   Biased sample of student weights.

able) for our first choice. If no student has that number, we next check 8293, then 0381, and so forth, continuing in this manner until our 30 weights have been chosen. If there is no current numbering of the members of the parent population from which we must draw our sample, then we must first assign numbers to the members (called an *enumeration*). Although many types of enumeration are permissible, we need to use only two digits apiece (and start with 01, 02, . . .) if there are fewer than 100 members in the population, three digits each for fewer than 1,000 members, and so on. Consider Example 1.

50288	29303	81597	12735	53489	67908	85341
69014	84784	93814	93900	62821	00441	82556
12569	35356	15683	44225	53161	10650	72436

FIGURE 8.4.2   Random digit table.

❓ Example 1

At the local zoo, 10 animals are to be randomly chosen by a state veterinarian for a health checkup. Use the table in Figure 8.4.2 to pick a random sample from the following list.

tiger	polar bear	chameleon	armadillo
tortoise	zebra	rhinoceros	flamingo
alligator	kangaroo	giraffe	toucan
gorilla	chimpanzee	lion	panther
kingsnake	African bullfrog	camel	orangutan
golden eagle	antelope	elephant	giant anteater
sea lion	manatee	snow owl	Gila monster

⚙ Solution

After enumerating the animals with two-digit numbers, we use the table to make our selections. If we ignore pairs of digits greater than 28, the choices result in a sample consisting of the alligator, golden eagle, polar bear, manatee, rhinoceros, lion, snow owl, panther, giant anteater, and Gila monster. You might think of this selection process as being roughly equivalent to throwing darts at a dartboard. (See **Figure 8.4.3**.)

50**288**   29**303**   81597   12**735**   53489   67**908**   85341
69**014**   84784   93814   93900   **62821**   00**441**   **82556**

01 tiger	09 zebra	• 16 rhinoceros	24 toucan
02 tortoise	10 kangaroo	17 giraffe	• 25 panther
• 03 alligator	11 chimpanzee	• 18 lion	26 orangutan
04 gorilla	12 African bull	19 camel	• 27 giant anteater
05 kingsnake	frog	20 elephant	• 28 Gila
• 06 golden eagle	13 antelope	• 21 snow owl	monster ◆
07 sea lion	• 14 manatee	22 armadillo	
• 08 polar bear	15 chameleon	23 flamingo	

A polling procedure used in pseudonews programs on television and some popular magazines consists of having viewers and readers call in their responses to certain questions. In the next episode or issue, the viewer or reader is led to believe the results of this survey are indicative of common public opinion, but this is definitely not true. In fact, this method fails to be a random selection process in several ways. One of the main problems is that the population from which this sample is being drawn is not society at large, but rather is that very small portion who happen to have read that particular copy of the magazine or have been watching that specific program. The person must also own a telephone and be willing to pay any associated cost. These barriers systematically exclude certain groups of people from the survey, and the problem is known as selection bias. Furthermore, only people who happen to be feeling passionate enough about the topic bother to take the time and energy to participate. Because many people will not respond, this effect on the sample result is known as nonresponse bias. Both of these problems are violations of the core principle that every population member must have an equal chance for selection, and they keep this type of polling from producing reliable results.

FIGURE 8.4.3   Selections must be made randomly.

We now assume that all of our sampling is done in a random manner, and we return to the concept of using a statistic such as the value of a sample mean $\bar{x}$ to estimate the value of a parameter such as the mean μ of a population. We recall the second of the three questions asked at the outset: How do we express our estimation in a meaningful way?

We illustrate a procedure for doing this by again examining the problem of determining the average weight of the 8,000 students at your school. Suppose you have randomly chosen 30 people and computed the mean weight of this sample to be $\bar{x}_1 = 156.3$ pounds (lb). Although we have no way of knowing for certain how close 156.3 is to the actual mean μ, it seems reasonable that this number is in some sense a more reliable estimate of μ than would be an average based on a sample of size 10 or 20 students, but perhaps it is less trustworthy than the results of weighing 50 people. In the extreme case, if our sample included the entire student body of 8,000, our estimate would be absolutely correct! Of course, the whole point of sampling and polling in the first place is to avoid such time-consuming tasks, and so the trade-off between accuracy and time efficiency is a standard problem for every pollster. This somewhat obvious premise is still worth stating as a general principle:

Larger sample sizes produce more reliable statistics but require greater effort.

Adopt a consistent sample size of 30, and then suppose that several friends have randomly chosen their own samples of 30 weights each and have computed these sample means to be $\bar{x}_2 = 159.2$, $\bar{x}_3 = 154.7$, $\bar{x}_4 = 152.6$, and $\bar{x}_5 = 168.5$. Clearly, we cannot claim any of these values to be a best estimate of μ. However, it must be true that any average of a number of randomly selected weights stands a better chance of lying close to μ than does just a single weight drawn at random from the entire population. If you had an infinite number of friends who each had conducted a survey and calculated an $\bar{x}$, this collection would comprise a *distribution of sample means*, also called a **sampling distribution**. What we have just observed is that these $\bar{x}$ values would cluster closer to μ than the members of the parent population and so would have a smaller standard deviation. Indeed, three important facts are proved in advanced statistics books concerning any sampling distribution and the relationship of its mean $\mu_{\bar{x}}$ and standard deviation $\sigma_{\bar{x}}$ to the corresponding parent population parameters.

sampling distribution
A population of means of samples, all of the same size.

1. A sampling distribution is approximately normal.
2. The mean of a sampling distribution is equal to the mean of the parent population. In symbols, this translates to

$$\mu_{\bar{x}} = \mu.$$

3. The standard deviation of a sampling distribution generated by means of samples of size n is equal to the standard deviation of the parent population divided by $\sqrt{n}$. In symbols, this translates to

$$\sigma_{\bar{x}} = \frac{\sigma}{\sqrt{n}}.$$

This corresponds to our earlier principle about sample sizes. The greater the sample size, the greater the chance the mean of the sample will lie close to μ. To know these truths about the mean and standard deviation is all well and good, but what about the shape of the distribution itself? To put the icing on the cake here, we recall that the *Central Limit Theorem* of the last section guarantees that the results of summing over independent factors will be approximately normally distributed if the number of factors is large. Certainly a population of sample means involves summing, and so for a large enough sample size (30 or more will suffice for our purposes), the *sampling distribution is approximately normal* regardless of the distribution of the parent population. A set of overlaid graphs gives a visual summary of these facts in **Figure 8.4.4**.

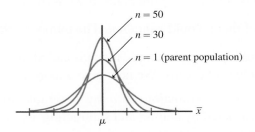

FIGURE 8.4.4　Sampling distributions generated by samples of increasing size.

It is important to remember that all members of a sampling distribution are generated by computations from samples *of the same size*. That is why each curve in Figure 8.4.4 is tagged with its own value for n. As n increases, the standard deviation of the associated normal distribution decreases, and so the shape of the associated curve narrows.

We are now in a position to numerically assess the likelihood of μ lying in an *interval* known as a **confidence interval** centered on a mean produced from a sample of a specified size. Because we can never be certain of the exact value of μ, every such interval must be accompanied by a percentage giving the **level of confidence** that it contains μ. To demonstrate the construction of a 95% confidence interval for the mean student weight problem, let's use our first sample statistic, $\bar{x}_1 = 156.3$ lb, based on the size $n = 30$; let us also assume a standard deviation for the parent population of $\sigma = 20$ lb.

In an example from the previous section, we found that 95% of a normal distribution lies within 1.96 standard deviations of the mean. Because we are dealing with a normal distribution of $\bar{x}$ values, this implies that there is a 95% chance of 156.3 lying within 1.96 standard deviations of μ. However, distance is a symmetric property: If Fred lives two blocks from Barney, then Barney lives

confidence interval
An interval centered on a sample statistic that contains a population parameter with a specified level of confidence.

level of confidence
A percentage that gives the portion of the time a confidence interval will contain the population parameter.

two blocks from Fred. We might say that if Fred lives in Barney's neighborhood, then Barney lives in Fred's neighborhood. In a similar vein, we can state that there is a 95% chance that μ lies in the neighborhood of 156.3—that is, within 1.96 standard deviations. (See **Figure 8.4.5**.) Remembering that the standard deviation $\sigma_{\bar{x}}$ of the sampling distribution is always equal to $\dfrac{\sigma}{\sqrt{n}}$, the endpoints of a 95% *confidence interval* (for μ) computed from a sample of size n and having mean $\bar{x}$ must be

$$\bar{x} \pm 1.96\left(\frac{\sigma}{\sqrt{n}}\right),$$

where σ is the standard deviation of the parent population. In the present case, we get

$$156.3 \pm 1.96\left(\frac{20}{\sqrt{30}}\right) = 156.3 \pm 7.2,$$

giving the interval [149.1, 163.5]. Statistically speaking, we are 95% confident that the true population mean μ lies between 149.1 and 163.5. Of course, different intervals must result from using the statistics of other samples, implying that the interval is a *function* of the sample mean. Here the intervals based on the four surveys by our friends have different centers at $\bar{x}_2 = 159.2$, $\bar{x}_3 = 154.7$, $\bar{x}_4 = 152.6$, and $\bar{x}_5 = 168.5$, but all have the same width of 14.4. An idea of where these intervals might be located (using a plausible μ) in the sampling distribution is shown in Figure 8.4.5. If we were to keep generating intervals in this manner, all but 5% of them would contain μ. The number $1.96\left(\dfrac{\sigma}{\sqrt{n}}\right)$, in this case 7.2, that is added and subtracted from each sample mean is called the *margin of error* and should accompany all respectable poll results. We now give a more explicit definition:

> Given an estimation of μ by a sample mean $\bar{x}$, a confidence interval is constructed by adding and subtracting a *margin of error*. If the sampling process were to be repeated indefinitely, the percentage of intervals that would contain μ is called the *level of confidence*.

So in our example,

$$159.2 \pm 7.2 \rightarrow [152.0, 166.4] \quad 154.7 \pm 7.2 \rightarrow [147.5, 161.9]$$
$$152.6 \pm 7.2 \rightarrow [145.4, 159.8] \quad 168.5 \pm 7.2 \rightarrow [161.3, 175.7].$$

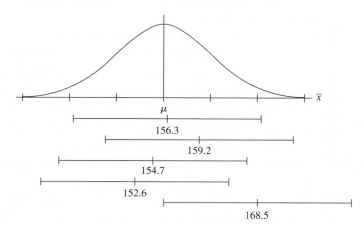

FIGURE 8.4.5   Several 95% confidence intervals based on different samples.

Therefore, any estimate of a population mean must be accompanied by two key elements: the level of confidence and the margin of error. If the level of confidence is absent from poll results given in a newspaper or magazine, it is usually assumed to be 95%, but any estimate given without a margin of error is essentially meaningless. As we have seen, the margin of error is a *decreasing* function of the sample size n, as is readily observed from its formula, $1.96\left(\dfrac{\sigma}{\sqrt{n}}\right)$, for a 95% level of confidence. This formula is also indicative of the connection between the margin of error and the confidence level. If a higher level of confidence is needed, a number bigger than 1.96 must be used in this formula, because only a wider interval could claim the presence of μ with greater confidence. Naturally, the opposite is also true. A lower level of confidence would use a number smaller than 1.96 for its margin of error, because a narrower interval would suffice. These variations are explored in greater detail in the exercises.

? Example 2

If we desire a smaller margin of error for our 95% confidence interval for the average weight of students, then we must increase our sample size. By surveying 50 students instead of only 30, we find that our margin of error is reduced to

$$1.96\left(\frac{20.0}{\sqrt{50}}\right) = 5.5.$$

If the mean of this sample is $\bar{x} = 154.9$, the confidence interval endpoints are 154.9 ± 5.5, giving a narrower interval of $[149.4, 160.4]$. ◆

? Example 3

Oliver works for an independent testing agency and has been told to estimate the mean speed in miles per hour that a particular Lexus will attain in 10 seconds(s) from a motionless start. In the past, the standard deviation of a distribution of such tests has been $\sigma = 1.6$ mi/h. If he has been asked to produce a 95% confidence interval with a margin of error of only 0.25 mi/h, how many times does Oliver need to test his Lexus?

⚙ Solution

Here, the margin of error is given, and the sample size needs to be determined. We need only substitute in the required formula and solve for n.

$$0.25 = 1.96\left(\frac{1.6}{\sqrt{n}}\right)$$

$$0.1276 = \frac{1.6}{\sqrt{n}}$$

$$\sqrt{n} = \frac{1.6}{0.1276}$$

$$= 12.54$$

$$n = (12.54)^2$$

$$= 157.3$$

To guarantee the desired margin of error, we round n up to the next whole number, 158. ◆

We conclude by addressing the topic of survey results used to introduce this section. The results of surveys and polls are used extensively to convey the "pulse of the public" in the writing of news articles in daily newspapers and weekly magazines such as *Newsweek, Time,* and *U.S. News & World Report.* The fictional survey shown in **Figure 8.4.6** is typical.

Campus Daily Register

Yesterday, 238 students were polled and asked whether they favor the selling of beer to patrons of legal age in the student union.

Yes	59%
No	41%

FIGURE 8.4.6   Results of a survey.

We restrict our attention to the gathering of answers from a random sample of people responding to a binary question—that is, a question for which yes or no is the only possible response. If we count each yes as a 1 and each no as a 0 in a sample of size n, then the sample mean is

$$\frac{1+1+0+1+0+\cdots+1}{n}$$

and it represents the fraction of yes answers in the survey. We multiply this fraction by 100 to convert it to a percentage and denote it by $\hat{p}$. Because $\hat{p}$ is computed from a sample, it is a *statistic* and is used to estimate the *parameter p*, the actual percentage of yes answers in the population from which the survey is taken.

The procedure for estimating p exactly parallels the previous use of the statistic $\bar{x}$ to estimate a mean μ. A large number of sample percentages—all generated in like fashion—form a sampling distribution that is approximately normal if the sample size n is large enough ($n \geq 30$ is usually sufficient). This distribution has mean p and a standard deviation that can be effectively approximated by

$$\sqrt{\frac{\hat{p}(100-\hat{p})}{n}}$$

for any $\hat{p}$ computed from a survey where the participants have been randomly selected. Because there is a 95% chance of $\hat{p}$ lying within 1.96 standard deviations of p, the margin of error is 1.96 times the above given square root expression, and we find the endpoints of a 95% confidence interval (for p) according to

$$\hat{p} \pm 1.96 \sqrt{\frac{\hat{p}(100 - \hat{p})}{n}}.$$

? Example 4

A survey is conducted to determine the popularity of a new student activities fee being added to the tuition of Gilbert University to help sponsor more concerts on campus. In a sample of 340 students, 57% favored the fee, while 43% opposed it. Determine a 95% confidence interval for the actual percentage of students who favor the fee. What is the margin of error?

⚙ Solution

In this case, $\hat{p} = 57$ and $n = 340$. Substituting in the previously given formula gives us a margin of error of

$$1.96 \sqrt{\frac{57(43)}{340}} = 1.96\sqrt{7.21} = 5.3\%$$

and a 95% confidence interval with endpoints of 57 ± 5.3. This results in the interval [51.7, 62.3]. In other words, we are 95% sure that the real percentage of students who favor an increased activities fee is between 51.7% and 62.3%. ◆

---

significance tests
A test involving a sample mean $\bar{x}$. If $\bar{x}$ lies outside the acceptance interval, it is said to be significant.

acceptance interval An interval centered on an assumed population mean that is used to test for whether it contains a mean computed from a sample.

significance level The percentage of time that a randomly collected sample mean would lie outside a prescribed acceptance interval.

We conclude our journey through statistics with a brief look at **significance tests** for statistics computed from samples of a population. People are often interested in determining whether the mean $\bar{x}$ of a randomly collected sample lies within the boundaries of a range of values, usually referred to as the **acceptance interval**. Every acceptance interval has a **significance level** defined by the percentage of sample means that would lie outside the prescribed acceptance interval if the assumed population mean μ were the true mean. By adding and subtracting $1.96\left(\dfrac{\sigma}{\sqrt{n}}\right)$ from μ, we have an interval that would contain 95% of the sample means and therefore have a significance level of 5%. If $\bar{x}$ lies outside that interval, then there may be reason to suspect whether the true mean of the entire population is actually close to μ, and, in that case, $\bar{x}$ is said to be significant.

? Example 5

Every bar of Whole Earth nutrition bars is supposed to contain 14.0 grams (g) of protein. If the historical standard deviation is $\sigma = 0.5$ g and 30 bars are sampled for the amount of protein in each bar, would a sample mean of $\bar{x} = 13.7$ g be considered significant at the 5% level?

⚙ Solution

We construct an acceptance interval centered at an assumed mean of 14.0 and check to see if 13.7 lies outside that interval. Because

$$1.96\left(\frac{\sigma}{\sqrt{n}}\right) = 1.96\left(\frac{0.5}{\sqrt{30}}\right) = 0.18,$$

the desired interval is $[14 - 0.18, 14 + 0.18] = [13.82, 14.18]$. Therefore, $\bar{x} = 13.7$ is considered significant at the 5% significance level. Whole Earth should reexamine the process being used to make these bars. ◆

Acceptance intervals are also constructed for analyzing the result of a single number chosen from a normal population. For example, the levels of many of the components detectable in a blood test are routinely checked as part of a medical checkup. In particular, the range for a normal red blood cell count (RBC) for middle-aged women is 4.2 to 5.4 million per microliter of blood, and these numbers correspond to the endpoints of an acceptance interval. *Polycythemia* is the medical name given to an RBC greater than 5.4 and is indicative of several conditions, including congenital heart disease, dehydration, pulmonary fibrosis, and other ailments. An RBC less than 4.2 implies anemia and is associated with internal bleeding, kidney disease, and so on. If the RBC of a specific patient lies outside the defined interval, then it is a significant result and therefore cause for concern.

In Figure 8.1.1 at the beginning of this chapter, a table was shown displaying the changes in employment for several industries between the months of March and April 2010. For each occupation, the rightmost column contained the result of the significance test. A yes meant that the (seasonally adjusted) monthly change was greater than the minimum significant change—that is, it is higher than an upper limit for an acceptance interval.

Significance tests are at the core of maintaining quality control in manufacturing and engineering research done in a huge number of vastly different fields, including medicine, biology, psychology, economics, chemistry, food production, sociology, environmental science, and many others. These tests are essential in maintaining quality standards of food, goods, and services, and in advancing the understanding of many facets of our society and the physical world.

Name _____

Exercise Set **8.4** ⬭⬭⬭⬤

1. You wish to conduct a poll of your fellow college students on whether they like the results of the recent construction projects on campus. You decide to stand in front of the library between 5:00 P.M. and 7:00 P.M. and question at least 35 students. Does this constitute a random sampling process? Why or why not?

2. This opinion poll question was published in a local newspaper. Explain why this method of conducting a survey cannot be labeled a random sampling.

> Should the sale of the handgun known as the *Saturday Night Special* be banned in our county?
> CALL: 626–HEAL for YES and 626–HURT for NO.

3. You wish to randomly select 8 people from the following list. Enumerate this list, and make your choices by starting with the first line in the table in Figure 8.4.2. Pick another sample of 8 by starting with the second line.

Chan	Stephanie	Peter	Subra	Kadhri
Jana	Kam	David	Ramona	Hadly
Cameron	Keon	Justin	Maurice	Rachel
Maria	Enrico	Rodolfo	Marilyn	Alma
Earvin	Donyell	Akeem	Lisa	Rabin
Lon	Juanita	Kevin	Gabriel	Kelly

4. Enumerate the list in Exercise 3 again, but this time assign 100 to Chan, 101 to Jana, and so on. Use Figure 8.4.2 to make a sample of 5 people. Would this be a random sample? Why? Do you think it matters how you enumerate your population?

Find the 95% confidence intervals for μ corresponding to the following values for a mean $\bar{x}$ of a sample of size n taken from a parent population having standard deviation σ. Round all your numbers to the same place value as given for $\bar{x}$.

5. $\bar{x} = 24.63$, $n = 30$, $\sigma = 3.50$

6. $\bar{x} = 54.7$, $n = 75$, $\sigma = 6.3$

7. $\bar{x} = 8.54$, $n = 36$, $\sigma = 1.26$

8. $\bar{x} = 8.54$, $n = 49$, $\sigma = 1.26$

9. $\bar{x} = 0.0625$, $n = 40$, $\sigma = 0.01$

10. $\bar{x} = 5,400$, $n = 100$, $\sigma = 300$

Because 90% of any normal distribution is contained within 1.65 standard deviations of the mean, the margin of error is $1.65\left(\dfrac{\sigma}{\sqrt{n}}\right)$ for an estimate in a 90% confidence interval from a sample of size n.

Find the 90% confidence intervals for μ corresponding to the following values for a mean $\bar{x}$ of a sample of size n taken from a parent population having standard deviation σ. Round all your numbers to the same place value as given for $\bar{x}$.

11. $\bar{x} = 24.63$, $n = 30$, $\sigma = 3.50$

    (Same values as Exercise 5.)

12. $\bar{x} = 54.7$, $n = 75$, $\sigma = 6.3$

    (Same values as Exercise 6.)

13. $\bar{x} = 8.54$, $n = 36$, $\sigma = 1.26$

    (Same values as Exercise 7.)

14. $\bar{x} = 8.54$, $n = 49$, $\sigma = 1.26$

    (Same values as Exercise 8.)

15. If 90% and 95% confidence intervals are computed from the same sample, which must be the wider interval? Which has a larger margin of error?

16. For a given sample mean, is the associated margin of error an increasing or decreasing function of the level of confidence?

The following values are sample percentages $\hat{p}$ of yes responses to a binary question given in a survey involving n participants. Find the 95% confidence interval and associated margin of error for the parameter p. Round all your numbers to the same place value as given for $\hat{p}$.

17. $\hat{p} = 46\%$, $n = 80$

18. $\hat{p} = 46\%$, $n = 176$

19. $\hat{p} = 59.2\%$, $n = 245$

20. $\hat{p} = 35.1\%$, $n = 542$

21. $\hat{p} = 34.8\%$, $n = 633$

22. $\hat{p} = 81.0\%$, $n = 72$

23. Cynthia is a quality control expert who wishes to alter the sample size n of the samples she uses to reduce the margin of error associated with the 95% confidence interval. Should Cynthia increase or decrease n? Why?

24. The fish company For the Halibut sells frozen fish in packages of 36 fillets each. A consumer protection group weighs all the pieces in one package and computes the mean weight per piece to be $\bar{x} = 4.1$ ounces (oz). If the standard deviation of the population of *all* fillets is $\sigma = 0.3$ oz, determine the 95% confidence interval for the mean μ. Find the 90% confidence interval. (Round to the nearest hundredth.)

25. Your mathematics class wonders about the mean grade point average (GPA) at the college and conducts a random survey of 45 students. If the mean of your sample is 2.45, construct a 95%

confidence interval for the mean GPA of the whole college population. (Assume the standard deviation is $\sigma = 0.32$.) What is the margin of error?

26. Nuts2U is a manufacturing company that produces mass quantities of nuts, bolts, screws, nails, and fasteners. Its quality control team needs to test the mean length of the bolts coming off the assembly line and has set the margin of error to be 0.25 centimeter (cm) for a 95% level of confidence. If the standard deviation for a particular type of bolt has been $\sigma = 0.8$ cm, find the sample size needed to test the length of this bolt.

27. In a poll conducted by American Opinion following the 2012 presidential election, 38% of 752 people responded affirmatively when asked if President Obama now has a mandate to develop new social and economic programs. Determine the margin of error of this poll by using the method of Example 4. (Assume a 95% level of confidence.)

28. Find the 95% confidence interval associated with the poll considered in Exercise 27.

29. In a poll taken after the 2008 presidential election, 69% of 752 people responded affirmatively when asked if President Obama should appoint Republicans to important positions. Determine a 95% confidence interval for this poll by using the method of Example 4.

30. What is the margin of error associated with the poll in Exercise 29?

31. Determine the formula for the margin of error as a function of sample size n that is associated with a 99% confidence interval for the mean μ in a population with standard deviation σ.

32. If we increase our sample size by a factor of 4, the margin of error of the new sampling distribution is decreased by what factor? What if we increase the sample size by a factor of 9?

33. Historically, the mean percentage grade of the mathematics students of Professor Gauss is 72.6%, and the standard deviation is 11.4%. What is the probability that the mean score of his 25 current students on the first test is at least 75%? (Remember that the distribution for sample means of sample size 25 has a standard deviation of $11.4 / \sqrt{25}$.)

34. Over the last month, Corinne has sometimes had trouble sleeping at night. Her motto is, "I'm irate with less than 8." If the mean and standard deviation for Corinne's nightly sleep are $\mu = 8.3$ and $\sigma = 0.7$, what percentage of the time is she irate? She recorded her amounts of sleep in hours for the last 30 nights, as shown in the following table.

6.5	7.2	8.6	8.3	7.6	4.5	6.4	8.0	7.5	9.5
6.0	7.8	5.5	9.5	8.6	7.4	5.9	7.2	10.5	8.2
7.5	6.3	6.0	8.3	7.3	8.6	7.4	8.3	8.4	7.6

Assuming this is a sample from a normal population of amounts of sleep for Corinne, what is the probability that the mean of the next 30 days will be less than the mean of this sample?

35. Two survey percentages were reported in the following newspaper article. Determine the margin of error corresponding to each value of $\hat{p}$.

Teens Find Cigarettes Easy Buy

The Associated Press

Cannaboro County—More than half of Cannaboro retailers are currently selling cigarettes to minors, according to a county survey released last week. More than 230 teenagers participated in the survey and asked to purchase cigarettes at 1,600 grocery stores, convenience stores, and motels chosen at random. Fifty-eight percent of the sales clerks did not ask for proof of age when the teens requested the tobacco product. Additionally, 73 percent of the time, nobody tried to stop them from buying cigarettes from a vending machine.

36. The normal range for thyroid stimulating hormone (TSH) in adults is defined by an acceptance interval of 0.4–4.5 milli-International units per liter (mIU/L) with a significance level of 5%. Explain the statistical meaning of an individual TSH reading of 4.8 mIU/L.

37. The length of a particular type of bolt produced by Nuts2U is supposed to be 7.0 cm. If the standard deviation is known to be $\sigma = 0.8$ cm, would the mean $\bar{x} = 7.4$ cm of a sample of 25 bolts be considered to be significant at the 5% significance level?

38. In Exercise 37, would a sample mean of $\bar{x} = 7.2$ cm be considered significant at the 5% significance level?

39. The miles per gallon figure achieved by a new automobile is usually part of most advertisements and is a key consideration for most car buyers. This value is, in fact, meant to be the mean of a large population of that same model, and no guarantee is given that every individual car will obtain that gas mileage. Suppose a new type of Toyota is billed as getting 28.0 miles per gallon (mi/gal) and a consumer rights organization tests 36 of these cars, obtaining a mean of $\bar{x} = 27.7$ mi/gal. Would this be considered significant at the 5% significance level? (Assume $\sigma = 2.4$ mi/gal.)

40. In Exercise 39, would a sample mean of $\bar{x} = 26.5$ mi/gal be considered significant at the 5% significance level?

41. An acceptance interval associated with a 10% significance level is determined by adding and subtracting $1.65\left(\dfrac{\sigma}{\sqrt{n}}\right)$ to/from the assumed mean. Would a sample mean $\bar{x}$ that is classified as significant at the 10% level also be significant at the 5% level?

42. Refer to Exercise 41. Would a sample mean $\bar{x}$ that is classified as significant at the 5% level also be significant at the 10% level?

Name _____

Chapter Review Test **8** ◯◯◯ ←⎯⎯⎯⎯⎯⎯⎯⎯⎯⎯⎯⎯⎯⎯

1. The standard deviation of a collection of values is a measure of

   (a) dispersion.   (b) range.   (c) average.   (d) symmetry.   (e) none of these.

2. This histogram displays a distribution whose mean is

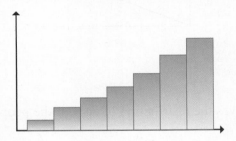

   (a) about the same as the median.   (b) less than the median.
   (c) greater than the median.   (d) about the same as the standard deviation.
   (e) none of these.

3. The mean in each of the distributions displayed in these two histograms is 25. In which of these distributions does 30 have the larger z-score?

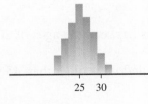 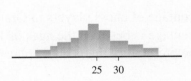

25   30   25   30

   (a) The one on the left   (b) The one on the right
   (c) Impossible to determine from just the histograms

4. Let 19.5 be the standard deviation of a parent population. Then the standard deviation of a sampling distribution generated from samples of size 40 equals (to the nearest hundredth) which of the following?

   (a) 0.49   (b) 0.11   (c) 3.08   (d) 0.70   (e) none of these

5. The graph of the density function $f(x) = 0.08x$ for $0 \leq x \leq 5$ gives the distribution curve shown here. Shade the appropriate region on the graph, and find the portion of the distribution that is between 2 and 4.5.

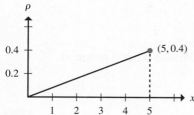

6. The following histogram gives the distribution of ages of the people who play chess regularly in Orange County.

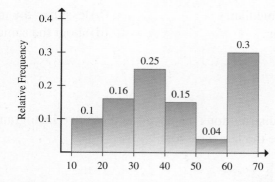

(a) What percentage of chess players in Orange County is between the ages of 20 and 30?
(b) What percentage is between the ages of 15 and 25?
(c) What is the probability that a chess player chosen at random is older than 45 years?

7. Here are 10 prices of video cameras found at various stores in town. Determine the mean and median price. If the most expensive camera goes on sale for $840, does the mean of the new list of prices change? Does the median?

    $525 $650 $939 $730 $489 $1,040 $810 $695 $705 $900

8. State the Central Limit Theorem.

9. Who is called the "Prince of Mathematicians" and is primarily responsible for the discovery of the normal distribution?

10. The relative frequency distribution showing the annual amount of money saved per person in the country of Philacornia is given here. Estimate the mean, median, and standard deviation of this distribution.

Interval ($)	Relative Frequency
0–200	0.23
200–400	0.35
400–600	0.18
600–800	0.16
800–1,000	0.05
1,000–1,200	0.03

11. The number of hours per week that the population of full-time students at Summa University devotes to study is normally distributed with a mean of 25 hours (h) and a standard deviation of 6 h. What percentage of students will study more than 20 h per week?

12. An educational research institution wishes to estimate the mean number of years of education of U.S. adults. A random sample of 520 adults is taken whose mean number of years of education is 13.4. Assuming a standard deviation $\sigma = 2.4$ years of the parent population, determine a 95% confidence interval for the parameter μ. What is the margin of error?

Chapter Objectives ☑◻◻ ← check off when you've completed an objective

- ◻ Construct residue classes modulo any positive integer.
- ◻ Know the pitch classes.
- ◻ Identify note names and positions on the treble clef staff.
- ◻ Identify variations of a primary chord.
- ◻ Identify and apply translations, rotations, and reflections.
- ◻ Determine a position sequence and note graph from a music score.
- ◻ Perform modular addition and multiplication.
- ◻ Construct a major scale and identify the major key from the key signature.
- ◻ Construct the circle of fifths.
- ◻ Apply the Russian peasant method for finding residues.
- ◻ Solve equations in modular algebra.
- ◻ Use the Caesar cipher to encrypt and decipher messages.
- ◻ Learn about Andrew Wiles.

Navigate Companion Website

go.jblearning.com/johnson

Visit go.jblearning.com/Johnson to access a Laboratory Manual, Student Solutions Manual, Interactive Glossary, and Interactive Flashcards.

9 Mathematics in Music and Cryptology

9.1 Residue Classes and Notes

9.2 Transformations

9.3 The Circle of Fifths

9.4 Modular Arithmetic and Cryptology

> One thing he discovered ...was that music held more for him than just pleasure. There was meat to it. What the music said was that there is a right way for things to be ordered so that life might not always be just tangle and drift but have a shape, an aim. It was a powerful argument against the notion that things just happen.
>
> —Charles Frazier, *Cold Mountain*

© pedrosala/ShutterStock, Inc.

When people are asked how mathematics is generally used in the world, their responses usually have to do with the practical side of the discipline related to engineering, business, aeronautics, statistics, and similar fields. Rarely do people think of the arts or music, since appreciation and invention within these domains are often considered to spring from the soul. They are always surprised to learn that most mathematicians would heartily agree that this is the exact same wellspring of mathematical thought. Indeed, one might argue that the richness of the soul offers the most ideal environment for mathematics to send down its deepest roots. Why? As the musician Edward Rothstein said of his college days, "Music and math together satisfied a sort of abstract appetite, a desire that was partly intellectual, partly esthetic, partly emotional, partly even, physical." The occurrence of patterns is the lifeblood of how mathematics permeates our existence, and nowhere are the creation and so-called feel for patterns more prevalent than in the composition and enjoyment of . . . *music*.

9.1 Residue Classes and Notes

pitch Property of sound that identifies it as high or low compared to other sounds. It is a function of the frequency of the sound waves producing it.

frequency The number of sound waves or vibrations per second reaching the ear that produce a particular pitch.

half-step Interval between two adjacent piano keys.

whole step Interval equivalent of two consecutive half-steps.

octave Interval between pitches that are 12 half-steps apart. This implies that the frequency of the higher pitch is twice that of the lower pitch.

Music is the result of an orderly arrangement of different types of sound, which, in turn, result from waves through the air causing vibrations of small bones in our ears. The perception of one sound as being either higher or lower than another sound is a function of the specific rapidity of these vibrations, which produce a property of sound referred to as **pitch**. The number of vibrations per second is known as the **frequency** of the pitch, and the greater the frequency, the higher the pitch sounds to the ear. When you press a piano key, a hammer inside the piano strikes a string that vibrates at a specific frequency based on the length and composition of the string. Indeed, a piano keyboard, a portion of which is pictured in **Figure 9.1.1**, provides a very useful means of displaying pitch relationships.

The interval between any two adjacent keys—whether black or white—is called a **half-step**, and two consecutive half-steps are called a **whole step**. The basic tenet of the keyboard is that any key that is 12 half-steps above another key produces a pitch whose frequency is exactly double that of the lower key. These two sounds are defined as being one **octave** apart and sound related to the ear. In Figure 9.1.1, for example, the key in position 4 is one octave higher than the key in position −8, position 15 is one octave higher than position 3, and position 31 is two octaves higher than position 7. In fact, we see position y is k octaves higher than position x if

$$y - x = 12k$$

for $k = 1, 2, 3, \ldots.$

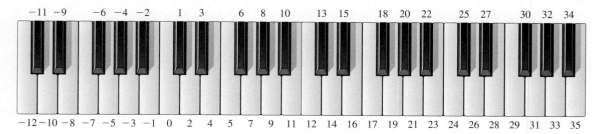

FIGURE 9.1.1   Part of a piano keyboard.

number theory The study of integers and associated properties.

This is a specific example of a relation between two numbers known as a *congruence*. Congruence is an important concept in **number theory**, the study of the properties of integers. Number theory has attracted the attention of great mathematicians since the time of Pythagoras, and research continues unabated even today. Not only does number theory possess an intrinsic beauty equal to any of the other branches of mathematics, but number-theoretic results have had significant applications in such practical areas as computer science and cryptography.

In general, if m is a positive integer and the difference $y - x$ of two integers is divisible by m, we say that y is congruent to x modulo m, and we write

modulus The positive integer m used in defining congruence modulo m.

$$y \equiv x \pmod{m}.$$

The integer m is called the **modulus**. For example,

$39 \equiv 4 \pmod{5}$	because	$39 - 4 = 35$, which is divisible by 5.
$47 \equiv 23 \pmod{8}$	because	$47 - 23 = 24$, which is divisible by 8.
$109 \equiv 21 \pmod{11}$	because	$109 - 21 = 88$, which is divisible by 11.
$2{,}374 \equiv 374 \pmod{1{,}000}$	because	$2{,}374 - 374 = 2{,}000$, which is divisible by $1{,}000$.
$-46 \equiv 10 \pmod{7}$	because	$-46 - 10 = -56$, which is divisible by 7.
$36 \equiv 0 \pmod{9}$	because	36 is divisible by 9.

Further examination of the above congruences shows that an equivalent way of defining $y \equiv x \pmod{m}$ is to say that y and x differ by a multiple of m—that is, $y - x = mk$, or

$$y = x + mk,$$

residue (or congruence) class Set of numbers such that the difference between any two members of the set is a multiple of a common integer m. Denoted by $[x]_m$.

m-modularization Grouping of the integers according to residue classes modulo m.

for some integer k. Given any integer x, the set of all integers that are congruent to x modulo m is called its **residue (or congruence) class**, and it is denoted by $[x]_m$. Note that if x is any one particular member of a residue class, the remaining members can be generated by adding successive multiples of m to x. It is a fact that every integer must be a member of one and only one of these classes. The following example illustrates the additional fact that there must be precisely m nonintersecting residue classes as a result of any particular *m-modularization*, or grouping of the integers modulo m.

❓ Example 1

A 6-modularization of the integers divides them into the following six residue classes.

$$[0]_6 = \{\ldots, -12, -6, \mathbf{0}, 6, 12, 18, 24, 30, \ldots\},$$
$$[1]_6 = \{\ldots, -11, -5, \mathbf{1}, 7, 13, 19, 25, 31, \ldots\},$$
$$[2]_6 = \{\ldots, -10, -4, \mathbf{2}, 8, 14, 20, 26, 32, \ldots\},$$
$$[3]_6 = \{\ldots, -9, -3, \mathbf{3}, 9, 15, 21, 27, 33, \ldots\},$$
$$[4]_6 = \{\ldots, -8, -2, \mathbf{4}, 10, 16, 22, 28, 34, \ldots\},$$
$$[5]_6 = \{\ldots, -7, -1, \mathbf{5}, 11, 17, 23, 29, 35, \ldots\}.$$

least residue The smallest nonnegative integer in a residue class.

These six classes collectively contain every integer yet are nonoverlapping. The smallest nonnegative integer (denoted in boldface in the six residue classes shown here) is called the **least residue** and is typically used as the representative of each class, although any member can be used. Hence,

$$[72]_6 = [12]_6 = [6]_6 = [0]_6, \quad [37]_6 = [19]_6 = [7]_6 = [1]_6,$$
$$[63]_6 = [33]_6 = [9]_6 = [3]_6, \quad [83]_6 = [29]_6 = [11]_6 = [5]_6. \quad \blacklozenge$$

How does one find the least residue of a given number x modulo m? Suppose we wish to find the least-residue representation of $[51]_4$. Dividing 51 by 4 yields a quotient of 12 and a remainder of 3, and so we can write $51 = 4 \cdot 12 + 3$. This implies that $51 - 3$ is a multiple of 4, and so

$$51 \equiv 3 \pmod 4.$$

This illustrates the fact that if r is the remainder after dividing the integer x by the integer m, then $x = mq + r$, where q is the quotient and $0 \leq r < m$. Therefore, $x - r$ is divisible by m, and so $x \equiv r \pmod m$. Hence, the least residue of the class $[x]_m$ must be r, and we can write

$$[x]_m = [r]_m.$$

❓ Example 2

Find least-residue representations for the residue classes $[207]_6$, $[415]_8$, and $[225]_5$.

⚙ Solution

$$[207]_6 = [3]_6 \text{ because } \frac{207}{6} = 34 \text{ with a remainder of } 3, \text{ or } 207 = 6 \cdot 34 + 3.$$

$$[415]_8 = [7]_8 \text{ because } \frac{415}{8} = 51 \text{ with a remainder of } 7, \text{ or } 415 = 8 \cdot 51 + 7.$$

$$[225]_5 = [0]_5 \text{ because } \frac{225}{5} = 45 \text{ with a remainder of } 0, \text{ or } 225 = 5 \cdot 45. \quad \blacklozenge$$

❓ Example 3

The separation of the integers into even and odd numbers can be thought of as the two classes resulting from a 2-modularization.

$$[0]_2 = \{\ldots, -4, -2, \mathbf{0}, 2, 4, 6, 8, \ldots\}, \quad [1]_2 = \{\ldots, -3, -1, \mathbf{1}, 3, 5, 7, 9, \ldots\} \quad \blacklozenge$$

? Example 4

Given a 4-modularization of the integers, what would be the smallest member of $[1]_4$ that is greater than 98? The smallest member of $[3]_4$ that is greater than 250?

⚙ Solution

We are seeking the smallest value x such that $x = 1 + 4k$ (for k a positive integer) and $x > 98$. In other words, we need the smallest integer k that satisfies the following inequality.

$$1 + 4k > 98$$
$$4k > 97$$
$$k > 24.25.$$

So $k = 25$, and then the required value is $x = 1 + 4(25) = 101$.
    For the second value, we need the smallest integer n such that

$$3 + 4n > 250$$
$$4n > 247$$
$$n > 61.75.$$

So $n = 62$, and the correct number is $3 + 4(62) = 251$. ◆

? Example 5

The dates of a month have a natural grouping contained in a 7-modularization that uses common colloquial names for the residue classes: Monday, Tuesday, Wednesday, etc. Each month, you need know only the weekday name of one date to determine the names of the remaining dates. For instance, suppose Thursday is the first day of the month.

S	M	T	W	Th	F	S
				1	2	3
4	5	6	7	8	9	10
11	12	13	14	15	16	17
18	19	20	21	22	23	24
25	26	27	28	29	30	31

Thursday	$= \{1, 8, 15, 22, 29\}$, which is contained in $[1]_7$.
Friday	$= \{2, 9, 16, 23, 30\}$, which is contained in $[2]_7$.
Saturday	$= \{3, 10, 17, 24, 31\}$, which is contained in $[3]_7$.
Sunday	$= \{4, 11, 18, 25\}$, which is contained in $[4]_7$.
Monday	$= \{5, 12, 19, 26\}$, which is contained in $[5]_7$.
Tuesday	$= \{6, 13, 20, 27\}$, which is contained in $[6]_7$.
Wednesday	$= \{7, 14, 21, 28\}$, which is contained in $[0]_7$.

Suppose the 1st day of the month was a Tuesday. What day of the week is the 25th?

⚙ Solution

Since $25 \equiv 4 \pmod 7$ and the 4th day of that month is a Friday, the 25th of the month is also a Friday.   ♦

❓ Example 6

Suppose the 1st day of the *year* is a Saturday. What day of the week is the 187th day of the year?

⚙ Solution

By the same reasoning as in Example 5, the 187th day of the year is a Wednesday, because $187 \equiv 5 \pmod 7$ and the 5th day of the year is a Wednesday.   ♦

❓ Example 7

Many phenomena of nature repeat their occurrences with systematic regularity—sunrise every morning, blooming flowers every spring, full moon every month. The time between repetitions is known as the **period**. Long before

period Time between repetitions of a repeating phenomenon.

the causes of solar and lunar eclipses were known, ancient peoples kept records accurate enough to discover that the same pattern of eclipses repeats itself every 18 years, 11.3 days. This is called the *Saros cycle*. There are various types of eclipses based on the precise distance of the moon from Earth during mideclipse. If you know the date x of one type of eclipse, then the dates of that exact same type of eclipse can be found by adding multiples of the Saros cycle to x. This forms a grouping much like modularizations of the integers. For example, the dates of some of the solar eclipses for the Saros series 131 are given here.

December 3, 1918	January 26, 2009
December 13, 1936	February 6, 2027
December 25, 1954	February 16, 2045
January 4, 1973	February 28, 2063
January 15, 1991	March 10, 2081 ♦

❓ Example 8

Most computer programming languages contain a set of intrinsic functions as part of the architecture of the language. The use of **mod** is an example of such a function.

Let the value of $x \bmod m$ for integers x and m be the least residue of the class $[x]_m$, and let the symbol := stand for an assignment of a value to a variable. If we write

$$y := x \bmod m,$$

then each row of the following table shows the value assigned to y after execution of the previous programming statement that corresponds to the given values of x and m.

x	m	y
81	7	4
141	21	15
532	3	1
922	50	22
1,674	2	0
1,675	2	1 ◆

note Symbol written on the music staff whose location indicates to the musician the pitch to play and the length of time the pitch is to be maintained. Often used interchangeably with the word *pitch*.

We now turn our attention to the connections of number theory to the world of music. In order for a composer to communicate the directions for the playing of a musical instrument to the musician, we need a musical language to explain when to play what pitch and for how long. The alphabet of this language consists of *notes*, whose names are based on the letters A, B, C, D, E, F, and G. A **note** is an oval-shaped symbol, often with a stem or flag attached, whose position on the music staff indicates what pitch is to be played by the musician (see **Figure 9.1.2**).

FIGURE 9.1.2   Notes on the music staff.

The specific type of note indicates the length of time the note is to be played. (You should be aware that even though *note* and *pitch* have different meanings, they are often used interchangeably in the common vernacular.) Two notes representing pitches that are a whole number of octaves (12 half-steps) apart are designated by the same letter. This necessarily means that each letter is the name of a residue class of key positions created by a 12-modularization of the integers. Each of these classes is called a **pitch class**, and the primary note names are defined by

pitch class Residue class modulo 12 containing the key positions of all pitches that are one or more octaves apart.

© lem/ShutterStock, Inc.

$$C = [0]_{12}, \quad D = [2]_{12}, \quad E = [4]_{12}, \quad F = [5]_{12}, \quad G = [7]_{12}, \quad A = [9]_{12}, \quad B = [11]_{12}.$$

The corresponding positions on the keyboard consist of all the white keys, as shown in **Figure 9.1.3**.

accidental Symbol preceding a note or note name, indicating the pitch to be played higher or lower by one half-step.

sharp Accidental indicating a note to be played higher by a half-step, signified by ♯.

The key in position 0 is called *middle C* and signifies the approximate middle of the full keyboard of an actual piano. Notice that the absence of a black key between E and F and between B and C correlates to the fact that these are the only pairs of white keys that are just a half-step apart. In general, the note name corresponding to a pitch located one half-step above or below the pitch of any primary note is formed by appending a symbol known as an **accidental** to the appropriate letter. We will be concerned with two accidentals. The **sharp**, ♯, refers to the

C	D	E	F	G	A	B	C	D	E	F	G	A	B	C	D	E	F	G	A	B	C	D	E	F	G	A	B
−12	−10	−8	−7	−5	−3	−1	0	2	4	5	7	9	11	12	14	16	17	19	21	23	24	26	28	29	31	33	35

FIGURE 9.1.3   Location of the primary notes on a portion of the keyboard.

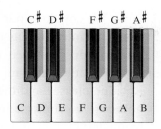

FIGURE 9.1.4   Names of pitches using accidentals.

flat Accidental indicating a note to be played lower by a half-step, signified by ♭.

enharmonic spellings Different names for the same pitch.

pitch located a half-step above the pitch corresponding to the letter; and the **flat**, ♭, indicates a half-step below. **Figure 9.1.4** shows the names of the black keys according to this scheme.

Observe that this means a pitch can have more than one name. Different names for the same pitch are called **enharmonic spellings**. The pairs C♯ and D♭, D♯ and E♭, F♯ and G♭, and even E♯ and F are all examples of enharmonic spellings. Recalling Figure 9.1.1, we see that the corresponding pitch classes are

$$C\sharp = D\flat = [1]_{12}, \qquad D\sharp = E\flat = [3]_{12}, \qquad F\sharp = G\flat = [6]_{12},$$

$$G\sharp = A\flat = [8]_{12}, \qquad A\sharp = B\flat = [10]_{12}.$$

staff Set of five parallel lines forming the grid on which notes are written to compose music.

clef Symbol preceding a staff that identifies the correspondence between lines and spaces and the pitches they represent.

treble clef Clef identifying the staff for those notes that are to be played primarily by the right hand on the piano.

bass clef Clef identifying the staff for those notes that are to be played primarily by the left hand on the piano.

Now that we have our alphabet, we need the correct type of ruled paper upon which to write our musical statements. These rulings consist of groups of five evenly spaced parallel lines known as the musical **staff**. Each line and each space between the lines correspond to a different pitch according to the **clef** that appears at the beginning of the staff. The two staffs most commonly used for piano music use the **treble clef** (as seen in **Figure 9.1.5**), which holds mainly the notes played by the right hand, and the **bass clef**, which contains primarily the notes played by the left hand. We will be concerned here with only the treble clef staff whose pitch correspondences are shown by the note placements in Figure 9.1.5. Notice that each line and each space correspond to a white piano key, and so the musical intervals between notes on adjacent spaces and lines vary depending on the pitches they represent. Recall that E and F are separated only by a half-step, as are B and C. Middle C occurs two whole steps below the staff, and so a small additional line is added. This is needed whenever a note occurs above or below the standard staff.

It is helpful to look at the locations of members of the same pitch class. By definition, any two members of a single class must lie one or more octaves apart on the staff, as shown in **Figure 9.1.6**.

An accidental is appended to a note by placing it directly in front of it. If we sharp each of the notes on the staff in Figure 9.1.5, then the key position of each of the resulting notes is simply increased by 1 (**Figure 9.1.7**). Observe how note F (positions 5 and 17 in Figure 9.1.5) occurs on both staffs since it lies just a half-step above E and therefore is enharmonic with E♯. The same can be said for note C at position 12, which is enharmonic with B♯. Flatting every note would result in a similar situation.

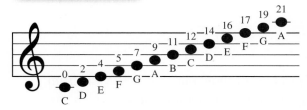

FIGURE 9.1.5   Note names and key positions on the treble clef staff.

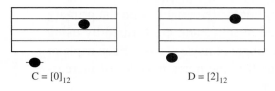

$$C = [0]_{12} \qquad\qquad D = [2]_{12}$$

FIGURE 9.1.6   Two members of the pitch classes for C and D.

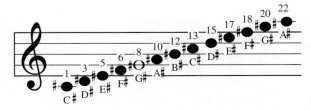

FIGURE 9.1.7   Key positions for sharped notes on the treble clef staff.

We conclude our introduction to the language of music by mentioning the ancient observations of the relationships between mathematics and music by the Greek mathematician Pythagoras (c. 570–495 BC). Residue classes and their association with pitches and notes are a prime example of the connections implied by the left branch of the *Quadrivium of Knowledge* (see **Figure 9.1.8**).

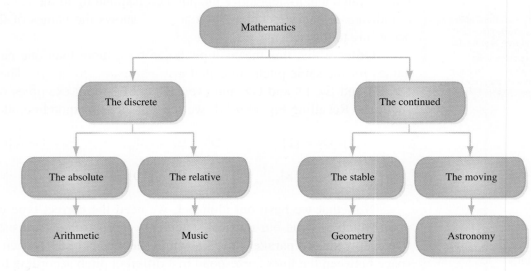

FIGURE 9.1.8   The Quadrivium of Knowledge.

Pythagoras would have been pleased with the notion of pitch classes as a categorization of notes and pitches insofar as it allows us to conveniently analyze a common technique used in musical composition. A **chord** is a set of three or more notes played at the same time on the piano whose combined tones produce a richer sound than that produced by single notes. We define a **primary chord** as one whose notes are at key positions between 0 and 11 inclusive. A composer wishing for a variation in the sound produced by a primary chord can create that by substituting for any individual pitch at key position x any member of the pitch class $[x]_{12}$. We know that this simply changes the octave of that pitch and so leaves the essential harmony of the primary chord unaltered.

chord A set of three or more notes to be played simultaneously.

primary chord Chord whose notes lie between key positions 0 and 11 inclusive.

❓ Example 9

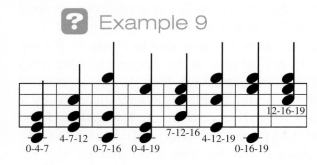

FIGURE 9.1.9   Variations of a primary chord.

The chord C-E-G shown in **Figure 9.1.9** is from a category known as *major triads*, which are used extensively. Since the corresponding pitch classes for this triad are $[0]_{12}$, $[4]_{12}$, and $[7]_{12}$, all of the other chords are variations on the primary chord located at 0-4-7 and are formed by using alternate members of these classes. Each triple of key positions lists the notes in order of increasing pitch. ◆

❓ Example 10

FIGURE 9.1.10 Chords 9-14-18 (A-D-F♯) and 2-6-9 (D-F♯-A).

It is useful to be able to identify the primary chord from any of its variations. This is done by finding the least residue of the pitch class of each note in the chord and then listing those positions in ascending order. The primary chord of the variation 9-14-18 (A-D-F♯) is found by observing that $[14]_{12} = [2]_{12}$ and $[18]_{12} = [6]_{12}$ and forming the chord 2-6-9 (D-F♯-A). These two chords are pictured in **Figure 9.1.10**. ◆

❓ Example 11

Which chords in the following measures from the commencement favorite *Pomp and Circumstance* by Edward Elgar (**Figure 9.1.11**) appear in variations of the same primary chord?

allarg.

FIGURE 9.1.11 A portion of *Pomp and Circumstance* by Edward Elgar.

⚙ Solution

The key positions of the chords, in order, are

1-9-13, 2-14, 4-12-16, 9-14-21, 2-9-14, 12-16-19-24, 16-19.

The notes of the third chord 4-12-16 and the last two chords 12-16-19-24 and 16-19 all come from the pitch classes of the primary chord 0-4-7 (C-E-G). Additionally, the notes of the second chord 2-14, fourth 9-14-21, and fifth 2-9-14 all come from the same pitch classes as the two-note primary chord 2-9 (D-A). ◆

> They [the Pythagoreans] supposed the whole heaven to be a harmonia and a number.
>
> —Aristotle

Name _____

Exercise Set **9.1** ◯◯◯←――――――――――――――――――

1. Two residue classes of a 5-modularization of the integers are given here in set notation.

$$[0]_5 = \{\ldots, -10, -5, 0, 5, 10, 15, \ldots\},$$
$$[1]_5 = \{\ldots, -9, -4, 1, 6, 11, 16, \ldots\}.$$

Give the residue classes denoted by $[2]_5$, $[3]_5$, and $[4]_5$, using set notation.

2. Three residue classes of an 8-modularization are given here in set notation.

$$[3]_8 = \{\ldots, -13, -5, 3, 11, 19, 27, \ldots\},$$
$$[5]_8 = \{\ldots, -11, -3, 5, 13, 21, 29, \ldots\},$$
$$[6]_8 = \{\ldots, -10, -2, 6, 14, 22, 30, \ldots\}.$$

Give the residue classes denoted by $[0]_8$, $[1]_8$, $[2]_8$ $[4]_8$, and $[7]_8$, using set notation.

3. Find the least-residue representations of the following classes.

$$[56]_6 \quad [69]_7 \quad [2{,}893]_2 \quad [99]_{11} \quad [211]_{18}$$

4. Find the least-residue representations of the following classes.

$$[49]_5 \quad [75]_8 \quad [310]_3 \quad [750]_{25} \quad [5{,}667]_{41}$$

5. A quick way to find the least residue modulo m of a large number x with your calculator is to divide x by m, subtract the whole-number part of the result on your display, and then multiply the remaining decimal portion by m. That result is the least residue. For instance, upon dividing 59,238 by 73, we get 811.4794521 (to 10 places). By subtracting 811 and multiplying by 73, we get 35. This means that $59{,}238 = 73 \cdot 811 + 35$ that is, $59{,}238 \equiv 35 \pmod{73}$ and $[59{,}238]_{73} = [35]_{73}$.

Find the least-residue representations of the following classes.

$$[37{,}497]_{39} \quad [75{,}898]_{74} \quad [310{,}936]_{528} \quad [8{,}206{,}851]_{463}$$

6. Find the least-residue representations of the following classes.

$$[67{,}309]_{82} \quad [44{,}625]_{19} \quad [803{,}242]_{907} \quad [12{,}774{,}953]_{188}$$

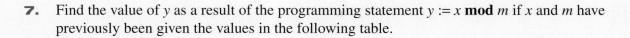

7. Find the value of y as a result of the programming statement $y := x \bmod m$ if x and m have previously been given the values in the following table.

x	m	y
24	5	
44	9	
109	16	
2,463	2	
3,778	2	
45,785	91	
296,689	403	
78,239	1,000	
78,239	100	
78,239	10	

8. Suppose you were designing a computer program in which you wished to use the functional assignment $y := x \bmod m$ to assign the last two digits of the value stored in x to the variable y. What value would you first have to assign to m? What if you just wanted the last digit of the value in x assigned to y?

9. Suppose the 1st day of the year is a Tuesday. What day of the week is the 304th day of the year?

10. Suppose the 1st day of the year is a Wednesday. What day of the week is the 168th day of the year?

11. Given a 9-modularization of the integers, find the smallest member of $[2]_9$ that is greater than 35, the smallest member of $[5]_9$ that is greater than 114, and the smallest member of $[8]_9$ that is greater than 200.

12. Given a 3-modularization of the integers, find the smallest member of $[0]_3$ that is greater than 17, the smallest member of $[1]_3$ that is greater than 29, and the smallest member of $[2]_3$ that is greater than 168.

13. Suppose you are giving a masquerade party, and you wish to seat 42 guests at 7 tables so that nobody sits at a table with a person with whom he or she arrived. (Assume no groups arrived with more than 6 people.) Explain how you might use congruences to seat your guests.

14. Find an integer that belongs to both classes $[4]_7$ and $[9]_{10}$.

15. Given the classes $[0]_2$ and $[0]_4$, which is a subset of the other? What about $[1]_2$ and $[1]_4$? Is there a connection between $[0]_2$ and the combined classes of $[0]_4$ and $[2]_4$? What is it? (A set A is a *subset* of a larger set B if every member of A is also a member of B.)

16. Given the classes $[0]_n$ and $[0]_{2n}$ for any positive integer n, which is a subset of the other? Is there a connection between $[0]_n$ and the combined classes of $[0]_{2n}$ and $[n]_{2n}$? What is it?

17. Give the note names of the following pitch classes.

$$[0]_{12}, \quad [4]_{12}, \quad [7]_{12}, \quad [3]_{12}, \quad [8]_{12}, \quad [13]_{12}, \quad [19]_{12}, \quad [24]_{12}, \quad [30]_{12}.$$

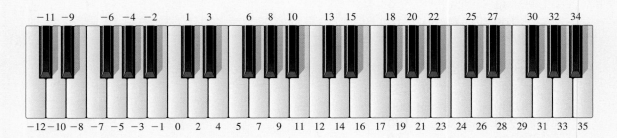

18. What is the musical connection between two pitches belonging to the same pitch class?

19. Name the pitch classes (using least-residue representatives) corresponding to C, D♯, F, G♭, A♯, B♭, and B♯.

20. What is an enharmonic spelling for C♯, E♭, G♭, A♯, and C?

21. What is an enharmonic spelling for D♭, F♯, A♯, C♭, and F?

22. Give the key positions between 0 and 11 of the notes in the chord C♯-E-A, add 2 to each number to obtain a new chord, and write the note names of this chord.

23. Give the key positions between 0 and 11 of the notes in the chord E♭-F-B♭, add 1 to each number to obtain a new chord, and write the note names of this chord.

24. Give the note names of the chord with positions 7-11-14. Subtract 7 from each number to obtain a new chord, and write the note names of this chord.

25. Give the names and key positions of the following notes on the treble clef staff.

26. Give the names and key positions of these notes on the treble clef staff:

27. Draw the correct note on the treble clef staff corresponding to the given key position, and write the note name.

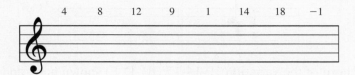

28. Draw the correct note on the treble clef staff corresponding to the given key position, and write the note name.

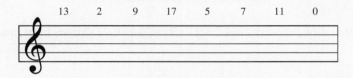

29. Although 12 notes are being played in this melody, they belong to just 6 pitch classes.

Name the 4 classes displaying more than one member.

30. This is a segment from *Moonlight Sonata* by **Ludwig van Beethoven** (1770–1827). The three notes in each set belong to the same three pitch classes. Name the classes and the associated note names. (The F, C, G, and D notes are all played with a sharp due to the key signature.)

31. Find the primary chord of the variation 7-11-13 (G-B-C♯) shown here, and draw it on the staff.

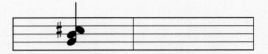

32. Find the primary chord of the variation 10-14-17 (B♭-D-F) shown here, and draw it on the staff.

33. Find the primary chord of the variation 13-17-20 (C♯-F-G♯) shown here, and draw it on the staff.

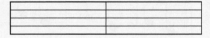

34. Create a variation of the primary chord 4-8-11 (E-G♯-B), and draw both of them on the staff.

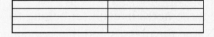

35. Create a variation of the primary chord 2-7-11 (D-G-B), and draw both of them on the staff.

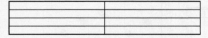

36. Create a variation of the primary chord 0-3-7 (C-E♭-G), and draw both of them on the staff.

(This is known as a *minor triad.*)

37. What primary chord in the following segment from *Bridal Chorus* by Richard Wagner appears in four variations? The notes for the left hand are shown on the lower staff. Observe that the last three notes are also from this same chord. (The key signature, which is not shown, requires that B be played in B♭.)

38. What primary chord in the following measure from *Pine Apple Rag* by Scott Joplin appears in two variations?

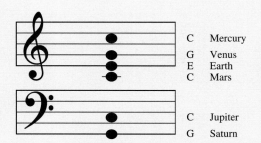

39. The astronomer Johannes Kepler felt that the principles of musical harmony were the keys to understanding the motions and arrangement of the planets. The central theme of his greatest work, *Harmonice Mundi*, was that the heavens resounded to the same scales and chords used by earthly musicians. He felt that the different speeds of the planets should be in the same proportions as the frequencies of the basic intervals that make up chords melodious to the ear. (For example, the frequency ratio of the two pitches of an octave is 1 : 2, a fifth is 2 : 3, and of a fourth is 3 : 4.) In 1599, he presented the following *planetary chord* in a letter to Edmond Bruce, an English colleague of Galileo.

C	Mercury
G	Venus
E	Earth
C	Mars
C	Jupiter
G	Saturn

Beginning with Mercury, the proportions of the frequencies of these pitches are 16 : 12 : 10 : 8 : 4 : 3. Using these ratios and assuming circular orbits, Kepler computed the distances from the sun to the planets relative to Earth. For example, the ratio of the velocities of Mars and Earth is $\dfrac{v_m}{v_E} = \dfrac{8}{10}$.

We then multiply the top and bottom of this fraction by each planet's period (1 year for Earth and 1.88 years for Mars) to compute the ratio of the circumferences of their orbits such that

$$\frac{C_m}{C_E} = \frac{v_m \cdot 1.88}{v_E \cdot 1} = \frac{8}{10} \cdot \frac{1.88}{1} = 1.5.$$

Because the ratio of their orbital radii must be the same as that of their circumferences, we have determined that the radius of Mars' orbit is 1.5 times that of Earth or, in modern units, 1.5 astronomical units (AU).

If the period of Venus is 0.62 year, use Kepler's chord method to find the planet's distance from the sun. How do these values compare to the modern distances of 1.52 AU for Mars and 0.72 AU for Venus?

40. If the period of Mercury is 0.24 year, use Kepler's chord method from Exercise 39 to find its distance from the sun. How does this compare to the modern distance of 0.39 AU?

41. If the period of Jupiter is 11.9 years, use Kepler's chord method from Exercise 39 to find its distance from the sun. How does this compare to the modern distance of 5.20 AU?

9.2 Transformations

The movement of pieces around most board games, such as Monopoly, involves the execution of a sequence of straight-line motions punctuated by stops and changes in direction. When these motions are carried out mainly on a flat surface (which we call a *plane*), they are referred to as geometric **transformations**. Here we shall see how the notion of transformation is rooted as deeply in music as it is in mathematics. A musical transformation affects a musical figure (e.g., a melody) on the musical scale in much the same way as a transformation affects a geometric figure (e.g., a triangle) in a plane. In both cases, a new figure is derived that has the same basic shape as the original but that has been moved to a different position. We will be restricting our attention to three types of motion known as *translations, rotations,* and *reflections*.

> **transformations** A change in location of the points in the Cartesian coordinate system resulting from a well-defined function.

When you push a plate of potatoes across the dinner table by extending your arm in a straight line toward a friend, you are imparting a rigid motion to that plate known as a **horizontal translation**. If the length of your arm defines how far you push any other plate of food, then you have the necessary characteristics for the definition of a *function*. Defining yourself as the origin, you can see that every point on any object you push is moved a distance k units farther from the origin, where k is the length of your arm. By using a Cartesian coordinate system, we see that a triangle starting in position 1 in **Figure 9.2.1** is affected by a horizontal translation of k units by being pushed over to position 2 and that any point (a, b) in the plane gets moved to the point $(a + k, b)$. Figure 9.2.1 illustrates this for a positive k.

> **horizontal translation** Horizontal movement of a point by k units in the Cartesian plane and having the functional representation $(a, b) \to (a + k, b)$.

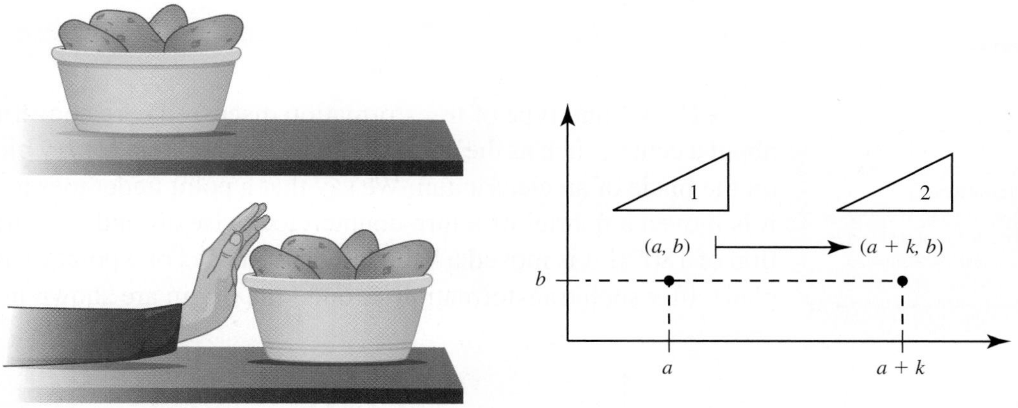

FIGURE 9.2.1 Horizontal translations. If $k > 0$, the triangle in position 1 is moved to position 2, and every point in the triangle is moved k units to the right.

Since every point is directed toward a unique ending point, every transformation is in fact a *function* with domain and range both consisting of all the points in the Cartesian plane. We write $(a, b) \to (a + k, b)$ to represent the function that defines a horizontal translation of k units, and we note that the sign of k determines whether the motion is directed to the left ($k < 0$) or to the right ($k > 0$). Similarly, a **vertical translation** is a transformation that moves an object up or down. In the Cartesian plane, such a translation can be denoted by the functional representation $(a, b) \to (a, b + k)$, where, again, the sign of k determines the direction of motion. If $k < 0$, then each point (a, b) is moved down k units; and if $k > 0$, then each point is moved up by that amount. (See **Figure 9.2.2**.)

> **vertical translation** Vertical movement of a point by k units in the Cartesian plane and having the functional representation $(a, b) \to (a, b + k)$.

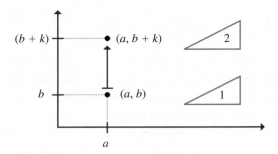

FIGURE 9.2.2   Vertical translation. If $k > 0$, the triangle in position 1 is moved to position 2, and every point in the triangle is moved up by k units.

? Example 1

Find the images of the points (3, 6), (−7, 1), and (4, −11) after applying

(a)   a horizontal translation of 5 units;
(b)   a vertical translation of −2 units;
(c)   a horizontal translation of −3 units *followed* by a vertical translation of 8 units.

⚙ Solution

(a)   Adding 5 to each x-coordinate yields (8, 6), (−2, 1), and (9, −11), respectively.
(b)   Adding −2 to each y-coordinate yields (3, 4), (−7, −1), and (4, −13).
(c)   This sequence of translations results in (0, 14), (−10, 9), and (1, −3).   ♦

rotation A rotation of 90° in the Cartesian plane has the functional representation $(a, b) \rightarrow (-b, a)$. A rotation of 180° in the Cartesian plane has the functional representation $(a, b) \rightarrow (-a, -b)$.

The second type of transformation discussed here concerns circular motion about a center, such as the spokes on the wheel of a stationary bicycle or the points on the blade of an electric fan. We say that a point undergoes a **rotation** of 90° if it is moved a quarter of a turn counterclockwise around a central point or a rotation of 180° if it is moved a half-turn. The images of a point (a, b) in the Cartesian plane after such transformations around the origin are shown in **Figure 9.2.3**.

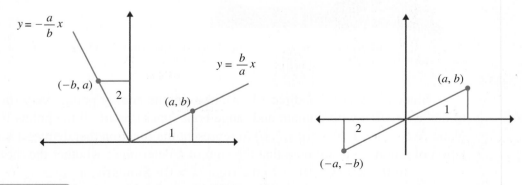

FIGURE 9.2.3   The Cartesian plane after a rotation of (a) 90°, and (b) 180°.

In Figure 9.2.3(a), note that triangles labeled 1 and 2 are congruent. This implies that the lengths of the corresponding sides are equal. Therefore, the point (a, b) in the first quadrant is moved by a 90° rotation to the point $(-b, a)$ in the second quadrant.

(Note also that the line from the origin through the point (a, b) must have slope $\dfrac{b}{a}$ and therefore has equation $y = \dfrac{b}{a}x$. The point $(-b, a)$ is on the line $y = -\dfrac{a}{b}x$, which we know from algebra is perpendicular to $y = \dfrac{b}{a}x$.) So the function producing a 90° rotation is represented by

$$(a, b) \rightarrow (-b, a)$$

In Figure 9.2.3(b), we see that any point (a, b) and its image after a rotation of 180° must lie equidistant from the origin and on a straight line through the origin. This forces the coordinates of the image to be $(-a, -b)$, and so the representation of this function is therefore

$$(a, b) \rightarrow (-a, -b).$$

Our last transformation is perhaps the most familiar to you, for it concerns that cheerful person who smiles back at you from your mirror every morning as you prepare for the day. If you and your mirror image could be viewed from the side, it might look something like the one in **Figure 9.2.4**.

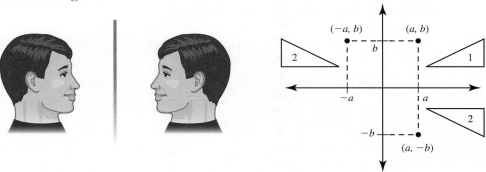

FIGURE 9.2.4   Reflections in a mirror and in the Cartesian plane.

reflection across a vertical A reflection across any line parallel to the y-axis in the Cartesian plane. The reflection across the y-axis has the functional representation $(a, b) \rightarrow (-a, b)$.

reflection across a horizontal A reflection across any line parallel to the x-axis in the Cartesian plane. The reflection across the x-axis has the functional representation $(a, b) \rightarrow (a, -b)$.

Note that the ear, eye, and nose of your reflected self appear to be the same distance behind the you mirror as your real ear, eye, and nose are in front of the mirror. If you align the y-axis of the Cartesian plane with the mirror (Figure 9.2.4), you can see how every point in the plane is affected by a **reflection across a vertical** (the y-axis). Similar analogies can be made for a **reflection across a horizontal** (the x-axis), and the corresponding functional representations are shown here.

Reflection across the y-axis   Reflection across the x-axis

$$(a, b) \rightarrow (-a, b) \qquad\qquad (a, b) \rightarrow (a, -b)$$

? Example 2

Find the images of the points $(7, 4)$, $(6, -3)$, and $(-12, -9)$ after applying

(a)   a rotation of 90°;

(b)   a rotation of 180°;

(c)   a reflection across the y-axis;

(d)   a reflection across the x-axis;

(e) a rotation of 90° followed by a reflection across the *x*-axis;

(f) a reflection across the *y*-axis followed by a rotation of 180°.

⚙ Solution

(a) By using the given functions, we find that the images after a rotation of 90° are

$$(7,4) \rightarrow (-4,7) \quad (6,-3) \rightarrow (3,6) \quad (-12,-9) \rightarrow (9,-12).$$

(b) A rotation of 180° results in

$$(7,4) \rightarrow (-7,-4) \quad (6,-3) \rightarrow (-6,3) \quad (-12,-9) \rightarrow (12,9).$$

(c) A reflection across the *y*-axis results in

$$(7,4) \rightarrow (-7,4) \quad (6,-3) \rightarrow (-6,-3) \quad (-12,-9) \rightarrow (12,-9).$$

(d) A reflection across the *x*-axis results in

$$(7,4) \rightarrow (7,-4) \quad (6,-3) \rightarrow (6,3) \quad (-12,-9) \rightarrow (-12,9).$$

(e) We call the function that results from applying consecutive transformations a **composite** of transformations. A picture helps to realize the final image. (See **Figure 9.2.5**.)

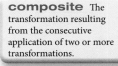

composite The transformation resulting from the consecutive application of two or more transformations.

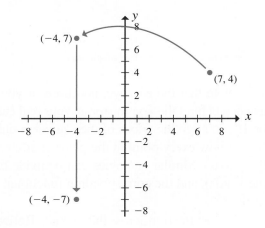

FIGURE 9.2.5 Composite of transformations.

We see that

$$(7,4) \rightarrow (-4,7) \rightarrow (\mathbf{-4,-7}),$$

and similarly,

$$(6,-3) \rightarrow (3,6) \rightarrow (\mathbf{3,-6})$$
$$(-12,-9) \rightarrow (9,-12) \rightarrow (\mathbf{9,12})$$

(f)   This composite of transformations yields

$$(7, 4) \rightarrow (-7, 4) \rightarrow (\mathbf{7, -4}),$$
$$(6, -3) \rightarrow (-6, -3) \rightarrow (\mathbf{6, 3}),$$
$$(-12, -9) \rightarrow (12, -9) \rightarrow (\mathbf{-12, 9}).$$

❓ Example 3

Which single transformation is equivalent to a reflection across the y-axis followed by a rotation of 180°?

⚙ Solution

The dotted line in **Figure 9.2.6** displays the transformation that is equivalent to the composite of this reflection and rotation.

Similarly, we can analyze the functional representations

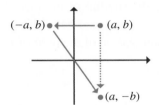

$$(a, b) \rightarrow (-a, b) \rightarrow (a, -b)$$

to see that the composite transformation consists of pairing the first point with the last point,

$$(a, b) \rightarrow (a, -b)$$

which we know to be a reflection across the x-axis.   ◆

FIGURE 9.2.6   A composite transformation.

Now that we know about how geometric transformations affect points in the Cartesian plane, what does any of this have to do with music? Well, as you might have guessed, we will consider the results of pushing, rotating, or reflecting a pattern of notes, To do this, however, we must first have a figure on paper that corresponds to any particular pattern. Our purposes will be better served here not by the usual musical staff with bass and treble clef, but with a *graph* of notes using piano key positions for the vertical axis according to the scheme we adopted in the last section. These are recalled in **Figure 9.2.7**.

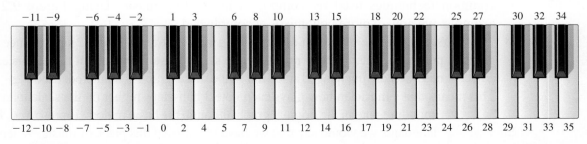

FIGURE 9.2.7   Piano key positions.

For example, we consider the graphical representation of the two measures in **Figure 9.2.8**.

FIGURE 9.2.8   Sample pattern of notes.

The pattern of notes is C-E-G-B♭-A-G-A-C-F. This translates to a numerical sequence of key positions given by

$$0\text{-}4\text{-}7\text{-}10\text{-}9\text{-}7\text{-}(-3)\text{-}0\text{-}5$$

position sequence A sequence of positions on the piano keyboard corresponding to a segment or sequence of musical notes.

note graph The graph of a position sequence as a function of time.

with the corresponding graph given in **Figure 9.2.9** below. For the sake of discussion we will refer to a sequence like the one above as a **position sequence** and to its graph as a **note graph**. The first coordinate in each point of a note graph is the order of the corresponding note in the sequence, and the second coordinate is the key position. So the points that are graphed in Figure 9.2.9 are (1, 0), (2, 4), (3, 7), (4, 10), and so on. (Observe that we have chosen to ignore the rhythm of the piece in the note graph since that is irrelevant to the concepts being developed here. Also, the lines connecting the points serve only to enhance the visualization of the pattern.)

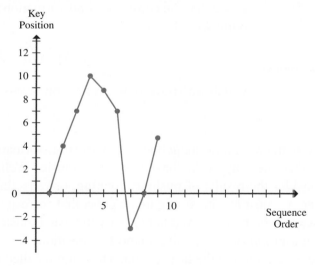

FIGURE 9.2.9   Note graph.

The application of geometric transformations to a note graph is equivalent to several common techniques used by composers in creating music. Using Figure 9.2.9 as our example, we now explore how this is accomplished.

First we observe the effect of applying a vertical translation of 2 units. Since the entire graph is pushed up, each note is raised by the same constant value, but the distances between adjacent notes remain unchanged since the shape of the graph is unaltered. (See **Figure 9.2.10**.) This means that a musician now plays each note 2 half-steps higher on the scale than the corresponding note in the original graph although the basic melody is preserved. In musical language, we would say that we have *transposed* this piece of music. When a vertical translation has been applied to an entire composition, it has undergone what is known musically as a **transposition**, or a change in key. We will explore the key signature of a piece of music in the next section.

transposition The changing of the key of a piece of music. It is the musical analog of a vertical translation of a note graph.

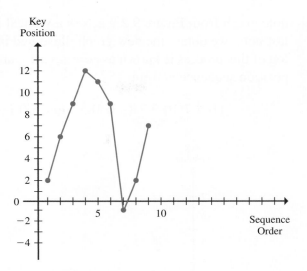

FIGURE 9.2.10   Vertical translation of 2 notes.

Recall that a vertical translation affects each point in the plane by adding a constant to its *y*-coordinate. Since the position sequence corresponding to a note graph is the set of *y*-coordinates of the points of the graph, the new sequence is easily obtained by simply adding 2 to each number in the original sequence. So,

$$0\text{-}4\text{-}7\text{-}10\text{-}9\text{-}7\text{-}(-3)\text{-}0\text{-}5$$

becomes

$$2\text{-}6\text{-}9\text{-}12\text{-}11\text{-}9\text{-}(-1)\text{-}2\text{-}7.$$

The associated sequence of note names is

$$\text{D-F\#-A-C-B-A-B-D-G,}$$

which gives us the new segment of notes on the staff shown in **Figure 9.2.11**.

FIGURE 9.2.11   New segment after the vertical translation.

We observe that a vertical translation has been applied 3 times consecutively to a group of 6 notes in the following segment from *Sonata No. 8, Op. 13* by Ludwig van Beethoven.

Have you ever heard the notes of the chorus of a song repeated in the reverse order? The transformation that creates this effect in a note graph is a *reflection*. By reflecting our

retrograde Musical terminology for the sequence of notes obtained by reflecting the note graph of an initial sequence across a vertical.

note graph from Figure 9.2.9 across a vertical line located immediately after the last note, we obtain the new graph illustrated in **Figure 9.2.12**. The musical analog of this process is known as **retrograde**, and the corresponding change in the position sequence is from

$$0\text{-}4\text{-}7\text{-}10\text{-}9\text{-}7\text{-}(-3)\text{-}0\text{-}5 \quad \text{to} \quad 5\text{-}0\text{-}(-3)\text{-}7\text{-}9\text{-}10\text{-}7\text{-}4\text{-}0$$

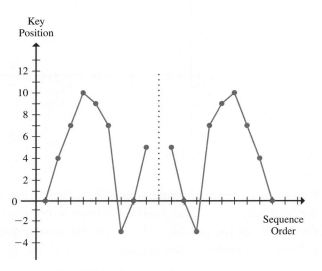

FIGURE 9.2.12   Reflection of a note graph across a vertical.

The effect of this reflection in the new note segment is made visually clear by comparing it to the original segment as seen in **Figure 9.2.13**.

FIGURE 9.2.13   Reflected segment of notes (retrograde).

Retrograde can be observed in the third movement of Beethoven's *Symphony No. 4* (**Figure 9.2.14**) when the progression up the staff at the end of the first line is reversed.

retrograde inversion Musical terminology for the sequence of notes obtained by the rotation of the note graph of an initial sequence through an angle of 180°.

note of inversion The note on the musical staff about which a segment of notes is rotated 180°.

Next we look at the result of rotating a note graph through an angle of 180°. This produces a new graph known musically as the **retrograde inversion** of the original. Here we must be careful about choosing the center of our rotation since rotating about different centers leads to different results. In **Figure 9.2.15**, the initial note graph has been rotated 180° about position 0 (middle C). This means that we place the center of the rotation (marked by an *I*) at level 0 and immediately following the last note of the graph. The point *I* now serves the same role as the origin did for our earlier rotations in the Cartesian plane, and the note C is known in musical language as the **note of inversion**.

To obtain the new position sequence, we first observe that a rotation of 180° is the composite of a reflection across a vertical followed by a reflection across a horizontal. (See the exercises.) We have already seen that reflecting across the vertical correlates to reversing the order. Reflecting across the horizontal line through the 0 key position (C)

FIGURE 9.2.14   Retrograde in Beethoven's *Symphony No. 4*.

implies that each position in the sequence must be replaced by its negative, since a number is required that is equidistant from 0 but in the opposite direction. Hence, the sequence that corresponds to the rotated note graph in **Figure 9.2.15** must be the result of applying these two operations to the original sequence.

$$(-5)\text{-}0\text{-}3\text{-}(-7)\text{-}(-9)\text{-}(-10)\text{-}(-7)\text{-}(-4)\text{-}0$$
$$\text{G-C-D\#-F-D\#-D-F-G\#-C}$$

In practice, it is usually advantageous to first obtain the new position sequence before graphing it.

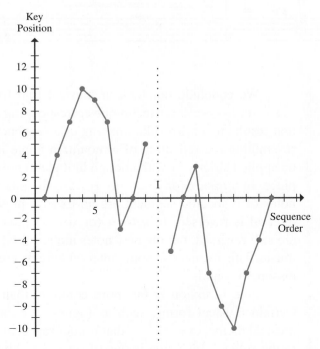

FIGURE 9.2.15   Rotation of 180° about position 0.

Good examples of retrograde and retrograde inversion can be seen in the accompanying piece by Maurice Ravel. Note that Ravel ensured that no one would miss these techniques by writing the letters HAYDN right in the score in the first few measures. Later in the piece, the letters are repeated in reverse order next to the corresponding retrograded notes, and still later, they reappear upside down and reversed next to the retrograde-inversion.

melodic inversion
A reflection of a segment of notes across a horizontal line through a pivot note.

strict (or real) inversion A melodic inversion in which the intervals between consecutive notes are preserved but in the opposite direction.

free (or tonal) inversion A melodic inversion in which some notes may be slightly altered to maintain the tonality of the music.

answer The segment of notes altered by a transformation of the subject that appears later in a fugue or canon.

We conclude our look at musical transformations with the procedure of **melodic inversion**, a technique corresponding to a reflection across a horizontal that results in a mirrorlike imaging of a sequence of notes on the staff somewhat resembling the reflection of a mountain range in a lake (**Fig. 9.2.16**). This may be applied strictly or with enough latitude to prevent veering out of the key of the piece. In a **strict** (or **real**) **inversion**, the interval lengths between consecutive notes of a specific segment are all exactly preserved, but the direction of each interval is reversed. In a **free** (or **tonal**) **inversion**, the interval progressions are still reversed, but the new notes are required to stay within the tonality of the music. This means that some interval lengths are altered by a half-step to prevent dissonant sounds.

Free inversion is far more common than strict inversion in tonal music. Certain musical forms, such as fugues and canons, begin with a fundamental musical theme, or **subject**, that is later repeated in an altered form as an **answer** to the subject. Inversion is one of several techniques used to implement such an

FIGURE 9.2.16   Lake reflection of a mountain range.

alteration. Free inversion is exemplified in **Figure 9.2.17** in a passage from J. S. Bach's *The Art of the Fugue*. The subject is shown in the first staff. Later, this is inverted to form the answer in *Contrapunctus VII*, appearing as shown in the second staff.

FIGURE 9.2.17   *Art of the Fugue.*

> **pivot note** The note on the staff across which an inversion is applied.
>
> **perfect fifth** An interval between two notes that spans 7 half-steps. This corresponds to 5 degrees on the staff.

In this particular segment, the initial note D is transposed up by 7 half-steps to the note A prior to the inversion process across the **pivot note** A. (An interval consisting of 7 half-steps is known as a **perfect fifth**, which is discussed in greater detail in the next section.) Alternately, we could think of the answer being formed by inverting across the pivot D first and then transposing the entire segment up by a fifth. By looking at the corresponding position sequences, we can more easily determine the lengths of the intervals between consecutive notes. The subject sequence is

$$2\text{-}9\text{-}7\text{-}5\text{-}4\text{-}2\text{-}1\text{-}2\text{-}4\text{-}5\text{-}7\text{-}5\text{-}4\text{-}2.$$

By subtracting consecutive positions, we get the corresponding sequence of interval lengths. For instance, the first interval length is $9 - 2 = 7$, the second is $7 - 9 = -2$, and so on.

$$7\text{-}(-2)\text{-}(\mathbf{-2})\text{-}(\mathbf{-1})\text{-}(-2)\text{-}(-1)\text{-}1\text{-}2\text{-}\mathbf{1}\text{-}\mathbf{2}\text{-}(\mathbf{-2})\text{-}(\mathbf{-1})\text{-}(-2).$$

The position sequence of the answer is

$$9\text{-}2\text{-}4\text{-}5\text{-}7\text{-}9\text{-}10\text{-}9\text{-}7\text{-}5\text{-}4\text{-}5\text{-}7\text{-}9.$$

This has a sequence of interval lengths given by

$$(-7)\text{-}2\text{-}\mathbf{1}\text{-}\mathbf{2}\text{-}2\text{-}1\text{-}(-1)\text{-}(-2)\text{-}(\mathbf{-2})\text{-}(\mathbf{-1})\text{-}\mathbf{1}\text{-}\mathbf{2}\text{-}2.$$

When we compare the two sequences of interval lengths, we see that every length has its sign reversed, but the boldface numbers indicate where an interval has been increased or decreased to preserve tonality. In this case, F has been substituted for the note F♯ that would have resulted from a strict inversion. In fact, F is the key of the piece, and if F♯ were to be played in its place, a truly awful assault on the ears would have resulted.

? Example 4

Write the sequences that result from performing each of the following transformations on the sequence 5-2-(−3)-(−1)-9-14-7.

(a)   Vertical translation of three notes (transposition)
(b)   Reflection across the vertical after the last note (retrograde)
(c)   Rotation of 180° (retrograde inversion)
(d)   Horizontal reflection across $p = 2$ (strict inversion)

⚙ Solution

(a)   8-5-0-2-12-17-10
(b)   7-14-9-(−1)-(−3)-2-5
(c)   (−7)-(−14)-(−9)-1-3-(−2)-(−5)
(d)   (−1)-2-7-5-(−5)-(−10)-(−3)

One way to find the inverted sequence in (d) is to observe that the pivot must be the midpoint note between each number in the original sequence and its corresponding position in the inversion. Therefore, the corresponding positions must always add up to $2p$. In this case $2p = 4$, and so $5 + (−1) = 4$, $2 + 2 = 4$, $(−3) + 7 = 4$, and so on.   ◆

? Example 5

Identify the transformation on the sequence 2-6-(−5)-0-9-12-9 that results in each of the following sequences.

(a)   (−2)-2-(−9)-(−4)-5-8-5
(b)   9-12-9-0-(−5)-6-2
(c)   (−9)-(−12)-(−9)-0-5-(−6)-(−2)
(d)   8-4-15-10-1-(−2)-1

⚙ Solution

(a)   Vertical translation of −4 notes (transposition)
(b)   Reflection across the vertical after the last note (retrograde)
(c)   Rotation of 180° (retrograde inversion)
(d)   Horizontal reflection across $p = 5$ (strict inversion)

Again, we determined that (d) was an inverted sequence by noting that each sum of corresponding positions was 10 and then dividing by 2. ◆

You can probably remember listening to songs in which you have heard a refrain that seems somehow to imitate a previous part of the song but in a manner that you cannot consciously identify. In fact, you may have heard the result of a transformation performed on the basic melody. Equivalence classes too are used in the experimental process of molding music that is satisfying to the human ear. Once a melodic strand has been decided upon, not only can a transformation provide compositional variation, but also any note can be replaced with another member of its equivalence class without affecting the integrity of the strand. Transformations and equivalence classes are mathematical tools that help to preserve the basic patterns and structure of a piece of music, but the amount and the manner in which they are applied are a part of the imagination that is at the heart of the creative process. This combination—structure and imagination—fuses the intellect with the soul and is a vital part of the essence of being human.

> Tuning [a piano] is always an approximation, an attempt to reconcile two notions: that of the mathematically pure with that of the musically appealing, the one empirical, the other intuitive. What the tuner is above all aiming to achieve is balance, a mean between the dissonance of theory and the pleasing tone of what the ear is used to hearing
>
> —Thad Carhart, *The Piano Shop on the Left Bank.*

Name _____

Exercise Set **9.2** ▢▢▢←

Identify each of the following transformations as a vertical translation, horizontal translation, rotation of 90°, rotation of 180°, reflection across the x-axis, or reflection across the y-axis.

1.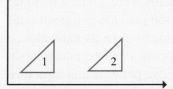

2.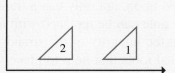

3.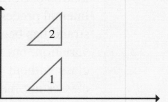

4.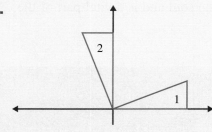

5.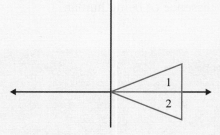

6.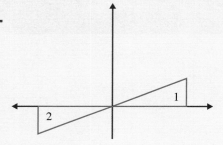

7.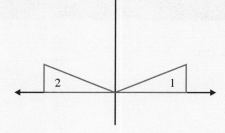

What transformation in the Cartesian plane results from applying the function given in each of following exercises to every point (a, b)?

8. $(a, b) \rightarrow (a, b + 5)$

9. $(a, b) \rightarrow (a + 10, b)$

10. $(a, b) \rightarrow (-b, a)$

11. $(a, b) \rightarrow (-a, -b)$

12. $(a, b) \rightarrow (-a, b)$

13. $(a, b) \rightarrow (a, -b)$

14. Find the image of the point (1, 7) after applying a horizontal translation of 4 units followed by a vertical translation of 5 units.

15. Find the image of the point (2, 9) after applying a vertical translation of 3 units followed by a rotation of 90°.

16. Find the image of the point (−3, 4) after applying a rotation of 90° followed by a reflection across the y-axis.

17. Find the image of the point $(-6, -4)$ after applying a reflection across the x-axis followed by a horizontal translation of -20.

18. Find the image of the point $(8, -10)$ after applying a rotation of $180°$ followed by a reflection across the x-axis.

19. Find the image of the point $(7, 13)$ after applying first a vertical translation of -10 units, then a reflection across the y-axis, and finally a rotation of $180°$.

20. Which single transformation is equivalent to the composite of a reflection across the x-axis followed by a reflection across the y-axis?

21. Find two transformations whose composite takes $(a, b) \rightarrow (b, a)$.

22. Determine the functional representation of a horizontal translation of 4 units followed by a reflection across the x-axis. Determine the functional representation of a reflection across the x-axis followed by a horizontal translation of 4 units. Does the order in which you perform these transformations make a difference?

23. Determine the functional representation of a vertical translation of 4 units followed by a reflection across the x-axis. Determine the functional representation of a reflection across the x-axis followed by a vertical translation of 4 units. Does the order in which you perform these transformations make a difference?

24. What points in the Cartesian plane are left unmoved by a reflection across the x-axis? Across the y-axis? What point would be left unmoved by a rotation?

25. Which single transformation is equivalent to the composite of a reflection across the x-axis followed by a rotation of $180°$?

26. The opening melody of *Canon in D* by Johann Pachelbel and accompanying position sequence are given here.

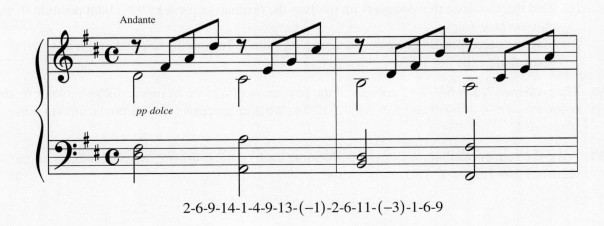

2-6-9-14-1-4-9-13-(−1)-2-6-11-(−3)-1-6-9

Draw the corresponding note graph with key position as the vertical axis and time as the horizontal axis. Notice the repetition of a 4-note pattern.

27. Find the sequence that results from applying a vertical translation of 6 notes to the sequence in the previous exercise. Draw the corresponding note graph on the same set or axes.

28. Here is the opening measure of *The Entertainer* by Scott Joplin. The position sequence of the top notes is 26-28-24-21-23-19.

What is the position sequence of the bottom notes? What is the interval between the top and bottom notes? Plot both the corresponding note graphs. Find the sequence resulting from a 180° rotation about position 0, and draw the note graph.

29. The opening measure of *Jesu, Joy of Man's Desiring* by Johann Sebastian Bach is given by the following position sequence:

$$11\text{-}7\text{-}9\text{-}11\text{-}14\text{-}12\text{-}12\text{-}16\text{-}14\text{-}14\text{-}19\text{-}17.$$

(a) Draw the corresponding note graph.
(b) Find the sequence that results from applying a vertical translation of −5 notes, and draw the new note graph.
(c) Find the sequence that results from rotating the original sequence 180° about position 0, and plot the new graph.
(d) Find the corresponding sequence that results from reflecting the original graph across the vertical after the last note.

30. The sequence 2-2-6-6-9-9-11 appears at the beginning of a piece of music, followed later by the sequences 0-0-4-4-7-7-9 and 5-5-9-9-12-12-14. What connections do you notice among these segments?

31. Write the sequences that result from performing each of the following transformations on the sequence 2-6-7-9-5-4-(−1)-0.

(a) Vertical translation of 3 notes (transposition).
(b) Reflection across vertical after the last note (retrograde).
(c) Rotation of 180° (retrograde inversion).
(d) Horizontal reflection across $p = 2$ (strict inversion).

32. Write the sequences that result from performing each of the following transformations on the sequence 9-12-14-2-6-9-7.

(a) Vertical translation of −4 notes (transposition).
(b) Reflection across vertical after the last note (retrograde).
(c) Rotation of 180° (retrograde inversion).
(d) Horizontal reflection across $p = 5$ (strict inversion).

33. Identify the transformation on the sequence 2-6-9-14-13-11-9 that results in each of the following sequences.

(a) (−2)-2-5-10-9-7-5
(b) 9-11-13-14-9-6-2
(c) (−9)-(−11)-(−13)-(−14)-(−9)-(−6)-(−2)
(d) 12-8-5-0-1-3-5

34. Identify the transformation on the sequence (−3)-1-4-2-6-7-9 that results in each of the following sequences.

(a) 9-7-6-2-4-1-(−3)
(b) (−9)-(−7)-(−6)-(−2)-(−4)-(−1)-3
(c) 9-13-16-14-18-19-21
(d) 9-5-2-4-0-(−1)-(−3)

35. The first segment in the accompanying measures (having sequence 13-11-14-13-11-21) is a subject found in Brahms, *Intermezzo in A Major*, Op. 118, No. 2. The next segment (20-21-18-20-21-11) is a free inversion used as an answer a short time later. Create the sequences of interval lengths between consecutive notes for both segments. Which intervals were adjusted to preserve tonality? (The key signature implies that C, F, and G are to be played as C♯, F♯, and G♯.)

(a) Brahms, Op. 118, No. 2, measures 1–2.

(b) Brahms, Op. 118, No. 2, measures 35–36. Free inversion of (a).

36. The segment here is a strict inversion of the subject (a) in Exercise 35 and has position sequence 20-22-19-20-22-12. Create the associated sequence of interval lengths, and compare it to the sequence for the free inversion in Exercise 35. (The *X* in front of F is a *double sharp*, meaning that it is played as a G.)

37. Several measures of *Healed Heart* by C. Johnson are given below.

(a) Create the position sequence for this segment. (Use just the top note of the chords.)

(b) Create a new sequence by applying a vertical translation of +5 notes.

(c) Create a strictly inverted sequence by reflecting the sequence from (a) across the horizontal through the note A at position 9.

38. The segment below is from the *Minute Waltz*, Op. 64, No. 1, by Frederic Chopin. According to the key signature, A and B must be played as A♭ and B♭. The inversion that occurs between measures is across what pivot note? Is it a strict or free inversion?

39. Identify any transformations that you recognize in this segment from J. S. Bach's *A Musical Offering*.

40. These are three measures from *Sonata No. 3* by Wolfgang Amadeus Mozart. What transformations are being applied within each measure and between consecutive measures?

Statue of Mozart in Salzburg, Austria

9.3 The Circle of Fifths

circle of fifths A geometric scheme for identification of the major keys with the corresponding number of accidentals.

The magic **circle of fifths** that is presented in any textbook on music theory is a wonderful example of a pattern of numbers occurring in the natural world. Before we can examine the components of this pattern, first we must discuss in greater depth both modulo arithmetic and musical structure.

First recall that an m-modularization of the integers partitions them into m distinct residue classes. If $[a]_m$ and $[b]_m$ are any two of these classes, the question naturally arises about whether we can add them in a reasonable way. The answer turns out to be yes. Suppose that a and c are two different representatives of the same residue class and that the same is true for b and d. In other words, let $a \equiv c \pmod{m}$ and $b \equiv d \pmod{m}$. Then

$$a = c + mn \quad \text{and} \quad b = d + mk$$

for some integers n and k, and so

$$a + b = c + mn + d + mk$$
$$= c + d + m(n + k).$$

Because this may be written

$$a + b \equiv c + d \pmod{m},$$

we have

$$[a + b]_m = [c + d]_m.$$

Note carefully what this means. It says that the residue class of the sum of two representatives from different classes is the same regardless of the choice of representatives. This in turn implies that the following equation is a legitimate definition of the addition of two residue classes modulo m:

$$[a]_m + [b]_m = [a + b]_m.$$

❓ Example 1

(a) $[3]_5 + [19]_5 = [22]_5 = [2]_5$ but since $[19]_5 = [4]_5$, we also could have written $[3]_5 + [4]_5 = [7]_5 = [2]_5$.

(b) $[11]_8 + [26]_8 = [37]_8 = [5]_8$ but since $[11]_8 = [3]_8$ and $[26]_8 = [2]_8$, we also could have written $[3]_8 + [2]_8 = [5]_8$.

(c) $[68]_3 + [118]_3 = [2]_3 + [1]_3 = [3]_3 = [0]_3$.

(d) $[45{,}409]_{20} + [79{,}135]_{20} = [9]_{20} + [15]_{20} = [4]_{20}$. ◆

Typically, modular addition is done using the least residues as the representatives for the congruence classes. The set $\{0, 1, 2, 3, \ldots, m - 1\}$ of any m-modularization is known as

least-residue system modulo *m* A complete residue system consisting entirely of least residues.

the **least residue system modulo *m*.** It is interesting to observe the patterns of numbers that result from computing an addition table using a least-residue system. For $m = 5$, that system is $\{0, 1, 2, 3, 4\}$, and each slot in the table in **Figure 9.3.1** contains the sum modulo 5 of the numbers heading the row and column.

+	0	1	2	3	4
0	0	1	2	3	4
1	1	2	3	4	0
2	2	3	4	0	1
3	3	4	0	1	2
4	4	0	1	2	3

FIGURE 9.3.1   Addition table for the residues of a 5-modularization.

In ordinary arithmetic, one must add $-a$ to a to obtain 0, and so we call $-a$ the *additive inverse* of a. We note in Figure 9.3.1 that this is not the case in modular arithmetic. In a 5-modularization, 1 and 4 are additive inverses of each other because their sum is congruent to 0 modulo 5. Likewise, 2 and 3 are also additive inverses of each other.

? Example 2

+	even	odd
even	even	odd
odd	odd	even

+	0	1
0	0	1
1	1	0

FIGURE 9.3.2   Addition tables for odd and even integers.

We all know that the sum of two even integers is even, the sum of two odd integers is even, and the sum of an odd and an even integer is odd. Earlier, we discussed that $[0]_2$ and $[1]_2$ are the sets of even and odd integers, respectively, and so now we see how this same information is contained in the addition table for the residues of a 2-modularization. (See **Figure 9.3.2**.) ♦

major scale A specific arrangement of eight notes separated by steps according to whole-whole-half-whole-whole-whole-half.

We now return to the musical connections. Before we talk about the circle of fifths, we need to introduce the notion of a major scale. A **major scale** is a specific arrangement of eight notes separated by whole steps and half-steps such that the sound of the produced pitches is very pleasing to the ear. The familiar vocal pattern

do-re-mi-fa-so-la-ti-do

represents a major scale, and the steps separating the pitches in such a scale are whole-whole-half-whole-whole-whole-half.

tonic: The first note of a major scale

Key: The name of a major scale according to the first note. Also the name of an ordered pair of integers that provides the information needed to decipher an encrypted message.

Each major scale is named by the first note in the scale (known as the **tonic** and is called the **Key** of that scale. For example, the major scale in **Figure 9.3.3** is in the key of C. Observe the locations of the half-step separations between the third and fourth notes and the seventh and eighth notes. This is the only major scale consisting entirely of notes having no accidentals.

Every other major can simply be thought of as a *vertical translation* of this basic position sequence:

$$0 - 2 - 4 - 5 - 7 - 9 - 11 - 12.$$

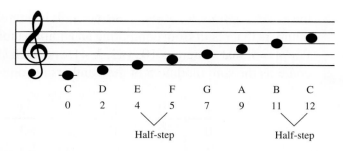

FIGURE 9.3.3   Major scale in the key of C.

❓ Example 3

To create a major scale in the key of D, we want a sequence beginning with 2, and so we simply add 2 to each position in the previous sequence for the scale of C to obtain 2-4-6-7-9-11-13-14. The corresponding notes are shown in **Figure 9.3.4**.

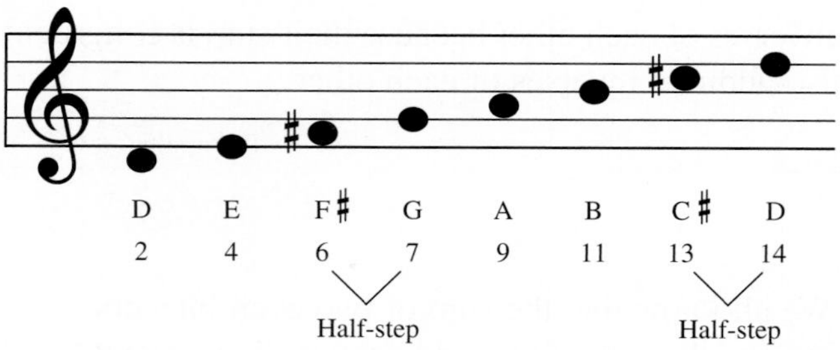

Notice that to maintain the half-steps in the correct locations, the two notes F and C had to be sharped. Similarly, we see that a scale exists for each of the 15 other possible starting keys, and each of these will need to have accidentals appended to certain notes to retain the proper structure. ◆

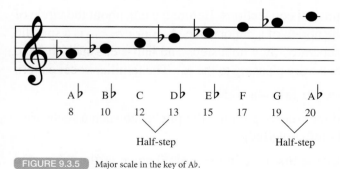

FIGURE 9.3.4   Major scale in the key of D.

❓ Example 4

Because A♭ is located 8 positions above C, the corresponding position sequence is 8-10-12-13-15-17-19-20, giving the scale in **Figure 9.3.5**.

The accidentals needed in this situation turn out to be flats as a result of the requirement for the notes of any major scale to be listed in alphabetical order, with every letter name appearing exactly once. For instance, position 13 is named by D♭ rather than by the enharmonic C♯ so that C is not listed twice. ◆

FIGURE 9.3.5   Major scale in the key of A♭.

Key signature A group of one or more accidentals at the beginning of a composition to identify the key.

Most compositions utilize a major scale as the underlying structure upon which the main musical themes are constructed. Rather than supply an accidental on the staff before every note throughout the piece, which needs to be altered for a particular scale, it is less cumbersome to use a device called a **Key signature**. A key signature consists of a group of one or more accidentals placed at the

FIGURE 9.3.6 Key signatures for D (on the left) and A♭.

beginning of the composition in positions on the staff that indicate which notes are to be played a half-step higher or lower. The major keys of D and A♭, for example, would be indicated by key signatures as shown in **Figure 9.3.6**, because C and F need to be sharped for the key of D and the notes A, B, D, and E likewise need to be flatted for the key of A♭.

As it turns out, not only is the number of sharps or flats a function of the key (as you would expect), but also the opposite is true—the key is a function of the type and number of accidentals. The precise correspondence is given in **Figure 9.3.7**.

Number of sharps or flats	Key	Pitch Class
0♯	C	0
1♯	G	7
2♯	D	2
3♯	A	9
4♯	E	4
5♯	B	11
6♯	F♯	6
7♯	C♯	1
7♭	C♭	11
6♭	G♭	6
5♭	D♭	1
4♭	A♭	8
3♭	E♭	3
2♭	B♭	10
1♭	F	5
0	C	0

FIGURE 9.3.7 The major keys as a function of the number and type of accidentals.

It is the third column that provides us with a most intriguing string of numbers. Our initial observation shows first that positions 11, 6, and 1 are repeated because B and C♭, F♯ and G♭, and C♯ and D♭ are all enharmonic pairs. Further inspection reveals that each successive number is the pitch class obtained by adding 7 (mod 12) to the previous number in the list.

$$[0]_{12} + [7]_{12} = [\mathbf{7}]_{12},$$
$$[7]_{12} + [7]_{12} = [14]_{12} = [\mathbf{2}]_{12},$$
$$[2]_{12} + [7]_{12} = [\mathbf{9}]_{12},$$
$$[9]_{12} + [7]_{12} = [16]_{12} = [\mathbf{4}]_{12},$$
$$[4]_{12} + [7]_{12} = [\mathbf{11}]_{12}.$$

This information puts us in a position to construct the *circle of fifths*. This is a circle with 12 evenly spaced tick marks on it—just as a clock has—with each mark designated by a major key. We start with the first of two preliminary circles by listing C at the top and then proceeding around clockwise naming each tick mark with the key that has a signature possessing one more sharp than the preceding key. This must end with C♯, the key with the necessary maximum of 7 sharps. Similarly, we can produce a second circle by proceeding in a counterclockwise direction from C, naming each mark with the key obtained by successively adding 1-flat and terminating, as before, with the key corresponding to the scale with 7 flats. These two figures are displayed in **Figure 9.3.8**.

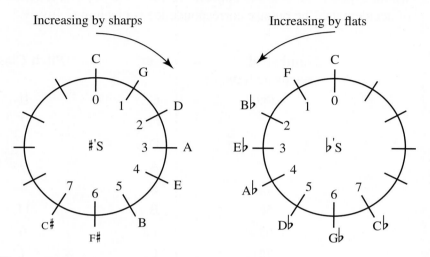

FIGURE 9.3.8   Circles showing the major keys according to increasing numbers of sharps and flats.

We have already noticed that the three pairs of keys at the bottom of each circle are enharmonically equivalent. This necessarily means that the corresponding scales are the same in each case, and so it makes perfect sense for these pairs of keys to coincide in a blending of these two circles into what is known as the *circle of fifths,* displayed in **Figure 9.3.9**. By also listing the position next to each key, we simultaneously have rendered a beautiful clockwise listing of the successive addition of 7 (mod 12) mentioned earlier, and herein lies the reason for the name of this particular scheme for remembering the signatures of the major keys.

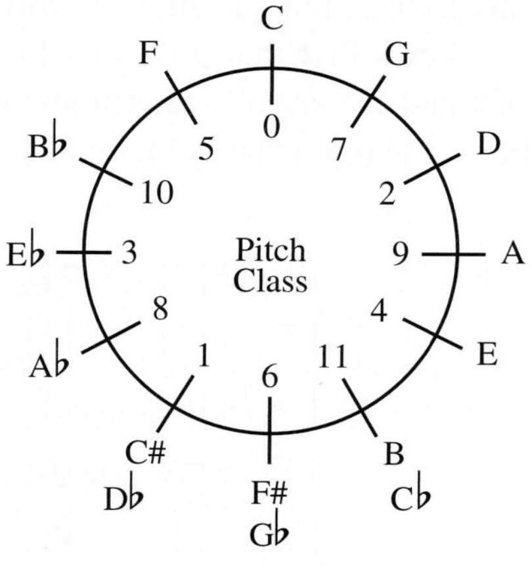

FIGURE 9.3.9   The circle of fifths.

In music theory, a *perfect fifth* is an interval on the staff between two notes that spans 5 degrees (spaces and lines) and therefore must encompass 7 half-steps. For instance, C and G form a perfect fifth, since G is located 5 degrees (7 half-steps) *above* C. Two other pairs are shown in **Figure 9.3.10**.

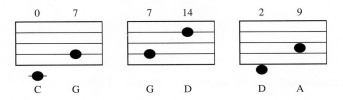

| FIGURE 9.3.10 | Three pairs of notes that form a perfect fifth interval.

Starting with C at the top of our circle and proceeding clockwise, we find that each successive note name on the circle can be obtained by *ascending* the piano keyboard in perfect fifth intervals, since this corresponds to the repeated addition of 7 half-steps. It is important to note that the numbers in our circle in Figure 9.3.9 represent pitch classes and not key positions.

Example 5

What major key has a signature containing 4 sharps?

Solution

Since the unknown key must be 4 perfect fifth intervals above C, we just add $[7]_{12}$ four times to $[0]_{12}$. This simply amounts to

$$[0]_{12} + 4 \cdot [7]_{12} = [28]_{12} = [4]_{12} = \text{E.} \quad \blacklozenge$$

If the key signature of musical composition involves sharps, Example 5 indicates that the major key is a function of the number n of sharps and can be represented by

$$k^{\#}(n) = \overbrace{[7]_{12} + [7]_{12} + \ldots + [7]_{12}}^{n\,\text{times}}$$
$$= n \cdot [7]_{12}$$
$$= [7n]_{12}.$$

Clearly, the domain of this function is just {0, 1, 2, 3, 4, 5, 6, 7}.

Alternately, note that we could generate the pitch classes on the circle of fifths by proceeding *counter* clockwise from C and repeatedly adding 5 (mod 12). Since adding 5 half-steps to a pitch corresponds to ascending 4 degrees on the staff, every successive note in the counterclockwise direction is an interval of a **perfect fourth** above the previous one. When presented with a signature consisting of n flats, we count counterclockwise n pitches from C to find the major key. Therefore, the key function now becomes

perfect fourth An interval between two notes that spans 5 half-steps. This corresponds to 4 degrees on the staff.

$$k^{\flat}(n) = \overbrace{[5]_{12} + [5]_{12} + \ldots + [5]_{12}}^{n \text{ times}}$$
$$= n \cdot [5]_{12}$$
$$= [5n]_{12}.$$

? Example 6

The major key whose signature contains 3 sharps is

$$k^{\#}(3) = [7 \cdot 3]_{12} = [21]_{12}$$
$$= [9]_{12} = \text{A}.$$

The major key whose signature contains 6 flats is

$$k^{\flat}(5) = [5 \cdot 6]_{12} = [30]_{12}$$

$$= [6]_{12} = \text{G}\flat. \quad \blacklozenge$$

> The fugue is like pure logic in music.
>
> —Frederic Chopin

Name _____

Exercise Set **9.3** 000←

In Exercises 1-8 perform the modular arithmetic. List your answers, using least residues.

1. $[26]_7 + [43]_7 =$

2. $[59]_{11} + [85]_{11} =$

3. $[39]_9 + [152]_9 =$

4. $[26]_5 + [32]_5 + [43]_5 =$

5. $[126]_{10} + [673]_{10} =$

6. $[2]_4 + [2]_4 =$

7. $[2,012]_{25} + [3,054]_{25} =$

8. $[27,784]_2 + [89,113]_2 =$

9. Construct an addition table, using the least residues of a 4-modularization. What is the additive inverse of 1? Of 2?

10. Construct an addition table, using the least residues of a 7-modularization. What is the additive inverse of 1? Of 2? Of 3? Of 5?

11. Construct an addition table, using the least residues of a 10-modularization. What is the additive inverse of 3? Of 6? Of 8?

12. Construct an addition table using the least residues of a 12-modularization. What is the additive inverse of 4? Of 7? Of 9?

The position sequence for the major scale in the key of C is 0-2-4-5-7-9-11-12. Find the position sequence for the major scale of each of the keys listed below, and draw each one on a musical staff.

13. G

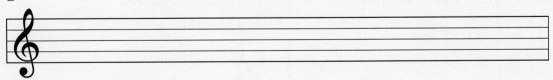

14. F♯

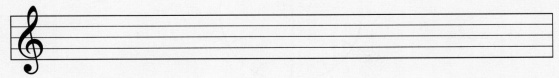

15. B♭

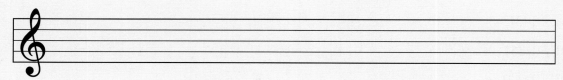

16. D♭

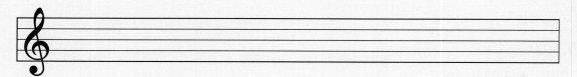

17. Fill in the missing pitch classes and note names in the following circle of fifths.

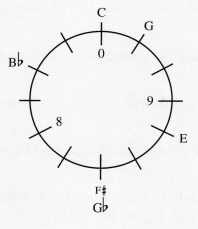

18. Starting with C (pitch class 0) on the circle of fifths shown here, compute the class of each of the given key names by successively *adding* 5 in a *counterclockwise* direction.

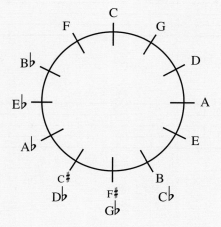

19. A is a perfect fifth interval above what note? A is a perfect fourth interval above what note?

20. B♭ is a perfect fifth interval above what note? B♭ is a perfect fourth interval above what note?

21. What are the two keys that name the same major scale having either 5 sharps or 7 flats?

22. What are the two keys that name the same major scale having either 7 sharps or 5 flats?

Use the function $k^\sharp(n) = [7n]_{12}$ and $k^\flat(n) = [5n]_{12}$ to determine the major key corresponding to each of the following signatures.

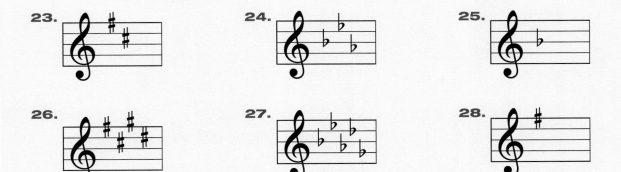

23.   **24.**   **25.**

26.   **27.**   **28.**

29. Consider the pairs of pitch classes that are reflections across a vertical line through the center of the circle of fifths. They are additive inverses of each other modulo 12. Using their note names, we could say, for example that F and G are additive inverses of each other. Use this scheme to determine the additive inverse of E♭ and of C♯.

30. What is the note name of the pitch class that is the additive inverse modulo 12 of B♭? Of E?

31. A **complete residue system modulo *m*** is a set of *m* integers, no two of which are congruent to modulo *m*. It is called *complete* because every integer must be congruent modulo *m* to exactly one number in such a collection. We form such a system by choosing one number from each residue class in an *m*-modularization of the integers. For example, any least-residue system is complete. For $m = 4$,

> **complete residue system modulo *m*** A set of *m* integers, no two of which are congruent to modulo *m*.

$$[0]_4 = \{\dots, -8, -4, 0, 4, 8, 12, \dots\},$$
$$[1]_4 = \{\dots, -7, -3, 1, 5, 9, 13, \dots\},$$
$$[2]_4 = \{\dots, -6, -2, 2, 6, 10, 14, \dots\},$$
$$[3]_4 = \{\dots, -5, -1, 3, 7, 11, 15, \dots\}.$$

So each of the sets {0, 1, 2, 3}, {4, 5, 6, 7}, and {−8, 9, 14, 3} is a complete system modulo 4. Is the set {12, −3, −6, 19} a complete system modulo 4? The set {14, 5, −1, 0}? The set

{3, 17, 12, −5}?

32. Any number in a complete residue system can be replaced by a number congruent to it, and that new system must also be complete. Since any least-residue system is such a system, we can show that {13, 15, 86, −8, 29} is a complete residue system modulo 5 by listing the congruences $13 \equiv 3 \pmod 5$, $15 \equiv 0 \pmod 5$, $86 \equiv 1 \pmod 5$, $-8 \equiv 2 \pmod 5$, and $29 \equiv 4 \pmod 5$. Show that {101, −13, 60, 19, 23} is a complete residue system modulo 5.

33. Show that {97, 18, 10, 87, −25, 4, 35, −31, 48} is a complete residue system modulo 9.

34. Show that {9, −11, 22, 17, 2, 42} is a complete residue system modulo 6.

35. Show that {0, 7, 14, 21, 28, 35, 42, 49} is a complete residue system modulo 8.

9.4 Modular Arithmetic and Cryptology

Now that we have learned some of the main concepts involved in modular addition of members of the same least-residue system, we now wish to expand our understanding and skills to include the other operations involved in **modular arithmetic**. We begin by revisiting addition with the help of a visual representation. For any positive integer m, we identify every least-residue x modulo m with every other integer $x + km$ for all integers k. Therefore, we can replace the standard number line with one constructed of repeated replications of the least residues. **Figure 9.4.1** illustrates this for $m = 8$.

> **modular arithmetic** Arithmetic done with residue classes.

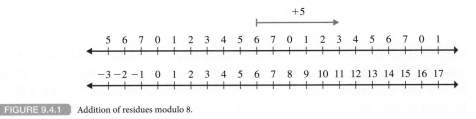

FIGURE 9.4.1   Addition of residues modulo 8.

This figure also demonstrates a hands-on method for doing addition modulo 8. From the last section, we know that $[6]_8 + [5]_8 = [11]_8 = [3]_8$ or, dispensing with the brackets,

$$6 + 5 \equiv 3 \,(\mathrm{mod}\,8).$$

In elementary school, addition of a positive number n is sometimes introduced as moving a distance n units to the right on the number line. If we perform this same motion on the *modular* number line, we arrive at the correct result, $6 + 5 \equiv 3 \pmod 8$. This suggests an identical procedure for the related operation of subtraction. Subtracting 5 from 3 corresponds to starting at 3 and moving 5 units to the *left*, arriving at 6 as our result. (See **Figure 9.4.2**.) So we see that $3 - 5 \equiv 6 \pmod 8$. This matches the more rote process,

$$3 - 5 = -2$$
$$\equiv 6 \,(\mathrm{mod}\,8).$$

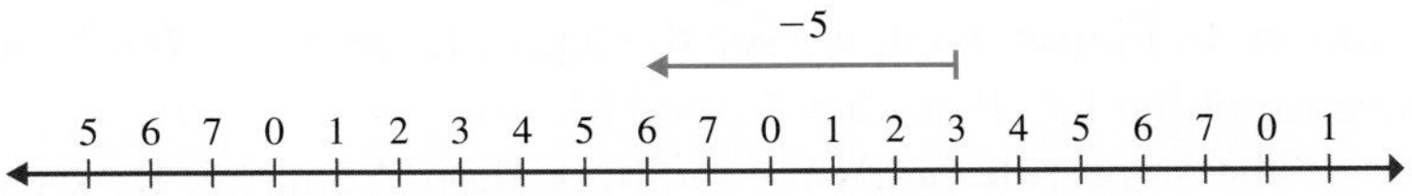

FIGURE 9.4.2   Subtraction on a modular number line.

To visualize this further, we utilize the repeated pattern of the modular number line by wrapping it around a circle marked with convenient units, as in **Figure 9.4.3**. Addition in this system now corresponds to a clockwise rotation around the circle, and subtraction corresponds to a counterclockwise rotation. (The use of this circle as a tool is why modular arithmetic is sometimes referred to as *clock arithmetic*.)

Recall from the last section that the addition of *any* residue class members may be done by first replacing them with their least residues. The so-called clock makes this more apparent. For instance, we can compute $6 + 21 = 27 \equiv 3 \pmod 8$. Alternately, adding 21 (modulo 8)

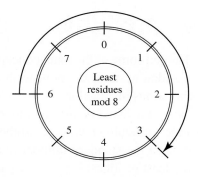

FIGURE 9.4.3   A clock for modular arithmetic. This represents this addition fact: $6 + 5 \equiv 3$ (mod 8).

is equivalent to traversing clockwise completely around the circle two times (16 units), followed by 5 more units. In other words, we could have skipped the two complete trips and simply moved 5 units. (See **Figure 9.4.4**.) In modular notation, $6 + \mathbf{21} \equiv 6 + \mathbf{5} \equiv 3$ (mod 8). In other words, we may first replace 21 by its *least residue* before performing the operation.

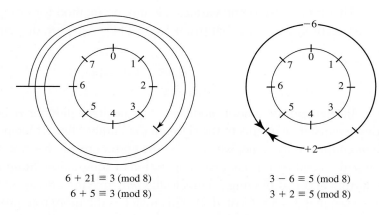

$$6 + 21 \equiv 3 \ (\text{mod } 8)$$
$$6 + 5 \equiv 3 \ (\text{mod } 8)$$

$$3 - 6 \equiv 5 \ (\text{mod } 8)$$
$$3 + 2 \equiv 5 \ (\text{mod } 8)$$

FIGURE 9.4.4   In modular arithmetic, we can replace any number with its least residue.

Note that this immediately implies that any subtraction can be replaced by an equivalent addition. In Figure 9.4.4, we see the equivalence of $3 - \mathbf{6} \equiv 5$ (mod 8) and $3 + \mathbf{2} \equiv 5$ (mod 8) because of the fact that $-6 \equiv 2$ (mod 8).

The same rule holds for multiplication. Suppose, for example, we wish to multiply 3 by 38. Then we can compute, $3 \cdot 38 = 114 \equiv 2$ (mod 8). Alternately, we could start at 0 on the clock and rotate clockwise around the circle a distance of 38 units, followed by another rotation of 38 units and finally one more rotation by this amount. But each traversal of 38 units takes us completely around 4 times, with 6 more units left over. So—exactly as in the situation with addition—this is equivalent to replacing 38 in the multiplication problem with its *least residue* 6 (modulo 8) and thereby reducing the complexity of the multiplication, as shown here:

$$3 \cdot 38 \equiv 3 \cdot 6$$
$$= 18$$
$$\equiv 2 \ (\text{mod } 8).$$

We list this discovery as a general rule.

Any number in a modular arithmetic operation may be replaced by its least residue.

❓ Example 1

Perform the following operations in the least-residue system modulo 12.

(a) $8 + 7$ (b) $5 - 10$ (c) $359 + 873$

(d) $126 - 451$ (e) $46 \cdot 87$ (f) $206 \cdot 627$

⚙ Solution

Since $a = b$ always implies that $a \equiv b \pmod{m}$, for convenience we will henceforth use only the congruence symbol.

(a) $8 + 7 \equiv 15 \equiv 3 \pmod{12}$,

(b) $5 - 10 \equiv 5 + 2 \equiv 7 \pmod{12}$,

(c) $359 + 873 \equiv 11 + 9 \equiv 20 \equiv 8 \pmod{12}$,

(d) $126 - 451 \equiv 6 - 7 \equiv 6 + 5 \equiv 11 \pmod{12}$,

(e) $46 \cdot 87 \equiv 10 \cdot 3 \equiv 30 \equiv 6 \pmod{12}$,

(f) $206 \cdot 627 \equiv 2 \cdot 3 \equiv 6 \pmod{12}$.   ◆

We have been using properties of modular arithmetic that we now formally gather into a single declaration. Property (a) was proved in the last section by a method that could easily be adapted to prove the other statements concerning multiplication and exponentiation.

If $a \equiv c \pmod{m}$ and $b \equiv d \pmod{m}$ for any positive integer m, then

   a. $a + b \equiv c + d \pmod{m}$,

   b. $ab \equiv cd \pmod{m}$,

   c. $a^n \equiv c^n \pmod{m}$ for n any positive integral power.

We have not yet used property (c) above. Since $15 \equiv 3 \pmod{4}$, we can square both sides to get

$$225 \equiv 9 \pmod{4},$$

which you can check to be true. If we again square both sides, we get

$$50{,}625 \equiv 81 \pmod{4},$$

which is also true.

Property (c) suggests a way efficiently to reduce a large power of any integer to a least residue. For example, many calculators are unable to compute 11^{107}, because the power is too large. However, we can still find its least-residue modulo 13 by first resolving the exponent 107 into a sum of powers of 2. The first few powers of 2 are 1, 2, 4, 8, 16, 32, 64, 128, 256, 512,.... Since 107 is greater than 64 but less than 128, we start with 64 as the first summand. We then keep adding successively smaller binary powers as long as the current sum does not exceed 107:

$$107 = 64 + 32 + 8 + 2 + 1.$$

So now we can write

$$11^{107} = 11^{64+32+8+2+1}$$
$$= 11^{64} \cdot 11^{32} \cdot 11^8 \cdot 11^2 \cdot 11.$$

We find the least residue of each of these factors by performing the following sequence of squaring and reducing operations, all modulo 13.

$$11^2 = 121 \equiv 4$$
$$11^4 \equiv 4^2 = 16 \equiv 3$$
$$11^8 \equiv 3^2 = 9$$
$$11^{16} \equiv 9^2 = 81 \equiv 3$$
$$11^{32} \equiv 9$$
$$11^{64} \equiv 81 \equiv 3 \,(\text{mod } 13).$$

Finally, we substitute accordingly to obtain

$$11^{107} = 11^{66} \cdot 11^{32} \cdot 11^8 \cdot 11^2 \cdot 11$$
$$\equiv 3 \cdot 9 \cdot 9 \cdot 4 \cdot 11$$
$$= 10,692$$
$$\equiv 6 \,(\text{mod } 13).$$

For historical reasons, this procedure is sometimes called the *Russian peasant method for finding power residues.*

? Example 2

Use the Russian peasant method to find the least residue of 5^{149} modulo 7.

⚙ Solution

First express the exponent as the sum $149 = 128 + 16 + 4 + 1$. This implies we only need the reduced squares in the boxes below. All congruences are modulo 7.

$$5^2 \equiv 4$$
$$\boxed{5^4 \equiv 16 \equiv 2}$$
$$5^8 \equiv 4$$
$$\boxed{5^{16} \equiv 2}$$
$$5^{32} \equiv 4$$
$$5^{64} \equiv 2$$
$$\boxed{5^{128} \equiv 4}$$

Therefore, $5^{149} = 5^{128} \cdot 5^{16} \cdot 5^4 \cdot 5 \equiv 4 \cdot 2 \cdot 2 \cdot 5 = 80 \equiv 3 \,(\text{mod } 7)$. ◆

In certain cases, a power residue can be easily computed without using the Russian peasant method if the original base is congruent to -1, 0, or 1.

❓ Example 3

Find the least residue.

(a) 15^{820} modulo 7
(b) 71^{305} modulo 18
(c) $(113 \cdot 55^{137} + 4{,}576^{50} \cdot 27)$ modulo 9

⚙ Solution

(a) Since $15 \equiv 1 \pmod{7}$, we have $15^{820} \equiv 1^{820} \equiv 1 \pmod{7}$.
(b) Since $71 \equiv -1 \pmod{18}$, we have $71^{305} \equiv (-1)^{305} \equiv -1 \equiv 17 \pmod{18}$.
(c) Since $55 \equiv 1 \pmod{9}$ and $27 \equiv 0 \pmod{9}$, we have

$$\left(113 \cdot 55^{137} + 4{,}576^{50} \cdot 27\right) \equiv 113 \cdot 1^{137} + 4{,}576^{50} \cdot 0 \pmod{9}$$

$$\equiv 5 \pmod{9}. \quad \blacklozenge$$

We have saved division for our last modular arithmetic operation because it involves a few subtleties. Recall from elementary school that dividing 30 by 5 is equivalent to finding a value for x such that the multiple $5x$ is equal to 30—that is, $x = 6$. We follow the same procedure when dividing in a least-residue system, keeping in mind the additional restriction imposed by the modular structure. For example, to divide 2 by 7 modulo 8, we solve for x in

$$7x \equiv 2 \pmod{8}$$

by checking each number in the system $\{0, 1, 2, 3, 4, 5, 6, 7\}$. By trial and error, we find that $x = 6$ is the solution. Therefore, we may write

$$\frac{2}{7} \equiv 6 \pmod{8}.$$

However, one very important difference exists between this and ordinary division. Depending on the modulus, some congruences will have *no* solutions or possibly *more than one* solution. Attempting to divide 5 by 4 modulo 8, for instance, we substitute values into

$$4x \equiv 5 \pmod{8}$$

and soon discover that none of the least residues in the system satisfy this equation. So this division simply cannot be done. Furthermore, one may check that the equation

$$2x \equiv 4 \pmod{8}$$

has *two* solutions, 2 and 6, in the least-residue system.

In general, the lack of a unique solution creates additional difficulties when we try to solve any **linear congruence**—i.e., modular equation having the form $ax \equiv b \pmod{m}$. The problem stems from the coefficient of x and the modulus containing like factors. However, if the modulus is a **prime** (a positive integer having only itself and 1 as factors), the situation is greatly simplified.

> **linear congruence**
> The congruence equation $ax \equiv b \pmod{m}$.
>
> **prime** A positive integer having only itself and 1 as factors.

If p is a prime and a is not divisible by p, then $ax \equiv b \pmod{p}$ has a unique solution.

? Example 4

Find the least-residue solutions (if any) to the following divisions.

(a) $\dfrac{6}{7}$ modulo 12

(b) $\dfrac{8}{3}$ modulo 12

(c) $\dfrac{9}{3}$ modulo 12

(d) $\dfrac{10}{4}$ modulo 13

(e) $\dfrac{28}{11}$ modulo 13

⚙ Solution

(a) $7x \equiv 6 \pmod{12}$ has the solution $x = 6$.

(b) $3x \equiv 8 \pmod{12}$ has no solutions.

(c) $3x \equiv 9 \pmod{12}$ has the solutions $x = 3, 7,$ and 11.

(d) $4x \equiv 10 \pmod{13}$ has the solution $x = 9$.

(e) $\dfrac{28}{11} \equiv \dfrac{2}{11}$ because we can replace any number with its least residue. $11x \equiv 2 \pmod{13}$ has the solution $x = 12$. ◆

We can now find solutions to a host of linear congruences. Remember that you may add or subtract like quantities from both sides, and you may multiply both sides by any number not congruent to 0; but division can only be done according to the procedure demonstrated here. We, of course, may also replace any number, including a coefficient, with any other member of its residue class.

? Example 5

Solve for x in the least-residue system.

(a) $2x - 16 \equiv 27 \pmod{5}$ (b) $54x + 29 \equiv 35 \pmod{8}$ (c) $85x + 142 \equiv 71 \pmod{11}$

⚙ Solution

(a) $2x - 16 \equiv 27 \pmod 5$

$2x - 1 \equiv 2$

$2x \equiv 3$

$x \equiv 4$

(b) $54x + 29 \equiv 35 \pmod 8$

$6x + 5 \equiv 3$

$6x \equiv -2 \equiv 6$

$x \equiv 1, 5$

(c) $85x + 142 \equiv 71 \pmod{11}$

$8x + 10 \equiv 5$

$8x \equiv -5$

$x \equiv 9.$ ◆

In regular arithmetic, we call $\frac{2}{3}$ the *multiplicative inverse* of $\frac{3}{2}$ because the product of these two numbers is 1. In general, a is the multiplicative inverse of b (and vice versa) if $ab = 1$. An analogous definition exists in modular arithmetic. If p is prime, then the unique (least-residue) solution to the equation $ax = 1 \pmod{p}$ is called the **multiplicative inverse** of a in the least-residue system modulo p.

> **multiplicative inverse** The number that, when multiplied by a, yields 1 modulo m is the multiplicative inverse of a modulo m.

❓ Example 6

Find the multiplicative inverses of 3, 7, and 10 modulo 11.

⚙ Solution

Because $3 \cdot 4 \equiv 1 \pmod{11}$, 3 and 4 are multiplicative inverses of each other.
Because $7 \cdot 8 \equiv 1 \pmod{11}$, 7 and 8 are multiplicative inverses of each other.
Because $10 \cdot 10 \equiv 1 \pmod{11}$, 10 is its own inverse.   ◆

© corepics/ShutterStock, Inc.

> **cryptology** The study of techniques used for writing and reading secret messages.

Many applications of number theoretic concepts are used today in such diverse areas as computer science, random number generation, and **cryptology**—the study of the techniques involved in the writing and reading of secret messages. Such techniques are commonly used by governments to secure communications between the various government branches that require secrecy. Talented cryptologists have long been essential in the making and breaking of codes used in military operations—in both wartime and peacetime—and, in fact, at times they have played a major role in turning the tide of key military engagements. Recently, interest in cryptology has increased due to ongoing efforts to provide security for electronic business transactions and for messages transmitted on the ever-expanding Internet.

The process of creating or encoding a secret message is known as *encrypting* or *enciphering,* while any procedure used to recapture the original message is called *decrypting* or *deciphering.* The unaltered readable version of any message is known *as* **plaintext**, while the language of any encryption is called **ciphertext**.

plaintext The letters in a message before it is encrypted.

ciphertext The letters in an encrypted message.

One of the inaugural events in the history of cryptology was a military communiqué from Julius Caesar during the Gallic Wars that was encrypted by shifting each letter three positions to the right in the standard alphabet. For instance, the word *ARMY* would be disguised by writing *DUPB* since *D* is three positions to the right of *A, U* is three to the right of *R*, and so forth. Note that we consider *B* to be three places to the right of *Y* by simply wrapping the end of the alphabet back around

A B C D E F G H I J K L M N O P Q R S T U V W X Y Z
D E F G H I J K L M N O P Q R S T U V W X Y Z A B C.

to the beginning. By aligning the plain sequence of letters with the cipher sequence, we get what is called a *substitution alphabet:*

Caesar cipher An encryption technique in which the ciphertext is created from the plaintext according to $C = (P + s)$ **mod** 26.

This encryption technique is called a **Caesar cipher** with shift constant 3. Since the same transformation is applied to each plaintext letter, we can easily create a function using modular arithmetic to implement this procedure. To do this, we first enumerate the standard alphabet, starting with 0 for A, 1 for B, 2 for C, and so forth:

A B C D E F G H I J K L M N O P Q R S T U V W X Y Z
0 1 2 3 4 5 6 7 8 9 10 11 12 13 14 15 16 17 18 19 20 21 22 23 24 25.

We then add 3 to each number *P* corresponding to the letters of the plaintext message and reduce the sum to the least-residue modulo 26. Calling the result *C* (for ciphertext), we write this function as

$$C = (P + 3) \textbf{ mod } 26,$$

where the domain and range are assumed to be the least-residue system mod 26. (The boldface **mod** is used to emphasize that a functional assignment is being made here.) The encrypted message is formed by writing down the letters corresponding to the output *C* values.

? Example 7

A Caeser cipher with shift constant *s* is provided by the following function:

$$C = (P + s) \textbf{ mod } 26.$$

Use a Caesar cipher with shift constant $s = 8$ to encrypt this message: "The east army is trapped."

Solution

We first determine the sequence of numbers corresponding to the following plaintext letters.

T H E	E A S T	A R M Y	I S	T R A P P E D
19 7 4	4 0 18 19	0 17 12 24	8 18	19 17 0 15 15 4 3.

We then apply the function $C = (P + 8) \bmod 26$ to transform the numerical sequence and obtain the encrypted message.

1 15 12	12 8 0 1	8 25 20 6	16 0	1 25 8 23 23 12 11
B P M	M I A B	I Z U G	Q A	B Z I X X M L.

Clearly, if we had been given this encrypted message and wished to decipher it, then we would have instead *subtracted* 8 (or, equivalently, added 18) from each ciphertext value to obtain the plaintext message. ◆

---

generalized Caesar cipher An encryption technique in which the ciphertext is created from the plaintext according to $C = (rP + s)$ mod 26.

A **generalized Caesar cipher** is an encryption algorithm obtained by replacing the numerical equivalents of plaintext P with cyphertext C according to the function

$$C = (rP + s)\bmod 26,$$

where we first multiply each plaintext number by a positive integer multiplier r before adding the shift constant s. To obtain a complete least-residue system modulo 26, it is necessary that r and 26 not share a common divisor. Therefore, we can choose r to be any odd, positive integer less than 26, except for 13. The ordered pair of integers (r, s) is called the **key** to the cipher scheme, because knowledge of the pair allows the recipient of any encrypted message to "unlock" or decipher it. One simply needs to apply the function

key The name of a major scale according to the first note. Also the name of an ordered pair of integers that provides the information needed to decipher an encrypted message.

$$P = r^{-1}(C - s)\bmod 26,$$

where r^{-1} is the multiplicative inverse of r modulo 26. The integer r^{-1} will be unique, providing r is chosen as described earlier. (*Note:* In practice, it is usually easier to add the least residue congruent to $-s$ than to subtract s.)

Example 8

Decipher the following message that is encrypted by a generalized Caesar cipher with key (3, 7). Note that the letters have been collected into groups of 5 each. This is a common encryption practice done to increase the difficulty of breaking the code.

C T O A V   F O O H G   G F S T J   X X U.

⚙ Solution

First, we repeat and enumerate the message:

C T O A V   F O O H G   G F S T J   X X U
2 19 14 0 21 5 14 14 7 6   6 5 18 19 9   23 23 20.

To decipher it, we first find the multiplicative inverse of $r = 3$ by solving $3x = 1$ (mod 26), giving us $x = r^{-1} = 9$. So the deciphering function is

$$P = 9(C - 7)\mathbf{mod}\,26$$
$$= 9(C + 19)\mathbf{mod}\,26,$$

and substituting the earlier sequence into this function gives us this plaintext message:

7 4 11 15 22   8 11 11 0 17   17 8 21 4 18   14 14 13
H E L P W   I L L A R   R I V E S   O O N.

Appropriate spacing of the letters then reveals the message as this one:

H E L P   W I L L   A R R I V E   S O O N. ◆

Interesting and useful properties of integers continue to be discovered today, including a recent solution to one of the most famous unsolved problems in number theory. The great mathematician and philosopher **Pierre de Fermat** (1601–1665) contributed to the folklore of mathematics when he conjectured in 1637 that no positive integers *x, y, z* exist that satisfy the equation

$$x^n + y^n = z^n$$

when *n* is an integer greater than 2. Quickly stated and easily understood, it is frighteningly difficult to prove. Fermat inadvertently created a legend around his conjecture—later to be named *Fermat's Last Theorem*—when he teased the world by writing in his notebook next to his proposition, "I have found for this a truly marvellous demonstration, but the margin is too narrow to contain it."

Fermat never published any such demonstration, and, in fact, the construction of a legitimate proof withstood the painstaking efforts of many of the world's greatest minds for over 350 years. Only in 1995, after 8 years of intensive work—including the correction of a flaw discovered in the original version announced in 1993—did **Andrew J. Wiles** (1953–), a professor of mathematics at Princeton University, shock the world with a bona fide 200-page proof. First exposed to the problem as a 10-year-old child, Wiles was intrigued by the challenge of solving it his entire life. In an interview he said, "Well mathematicians just love a challenge and this problem, this particular problem just looked so simple, it just looked as if it had to have a solution, and of course it's very special because Fermat said he had a solution."

After receiving his doctorate at Cambridge, England, Wiles eventually settled at Princeton and soon after decided to concentrate exclusively on solving Fermat's Last Theorem. He then confined himself primarily to the attic of his house for much of the next 7 years, separating himself from the rest of the mathematical community and telling no one (except his wife) the object of his labors. This was in keeping with a long-standing tradition of mathematicians finding their greatest form of satisfaction in single-minded devotion to solving puzzles. The essence of the joy of discovery can be detected in his thrill of a singular moment in his long odyssey,

"At the beginning of September I was sitting here at this desk when suddenly, totally unexpectedly, I had this incredible revelation. It was the most, the most important moment of my working life. Nothing I ever do again will… I'm sorry."

As is so often the case in scientific research of all kinds, the auxiliary mathematical tools and structures that Wiles had to invent to help solve this deceptively perplexing problem have opened the doors to several other fields of rich mathematics. The British-born Wiles has since been the recipient of numerous honors and awards from prestigious institutions all over the world. Such current discoveries continue today to make number theory one of the most intriguing and beautiful areas of higher mathematics. In conclusion, we consider one final example of a curious property of modular algebra.

FIGURE 9.4.5   Andrew J. Wiles.

Example 9

A common mistake by abusers of high school algebra is to equate $(a + b)^n$ with $a^n + b^n$. They should enjoy learning that if p is a prime, then the coefficient of every term in the expansion of $(a + b)^p$ except a^p and b^p is divisible by p. Therefore, it *is* true that

$$(a+b)^p \equiv a^p + b^p \pmod{p}.$$

Verify this fact by expanding the left side for $p = 3$ and 7.

Solution

Recall Pascal's triangle for the coefficents of biniomial expansions of $(a + b)^n$.

$n = 1$						1		1				
$n = 2$					1		2		1			
$n = 3$				1		3		3		1		
$n = 4$			1		4		6		4		1	
$n = 5$		1		5		10		10		5		1
$n = 6$	1		6		15		20		15		6		1
$n = 7$	1		7		21		35		35		21		7		1

$$(a+b)^3 = a^3 + 3a^2 b + 3ab^2 + b^3$$
$$= a^3 + b^3 \pmod{3} \; because \; 3 \equiv 0 \pmod{3}.$$
$$(a+b)^7 = a^7 + 7a^6 b + 21a^5 b^2 + 35a^4 b^3 + 35a^3 b^4 + 21a^2 b^5 + 7ab^6 + b^7$$
$$= a^7 + b^7 \pmod{7} \; because \; 35 \equiv 21 \equiv 7 \equiv 0 \pmod{7}. \quad \blacklozenge$$

Name _____

Exercise Set **9.4** ◯◯◯←

Perform the indicated operations in the least-residue system modulo 8.

1. $6 + 7$

2. $177 + 892$

3. $27 - 39$

4. $378 - 699$

5. $433 \cdot 8{,}902$

6. $527 \cdot 1{,}193$

7. $\dfrac{6}{2}$

8. $\dfrac{5}{2}$

9. $\dfrac{1}{7}$

10. $\dfrac{48}{76}$

11. 65^{12}

12. 408^{25}

Perform the indicated operations in the least-residue system modulo 11.

13. $10 + 5$

14. $42 - 89$

15. $853 \cdot 931$

16. $6{,}027 \cdot 4{,}491$

17. $\dfrac{1}{2}$

18. $\dfrac{9}{4}$

19. $\dfrac{99}{73}$

20. $\dfrac{5}{8}$

21. $\dfrac{3}{10}$

22. $\dfrac{49}{100}$

23. 12^{800}

24. 55^{55}

Use the Russian peasant method to find the least residue.

25. 3^{41} modulo 7

26. 9^{107} modulo 13

27. 3^{106} modulo 5

28. 15^{302} modulo 17

29. 8^{257} modulo 10

30. 7^{70} modulo 9

31. 5^{49} modulo 6

32. $2^{1{,}040}$ modulo 5

Compute and reduce to the least residue without using the Russian peasant method.

33. $(31^{91} + 24 \cdot 929^{430})$ modulo 8

34. $(29 \cdot 42^{75} + 91 \cdot 88^{219} + 667 \cdot 14^{504})$ modulo 3

35. $(20^{10} + 21^{11} + 22^{12})$ modulo 21

36. $(1{,}001^{2{,}001} + 999^{2{,}001})$ modulo 10

Solve for x in the least-residue system.

37. $3x - 2 \equiv 6 \pmod{7}$

38. $6x + 9 \equiv 5 \pmod{11}$

39. $12x - 31 \equiv 45 \pmod{8}$

40. $51x - 30 \equiv 94 \pmod{12}$

The following enumeration of the alphabet is provided to assist with Exercises 41–47.

A	B	C	D	E	F	G	H	I	J	K	L	M	N	O	P	Q	R	S	T	U	V	W	X	Y	Z
0	1	2	3	4	5	6	7	8	9	10	11	12	13	14	15	16	17	18	19	20	21	22	23	24	25.

41. Use a Caesar cipher with shift constant $s = 8$ to encrypt the following message.

    T O   E R R   I S   H U M A N.

42. Use a generalized Caesar cipher with key $(7, 2)$ to encrypt the following message.

    B E R T   W O N   T H E   E L E C T I O N.

43. A Caesar cipher with shift constant $s = 12$ was used to create the following encrypted message. Find the plaintext message.

    F T Q O A   P Q U E G   Z N D Q M   W M N X Q.

44. Create the function needed to decipher any message encrypted by a generalized Caesar cipher with key $(3, 4)$. Use it to decipher the following message.

    W E L C L   Q U G E C   N C J O U   P Q G.

45. Create the function needed to decipher any message encrypted by a generalized Caesar cipher with key $(5, 2)$. Use it to decipher the following message.

    T L W W C   G F W L C   O F C P R   W R.

46. A generalized Caesar cipher with key $(11, 16)$ was used to create the following encrypted message. Find the plaintext message.

    S Q R P I   S Q R A M   G A G T C   D.

47. A generalized Caesar cipher with key $(15, 20)$ was used to create the following encrypted message. Find the plaintext message.

    U D D Q W   I H C C N   K E D W X   C.

48. Show that $(m - x)^2 \equiv x^2 \pmod{m}$ for any positive integer x and modulus m.

49. Verify the property in the previous exercise with $m = 9$ and $x = 1, 2, 3, 4$.

50. Recall from Example 9 that $(a + b)^p \equiv a^p + b^p \pmod{p}$. Verify this fact by using Pascal's triangle to expand the left side for $p = 2$ and 5.

51. Show that $a^3 \equiv a \pmod{3}$ for every integer a.

52. Show that $a^5 \equiv a \pmod{5}$ for every integer a. [*Hint:* $a^2 + 1 \equiv a^2 - 4 \pmod{5}$.]

53. Using $(a + b)^p \equiv a^p + b^p \pmod{p}$ with p a prime, we get successively

$$2^p = (1+1)^p \equiv 1^p + 1^p = 1 + 1 = 2 \pmod{p},$$

$$3^p = (2+1)^p \equiv 2^p + 1^p \equiv 2 + 1 = 3 \pmod{p},$$

$$4^p = (3+1)^p \equiv 3^p + 1^p \equiv 3 + 1 = 4 \pmod{p},$$

$$5^p = (4+1)^p \equiv 4^p + 1^p \equiv 4 + 1 = 5 \pmod{p}, \text{ and soon.}$$

Use these results to create a general congruence property about a^p if p is a prime.

> I think prime numbers are like life. They are very logical but you could never work out the rules, even if you spent all your time thinking about them.
>
> —Mark Haddon, *The Curious Incident of the Dog in the Night-time*

*use the extra space
to show your work*

Chapter Review Test **1**  ←

1. Give the residue classes denoted by $[2]_3$, $[4]_7$, and $[11]_{10}$. What number occurs in all three of these classes?

2. What is the smallest member of $[4]_7$ that is greater than 261?

3. Given the classes $[0]_7$ and $[0]_{14}$, which is a subset of the other?

4. Perform the following modular arithmetic. List your answers, using least residues.

   (a) $[3]_7 + [12]_7 =$ (b) $[85]_4 + [51]_4 =$
   (c) $[105]_{15} + [538]_{15} =$ (d) $[5,639]_{10} + [8,247]_{10} =$

5. Construct an addition table using the least residues of a 6-modularization. What is the additive inverse of 3? Of 5?

6. Give the note names of the following pitch classes.

   $[2]_{12}$ $[7]_{12}$ $[10]_{12}$ $[25]_{12}$

7. Give the name of each note on this treble clef staff directly below the staff, and give the key position directly above the staff.

8. Find the primary chord of the variation A-D-F shown here, and draw it on the staff.

9. What transformations in the Cartesian plane results from applying each of the following functions?

(a) $(a, b) \rightarrow (-a, -b)$ (b) $(a, b) \rightarrow (a, -b)$ (c) $(a, b) \rightarrow (a + k, b)$

10. Find the images of the following points after applying the given transformations.

(a) $(5, -1)$ after a reflection across the x-axis, followed by a rotation of $180°$.

(b) $(-3, 6)$ after a horizontal translation of 9 units, followed by a reflection across the y-axis.

11. What single transformation has the same net effect on a point (a, b) in the Cartesian plane as the composite of a reflection across the x-axis followed by a rotation of $180°$?

12. The opening melody of *Brian's Song* by Michel Legrand is given by the following sequence of key positions:

$$7\text{-}5\text{-}7\text{-}11\text{-}0\text{-}2\text{-}7\text{-}5\text{-}7\text{-}11\text{-}12.$$

(a) Find the position sequence that results from applying a reflection across a vertical line after the last note.
(b) Find the new position sequence that results from a vertical translation of -4 notes.
(c) Find the new position sequence that results from rotating the original sequence by $180°$.

13. F# is an interval of a perfect fifth above what note? F# is an interval of a perfect fourth above what note?

14. Use the functions $k^\sharp(n) = [7n]_{12}$ and $k^\flat(n) = [5n]_{12}$ to determine the major key corresponding to each of the following signatures.

(a)

(b)

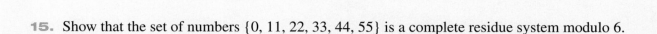

15. Show that the set of numbers {0, 11, 22, 33, 44, 55} is a complete residue system modulo 6.

16. Perform the indicated operations in the least-residue system.

(a) $(67 - 97)$ modulo 7

(b) $(204 \cdot 551)$ modulo 14

(c) $\dfrac{6}{11}$ modulo 13

(d) $4{,}029^{307}$ modulo 5

17. Use the Russian peasant method to find the least residue of 7^{74} modulo 9.

18. Solve for x in the least-residue system $19x + 32 \equiv 90 \pmod{7}$.

19. A generalized Caesar cipher with key (7, 5) was used to create the following encrypted message. Find the plaintext message.

C F G G R   A F R R Z   P F U H A   Z S H.

Paper airplane

Chapter Objectives

check off when you've
completed an objective

☐ Use linear interpolation.
☐ Compute the absolute error and relative error of a prediction.
☐ Form the Wald function from a data set.
☐ Identify a modeling type by examination of a scatter plot.
☐ Transform the data in order to compute the parameters of a modeling function.
☐ Use exponential averaging to find an exponential modeling function.

© SmallAtomWorks/ShutterStock, Inc.

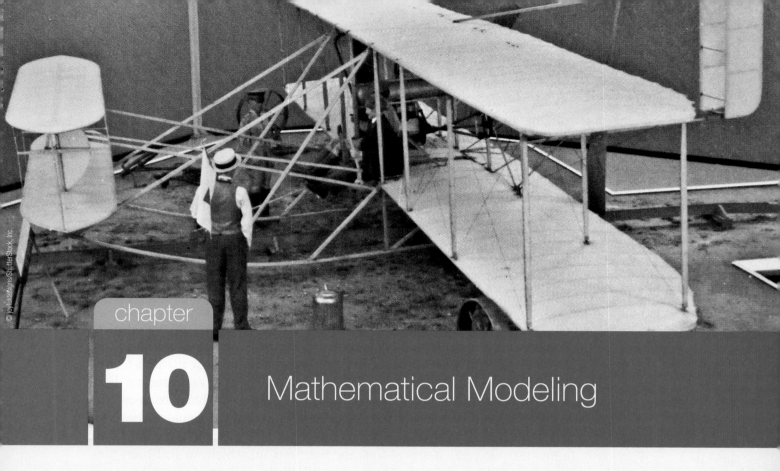

10 Mathematical Modeling

10.1 Linear Interpolation

10.2 Linear Prediction with the Wald Function

10.3 Nonlinear Modeling Functions

At school, I was cheated by my physics teacher... A bland statement of the Law of Universal Gravitation without any indication of how the trick was done.

—George Pólya

One more staple in the nose ought to make it fly better.

—Scott Johnson

Oh yeah, and bend the tail up too.

—Nicholas Johnson

To create a model with our own hands and to improve upon it is something known to many of us since childhood. The above statements were made by my sons as youngsters when observing the flight of one of their paper airplanes. A mathematician could not help noticing the essential elements of the *modeling process* present in this situation:

- Observation of a physical phenomenon (a bird or airplane)
- Creation of a model to simulate the behavior of the phenomenon (a paper airplane)
- Modification of the model in response to testing (adding a staple or fold)

mathematical modeling Simulating a real-world phenomenon through the creation of a mathematical structure.

model Function, system of equations, computer program, or other vehicle capable of processing quantitative information to make predictions.

Instead of paper, the type of model that we will be studying in this chapter is constructed using mathematics. **Mathematical modeling** is a process in which one attempts to simulate a physical phenomenon in the real world through the creation of some type of mathematical structure. This structure, or **model**, as it is called, approximately mimics the phenomenon by relating variable quantities; and the model can take the form of a function, system of equations, computer program, or other vehicle capable of processing quantitative information. Although modeling can require quite complicated analysis beyond the scope of this text, we can still explore some of the basic concepts involved by restricting our attention to creating a function that approximately fits a set of ordered pairs of experimental data. Note that this means we will be limiting ourselves to two-variable phenomena.

We have already seen examples of the ability of a function to compute values in its range, given a specific value in the domain *if* the function is expressible in a general form. Previously in this text we created expressions for functions for which we had been presented with convenient tables of values. These tables all displayed precise patterns that we trained ourselves to recognize, but often no such observable pattern is present. In that case a formula expression is not possible that fits the data exactly, and so then we enter the realm of mathematical modeling. A general procedure has similarities to the earlier discussion of folding a paper airplane.

1. Observe a real-world phenomenon involving two variables.
2. Record and plot paired data.
3. Create a function to fit the data.
4. Use the function to predict behavior.
5. Compare the predictions to further observations.

The last step in the procedure directs the process back to step 1 and thus closes the loop in an iterative process illustrated schematically in **Figure 10.0.1**.

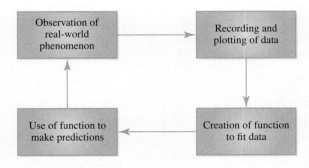

FIGURE 10.0.1   The modeling process as a closed loop.

Of course, the whole notion of making predictions of a natural event has a long and glorious history stemming from humankind's early and persistent desire to understand the universe and parallels the development of the *scientific method*, discussed elsewhere. Generally speaking, making a prediction in our present context means computing a value for a variable that depends on the value of an associated independent variable. Our ultimate goal is to construct a function that accomplishes this globally, that is, for a fairly large domain. To accomplish this, we will be studying the art of *curve fitting*. This entails the drawing of a curve approximating the graph of a function that we *can* formulate and, in some sense, comes reasonably close to the set of ordered independent-dependent pairs that have been gathered experimentally. The first attempt usually involves situating a straight line as close to the given points as possible. (See **Figure 10.0.2**.) We call such a graph a **linear fit**.

> **linear fit** A straight line that comes as close as possible to the points in a scatter plot.

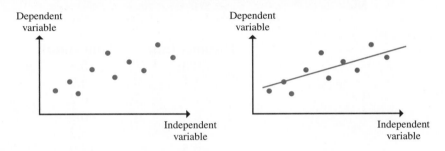

FIGURE 10.0.2 Fitting a straight line to a set of experimentally gathered ordered pairs.

To gain insight into the global situation, first we examine one technique for making a prediction locally, or in a small subset of the domain using only two of the data points at a time. The method is called *linear interpolation*.

10.1 Linear Interpolation

Webster defines *interpolate* as "to insert between or among others." This is a fair description of the process we are about to describe. Given the coordinates of two points in the Cartesian plane, we will locate a third point on the straight-line segment connecting them. Because the equation of that line gives a relationship between an independent variable (usually *x*) and a dependent variable (usually *y*), we typically wish to find the value of *y*, given a value for *x*. The use of a line explains why we call the process *linear* interpolation, and our method simply involves determining the equation of the desired line.

? Example 1

In Suffix County, a record was kept of the response times in minutes (min) of the ambulance at Fire Station #12 as a function of the distance in miles (mi) from the station. The values are shown in the table below.

Distance (mi)	Time (min)
0.8	5.3
1.1	5.9
1.5	6.2
2.2	6.8
3.4	8.2
6.5	12.9
8.8	14.7

We wish to estimate the response time for a call made from a distance of 5.0 mi from the fire station. Since 5.0 is between 3.4 and 6.5, we guess the time to be between 8.2 and 12.9 min. But how can we make this guess more precise? If we were to plot all these ordered pairs on a Cartesian coordinate system, they would not all lie on the same straight line; but between any two neighboring points we can assume the relationship is linear and can find the equation of the line connecting just those two points. To do so, we first determine the slope m through the ordered pairs (3.4, 8.2) and (6.5, 12.9):

$$m = \frac{12.9 - 8.2}{6.5 - 3.4} = 1.5 \text{ min/mi}.$$

We then use one of the points to get the point-slope form of the equation of the line through those pairs:

$$y - 8.2 = 1.5(x - 3.4).$$

Since we want a linear function, to make our estimate we let $y = L(x)$, and so

$$L(x) = 1.5(x - 3.4) + 8.2.$$

Then our estimate for the response time, given a distance of $x = 5.0$ mi, is

$$L(5.0) = 1.5(1.6) + 8.2 = 10.6 \text{ min}. \quad \blacklozenge$$

linear interpolation Prediction of a functional value based on the equation of the straight line connecting two ordered pairs.

This technique is known as **linear interpolation**. We compute the desired y-coordinate by formulating the linear function whose graph contains this straight-line segment.

Linear Interpolation Method

Given two points (x_0, y_0) and (x_1, y_1):

1. Find the slope m of the line between the points:

$$m = \frac{y_1 - y_0}{x_1 - x_0}.$$

2. Use either of the two points to form the linear interpolation function L:

$$L(x) = m(x - x_0) + y_0.$$

3. Substitute the desired input for x.

? Example 2

Kevin wants to go fishing at 6:15 A.M. on Saturday. His boat, *The Lost Debit*, is anchored in Morro Bay and has a keel that needs the level of the bay to be at least 1.5 feet (ft) above average in order to set sail. Consulting a local tide guide, he reads that low tide that day will be 1.3 ft below average, occurring at 4:02 A.M., followed later by a high tide at 5.8 ft above average and occurring at 10:26 A.M. Will Kevin be able to cast off at the desired time?

⚙ Solution

The level of all ocean bays everywhere is a function of the time of day, continuously alternating between a minimum at low tide and a maximum at high tide. Although the graph of that function would be nonlinear when the domain consisted of an entire day, it can be *approximated as linear* when we restrict ourselves to a short time span between consecutive extreme tides. It seems reasonable that since 6:15 A.M. is between 4:02 A.M. and 10:26 A.M., the tide level for 6:15 A.M. should be between −1.3 ft (with the negative sign indicating below average) and 5.8 ft. We can increase the precision of our guess by examining a straight line drawn through the graph of the two given ordered pairs. It will make life easier if we first convert the times to minutes past midnight (4:02 A.M. is 242 min and 10:26 A.M. is 626 min) and then graph the points (242, −1.3) and (626, 5.8), as shown in **Figure 10.1.1**. We can now estimate a prediction for the tide

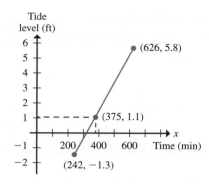

FIGURE 10.1.1   Using linear interpolation to approximate a functional value.

stakes The two data points used in linear interpolation whose first coordinates bracket the value used as the input for the prediction.

level at 375 min (6:15 A.M.) by using the y-coordinate of the point on the line having x-coordinate 375.

We will call the two chosen points (x_0, y_0) and (x_1, y_1) **stakes** since this method is rather like tying a rope between two stakes and finding a knot in the middle. For the current situation, we use the points $(242, -1.3)$ and $(626, 5.8)$ as our stakes.

The slope for the interpolation function is

$$m = \frac{5.8 - (-1.3)}{626 - 242} = \frac{7.1}{384} = 0.018.$$

So we get

$$L(x) = 0.018(x - 242) - 1.3$$
$$L(375) = 0.018(375 - 242) - 1.3$$
$$= 0.018(133) - 1.3$$
$$= 2.4 - 1.3$$
$$= 1.1 \text{ ft.}$$

In essence we have assumed a constant rate (0.018 ft/min) at which the tide level is rising, multiplied by the number of minutes (133) past the time of low tide, and added the result to -1.3. Since 1.1 ft is less than the desired level of 1.5 ft, Kevin will have to wait until a short time after 6:15 A.M. to set sail.   ◆

Significant Digits

Note in the above example that we have rounded the slope to two significant digits (cf. Appendix III). We will be forming a variety of functions from data tables throughout this chapter. We adopt the practice of using the same number of significant digits for the slope of any such function as that used for the *dependent* variable in the table. The same is true for all predictions or estimates to be computed with these functions.

? Example 3

Sharin is growing a very light-sensitive plant for a gardening experiment in Fresno, California, and needs to know the light duration times for several days in July not listed in the table below. (See **Figure 10.1.2**.) In the northern hemisphere the amount of daylight per day is a decreasing function of the date from summer solstice (June 22) to winter solstice (December 22) and an increasing function from winter to summer solstice. Although an astronomical function already exists to compute the amount of daylight for any given date, it is not needed if Sharin can be content with some fairly accurate approximations. Estimate the amount of light for July 7.

Date	Sunrise (A.M.)	Sunset (P.M.)	Light Duration (min)
July 1	5:43	8:22	879
July 9	5:47	8:20	873
July 31	6:03	8:06	843

FIGURE 10.1.2   Light duration in Fresno, California, in July 2008.

⚙ Solution

First, since there can only be one specific amount of daylight for any given day, we are assured that light duration is indeed a function of the date. There would be no curve for the graph of this particular function because the domain contains just whole numbers and is not an interval. The graph would consist of discrete points. However, by assuming the full set of data points over the interval [1, 9] would fall reasonably close to a line drawn through the stakes (1, 879) and (9, 873), we make an estimate for the amount of light on day 7 by using the y-coordinate of the point on the line having x-coordinate 7. See **Figure 10.1.3**.

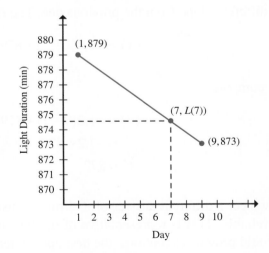

FIGURE 10.1.3   Using linear interpolation to approximate light duration.

The slope m of the line through the two stakes (1, 879) and (9, 873) is

$$m = \frac{873 - 879}{9 - 1} = \frac{-6}{8} = -0.75.$$

This slope tells us that between July 1 and July 9, the amount of daily light decreases by about 0.75 min/day. Using the point (1, 879), the L function is

$$L(x) = -0.75(x-1) + 879.$$

Then we compute

$$L(7) = -0.75(7-1) + 879$$
$$= -4.5 + 879$$
$$= 874.5$$
$$\approx 875.$$

So we conclude that there is about 875 min of daylight on July 7. In all ensuing computations, the "=" sign will be used consistently (instead of the "$\approx$" sign) even when the value has been rounded.   ◆

Clearly, the closer the actual function is to being linear over the particular interval, the more accurate the approximation. If we choose two stakes farther away from the point we wish to estimate, we increase our risk of the function deviating significantly from linearity and so the risk of the associated error increases as well. For instance, instead of (1, 879) and (9, 873), suppose we choose (1, 879) and (31, 843) from Figure 10.1.2 to use as our stakes. In this case,

$$m = \frac{843 - 879}{31 - 1} = \frac{-36}{30} = -1.2,$$

which is a different slope from the previous one. The L function this time is

$$L(x) = -1.2(x-1) + 879.$$

And so we compute

$$L(7) = -1.2(7-1) + 879$$
$$= -7.2 + 879$$
$$= 872.$$

Therefore we get a different answer of 872 min by using different stakes, and it is probably not as reliable. This is representative of the typical situation. Although certain exotic functions could provide exceptions, the best approximation to a functional value is usually achieved by using the closest available stakes to bracket the point in question.

? ## Example 4

Many factors influence the level of consumer spending, but statistical data suggest that the most important of these factors is, quite naturally, income. The following table gives (hypothetical) data in dollars for per capita household spending and disposable (after-tax) income of a particular country for several years. An economist wishes to estimate household spending when disposable income is $47,500, using linear interpolation. What is his estimate?

Disposable Income	Household Spending
45,900	45,620
46,500	46,150
47,100	46,700
47,800	47,450

Solution

First, we see that the stakes whose x-values bracket 47,500 the closest are (47,100, 46,700) and (47,800, 47,450). Thus, the slope for our interpolation function is

$$m = \frac{47{,}450 - 46{,}700}{47{,}800 - 47{,}100}$$

$$= \frac{750}{700}$$

$$= 1.071.$$

Note how we use four significant digits for the slope to match the number used in the data. The required function is

$$L(x) = 1.071(x - 47{,}100) + 46{,}700,$$

giving us

$$L(47{,}500) = 1.071(400) + 46{,}700$$

$$= 47{,}130. \quad \blacklozenge$$

Linear interpolation helps us to estimate values for functions that we know exist but for which we do not have a formula. What about those functions for which we *do* know an expression, but that are still difficult to compute (without electronic assistance)? In a world without calculators, how would you go about trying to determine $\sqrt{41}$, for instance? As you may have guessed, the assumption that our function is linear over a short interval again proves to be extremely useful.

Suppose f is our known function and we wish to determine $f(x)$ for $x = a$. Now f may be very complicated or difficult to manipulate. But *if* we already know $f(x_0)$ and $f(x_1)$ for two numbers x_0 and x_1 such that $x_0 < a < x_1$, we can proceed just as before. We compute the slope

$$m = \frac{f(x_1) - f(x_0)}{x_1 - x_0}$$

of the line through the two stakes $(x_0, f(x_0))$ and $(x_1, f(x_1))$ and form the linear interpolation function

$$L(x) = m(x - x_0) + f(x_0).$$

absolute error The absolute value of the difference between a predicted value and the actual value.

Then $L(a)$ is a linear approximation to $f(a)$. Graphically, the situation is displayed in **Figure 10.1.4**. The **absolute error** of the approximation is defined to be $|f(a) - L(a)|$ and is equal to the length of the vertical shaded line segment.

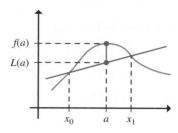

FIGURE 10.1.4   The error between $f(a)$ and $L(a)$.

? Example 5

Find linear approximations for $\sqrt{32}$ and using $\sqrt{77}$, using 4 significant digits.

⚙ Solution

Our function here is $f(x) = \sqrt{x}$. The *closest* perfect squares bracketing 32 are 25 and 36. These are numbers that have exact principal square roots, namely, $f(25) = 5$ and $f(36) = 6$. So the slope we want is

$$m = \frac{6-5}{36-25} = 0.0909$$

giving us

$$L(x) = 0.0909(x - 25) + 5$$

and an estimate of $\sqrt{32} \approx L(32) = 0.0909(32 - 25) + 5 = 5.636$. Using a calculator, $\sqrt{32} \approx 5.657$, giving an absolute error of 0.021, which only misses the actual value by 0.4%—not a bad estimate!

In **Figure 10.1.5**, we see a graphical magnification of the situation. The graph of $L(x)$ is the straight line through (25, 5) and (36, 6). Since that line has slope 0.0909, the y-coordinate of the point on the line with x-coordinate 32 is found by adding $0.0909(32 - 25) = 0.636$ to 5.

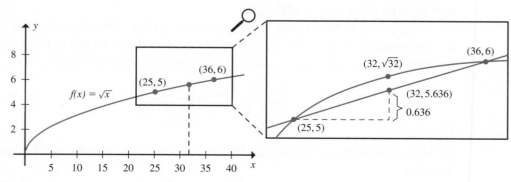

FIGURE 10.1.5   Linear approximation to $\sqrt{32}$ by computing $L(32)$ using (25, 5) and (36, 6) as stakes.

Similarly, we estimate $\sqrt{77}$ by using the stakes $(64, 8)$ and $(81, 9)$. The slope of the line through these two points is

$$m = \frac{9-8}{81-64} = 0.05882$$

giving us

$$L(x) = 0.05882(x - 64) + 8$$

and an estimate of $\sqrt{77} \approx L(77) = 0.05882(77 - 64) + 8 = 8.765$. Our calculator yields $\sqrt{77} \approx 8.775$, giving an absolute error of 0.010, missing the actual value by only 0.1%. ◆

Name _____

Exercise Set **10.1** ◯◯◯←

1. A vice-president for *Spritzer* soft drinks examined the market research to find the following correlation between money spent on TV advertising and the number of people who viewed the ads. Estimate the number of viewers who would be reached by an advertising expenditure of $250,000. Make an estimate of viewers reached by spending $300,000.

Cost ($1000)	Viewers (millions)
150	2.8
220	3.3
370	4.5
500	6.4

2. Use the above table to estimate the number of viewers who would be reached by an advertising expenditure of $400,000.

Use this table of tides for Morro Bay, California, to estimate the tide levels (to the nearest tenth) for the times and dates given in Exercises 3–8.

	Low Tide				High Tide			
Day	**A.M.**	**Ht.**	**P.M.**	**Ht.**	**A.M.**	**Ht.**	**P.M.**	**Ht.**
10	1:59	0.3	1:01	1.5	8:00	3.2	7:36	5.6
11	2:50	−0.2	1:49	1.7	9:06	3.3	8:18	5.8
12	3:35	−0.6	2:34	1.9	10:02	3.4	8:56	5.9
13	4:15	−0.8	3:16	2.0	10:47	3.4	9:35	5.9
14	4:54	−0.9	3:53	2.1	11:30	3.5	10:10	5.9

3. 5:00 A.M. on June 10

4. 5:00 P.M. on June 10

5. 7:15 A.M. on June 11

6. 3:30 P.M. on June 12

7. 12:02 A.M. on June 14

8. 8:42 A.M. on June 14

9. In August 1989, Voyager 2 became the first spacecraft from Earth to fly past the planet Neptune. Among many surprises found in the collected data was the fact that Neptunian wind speeds in kilometers per hour (km/h) in the outer atmosphere turned out to be the fastest in the solar system, although they vary according to latitude.

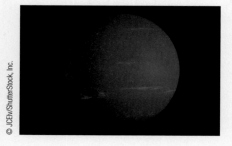

Latitude	Wind Speed (km/h)
−40°	360
−30°	700
−20°	1,420
−10°	1,880
0°	1,600
10°	1,500
20°	520

Estimate the wind speed at a latitude of −15°, using stakes with latitudes of −20° and −10°. (Round off to the nearest 10 km/h.) Form another estimate for the wind speed at the same latitude, but using stakes at −20° and 10° this time. Repeat the process one more time, staking at −40° and 20°. What accounts for the differences in your answers? (It might help to plot the points.) Which answer would you trust the most?

Use the following table to estimate the light duration for the days given in Exercises 10–13. These data are typical of any location at a latitude of about 36°45′.

Date	Light Duration (min)
August 1	841
August 5	834
August 7	830
August 20	804
August 27	789
August 31	780

10. August 6, using stakes obtained from August 1 and August 20

11. August 6, using stakes obtained from August 5 and August 7

12. August 30, using stakes obtained from August 1 and August 31

13. August 30, using stakes obtained from August 27 and August 31

Over a relatively short time span, the supply of a readily made commodity typically goes up as its price increases but the demand goes down. Suppose the supply and demand of bags of peanuts in a particular region fluctuated with price over several months according to the following table.

Price ($)	Supply (1000s)	Demand (1000s)
1.75	850	1,150
1.83	912	1,105
1.90	960	1,039
1.94	991	1,018
2.05	1,015	1,002
2.21	1,027	989

14. Estimate the supply and demand of these peanuts for a price of $1.80.

15. Estimate the supply and demand of these peanuts for a price of $2.10.

16. Use the last two rows to predict the supply and demand for a price of $2.25.

17. The price at which the supply is equal to the demand is known as the *equilibrium price*. Between what two prices is the equilibrium price for these peanuts?

The following chart gives some of the values for air pressure in atmospheres (atm) in our atmosphere as a function of altitude. Estimate the pressure at the given altitudes by using the two closest stakes.

Altitude (1,000 ft)	Pressure (atm)
0	1.00
10	0.69
20	0.46
30	0.30
40	0.19
50	0.11

18. 24,000 ft   **19.** 35,000 ft   **20.** 48,000 ft

21. Matthew is a 7-year-old who just joined the city swim team. At his first meet he recorded 32.46 seconds (s) in the 25-yard (25-yd) freestyle. Hard work and dedication helped him lower that time to 30.37 s at the second meet, and by the third meet he had improved to 28.25 s. If these reductions continue in a linear fashion, use Matthew's last two times to predict his performance in the next swim meet.

22. The debt-to-equity ratios of the U.S. auto company General Motors are given below for six dates during 2010–2012.

Date	Debt/Equity Ratio
January 1, 2011	0.3119
March 31, 2011	0.3194
June 30, 2011	0.2860
September 30, 2011	0.2557
December 31, 2011	0.3548
March 31, 2012	0.3544

(a) Estimate the debt-to-equity ratio for General Motors on April 30, 2011, using the two closest stakes. (*Hint:* Let x represent the number of days in the year, starting with January 1 corresponding to $x = 1$.)

(b) Estimate the debt-to-equity ratio for General Motors on July 31, 2011, using the two closest stakes. (*Hint:* Let x represent the number of days in the year, starting with January 1 corresponding to $x = 1$.)

23. Use January 1, 2011, and March 31, 2012, to make a ratio estimate for July 31, 2011. Compare this to your answer for the previous exercise. Why are they different?

Use the function $f(x) = \sqrt[3]{x}$ to estimate the following numbers to 4 significant digits by the method of linear interpolation with the given ordered pairs as stakes. Also, compute the absolute error by computing the actual value with a calculator.

24. $\sqrt[3]{100}$; (64, 3) and (125, 5)   **25.** $\sqrt[3]{15}$; (8, 2) and (27, 3)

26. $\sqrt[3]{22}$; (8, 2) and (27, 3)

27. $\sqrt[3]{50}$; (27, 3) and (64, 4)

28. Estimate $3^{5.7}$ to 4 significant digits by using linear interpolation with the function $f(x) = 3^x$ between $x = 5$ and $x = 6$. Compute the absolute error by comparing your answer with the calculator value.

29. Estimate $5^{2.4}$ to 4 significant digits by using linear interpolation with the function $f(x) = 5^x$ between $x = 2$ and $x = 3$. Compute the absolute error by comparing your answer with the calculator value.

30. Give a linear approximation of $100^{1.75}$ to 4 significant digits. Compute the absolute error by comparing your answer with the calculator value. Why is the error so large?

31. Estimate $\sqrt[4]{30} + \sqrt{30}$ to 4 significant digits by using linear interpolation with the function $f(x) = \sqrt[4]{x} + \sqrt{x}$ between $f(16) = 6$ and $f(81) = 12$. Compute the absolute error by comparing your answer with the calculator value.

32. The base-10 logarithm of a number x is defined as the exponent a for which $x = 10^a$. We write this as

$$\log_{10} x = a.$$

For example, $\log_{10} 100 = 2$ because $100 = 10^2$ and $\log_{10} 1{,}000 = 3$ because $1{,}000 = 10^3$. Therefore, the points (100, 2) and (1,000, 3) would be on the graph of the function $f(x) = \log_{10} x$. Use these two points as stakes to estimate $\log_{10} 500$. Your calculator contains a key for the logarithm function. Compute the absolute error by comparing your answer with the calculator value.

33. The base-10 logarithm of a number x is defined in the previous exercise. Use the fact that $\log_{10} 10 = 1$ and $\log_{10} 100 = 2$ to estimate $\log_{10} 72$. Compute the absolute error by comparing your answer with the calculator value.

10.2 Linear Prediction with the Wald Function

In the previous section we learned how to make a prediction for the dependent variable of a two-variable phenomenon, upon being given a value for the independent variable. This prediction was made by creating a linear function using the coordinates of only two "nearby" ordered pairs. Therefore the prediction is based on very little information about the behavior of the phenomenon outside this small neighborhood. We now wish to create a linear function that is based on a larger table of ordered pairs collected from extended observation of the phenomenon and therefore can be used more globally for estimating outcomes. Not only will such a function be more representative of the broader nature of the phenomenon in question, but also it will provide us with a single function for making multiple predictions.

    The essence of the construction process of this function is just the idea of fitting a straight line to the graph of a collection of experimentally gathered data points, called the **scatter plot**. A formal definition of a best-fitting straight line requires that the sum of the squares of the vertical distances of the line from the data points be less than the sum obtained using any other line. Intuitively we see that the sum for the line shown in **Figure 10.2.1**(d) is less than that for the lines in (b) and (c).

> **scatter plot** The graph of a set of data points obtained from an experiment or observation of a phenomenon.

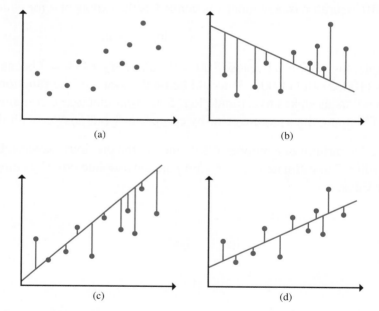

(a)   (b)

(c)   (d)

FIGURE 10.2.1   Fitting a line to data points. The scatter plot is shown in (a). Two attempts to fit a line to the points are shown in (b) and (c). The sum of the vertical distances from the points to the line is a minimum in (d).

Although there exist formulas to determine the slope and y-intercept of *the* best-fitting line, they are rather tedious and cumbersome to use. Wishing to keep our approach light and intuitive, we examine an approach that gives us a linear function that makes reasonably accurate predictions. We can thank the statistician **Abraham Wald** for providing us with this simpler approach.

    Suppose we have observed a two-variable phenomenon long enough to have recorded an even number of ordered pairs

$$(x_1, y_1), (x_2, y_2), \ldots, (x_{2n}, y_{2n}),$$

where n is some positive integer and the x-coordinates are in ascending order $x_1 \leq x_2 \leq \cdots \leq x_{2n}$. (If we have an odd number of data points, we disregard the one in the middle.) Our method uses the reasonable assumption that the graph of the linear function we wish to construct should contain $(\bar{x}, \bar{y})$ where $\bar{x}$ is the mean of the x-coordinates and $\bar{y}$ is the mean of the y-coordinates. (The *mean* of a set of values is the sum of the values divided by the number of values in the set.) Since we need two points to determine a line, we compute two such average ordered pairs (x', y') and (x'', y'')—done for each half of the collection of data points—and find the function whose graph contains these two points.

Wald Method for Linear Prediction

1. Find the average ordered pair $P' = (x', y')$ of the first half of the data points:

$$x' = \frac{x_1 + x_2 + \cdots + x_n}{n} \quad \text{and} \quad y' = \frac{y_1 + y_2 + \cdots + y_n}{n}.$$

2. Find the average ordered pair $P'' = (x'', y'')$ of the second half of the data points:

$$x'' = \frac{x_{n+1} + x_{n+2} + \cdots + x_{2n}}{n} \quad \text{and} \quad y'' = \frac{y_{n+1} + y_{n+2} + \cdots + y_{2n}}{n}.$$

3. Determine the slope m of the line through P' and P'':

$$m = \frac{y'' - y'}{x'' - x'}.$$

4. Use either of the two points to form the **Wald function**:

$$w(x) = m(x - x') + y'.$$

Wald function
The linear function created by using two average ordered pairs of points from a set of ordered pairs observed from some experiment or phenomenon.

Wald line The graph of the Wald function.

It is important to realize that the graph of any Wald function—called the **Wald line**—is *not* a best-fitting line and that the predictions it allows are only estimates. One of the main advantages of the Wald function is the simplicity of its formulation. Therefore it affords a "quick and dirty" method for the purpose of making predictions.

? Example 1

Darron wants to be able to predict his monthly natural gas bill by observing the high temperature on the first day of the month. His records for last year are given below. Find the Wald function, and use it to predict Darron's gas bill for a particular month if the first-day high temperature was 20°F.

Month	First Day High Temperature (°F)	Gas Bill ($)
Jan.	12	258
Feb.	15	266
Mar.	27	191
Apr.	51	133
May	67	107
June	78	81
July	85	44
Aug.	92	43
Sept.	88	52
Oct.	61	120
Nov.	49	142
Dec.	39	188

© K. Geijer/ShutterStock, Inc.

⚙ Solution

Note that the first-day high temperature will be the independent variable of our function. We order the data points with respect to the independent (first) coordinate and split them into two groups. The first group is

$$(12, 258), (15, 266), (27, 191), (39, 188), (49, 142), (51, 133).$$

The second group is

$$(61, 120), (67, 107), (78, 81), (85, 44), (88, 52), (92, 43).$$

We now find the average ordered pair of each group, using the same number (3) of significant digits for each coordinate as that used for the given dependent variable values. We will continue to use this convention for all problems.

$$x' = \frac{12 + 15 + 27 + 39 + 49 + 51}{6} = 32.2 \quad y' = \frac{258 + 266 + 191 + 188 + 142 + 133}{6} = 196$$

$$x'' = \frac{61 + 67 + 78 + 85 + 88 + 92}{6} = 78.5 \quad y'' = \frac{120 + 107 + 81 + 44 + 52 + 43}{6} = 74.5$$

The Wald line contains the points $P' = (32.2, 196)$ and $P'' = (78.5, 74.5)$, and so the required slope (also using 3 significant digits) is

$$m = \frac{74.5 - 196}{78.5 - 32.2} = \frac{-121.5}{46.3} = 2.62.$$

Using the first of the above ordered pairs (either one is fine) to determine the Wald function, we obtain

$$w(x) = -2.62(x - 32.2) + 196.$$

Or simplifying, we get

$$w(x) = -2.62x + 280.$$

Note that we use 3 significant digits for both parameters in the final Wald function. If the temperature on the first day of a particular month reaches a high of 20°F, this function gives Darron a predicted gas bill for that month of

$$w(20) = -2.62(20) + 280$$
$$= \$228$$

where we continue to round to 3 significant digits. The scatter plot and the Wald line are displayed in **Figure 10.2.2**. Although it is not guaranteed to be the best-fitting line, we see that in this case it comes fairly close to the data. ◆

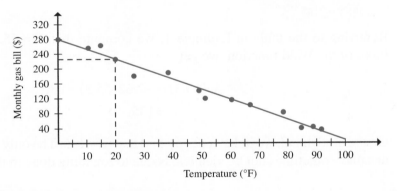

FIGURE 10.2.2   The data of Example 1 and the graph of the Wald function.

When we make predictions, there is always an associated error. The absolute error of a prediction is defined to be the absolute value of the difference between the predicted value and the actual observed value.

$$\text{Absolute error} = |\text{Wald value} - \text{observed value}|$$

relative error
The result of dividing an absolute error by the observed value or actual value of a function and converting to a percentage.

The **relative error** of a prediction is then defined to be the quotient of the absolute error divided by the observed value, converted to a percentage.

$$\text{Relative error} = \frac{\text{absolute error}}{\text{observed value}} \cdot 100\%$$

We will round all relative errors to the nearest tenth of a percent.

? Example 2

At the end of a month that started with a high temperature of 20°F, Darren paid a gas bill of $241. What are the absolute and relative errors of the prediction of $228 made in Example 1?

⚙ Solution

$$\text{Absolute error} = |\$228 - \$241| = \$13,$$

$$\text{Relative error} = \frac{13}{241} \cdot 100\% = 5.4\%. \quad \blacklozenge$$

As mentioned above, given a set of data points $(x_1, y_1), (x_2, y_2), \ldots, (x_{2n}, y_{2n})$ with first coordinate mean $\bar{x}$ and second coordinate mean $\bar{y}$, one should expect any good predictor function to produce a value near $\bar{y}$, given an input value of $\bar{x}$. Indeed, for the Wald function w one can prove that it is always the case that $w(\bar{x}) = \bar{y}$.

? Example 3

Referring to the table in Example 1, we compute that $\bar{x} = 55.3$. Using this for the input of the Wald function, we get

$$w(55.3) = -2.62(55.3) + 280$$
$$= 135.$$

On the other hand, we independently find that $\bar{y} = 135$. The only reason these two values may sometimes not be identical is due to rounding done in the computations. ♦

? Example 4

The following table gives the voter turnout (as a percentage of voting age population) for the U.S. House of Representatives elections in 8 nonpresidential years along with the percentage of seats won by Democratic candidates. Form the appropriate Wald function to predict the percentage of Democratic seats won in the next election if there is a voter turnout of 40.0%.

Voter Turnout (%)	Seats Won by Democratic Candidates (%)
41.1	54.0
41.7	53.3
43.0	64.9
45.4	59.4
45.4	57.0

Voter Turnout (%)	Seats Won by Democratic Candidates (%)
43.5	58.6
36.1	66.9
35.1	63.7

⚙ Solution

The percentage of seats won by Democrats is to be a Wald function of the independent variable of voter turnout. We order the data points according to the first coordinate and split them into two groups:

$$(35.1,\ 63.7),\ (36.1,\ 66.9),\ (41.1,\ 54.0),\ (41.7,\ 53.3)$$

and

$$(43.0,\ 64.9),\ (43.5,\ 58.6),\ (45.4,\ 59.4),\ (45.4, 57.0).$$

The average ordered pairs of these two groups are (38.5, 59.5) and (44.3, 60.0). The slope through these two points is

$$m = \frac{60.0 - 59.5}{44.3 - 38.5} = \frac{0.5}{5.8} = 0.0862.$$

The Wald function is given by

$$w(x) = 0.0862(x - 38.5) + 59.5$$

or

$$w(x) = 0.0862x + 56.2.$$

Using this function to predict the percentage of House seats won by Democrats if the voter turnout is 40.0%, we get

$$w(40.0) = 59.6\%. \quad ♦$$

This example raises several important issues associated with prediction and estimation:

1. In Example 4 the slope of 0.0862 is close enough to 0 to render the knowledge of voter turnout practically useless when making a prediction. We note that it would be equally valid in this case to predict the percentage of seats won by Democrats to be simply the average of those eight given values, namely, 59.7%. In more-advanced courses on this subject, a number known as the *correlation coefficient* is computed and tested to provide a measure of the strength of the correlation between the two variables. Depending on the units being used for the independent and dependent variables, the closer the slope of the Wald line is to 0, the weaker is the correlation between the identified variables.

2. More data points provide you with a more reliable Wald function. It is possible, for instance, that the results of 25 elections would reveal a significantly different Wald slope from that determined from the eight pairs given in this example. However, even a great many data points do not *guarantee* the determination of completely dependable predictions regardless of the technique being used. Any table of paired values is a sample from a much larger collection, and therefore it is always possible that the predicting function produced from that sample is, by random chance, quite different from a more realistic model. That possibility decreases as the number of data points increases.

3. This technique does nothing to identify *causality* between variables. A large slope for our function does not imply that a change in the independent variable causes a change in the functional value. It means only that a *trend* exists that we may utilize for the sole purpose of making predictions, as in the following example.

? Example 5

Many species of plants and animals show a regular variation in certain characteristics that accompany changes in geographic location. The following table displays one study of the heights in centimeters (cm) of the plant *Achillea lanulosa* as measured at several sites throughout the Sierra Nevada mountain range. Determine the Wald function that fits these data.

Altitude (1,000 ft)	Height (cm)
4.3	75
6.0	50
6.8	49
7.5	32
8.0	21
10.2	15
9.2	20
7.8	25
6.2	43

⚙ Solution

Clearly there are many environmental factors (e.g., soil composition, average rainfall, locally competing species) that contribute to the height of any species of plant. These factors can vary even at the same altitude. Therefore you cannot precisely predict height based solely on altitude, but you can make an estimate using the Wald method that takes advantage of a trend between these two variables.

Note that we have an odd number of data points in this case. After arranging the 9 data points according to the first coordinate, we form two groups of 4 each by simply disregarding the point in the middle. The middle point turns out to be $(7.5, 32)$, and so the two groups are

$$(4.3, 75), (6.0, 50), (6.2, 43), (6.8, 49)$$

and

$$(7.8, 25), (8.0, 21), (9.2, 20), (10.2, 15).$$

The average ordered pairs of these two groups are (5.8, 54) and (8.8, 20). The slope through these two points is

$$m = \frac{20 - 54}{8.8 - 5.8} = \frac{-34}{3.0} = -11.$$

The Wald function is given by

$$w(x) = -11(x - 5.8) + 54$$

or

$$w(x) = -11x + 118.$$

Any prediction made using this function should be rounded to 2 significant digits. As an example we compute $w(7.5) = 36$, corresponding to our omitted data point (7.5, 32). The prediction of 36 yields a relative error $= \frac{4}{32} \cdot 100\% = 12.5\%$. ◆

The above example raises one final precautionary comment. One must be careful about inputting a value to the Wald function from an interval domain wider than the one implied by the given data. In other words, the largest allowable domain is typically $[L, G]$, where L is the least value of the independent variable and G is the greatest value. In Example 5, this means that $w(x) = -11x + 118$ has an assumed domain of $[4.3, 10.2]$. Allowing x to assume a value outside this interval should involve additional preliminary analysis. For instance, this function gives a predicted height at sea level ($x = 0$) of 118 cm, or worse, a height of 129 cm at 1,000 ft below sea level ($x = -1$), a highly unlikely occurrence!

Name _____

Exercise Set **10.2** ☐☐☐←──────────────────────

1. Determine the Wald function that corresponds to the data in the following table. What is the assumed domain of this function?

x	y
13.2	31.4
10.5	27.0
17.7	38.8
21.3	47.5
22.4	52.1
11.9	27.6
18.8	39.9
25.0	53.7

2. Graph both a scatter plot of the data points from the previous exercise and the associated Wald function on the same set of axes. Use the function to predict a value for y when x = 20.3. If an actual observation reveals that y = 46.1 when x = 20.3, what are the absolute and relative errors of the prediction?

3. In Exercise 1, verify that $w(\bar{x}) = \bar{y}$, where w is the Wald function, $\bar{x}$ is the mean of the first coordinate values, and $\bar{y}$ is the mean of the second coordinate values.

4. Samuel owns an apartment complex with over 300 identical units that all rent for the same amount. As the rental price increased over the last 6 years, he computed the average daily vacancy percentage for each period in which the rent remained constant. Using the Wald method, what might Samuel expect to be the average vacancy percentage if he increases the rent to $625 per month?

Rent ($ per month)	Average Daily Vacancy (%)
475	2.5
500	2.7
530	3.2
550	2.9
585	3.2
600	3.4

5. Graph both a scatter plot of the data points from the previous exercise and the associated Wald function on the same set of axes. Samuel makes a prediction for a raise in the rent to $625 per month. He then implements the raise and experiences an average daily vacancy of 3.4%. What are the absolute and relative errors of his prediction?

6. In Exercise 4, verify that $w(\bar{x}) = \bar{y}$, where w is the Wald function, $\bar{x}$ is the mean of the first coordinate (rent) values, and $\bar{y}$ is the mean of the second coordinate values.

7. Give an example in your major field of two variables such that a graph of 20 data points prepared from paired values of those variables would probably yield a Wald function with a positive slope. Be sure to identify the independent variable of the function.

8. Give an example in your major field of two variables such that a graph of 20 data points prepared from paired values of those variables would probably yield a Wald function with a negative slope. Be sure to identify the independent variable of the function.

9. The graph of the global land–ocean temperature index (anomaly with base 1951–1980) is given below for the years 1880 to 2011. The index is the difference in degrees Celsius (°C) between the year's mean global surface temperature and the mean for the base years. Would the Wald function for these data have a positive or a negative slope? Use a pencil and ruler to draw a straight line that you feel fits the data points reasonably well.
(a) Use your line to estimate the mean temperature index for the years 1920, 1930, and 1940.
(b) Use your line to estimate the mean temperature index for the years 1990, 2000, and 2010.
(c) Draw a second line that begins with the data for 1960. Use this line to again estimate the mean temperature index for 1990, 2000, and 2010. (*Source:* Goddard Institute for Space Studies, NASA, May 10, 2012.)

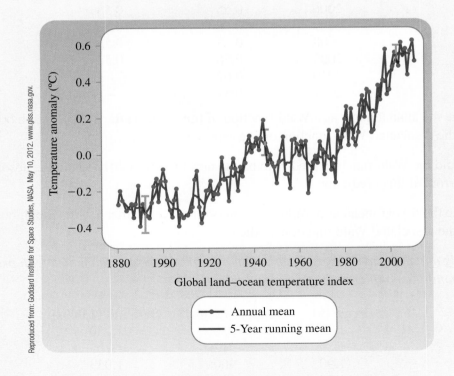

The annual mean global land–ocean temperature index and the 5-year mean (± 2 years) are given below for the years 1990–2011.

Year	Annual Mean	5-Year Mean
1990	0.36	0.27
1991	0.35	0.23
1992	0.12	0.24
1993	0.13	0.24
1994	0.23	0.23
1995	0.38	0.29
1996	0.29	0.38
1997	0.41	0.40
1998	0.58	0.39
1999	0.33	0.43
2000	0.35	0.46
2001	0.48	0.45
2002	0.56	0.48
2003	0.55	0.54
2004	0.48	0.55
2005	0.61	0.55
2006	0.55	0.53
2007	0.58	0.55
2008	0.43	0.55
2009	0.56	0.54
2010	0.62	–
2011	0.51	–

10. Determine the annual mean as a Wald function of time (years after 1990). Graph both a scatter plot and the associated Wald function on the same set of axes.

11. What would the Wald function predict to be the mean index in 2012? Find the absolute and relative errors of this prediction.

12. Determine the 5-year mean as a Wald function of time (years after 1990). Graph both a scatter plot and the associated Wald function on the same set of axes.

The supply and demand of bags of peanuts in a particular region fluctuated with price over several months according to the following table.

Price ($)	Supply (1,000s)	Demand (1,000s)
1.75	850	1,150
1.83	912	1,105
1.90	960	1,039
1.94	991	1,018
2.05	1,015	1,002
2.21	1,027	989

13. Using price as the independent variable, find the Wald function for the supply of peanuts.

14. Using price as the independent variable, find the Wald function for the demand of peanuts.

15. Use the functions from the previous two exercises to predict the supply and demand of peanuts for a price of $1.80. Compare your answers to those from Exercise 14 in the previous Exercise Set.

16. Use the functions from the previous two exercises to predict the supply and demand of disks at a price of $2.10. Compare your answers to those from Exercise 15 in the previous Exercise Set.

17. Maxine is a botanist who wishes to plant three members of the species *Achillea lanulosa* at an altitude of 7,200 ft. Use the Wald function from Example 5 to predict the height of each plant. If the mature heights of the three plants turn out to be 39, 42, and 45 cm, find the absolute and relative errors in each case.

18. Find the mean $\bar{x}$ of the altitudes given in Example 5. Compare $w(\bar{x})$ to the mean $\bar{y}$ of the plant heights, where w is the Wald function. Explain why they are different.

19. The following table matches the percentage score on the first test for each of 20 students in a liberal arts mathematics class with the final exam score for the same student. Use the Wald method to predict what score Arthur will obtain on the final exam if he scores a 78 on the first test in the same course from the same teacher.

First Test	Final Exam	First Test	Final Exam
89	86	52	49
62	70	80	77
65	57	67	64
91	94	73	79
84	92	82	93
47	40	98	97
76	83	90	94
95	98	85	79
73	70	74	74
79	81	58	70

20. If Arthur does score an 80 on the final examination, what are the absolute and relative errors of the prediction made in the previous exercise?

21. Lauren likes to take three books with her on her backpacking trips, but always keeps their total mass to less than 700 grams (g). Since she does not usually have access to a scale, she wants to formulate a Wald function to estimate the mass of a book based on the number of pages. Lauren brought in a selection of 10 books to her chemistry lab and obtained the following data. What will be her function?

Number of pages	Mass (g)
373	186
202	204
832	569
130	203
183	324

Number of pages	Mass (g)
355	443
105	193
107	133
230	224
110	185

22. In the previous exercise, graph a scatter plot of the data and the Wald function on the same set of axes.

23. In Exercise 21, on what condition would Lauren take a book of 635 pages based on her Wald estimate of the mass?

24. In Exercise 21, Lauren wishes to take three books of 213, 189, and 240 pages. Will she be able to take these three particular books based on her Wald estimates? Suppose she adds the pages of these three books together first and uses that number to estimate the total mass. What is wrong with that approach?

25. Omar owns a small cafe in Cairo and wishes to create a formula for the price he charges the customer based on the cost. He prepared a list of these numbers for the current menu. What would be his formula based on the Wald method?

Item	Cost	Price charged
Soft pretzel	0.12	0.50
Ice cream sandwich	0.17	0.50
Large ice cream cone	0.74	1.95
Small coffee	0.05	0.60
Juice box	0.50	1.00
Hot dog	0.41	1.00
Personal-size pizza	0.71	2.00
French fries	0.69	1.25
Small potato chips	0.16	0.60
Quarter-pounder	1.00	1.75
Fruit roll-up	0.24	0.75
Bagel with cheese	0.28	1.35

26. The length of a baseball game is certainly related to the total number of player at-bats for that game. A television network that broadcasts baseball games is interested in predicting the length of a game as a function of the number of at-bats. This list gives some times and numbers of at-bats for a set of major league games played over several weeks. Predict the length of a game with 70 at-bats.

At-bats	Game Duration (min)
63	171
61	172
77	199
56	160
76	182
68	184

At-bats	Game Duration (min)
64	169
68	190
105	316
82	271
61	176
68	165

27. A real estate agent thinks it would be a convenience for her customers if she could create an easy-to-use linear function for estimating the cost of a house based on the total square feet of living space. Her current book of listings gives her the following information. Find her function, and use it to estimate the cost of a house having 3,100 square feet (ft^2).

Square Footage	Cost ($100,000)
2,600	2.9
2,820	3.3
3,000	3.4
3,060	4.1
3,270	4.6
3,600	4.1
4,000	4.4
4,000	4.9
4,500	5.9
4,950	6.0
5,000	6.4
5,080	6.3

28. Diondre is in the market for a used car. He cannot afford a car that is less than 4 years old, but he does not want the car to have more than 50,000 mi on it. Based on the following data, he wants to know a reasonable estimate for the mileage of a car, given its age. Use the Wald method to see whether Diondre can expect to find a car that satisfies his criteria.

Age of Car	Odometer Mileage
1	7,600
1	11,400
2	35,000
2	40,500
4	37,000
5	16,300
5	57,200
5	86,000
5	125,200
6	49,000
6	78,100
8	115,300
9	86,000
11	143,400

29. Gretchen is a schoolteacher who wishes to explore the relationship between the heights of her students (which vary considerably) and certain athletic accomplishments. She found the average length of three efforts in the standing broad jump for 12 students. Find the Wald function for jump length as a function of height in feet and inches (in). What would Gretchen predict to be the jump length of a student who is 5 ft 6 in tall?

Student Height (in.)	Jump length (in.)
40	34
59	44
60	82
63	67
64	69
67	82
68	75
69	81
71	111
72	91
73	87
75	84

10.3 Nonlinear Modeling Functions

The construction of the Wald function in the previous section is based on the assumption of a linear model representing the relationship between two variables, and so the examples and situations we examined were chosen with that condition in mind. We were not concerned with the possibility of a better model, and so we did not seek alternative **modeling types**. However, you do not have to search very hard to encounter a phenomenon for which a close analysis strongly suggests that the assumption of a **nonlinear function**—one whose graph is not a straight line—would provide a more accurate model representation. We will be focusing on the third link in our modeling process displayed in the beginning of the chapter. The creation of the desired function actually involves first selecting one or more model types, then finding the values of specific constants, and finally testing the competing functions for the most appropriate choice (**Figure 10.3.1**).

> **modeling type**
> A function chosen from the choices:
> linear ($y = mx + k$),
> power ($y = kx^p$),
> hyperbolic $\left(y = \dfrac{k}{x} \right)$,
> or exponential ($y = kb^x$).
>
> **nonlinear function**
> A function whose graph is not a straight line.

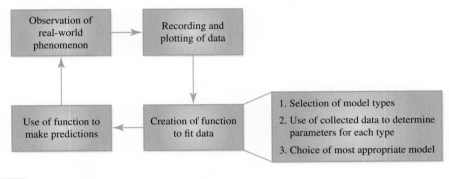

FIGURE 10.3.1   The modeling function could be one of several types.

Our goal here is to present a broad overview of some of the procedures used to select a model and to create a nonlinear function, given a collection of data. We will restrict our study to three very specific modeling types:

$$\text{Power:} \qquad y = kx^p \quad (k \text{ and } p \text{ are positive constants})$$

$$\text{Hyperbolic:} \quad y = \frac{k}{x} \quad (k \text{ is a positive constant})$$

$$\text{Exponential:} \quad y = kb^x \quad (k \text{ and } b \text{ are positive constants})$$

> **parameter** The constants k, p, and b used in a nonlinear modeling function such as $y = kx^p$, $y = \dfrac{k}{x}$, or $y = kb^x$

We refer to the above constants k, p, and b as **parameters**. Our task in each case will be to first select a model type by examining a scatter plot of the given data points and then determine the values of the parameters for the function associated with that type. We will examine two procedures for finding the parameters, called *transforming the data* and *exponential averaging*. It is important to note that the assumption of different modeling types or the use of alternate procedures for finding the parameters can lead to different modeling functions. So the final step, which is beyond the scope of this text, always consists of testing the competing functions and choosing the most appropriate one.

Transforming the data

? Example 1

Among the first experiments done in an attempt to quantitatively describe a phenomenon of motion were those performed by Galileo Galilei (1564–1642) concerning the relationship between the length x of a pendulum and its period of oscillation T. These attempts bore great historical significance as they were a break in procedure from the more qualitative approach of Aristotle that had been in use for centuries. In her physics class, Hsuan wished to duplicate the findings of Galileo and so experimentally determined the periods of several pendulums. She would like to estimate the period for a pendulum of length 100 cm. How can she estimate the period as a function of length?

Length x (cm)	Period T (s)
10	0.6
15	0.8
20	0.9
30	1.2
40	1.3
45	1.4
60	1.6
80	1.8

⚙ Solution

As always, a scatter plot of the data provides much insight. In **Figure 10.3.2**, the graph on the left shows a straight line that has been fitted to the data points. The Wald method was not used to determine this line, but another common technique

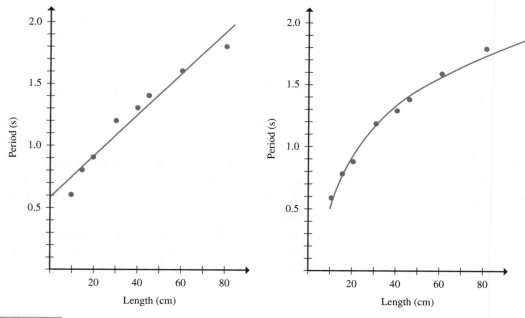

FIGURE 10.3.2   Fitting a linear function (left) and a nonlinear function (right) to a set of data points.

commonly called *linear regression*. Even though linear regression methods do find a best-fitting line, we see how the points fall far enough from this line to indicate the need to search for a nonlinear model.

On the right, an **approximating curve** has been sketched which comes closer to the points, and this helps to reveal a more feasible function to use for our model. The fact that the curve is concave down ("spills water") implies that raising the dependent variable (T in this case) to a power greater than 1, say T^2 and then graphing T^2 versus length may result in a group of data points for which a straight line provides a reasonable fit. Since we are altering the dependent values by squaring, we are **transforming the data**. Note that we rounded to two significant digits. A scatter plot of these data (**Figure 10.3.3**) shows that these points come much closer to a linear fit.

x	T^2
10	0.36
15	0.64
20	0.81
30	1.4
40	1.7
45	2.0
60	2.6
80	3.2

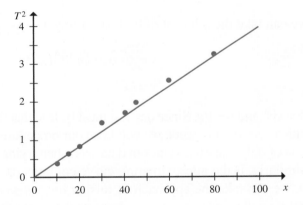

FIGURE 10.3.3 Fitting the Wald line to transformed data.

We now find the slope of the Wald line that would fit the transformed data set. The average data point for the group (10, 0.36), (15, 0.64), (20, 0.81), and (30, 1.4) is $P' = (19, 0.81)$ and the average point for the group (40, 1.7), (45, 2.0), (60, 2.6), and (80, 3.2) is $P'' = (56, 2.4)$. The slope of the line through P' and P'' is

$$m = \frac{2.4 - 0.81}{56 - 19} = 0.043.$$

Observing that the y-intercept is close enough to zero to be ignored, we can write

$$T^2 = 0.043x$$
$$T = 0.21\sqrt{x}$$

By taking the principal square root of both sides, we get T as a power function of x: $T = kx^p$, where $k = 0.21$ and power $p = 1/2$. Hsuan can now predict the period of a pendulum of length 100 cm to be

$$T = 0.21\sqrt{100} = 2.1\,\text{s}. \quad \blacklozenge$$

For small oscillations, the pendulum length and period are more accurately related by $T = 2\pi\sqrt{x/g}$, where g is acceleration due to gravity. Since x is being measured in centimeters here, we use $g = 980$ cm/sec^2 to see that $\dfrac{2\pi}{\sqrt{980}} = 0.20$ to two significant digits. This is a nice match to Hsuan's experimental data.

This example featured a set of data points for which the curve fitting them was:

   i.   concave downward ("spills water") and
  ii.   contained the point (0, 0).

These are characteristics associated with the graph of a power function $y = kx^p$, where $k > 0$ and $0 < p < 1$. In this case, we squared the dependent variable in an effort to "straighten the data." In general, if raising the dependent variable to the power $q > 1$ results in a transformed data set with a linear look to it, we can find the Wald slope m and obtain

$$y^q = mx.$$

Then we can take the qth root of both sides to get

$$y = \sqrt[q]{mx} = \left(m^{1/q}\right)\left(x^{1/q}\right)$$
$$y = kx^p$$

where $k = m^{1/q}$ and $p = 1/q$. Since $q > 1$, it must be true that $0 < p < 1$. Although we could experiment with a variety of powers, we will keep our procedure straightforward by maintaining the restriction of only squaring values in data sets when trying to construct a power function.

Similarly, if the approximating curve contains (0, 0) and is concave upward ("holds water"), we can take the approach of raising the *independent* variable to a power $p > 1$ to try to achieve the same result. It is important to note that, in both of these situations, the curves involved contain the point (0, 0). This implies that this technique can only be applied to a phenomenon in which a zero value for one variable implies a zero value for the other variable. See **Figure 10.3.4**.

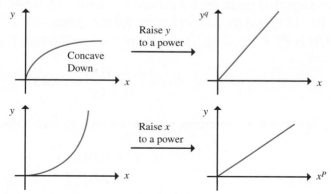

FIGURE 10.3.4   We find a power modeling function $y = kx^p$ by transforming the data. Note that (0, 0) is on each curve. In each case, the Wald slope leads to the value of k.

As before, we shall only be squaring values in either of the above cases. Also, we should pause at this point and reflect on the fact that there exist many functions whose graphs will contain the data points of a given experimental set. We are exploring only certain specific functions whose formulations are accessible to us in a text of this type. We should be aware that the functions we are creating may very well not be the best available and that using them for predictive purposes must be done with caution.

? Example 2

The following table (**Figure 10.3.5**) gives some measurements for the time t in seconds (s) and distance (d) in feet (ft) traveled by an object dropped from rest. Select a modeling type by examining the scatter plot of the data, and find a modeling function by transforming the data. Use three significant digits.

$t\,(\mathbf{s})$	$d\,(\mathbf{ft})$
2.0	63.2
4.2	286
3.5	199
7.3	849
5.1	417
6.4	652

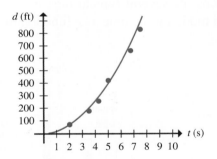

FIGURE 10.3.5 An approximating curve that is concave upward and contains (0, 0). It can be modeled by a power function with power greater than 1.

⚙ Solution

Since the approximating curve contains $(0, 0)$ and is concave upward this time, we assume a power modeling type $d = kt^2$. We proceed by squaring all the independent-variable values and listing them in ascending order.

t^2	d
4.0	63.2
12.3	199
17.6	286
26.0	417
41.0	652
53.3	849

A scatter plot of the transformed data would yield points lying quite close to a straight line. The two average data points of this set are (11.3, 183) and (40.1, 639). The Wald slope of the line through these two points is

$$\frac{639 - 183}{40.1 - 11.3} = 15.8$$

and provides us with the value of the constant k. So our modeling function is

$$d = 15.8t^2. \quad \blacklozenge$$

? Example 3

Pouring liquid mercury into the open end of the curved glass tube pictured in **Figure 10.3.6** compresses the gas trapped in the closed end. The pressure exerted on the air is measured by the height of the column of mercury and the relationship between the pressure, measured in atmospheres (atm), and the volume of gas, measured in liters (L), at a fixed temperature was first discovered by Robert Boyle in 1662. Some values for several typical measurements are given in the table below. Find an equation modeling volume as a function of pressure.

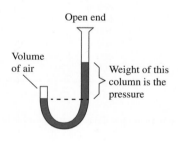

P (atm)	V (L)
0.013	0.150
0.016	0.124
0.020	0.100
0.024	0.084
0.033	0.060
0.066	0.030
0.099	0.020
0.132	0.015
0.164	0.012

FIGURE 10.3.6   Experimental data revealing Boyle's law.

⚙ Solution

We see clearly that the volume of air is a decreasing function of pressure. Fitting a curve to these data points (**Figure 10.3.7**) reveals a

asymptote A line used as a guide in graphing a function. The graph approaches it forever but does not intersect it.

beautiful curve known in analytic geometry as a *rectangular hyperbola.* Characteristically, the graph of this type of hyperbola

■ is concave upward and
■ approaches (but does not intersect) both axes.

An **asymptote** is a line that a graph approaches forever but never intersects.

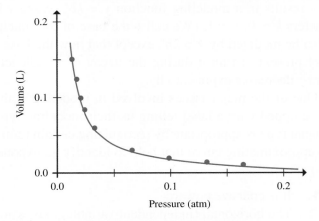

FIGURE 10.3.7 The graph of volume versus pressure is very close to a rectangular hyperbola.

A hyperbolic modeling function has the form

$$V = \frac{k}{p} = k\left(\frac{1}{p}\right).$$

Since V is a linear function of $\frac{1}{p}$ with a graph having slope k, we can determine the value of k by inverting the P-values and then finding the slope of the Wald line through the transformed data points. (See **Figure 10.3.8**.)

1/P	V
6.10	0.012
7.58	0.015
10.1	0.020
15.2	0.030
30.3	0.060
41.7	0.084
50.0	0.100
62.5	0.124
76.9	0.150

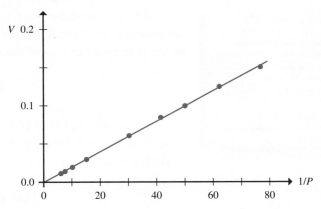

FIGURE 10.3.8 We find a hyperbolic function by inverting the independent variable and finding the Wald slope.

The average of the first four data points is (9.75, 0.019), and the average of the last four points is (57.8, 0.115). These two points yield a Wald slope = 0.096/48.1 = 0.0020. This value for k gives us

$$V = \frac{0.0020}{P}. \quad \blacklozenge$$

Exponential Averaging

In the absence of constraints, many quantities increase or decrease at a rate that is, at any time, proportional to the amount present at that time. Such quantities can be modeled by functions of an exponential type. One example of such an increasing quantity is the unrestrained growth of a collection of like organisms such as bacteria, rabbits, or people. This usually results in a modeling function $y = kb^t$, where y is the amount, t is time, and the parameters $k > 0$, $b > 1$. (We call b the base of the function.) Some decreasing quantities can also be modeled by $y = kb^t$, except that now the base $b < 1$. An example would be the amount present at time t during the decay of a radioactive element. As t increases, the amount y decreases exponentially.

One of the major issues involved in modeling is that raw experimental data do not come equipped with a label telling us their modeling type. Rather, we first have to determine what type is appropriate by recognizing certain telltale features. Three characteristics of the approximating curve that help to identify an exponential modeling type (see **Figure 10.3.9**) are

1. It is concave upward.
2. The horizontal (independent variable) axis is an asymptote.
3. It crosses the vertical (dependent variable) axis at a positive value.

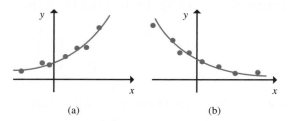

(a)   (b)

FIGURE 10.3.9   (a) Increasing exponential type ($b > 1$) and (b) decreasing exponential type ($b < 1$).

exponential averaging The technique for finding an exponential function whose graph forms an approximating curve to a set of data points.

geometric mean The nth principal root of the product of n values.

To describe the **exponential averaging** procedure for creating an exponential function, we first define the **geometric mean** $\hat{y}$ of n numbers $y_1, y_2, \cdots, y_n$ to be the principal nth root of their product:

$$\hat{y} = \sqrt[n]{y_1 y_2 \ldots y_n}.$$

Now, our goal is to compute the parameters k and b to be used in the modeling function $y = kb^t$. Let $(t_1, y_1), (t_2, y_2), \ldots, (t_n, y_n)$ be any n points on the graph of such a function. Then

$$y_1 y_2 \cdots y_n = \left(kb^{t_1}\right)\left(kb^{t_2}\right)\cdots\left(kb^{t_n}\right) = k^n\left(b^{t_1+t_2+\cdots+t_n}\right).$$

If we take the nth root of both sides, we get

$$\sqrt[n]{y_1 y_2 \cdots y_n} = kb^{(t_1+t_2+\cdots+t_n)/n}$$

In other words, if $\bar{t}$ is the arithmetic average of the first coordinates of the given points and $\hat{y}$ is the geometric mean of the second coordinates of the points, then we must have

$$\hat{y} = kb^{\bar{t}}.$$

Now suppose we have decided an exponential function would be the best fit to a given data set $(t_1, y_1), (t_2, y_2),\ldots, (t_{2n}, y_{2n})$. We find the parameters of the function according to this procedure:

1. As before with the Wald method, we divide the points into two groups (eliminating the middle point if there are an odd number of points) by ordering them according to the first coordinates.

2. We determine two average ordered pairs (t', y') and (t'', y'') for these two groups in which the first coordinate of each pair is still the arithmetic mean of the independent values, but the second coordinate is the geometric mean of the dependent values.

3. We require the graph of our modeling function to contain both of these average points. Therefore,

$$y' = kbt' \text{ and } y'' = kb^{t''}.$$

To find b, we calculate the ratio $\dfrac{y''}{y'}$ and note that

$$\frac{y''}{y'} = \frac{kb^{t''}}{kb^{t'}} = b^{t''-t'}.$$

4. Let $r = t'' - t'$. We obtain b by taking the rth root of both sides of the prior equation.

$$b^r = \frac{y''}{y'},$$

$$b = \sqrt[r]{\frac{y''}{y'}}.$$

5. With this value for b, we can then solve for k by substituting either (t', y') or (t'', y'') into $y = kb^t$. The following example helps to clarify this process.

❓ Example 4

Census counts for the population of Jolyndia are given in the table below along with a scatter (see **Figure 10.3.10**) plot. Find the modeling function and use it to predict the population of Jolyndia in the year 2021.

Year	t (years)	y (1000s)
1996	0	28.0
1998	2	32.3
2001	5	38.7
2002	6	42.2
2006	10	56.4
2009	13	68.2
2011	15	76.4
2012	16	82.1

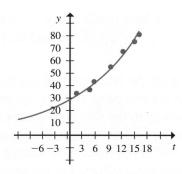

FIGURE 10.3.10   Population growth is typically of the exponential type.

⚙ Solution

Note that we chose 1996 to correspond to $t = 0$. Then each of the other t values corresponds to the number of years past 1996. This allows a more convenient domain to use for our function. The approximating curve to the scatter plot satisfies the above criteria for a modeling function of exponential type, and so we proceed with the method of exponential averaging. The data set splits into two groups of 4 each. The two average data points are

$$t' = \frac{0+2+5+6}{4} = 3.25 \qquad y' = (28.0 \cdot 32.3 \cdot 38.7 \cdot 42.2)^{1/4} = 34.9$$

$$t'' = \frac{10+13+15+16}{4} = 13.5 \qquad y'' = (56.4 \cdot 68.2 \cdot 76.4 \cdot 82.1)^{1/4} = 70.1.$$

Therefore,

$$r = 13.5 - 3.25 = 10.3 \quad \text{so} \quad b = \left(\frac{70.1}{34.9}\right)^{1/10.3} = 2.01^{0.097} = 1.07.$$

We now have $y = k \cdot 1.07^t$ for our function. To find k, we substitute either of our average data points and solve for k. Choosing $t' = 3.25$ and $y' = 34.9$, we get

$$34.9 = k \cdot 1.07^{3.25}$$
$$= k \cdot 1.25$$
$$k = \frac{34.9}{1.25}$$
$$= 27.9.$$

So we use $b = 1.07$ and $k = 27.9$ in our exponential modeling function to get

$$y = 27.9 \cdot 1.07^t.$$

In the year 2021 ($t = 25$), the value of y is

$$y = 27.9 \cdot 1.07^{25}$$
$$= 151.$$

So our prediction for the population of Jolyndia in 2021 is 151,000 people.   ◆

Name _____

Exercise Set **10.3** ☐☐☐←

Identify a modeling type as defined in this section that is indicated by the approximating curve drawn through each of the following scatter plots. Choose from linear, power, hyperbolic, exponential, or none of these.

1.

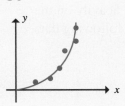

2.

3.

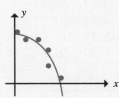

4.

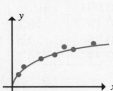

5.

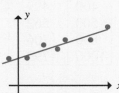

6.

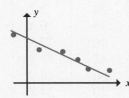

7.

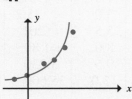

8.

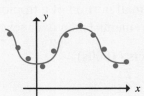

9.

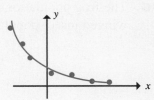

Each of the following tables of data consists of measurements from the observation of some two-variable phenomenon. Consider the left-hand column to be values for the independent variable. Perform the following:

(a)   *Draw a scatter plot. Be sure to use appropriate scaling on your axes.*

(b)   *Sketch an approximating curve on the same set of axes.*

(c)   *Select a modeling type.*

(d)   *Find a modeling Junction.*

10.

x	y
1.3	5.1
2.6	21.1
3.2	29.8
4.5	60.5
5.7	98.3
6.0	106.9

11.

x	y
200	355
300	436
400	499
500	562
800	706
1000	792

12.

x	y
0.021	477
0.044	225
0.091	107
0.18	56.2
0.25	39.8
0.37	26.2

13.

x	y
10	0.48
20	2.2
35	6.4
40	7.8
48	12.1
60	18.9
70	24.8
79	30.5

14.

x	y
0.01	8.4
0.10	0.81
1.00	0.085
10.0	0.0079
100	0.00088
1,000	0.00008

15.

t	y
1.0	9.3
3.0	35.2
5.5	201
8.0	1,1 80
9.5	3,230
13.0	36,600

16. The frog population in a certain small portion of a tropical rain forest depended heavily on the winged insect population. Measurements taken over several years produced the following data.

Insects (1000s)	Frogs
23	175
34	210
26	183
10	102
38	217
41	230

Model the frog population as a function of the insect population, and predict the number of frogs given 45,000 winged insects.

17. The wavelength at which an object radiates the maximum amount of energy (called λ_{max}) depends on the object's temperature. The following table gives some measurements for λ_{max} in nanometers (nm) at certain temperatures, in kelvins (K). Graph a scatter plot of the data, fit a curve to the points, and find a modeling function by transforming the data.

Temperature (K)	λ_{max}(nm)
10,000	310
7,900	380
7,100	420
6,500	470
5,500	530
4,100	720
3,500	870
3,100	980

18. Often the revenue generated by a manufacturing firm depends on the quantity of the produced item in a nonlinear fashion. The records of Sheila Shirts, Inc. show the following revenue was brought in by sales of a trendy new shirt for the last 4 months.

Quantity of Shirts Produced	Revenue (1,000 $)
800	19.4
875	22.6
950	26.7
1,075	35.2

Find a power function that relates revenue R to the quantity x of shirts produced. If this relationship holds true for larger numbers of shirts, what revenue would you predict as a result of increasing production to 1,200 shirts next month? Do you think this function could have an unlimited domain?

19. Drugs such as nitroglycerin can be used to enlarge the diameters of blood vessels. An experiment yielded the following data on the cross-sectional area A, in square centimeters (cm^2), of a blood vessel t hours (h) after nitroglycerin was taken. Find a power function that could be used to model this relationship.

t(h)	A(cm^2)
0.5	0.0026
1.0	0.0098
1.5	0.023
2.0	0.043
2.5	0.061
3.0	0.087

20. A cup of coffee is heated to a temperature of 180°F and left to cool outside. Find an exponential modeling function that will give the temperature T as a function of minutes x.

x(min)	T(°F)
0	180
5	152
10	134
15	117
20	97
25	82

21. The parallax angles p (in arc-seconds) and distances d (in light-years) of eight stars are given below. Graph a scatter plot of the data, fit a curve to the points, and create a modeling function by transforming the data.

p	d
0.08	39.1
0.10	33.4
0.15	21.2
0.22	14.2
0.25	13.8
0.38	8.9
0.46	7.2
0.53	5.7

22. The world population figures for 7 years are given below. Find the exponential model that gives population as a function of the number of years after 1980, and use it to predict the population of the world in the years 2015 and 2020.

Year	Population (billions)
1980	4.45
1985	4.85
1990	5.28
1995	5.69
2000	6.08
2005	6.46
2010	6.84

23. The table below lists the population figures for the United States for 7 years. Find the exponential model that gives U.S. population as a function of the number of years past 1980, and use it to predict the population in the years 2020 and 2025.

Year	Population (millions)
1980	227
1985	238
1990	250
1995	266
2000	282
2005	296
2010	309

24. The following chart displays the distances needed to stop a car traveling at various speeds in miles per hour (mi/h). Find a modeling function that gives stopping distance as a function of speed, and use it to predict the distance needed to stop a car traveling at a speed of 70 mi/h.

Speed (mi/h)	Stopping Distance (ft)
25	61.7
35	106
45	161.5
55	228
65	305.7

25. Bradford is a 7-year-old who just joined the city swim team. The following table shows his times in the 25-y freestyle at the first six swim meets. Model his freestyle times f in seconds (s) as an exponential function of the number of swim meets x, and predict Bradford's time at the 12th meet of the summer.

Meet	Time (s)
1	32.8
2	31.4
3	29.6
4	28.5
5	26.8
6	25.3

26. In the coffee shop of Marywood University 500 cupcakes were placed on a large table at 7:00 A.M. The cupcake counts at the end of each of the first 6 h were as follows.

Clock Time	Cupcakes
8:00	404
9:00	335
10:00	245
11:00	197
12:00	170
1:00	120

Model the number of cupcakes as a decreasing function $C(t)$ of t hours past 7:00 A.M. and predict the number of cupcakes left on the table at 6:00 P.M.

Name _____

Chapter Review Test **10** ⃞⃞⃞ ←

The following table gives the heights in inches (in) and weights in pounds (lb) of eight NBA basketball players. Use this table to answer questions 1–3.

Height (in.)	Weight (lb)
73	190
74	196
76	205
79	220
80	237
82	252
85	270

1. Use linear interpolation to estimate the weight of an NBA player who is 6 ft 5 in tall.

2. Find the linear function, using the Wald method for estimating the weight of an NBA player as a function of his height.

3. Use your function from the previous problem to predict the weight of an NBA player who is exactly 7 ft tall. If the actual weight of a 7-ft player is 290 lb, find the absolute and relative errors of your prediction.

4. Use linear interpolation to estimate $2^{5.3}$ to 3 significant digits by using the stakes on the graph of the function $f(x) = 2^x$ obtained by letting $x = 5$ and 6. Find the error between your estimate and the value of $2^{5.3}$ on your calculator.

5. Identify which modeling type is indicated by the approximating curve drawn through the following scatter plots.

A. Linear   B. Power   C. Hyperbolic   D. Exponential   E. None of these

(i)
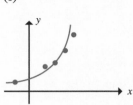

(ii)
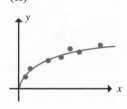

(iii)
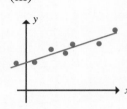

(iv)
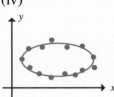

(v)
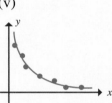

(vi)
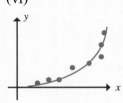

6. Michelle claims that $w(x) = 2.54x + 62.3$ is the Wald function constructed from a collection of data points for which the average of the independent values $\bar{x} = 70.1$ and the average of the dependent values $\bar{y} = 98.8$. Is Michelle's claim valid? Why or why not?

7. The heat-stress index measures the combined effects of temperature and humidity on physical exercise. (An index of 70 is comfortable while 110 is miserable.) The following table gives the heat-stress index for several levels of humidity at a temperature of 90°F. Find an exponential modeling function and use it to predict the index at a humidity of 65. (Use 3 significant digits.)

Humidity	Heat-Stress Index
0	83 5
20	87.5
40	93.0
60	100
80	113
90	122

Standardized Normal Distribution Table

Note: This table gives $p(z > \tau)$ for the standardized normal curve.

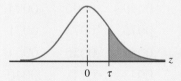

τ	.00	.01	.02	.03	.04	.05	.06	.07	.08	.09
0.0	.5000	.4960	.4920	.4880	.4840	.4801	.4761	.4721	.4681	.4641
0.1	.4602	.4562	.4522	.4483	.4443	.4404	.4364	.4325	.4286	.4247
0.2	.4207	.4168	.4129	.4090	.4052	.4013	.3974	.3936	.3897	.3859
0.3	.3821	.3783	.3745	.3707	.3669	.3632	.3594	.3557	.3520	.3483
0.4	.3446	.3409	.3372	.3336	.3300	.3264	.3228	.3192	.3156	.3121
0.5	.3085	.3050	.3015	.2981	.2946	.2912	.2877	.2843	.2810	.2776
0.6	.2743	.2709	.2676	.2643	.2611	.2578	.2546	.2514	.2483	.2451
0.7	.2420	.2389	.2358	.2327	.2296	.2266	.2236	.2206	.2177	.2148
0.8	.2119	.2090	.2061	.2033	.2005	.1977	.1949	.1922	.1894	.1867
0.9	.1841	.1814	.1788	.1762	.1736	.1711	.1685	.1660	.1635	.1611
1.0	.1587	.1562	.1539	.1515	.1492	.1469	.1446	.1423	.1401	.1379
1.1	.1357	.1335	.1314	.1292	.1271	.1251	.1230	.1210	.1190	.1170
1.2	.1151	.1131	.1112	.1093	.1075	.1056	.1038	.1020	.1003	.0985
1.3	.0968	.0951	.0934	.0918	.0901	.0885	.0869	.0853	.0838	.0823
1.4	.0808	.0793	.0778	.0764	.0749	.0735	.0722	.0708	.0694	.0681
1.5	.0668	.0655	.0643	.0630	.0618	.0606	.0594	.0582	.0571	.0559
1.6	.0548	.0537	.0526	.0516	.0505	.0495	.0485	.0475	.0465	.0455
1.7	.0446	.0436	.0427	.0418	.0409	.0401	.0392	.0384	.0375	.0367
1.8	.0359	.0352	.0344	.0336	.0329	.0322	.0314	.0307	.0301	.0294
1.9	.0287	.0281	.0274	.0268	.0262	.0256	.0250	.0244	.0239	.0233
2.0	.0228	.0222	.0217	.0212	.0207	.0202	.0197	.0192	.0188	.0183
2.1	.0179	.0174	.0170	.0166	.0162	.0158	.0154	.0150	.0146	.0143
2.2	.0139	.0136	.0132	.0129	.0125	.0122	.0119	.0116	.0113	.0110
2.3	.0107	.0104	.0102	.0099	.0096	.0094	.0091	.0089	.0087	.0084
2.4	.0082	.0080	.0078	.0075	.0073	.0071	.0069	.0068	.0066	.0064
2.5	.0062	.0060	.0059	.0057	.0055	.0054	.0052	.0051	.0049	.0048
2.6	.0047	.0045	.0044	.0043	.0041	.0040	.0039	.0038	.0037	.0036
2.7	.0035	.0034	.0033	.0032	.0031	.0030	.0029	.0028	.0027	.0026
2.8	.0026	.0025	.0024	.0023	.0023	.0022	.0021	.0021	.0020	.0019
2.9	.0019	.0018	.0017	.0017	.0016	.0016	.0015	.0015	.0014	.0014
3.0	.0013	.0013	.0013	.0012	.0012	.0011	.0011	.0011	.0010	.0010

Data Set A

Number of games won during a season by the Chicago Cubs baseball team from the years 1901 to 1940. (**Source**: *The Sports Encyclopedia: Baseball* by Neft/Johnson/ Cohen/Deutsch, Grosset & Dunlap, 1974.)

53	68	82	93	92	116	107	99	104	104
92	91	88	78	73	67	74	84	75	75
64	80	83	81	68	82	85	91	98	90
84	90	86	86	100	87	93	89	84	75

Data Set B

Absolute magnitudes of 25 of the nearest stars to Earth. (**Source**: *Horizons, Exploring the Universe*, Thirteenth Edition by Michael A. Seeds, Wadsworth, 2013.)

4.83	4.38	5.76	13.21	16.80	10.42	1.41	11.54	15.27	15.8
13.3	14.8	6.13	14.6	13.5	7.58	8.39	7.0	2.64	13.1
11.15	11.94	10.23	13.29	9.59					

Data Set C

Percentage of high school students in each state who graduated from high school in 1996 and 2006. (**Source**: EPE Research Center 2009, from U.S. Department of Education.)

State	1996	2006
Alabama	57.0	61.4
Alaska	66.1	65.9
Arizona	56.6	68.6
Arkansas	69.4	71.9
California	67.5	67.5
Colorado	71.5	72.7
Connecticut	76.1	78.9
Delaware	63.2	66.0
Florida	57.5	57.5
Georgia	55.1	55.9
Hawaii	59.0	63.9
Idaho	78.6	76.8
Illinois	78.7	74.1
Indiana	69.8	73.3
Iowa	79.1	80.7
Kansas	72.8	75.4
Kentucky	62.9	72.0
Louisiana	54.0	61.9
Maine	75.1	76.3
Maryland	73.5	73.5
Massachusetts	75.2	75.9
Michigan	69.0	69.6
Minnesota	77.1	79.2
Mississippi	54.5	60.5
Missouri	70.4	74.4
Montana	76.2	76.1
Nebraska	79.3	78.7
Nevada	70.5	47.3
New Hampshire	71.3	77.0
New Jersey	83.1	82.1
New Mexico	55.3	56.0
New York	61.1	68.3
North Carolina	61.9	63.3
North Dakota	80.9	79.0
Ohio	67.8	74.3
Oklahoma	67.0	70.6

State	1996	2006
Oregon	66.0	74.9
Pennsylvania	74.8	77.6
Rhode Island	69.0	72.8
South Carolina	53.2	66.3
South Dakota	77.6	77.1
Tennessee	56.7	69.5
Texas	58.5	65.3
Utah	78.5	72.2
Vermont	75.2	78.7
Virginia	73.4	69.2
Washington	68.0	62.4
West Virginia	75.7	71.8
Wisconsin	77.0	81.7
Wyoming	75.9	73.2

Data Set D

Number of Earthquakes in the United States during 2000–2009. (**Source**: U.S. Geological Survey; http://neic.usgs.gov/neis/eqlists/)

Number of Earthquakes in the United States for 2000 – 2009 □Located by the U.S. Geological Survey National Earthquake Information Center

Magnitude	2000	2001	2002	2003	2004	2005	2006	2007	2008	2009
8.0 to 9.9	0	0	0	0	0	0	0	0	0	0
7.0 to 7.9	0	1	1	2	0	1	0	1	0	0
6.0 to 6.9	6	5	4	7	2	4	7	9	9	0
5.0 to 5.9	63	41	63	54	25	47	51	72	85	25
4.0 to 4.9	281	290	536	541	284	345	346	366	432	155
3.0 to 3.9	917	842	1,535	1,303	1,362	1,475	1,213	1,137	1,485	622
2.0 to 2.9	660	646	1,228	704	1,336	1,738	1,145	1,173	1,578	1,000
1.0 to 1.9	0	2	2	2	1	2	7	11	14	12
0.1 to 0.9	0	0	0	0	0	0	1	0	0	1
No magnitude	415	434	507	333	540	73	13	22	20	10
Total	2,342	2,261	3,876	2,946	3,550	3,685	2,783	2,791	3,623	1,825
Estimated deaths	0	0	0	2	0	0	0	0	0	0

Data Set E

Average Annual Health Care Expenditures of All Consumer Units by Metropolitan Area: 2005–2006 (**Source**: U.S. Bureau of Labor Statistics, *Consumer Expenditures in 2006*; http://stats.bls.gov/cex/homr.html)

Metropolitan Area	Health Care
Atlanta	2,017
Baltimore	2,551
Boston–Lawrence–Salem	2,794
Chicago–Gary–Lake County	2,878
Cleveland–Akron–Lorain	3,035
Dallas–Fort Worth	3,075
Detroit–Ann Arbor	2,349
Houston–Galveston–Brazoria	3,259
Los Angeles–Long Beach	2,316
Miami–Fort Lauderdale	2,190
Minneapolis–St. Paul	3,322
New York–No. New Jersey–Long Island	2,607
Philadelphia–Wilmington–Trenton	2,188
Phoenix–Mesa	3,134
San Diego	3,421
San Francisco–Oakland–San Jose	2,820
Seattle–Tacoma	2,889
Washington, D.C.	2,505

Data Set F

Estimated fuel costs to drive new vehicles for 15,000 miles (**Source**: *Consumer Reports*, April 1993, vol. 58, no. 4, pp. 222–233), categorized by vehicle type.

Small Cars

Saturn: $660
Ford Escort: $680
Nissan Sentra: $650
Plymouth Sundance: $650
Hyundai Elantra: $715

Honda Civic: $620
Mercury Tracer: $630
Toyota Tercel: $510
Dodge Shadow: $650

Mid-Sized Cars

Toyota Camry V6: $985
Ford Taurus: $890
Nissan Maxima: $985
Dodge Dynasty: $860
Olds Cutlass Supreme: $895

Toyota Camry 4: $740
Mercury Sable: $890
Chrysler New Yorker: $860
Buick Regal: $895
Pontiac Grand Prix: $895

Chevrolet Lumina: $815
Infinity Q45: $1,230
BMW 5-Series: $1,230
Lincoln Continental: $1,020
Acura Legend: $1,045
Acura Vigor: $910

Lexus LS400: $1,105
Lexus ES300: $1,030
Volvo 940/960: $1,055
Mazda 929: $1,045
Mitsubishi Diamante: $1,030
Audi 100: $1,020

Large Cars

Pontiac Bonneville: $910
Dodge Intrepid: $850
Ford Crown Victoria: $905
Oldsmobile 88 Royale: $965
Chevrolet Caprice Classic: $990

Chrysler Concorde: $930
Eagle Vision: $930
Mercury Grand Marquis: $905
Buick Le Sabre: $965
Buick Roadmaster: $1,080

Vans

Mercury Villager: $895
Toyota Previa 2WD: $875
Dodge Caravan 2WD: $910
Ford Aerostar 2WD: $1,010
Dodge Grand Caravan: $1,075
Volkswagon EuroVan: $1,080
Chevrolet Astro: $1,370

Nissan Quest: $960
Toyota Previa 4WD: $975
Dodge Caravan 4WD: $960
Ford Aerostar 4WD: $1,040
Pontiac Trans Sport: $1,030
Mazda MPV: $1,105

Data Set G

National League team runs, batting averages, and home runs for the 2008 season. (**Source**: http://sports.espn.go.com/mlb/stats)

Team	Runs	Average	Home Runs
Chicago Cubs	855	.278	184
NY Mets	799	.266	172
Philadelphia	799	.255	214
St. Louis	779	.281	174
Florida	770	.254	208
Atlanta	753	.270	130
Milwaukee	750	.253	198
Colorado	747	.263	160
Pittsburgh	735	.258	153
Arizona	720	.251	159
Houston	712	.263	167
Cincinnati	704	.247	187
LA Dodgers	700	.264	137
Washington	641	.251	117
San Francisco	640	.262	94
San Diego	637	.250	154

Data Set H

Individual passing yards and ratings for 20 quarterbacks in the National Football League for the 2008 season. (**Source**: http://www.nfl.com/stats)

Name	Yards	Rating
Drew Brees	5,069	96.2
Kurt Warner	4,583	96.9
Jay Cutler	4,526	86.0
Aaron Rodgers	4,038	93.8
Philip Rivers	4,009	105.5
Peyton Manning	4,002	95.0
Donovan McNabb	3,916	86.4
Matt Cassel	3,693	89.4
Chad Pennington	3,653	97.4
David Garrard	3,620	81.7
Brett Favre	3,472	81.0
Tony Romo	3,448	91.4
Matt Ryan	3,440	87.7
Ben Roethlisberger	3,301	80.1
Jake Delhomme	3,288	84.7
Jason Campbell	3,245	84.3
Eli Manning	3,238	86.4
Matt Schaub	3,043	92.7
Kyle Orton	2,972	79.6
Joe Flacco	2,971	80.3

Appendix III

Scientific Notation

It is often inconvenient to express very large numbers or very small numbers by using ordinary decimal representation. For such numbers, the use of *scientific notation* allows a more concise form. A number written in scientific notation is expressed as a product

$$A \times 10^n$$

where A is a number between 1 and 10 (called the *leading part*) and n is an integer. The number of digits contained in the leading part is called the number of *significant digits*.

? Example 1

$$93,000,000 = 9.3 \times 10^7 \qquad 0.00000093 = 9.3 \times 10^{-7}$$

$$378,500,000,000 = 3.785 \times 10^{11} \qquad 0.000774 = 7.74 \times 10^{-4}$$

$$56,000 = 5.6 \times 10^4 \qquad 0.0000000000081 = 8.1 \times 10^{-12} \quad \blacklozenge$$

Note that to write a number in scientific notation, we shift the decimal point to the right or left, counting the decimal places as we go, until only one nonzero digit remains to the left of the point. The corresponding exponent n will be positive if you have moved the point to the left (in the case of a very large number) or negative if you have moved it to the right (in the case of a very small number). Additionally, $|n|$ must necessarily be equal to that number of shifted places.

When you are multiplying (or dividing) two numbers expressed in scientific notation, it is easiest to use your calculator to first multiply (or divide) only the leading parts and then determine the product of the powers of 10 by adding (or subtracting) their exponents. Then you usually express the final result in scientific notation. Since we can never claim greater accuracy in a final result than the least accurate input to the computation, the final leading part should be rounded off to a number of significant digits equal to the least number of sigificant digits of any input.

? Example 2

$$\left(5.34\times10^4\right)\left(2.17\times10^8\right)=11.6\times10^{12}=1.16\times10^{13}$$

$$\left(3.50\times10^{-5}\right)\left(6.379\times10^{23}\right)=22.3\times10^{18}=2.23\times10^{19}$$

$$\frac{4.93\times10^{15}}{7.2886\times10^7}=0.676\times10^8=6.76\times10^7$$

$$\frac{9.533\times10^6}{1.80591\times10^{32}}=17.21\times10^{-26}=1.721\times10^{-25}$$

$$\frac{\left(8.392\times10^{15}\right)\left(3.78\times10^{-28}\right)}{2.9961\times10^{-5}}=10.6\times10^{-8}=1.06\times10^{-7}$$

$$\frac{\left(4.881\times10^{56}\right)\left(5.5977\times10^{-12}\right)}{\left(7.462\times10^9\right)\left(6.7323\times10^{11}\right)}=0.5439\times10^{24}=5.439\times10^{23} \quad \blacklozenge$$

Name _____

Exercise Set

Convert to scientific notation.

1. 45,730,000

2. 966,200,000,000

3. 0.000238

4. 0.000000077831

Convert to ordinary decimal notation.

5. 6.71×10^5

6. 8.185×10^{10}

7. 3.0032×10^{-3}

8. 1.15×10^{-8}

Compute and express the result in scientific notation.

9. $(6.35 \times 10^6)(1.7 \times 10^{15})$

10. $(5.1259 \times 10^{23})(4.30993 \times 10^{-12})$

11. $\dfrac{8.46 \times 10^{41}}{2.00 \times 10^6}$

12. $\dfrac{4.6193 \times 10^{20}}{7.3325 \times 10^{55}}$

13. $\dfrac{\left(4.3 \times 10^{31}\right)\left(6.89 \times 10^{-18}\right)}{1.088 \times 10^6}$

14. $\dfrac{\left(4.338974 \times 10^{-23}\right)\left(6.5050 \times 10^{-11}\right)}{\left(3.4451 \times 10^{-17}\right)\left(2.0996 \times 10^{-8}\right)}$

Answers

1. 4.573×10^7

2. 9.662×10^{11}

3. 2.38×10^{-4}

4. 7.7831×10^{-8}

5. 671,000

6. 81,850,000,000

7. 0.0030032

8. 0.0000000115

9. 1.1×10^{22}

10. 2.2092×10^{12}

11. 4.23×10^{35}

12. 6.2998×10^{-36}

13. 2.7×10^8

14. 3.9021×10^{-9}

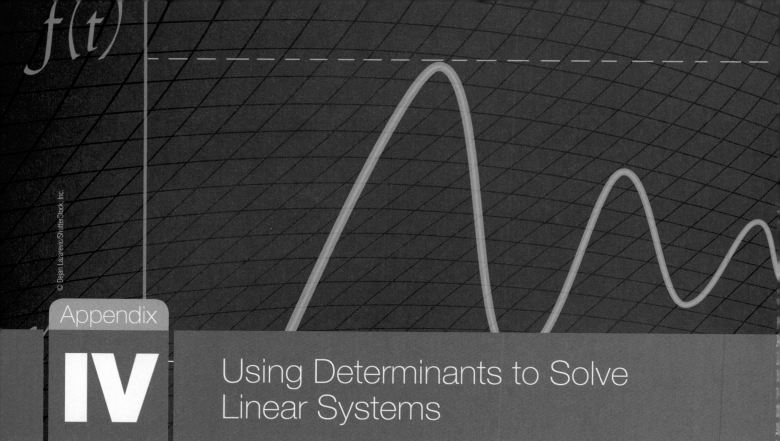

IV

Using Determinants to Solve Linear Systems

IV.1 Determinants

Definition 1 Let $A = \begin{bmatrix} a & b \\ c & d \end{bmatrix}$ be a 2×2 matrix. The determinant of A, written $|A|$, is an associated real number defined by

$$|A| = \begin{vmatrix} a & b \\ c & d \end{vmatrix}$$

$$= ad - bc.$$

? Example 1

(a) $\begin{vmatrix} 3 & 5 \\ 2 & 6 \end{vmatrix} = 3(6) - 2(5) = 18 - 10 = 8$

(b) $\begin{vmatrix} 4 & -2 \\ 8 & 11 \end{vmatrix} = 4(11) - 8(-2) = 44 + 16 = 60$

(c) $\begin{vmatrix} -7 & 1 \\ 5 & 3 \end{vmatrix} = (-7)(3) - 5(1) = -21 - 5 = -26$

(d) $\begin{vmatrix} 20 & 1.6 \\ -4 & -0.75 \end{vmatrix} = 20(-0.75)-(-4)(1.6) = -15+6.4 = -8.6$

(e) $\begin{vmatrix} 12 & 4 \\ 9 & 3 \end{vmatrix} = 36-36 = 0$

(f) $\begin{vmatrix} -38 & 22 \\ 0 & 0 \end{vmatrix} = 0-0 = 0$ ◆

We see that all possible real numbers may result from finding a determinant. In particular, note that the zero determinant in part (f) resulted from the presence of a row consisting entirely of zeros.

Definition 2 The determinant of a 3×3 matrix $\begin{bmatrix} a_{11} & a_{12} & a_{13} \\ a_{21} & a_{22} & a_{23} \\ a_{31} & a_{32} & a_{33} \end{bmatrix}$ is defined to be

$$\begin{vmatrix} a_{11} & a_{12} & a_{13} \\ a_{21} & a_{22} & a_{23} \\ a_{31} & a_{32} & a_{33} \end{vmatrix} = a_{11}\begin{vmatrix} a_{22} & a_{23} \\ a_{32} & a_{33} \end{vmatrix} - a_{21}\begin{vmatrix} a_{12} & a_{13} \\ a_{32} & a_{33} \end{vmatrix} + a_{31}\begin{vmatrix} a_{12} & a_{13} \\ a_{22} & a_{23} \end{vmatrix}.$$

This is known as *expanding along the first column*. Note the minus sign in front of the second term.

? Example 2

(a) $\begin{vmatrix} 2 & 8 & 7 \\ 3 & -5 & 0 \\ -4 & 3 & -1 \end{vmatrix} = 2\begin{vmatrix} -5 & 0 \\ 3 & -1 \end{vmatrix} - 3\begin{vmatrix} 8 & 7 \\ 3 & -1 \end{vmatrix} + (-4)\begin{vmatrix} 8 & 7 \\ -5 & 0 \end{vmatrix}$

$$= 2(5) - 3(-29) + (-4)(35)$$
$$= -43$$

(b) $\begin{vmatrix} 5 & 0 & -2 \\ -6 & 1 & 7 \\ 0 & -3 & 4 \end{vmatrix} = 5\begin{vmatrix} 1 & 7 \\ -3 & 4 \end{vmatrix} - (-6)\begin{vmatrix} 0 & -2 \\ -3 & 4 \end{vmatrix} + 0\begin{vmatrix} 0 & -2 \\ 1 & 7 \end{vmatrix}$

$$= 5(25) + 6(-6) + (0)(2)$$
$$= 89$$

(c) $\begin{vmatrix} 9.4 & -5 & 11 \\ 7 & 2.3 & 8 \\ 0 & 0 & 0 \end{vmatrix} = 9.4 \begin{vmatrix} 2.3 & 8 \\ 0 & 0 \end{vmatrix} - 7 \begin{vmatrix} -5 & 11 \\ 0 & 0 \end{vmatrix} + (0) \begin{vmatrix} -5 & 11 \\ 2.3 & 8 \end{vmatrix}$

$$= 9.4(0) - 7(0) + (0)(-65.3)$$
$$= 0$$

For the matrix $A = \begin{bmatrix} a_{11} & a_{12} & a_{13} \\ a_{21} & a_{22} & a_{23} \\ a_{31} & a_{32} & a_{33} \end{bmatrix}$, each of the *submatrices* $M_{11} = \begin{vmatrix} a_{22} & a_{23} \\ a_{32} & a_{33} \end{vmatrix}$,

$M_{21} = \begin{vmatrix} a_{12} & a_{13} \\ a_{32} & a_{33} \end{vmatrix}$, and $M_{31} = \begin{vmatrix} a_{12} & a_{13} \\ a_{22} & a_{23} \end{vmatrix}$ used in finding $|A|$ is called a minor of

the matrix A. So we could have written

$$|A| = a_{11}|M|_{11} - a_{21}|M|_{21} + a_{31}|M|_{31}.$$

We can define minors for any square matrix.

Definition 3 Let A be an $n \times m$ matrix. We define the minor M_{ij} to be the $(n-1)$ $\times (n-1)$ matrix obtained by deleting the ith row and the jth column from A.

Definition 4 Let $A = \begin{bmatrix} a_{11} & a_{12} & a_{13} & a_{14} \\ a_{21} & a_{22} & a_{23} & a_{24} \\ a_{31} & a_{32} & a_{33} & a_{34} \\ a_{41} & a_{42} & a_{43} & a_{44} \end{bmatrix}$ be a 4×4 matrix. In this case,

$$|A| = a_{11}|M_{11}| - a_{21}|M_{21}| + a_{31}|M_{31}| - a_{41}|M_{41}|.$$

Note the minus signs in front of both the second and the fourth terms.

? Example 3

$\begin{vmatrix} 2 & -3 & 1 & 5 \\ -3 & 1 & 0 & -1 \\ -5 & 2 & 3 & -4 \\ 4 & 6 & -2 & 0 \end{vmatrix} = 2 \begin{vmatrix} 1 & 0 & -1 \\ 2 & 3 & -4 \\ 6 & -2 & 0 \end{vmatrix} - (-3) \begin{vmatrix} -3 & 1 & 5 \\ 2 & 3 & -4 \\ 6 & -2 & 0 \end{vmatrix} + (-5) \begin{vmatrix} -3 & 1 & 5 \\ 1 & 0 & -1 \\ 6 & -2 & 0 \end{vmatrix}$

$$-4 \begin{vmatrix} -3 & 1 & 5 \\ 1 & 0 & -1 \\ 2 & 3 & -4 \end{vmatrix}$$

$$= 2(14) + 3(-110) - 5(-10) - 4(8)$$
$$= -284. \quad \blacklozenge$$

Name _____

Exercise Set **IV.1** ◯◯◯ ⟵

Find the following determinants.

1. $\begin{vmatrix} 3 & 5 \\ 2 & 6 \end{vmatrix}$

2. $\begin{vmatrix} 3 & -2 \\ 7 & -3 \end{vmatrix}$

3. $\begin{vmatrix} 13 & 45 \\ 0.2 & 0 \end{vmatrix}$

4. $\begin{vmatrix} -6 & 8 \\ -3 & 4 \end{vmatrix}$

5. $\begin{vmatrix} 2 & 5 & -2 \\ 3 & 0 & 1 \\ -1 & 4 & 3 \end{vmatrix}$

6. $\begin{vmatrix} -2 & 0 & -5 \\ -3 & 2 & 0 \\ 2 & 1 & 6 \end{vmatrix}$

7. $\begin{vmatrix} 8 & 3 & -10 \\ 0 & 0 & 0 \\ 5 & -7 & 54 \end{vmatrix}$

8. $\begin{vmatrix} 1 & 2 & 3 \\ 0 & 4 & 5 \\ 0 & 0 & 6 \end{vmatrix}$

9. $\begin{vmatrix} 3 & -1 & 7 \\ -6 & 2 & -14 \\ 5 & 4 & -2 \end{vmatrix}$

10. $\begin{vmatrix} 2 & 3 & -4 & 0 \\ 1 & -2 & 0 & 2 \\ -1 & 4 & 1 & -3 \\ 3 & 0 & 5 & 1 \end{vmatrix}$

11. $\begin{vmatrix} 2 & -6 & 4 & 7 \\ 0 & 1 & 5 & 12 \\ 0 & 0 & -4 & 3 \\ 0 & 0 & 0 & 15 \end{vmatrix}$

12. $\begin{vmatrix} -11 & 5 & 0 & 16 \\ 3 & -4 & 0 & 2 \\ 8 & 7 & 0 & 1 \\ -9 & 9 & 0 & 10 \end{vmatrix}$

Solutions

1. 8   2. 5   3. −9   4. 0   5. −82   6. 11

7. 0   8. 24   9. 0   10. 32   11. −120   12. 0

IV.2 Properties of Determinants

Theorem 1 Let A be an $n \times n$ matrix.

(a) The determinant of an upper triangular matrix is the product of the diagonal elements.

(b) If any row or column of A consists of only zeros, then $|A| = 0$.

(c) If any row or column of A is multiplied by the constant number c, then $|A|$ is multiplied by c.

(d) Interchanging any two rows or columns of A has the effect of multiplying $|A|$ by -1.

(e) If a multiple of one row (column) of A is added to or subtracted from any other row (column), then $|A|$ is unchanged.

(f) If two rows (columns) are identical, then $|A| = 0$.

(g) If one row (column) of A is a multiple of any other row (column), then $|A| = 0$.

? Example 1

(a) $\begin{vmatrix} 4 & -2 & 8 & 1 \\ 0 & 3 & -7 & 11 \\ 0 & 0 & -9 & 5 \\ 0 & 0 & 0 & -2 \end{vmatrix} = 216$

(b) $\begin{vmatrix} 3 & 0 & -7 \\ 5 & 0 & 12 \\ 6.2 & 0 & 4 \end{vmatrix} = 0$

(c) $\begin{vmatrix} 3 & -2 & 0 \\ 5 & 0 & 1 \\ 7 & -4 & 5 \end{vmatrix} = 48$. Therefore, $\begin{vmatrix} 3 & -2 & 0 \\ 5 & 0 & 1 \\ 14 & -8 & 10 \end{vmatrix} = 96$.

(d) $\begin{vmatrix} 3 & -2 & 0 \\ 5 & 0 & 1 \\ 7 & -4 & 5 \end{vmatrix} = 48$. Therefore, $\begin{vmatrix} 7 & -4 & 5 \\ 5 & 0 & 1 \\ 3 & -2 & 0 \end{vmatrix} = -48$.

(e) $\begin{vmatrix} 2 & -1 & 4 \\ -5 & 2 & -3 \\ -4 & 0 & 1 \end{vmatrix} = 19$. Add 4(row 1) to row 2. We get $\begin{vmatrix} 2 & -1 & 4 \\ 3 & -2 & 13 \\ -4 & 0 & 1 \end{vmatrix} = 19$.

(f) $\begin{vmatrix} 6 & -1 & 3 \\ 6 & -1 & 3 \\ -7 & 4 & 9 \end{vmatrix} = 0$

(g) $\begin{vmatrix} 6 & -1 & 3 \\ 12 & -2 & 6 \\ -7 & 4 & 9 \end{vmatrix} = 0$ ♦

Use of the above properties can reduce the difficulty of finding some determinants.

❓ Example 2

(a) $\begin{vmatrix} 0 & 4 & 5 \\ 0 & 0 & -2 \\ 3 & 7 & -6 \end{vmatrix} = (-1)\begin{vmatrix} 0 & 4 & 5 \\ 3 & 7 & -6 \\ 0 & 0 & -2 \end{vmatrix} = (-1)^2\begin{vmatrix} 3 & 7 & -6 \\ 0 & 4 & 5 \\ 0 & 0 & -2 \end{vmatrix} = (3)(4)(-2) = -24.$

(b) $\begin{vmatrix} 2 & -3 & 1 \\ 6 & -2 & -5 \\ -4 & -1 & 8 \end{vmatrix} = \begin{vmatrix} 2 & -3 & 1 \\ 0 & 7 & -8 \\ -4 & -1 & 8 \end{vmatrix}$ We added $(-3)(\text{row }1)$ to row 2.

$= \begin{vmatrix} 2 & -3 & 1 \\ 0 & 7 & -8 \\ 0 & -7 & 10 \end{vmatrix}$ We added $2(\text{row }1)$ to row 3.

$= \begin{vmatrix} 2 & -3 & 1 \\ 0 & 7 & -8 \\ 0 & 0 & 2 \end{vmatrix}$ We added $1(\text{row }2)$ to row 3.

Note: This process is called *triangulation* of the matrix.   ♦

Name _____

Exercise Set **IV.2** ⬜⬜⬜ ←

Find the following determinants.

1.
$$\begin{vmatrix} -3 & 1 & 2 \\ 0 & 2 & 4 \\ 0 & 0 & -7 \end{vmatrix}$$

2.
$$\begin{vmatrix} 0 & 2 & 5 \\ 4 & -7 & -1 \\ 0 & 0 & 11 \end{vmatrix}$$

3.
$$\begin{vmatrix} 9 & -2 & 7 \\ 0 & 0 & 6 \\ 0 & -5 & 3 \end{vmatrix}$$

4.
$$\begin{vmatrix} 1 & -3 & 6 \\ 0 & 2 & 7 \\ 3 & -17 & 5 \end{vmatrix}$$

5.
$$\begin{vmatrix} 5 & 2.8 & 0 \\ -1 & 3 & 0 \\ 9.4 & -5 & 0 \end{vmatrix}$$

6.
$$\begin{vmatrix} 2 & 9 & -3 \\ 8 & -5 & 1 \\ 2 & 9 & -3 \end{vmatrix}$$

7.
$$\begin{vmatrix} 6 & 1 & -1 \\ 2 & -3 & 3.5 \\ -4 & 6 & -7 \end{vmatrix}$$

8.
$$\begin{vmatrix} 5 & -3 & 1 & 3 \\ 0 & -2 & 4 & 1.8 \\ 0 & 0 & 5 & -6.1 \\ 0 & 0 & 0 & 7 \end{vmatrix}$$

9.
$$\begin{vmatrix} -4 & 3 & 6 & 2 \\ 0 & 0 & 8 & 1 \\ 0 & 2 & -1 & 4 \\ 0 & 0 & 0 & 3 \end{vmatrix}$$

10.
$$\begin{vmatrix} 1 & 2 & 3 & -1 \\ 0 & 4 & 5 & 2 \\ 0 & 0 & 1 & -4 \\ 5 & 6 & 0 & 7 \end{vmatrix}$$

Solutions

1. 42 **2.** −88 **3.** 270 **4.** 30 **5.** 0

6. 0 **7.** 0 **8.** −350 **9.** 192 **10.** −104

Section IV.3 Cramer's Rule

Definition 1 Consider the $n \times n$ linear system:

$$a_{11}x_1 + a_{12}x_2 + \cdots + a_{1n}x_n = b_1$$
$$a_{21}x_1 + a_{22}x_2 + \cdots + a_{2n}x_n = b_2$$
$$\cdots \quad \cdots \quad \cdots \quad \cdots \quad \cdots$$
$$a_{n1}x_1 + a_{n2}x_2 + \cdots + a_{nn}x_n = b_n$$

Define $A = \begin{bmatrix} a_{11} & a_{12} & \cdots & a_{1n} \\ a_{21} & a_{22} & \cdots & a_{2n} \\ \cdots & \cdots & \cdots & \cdots \\ a_{n1} & a_{n2} & \cdots & a_{nn} \end{bmatrix}$. This is called the *coefficient matrix* of the system.

Also, define

$$A_1 = \begin{bmatrix} b_1 & a_{12} & \cdots & a_{1n} \\ b_2 & a_{22} & \cdots & a_{2n} \\ \cdots & \cdots & \cdots & \cdots \\ b_n & a_{n2} & \cdots & a_{nn} \end{bmatrix}, A_2 = \begin{bmatrix} a_{11} & b_1 & \cdots & a_{1n} \\ a_{21} & b_2 & \cdots & a_{2n} \\ \cdots & \cdots & \cdots & \cdots \\ a_{n1} & b_n & \cdots & a_{nn} \end{bmatrix}$$

$$A_n = \begin{bmatrix} a_{11} & a_{12} & \cdots & b_1 \\ a_{21} & a_{22} & \cdots & b_2 \\ \cdots & \cdots & \cdots & \cdots \\ a_{n1} & a_{n2} & \cdots & b_n \end{bmatrix}.$$

In other words, the matrix A_j for $j = 1, 2, \ldots, n$ is defined by **replacing** the jth column of A with a column composed of the constants $b_1, b_2, \ldots, b_n$ from the system.

Theorem 2 Cramer's Rule The $n \times n$ linear system

$$a_{11}x_1 + a_{12}x_2 + \cdots + a_{1n}x_n = b_1$$
$$a_{21}x_1 + a_{22}x_2 + \cdots + a_{2n}x_n = b_2$$
$$\cdots \quad \cdots \quad \cdots \quad \cdots \quad \cdots \quad \cdots$$
$$a_{n1}x_1 + a_{n2}x_2 + \cdots + a_{nn}x_n = b_n$$

has a unique solution if and only if $|A| \neq 0$ where A is the matrix of coefficients. If the system has a unique solution, then it is given by

$$x_1 = \frac{|A_1|}{|A|}, x_2 = \frac{|A_2|}{|A|}, \ldots, x_n = \frac{|A_n|}{|A|}.$$

? Example 1

Solve the system by using Cramer's rule.

$$2x + 7y = -3,$$
$$5x - y = 4.$$

⚙ Solution

The coefficient matrix A is $A = \begin{bmatrix} 2 & 7 \\ 5 & -1 \end{bmatrix}$, and it has determinant $|A| = -37$.

The other two required matrices are

$$A_1 = \begin{bmatrix} -3 & 7 \\ 4 & -1 \end{bmatrix} \text{ with determinant } |A_1| = -25$$

and

$$A_2 = \begin{bmatrix} 2 & -3 \\ 5 & 4 \end{bmatrix} \text{ with determinant } |A_2| = 23.$$

By Cramer's rule our solution is given by

$$x = \frac{-25}{-37} = \frac{25}{37}, \quad y = \frac{23}{-37}. \quad ◆$$

Name _____

Exercise Set **IV.3** ☐☐☐←

Solve the following linear systems by using Cramer's rule.

1. $3x + 4y = 5$
  $2x + y = -6$

2. $5x - 8y = -3$
  $-6x + 3y = 1$

3. $-x + 7y = -2$
  $3x + 8y = 9$

4. $8x + 3y = 5$
  $x - 2y = 0$

5. $x_1 - 3x_2 + 6x_3 = 2$
  $2x_2 + 7x_3 = -4$
  $3x_1 - 17x_2 + 5x_3 = 1$

6. $4x_1 - x_2 - 2x_3 = -3$
  $5x_1 + 2x_2 = 4$
  $-x_1 + 3x_3 = -1$

7. $x_1 + 2x_2 + 3x_3 = -2$
  $4x_1 + 5x_2 + 6x_3 = 5$
  $3x_1 + x_2 - 2x_3 = 74$

8. $2x_1 + 2x_2 - x_3 + 3x_4 = 5$
  $x_1 - 2x_2 + x_4 = 1$
  $-3x_1 + x_3 + 6x_4 = 0$
  $4x_1 + x_2 - 5x_3 + x_4 = -7$

Solutions

1. $\left(\dfrac{29}{5}, \dfrac{28}{5} \right)$

3. $\left(\dfrac{47}{13}, \dfrac{3}{13} \right)$

5. $\left(\dfrac{191}{10}, \dfrac{29}{10}, -\dfrac{7}{5} \right)$

7. $\left(\dfrac{46}{3}, -\dfrac{65}{3}, \dfrac{26}{3} \right)$

An interval is an infinite set of real numbers of any of the following forms.

$[a, b] = \{x \mid a \leq x \leq b\}$

$(a, b) = \{x \mid a < x < b\}$

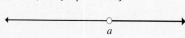

$[a, b) = \{x \mid a \leq x < b\}$

$(a, b] = \{x \mid a < x \leq b\}$

$[a, \infty) = \{x \mid x \geq a\}$

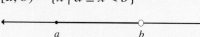

$(a, \infty) = \{x \mid x > a\}$

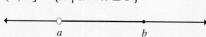

$(-\infty, a] = \{x \mid x \leq a\}$

$(-\infty, a) = \{x \mid x < a\}$

Exploring Mathematics: Investigations with Functions

Solutions to Odd-Numbered Exercises

C. M. Johnson

Chapter 1

Section 1.1

1. **D** = {Amsterdam, Calgary, Dublin, Lisbon, Manila, Oslo, Rome, Singapore, Toyko, Vienna}; **R** = {14, 24, 27, 36, 39, 41, 43, 66, 75}; f(Amsterdam) = 39, f(Calgary) = 14, f(Dublin) = 41.

3. No. It is likely that one weight will correspond to more than one player.

5. Every social security number corresponds to exactly one student.

7. No, because no unique assignment is possible for numbers which are midway between multiples of 4. Example: 22 could be assigned to either 20 or 24.

9. No. You could not define $f(2.25)$ to be unique.

11. Decreasing.

13. 7, 28, 20, 217

15. $\tau(p) = 2$; $\sigma(p) = 1 + p$

17. $f(-6) = 9, f(-1) = -6, f(0) = -3, f(2) = 9, f(10) = 137$

19. $h(-1) = 0.5, h(1) = 0.9, h(8) = 2.3, h(25) = 5.7, h(100) = 20.7$

21. $D(-3) = 24, D(-2) = 6\sqrt{11} \approx 19.9, D(0) = 6\sqrt{7} \approx 15.9, D(5) \approx 33.9$

23. $G(0) = \dfrac{7}{3}, G(1) = 2, G(2) = \dfrac{9}{7}, G(3) = \dfrac{5}{6}$

25. (a) $S(t) = 7300(1 + 0.045t)$

(b) \$8285.50;   (c) \$8942.50

27. $T(45,000) = 9400; T(52,500) = 11,200.$

29. $C(70) = \$31, C(100) = \$40, C(450) = \$145;$ Increasing.

31. Decreasing; $v(9.0 \times 10^6) = 6.7 \times 10^3$ m/sec, 5.0×10^3 m/sec, 4.5×10^3 m/sec.

33. 0.392; 0.81

35. Closer; Increase.

37. $\dfrac{49}{128}; \dfrac{9}{16}; \dfrac{75}{128}$

Section 1.2

1. $f(x) = x + 10$

3. $f(x) = 2x + 7$

5. $f(x) = 6x - 1$

7. $f(x) = -4x + 5$

9. $f(x) = 3.5x + 8$

11. $f(x) = -1.8x - 3$

13. $f(x) = 5.1x - 11.5$

15. $f(x) = -4.7x + 2$

17. $f(x) = 3 \cdot 2^x$

19. $f(x) = 4 \cdot 10^x$

21. $f(x) = 10 \cdot 5^x$

23. $f(x) = 3 \cdot (0.5)^x$

25. $f(x) = 1000 \cdot (0.2)^x$

27. $V(t) = 5200 - 225t; V(5) = \$4075; V(8) = \$3400$

29. $C(x) = 32.5x + 25; C(4.5) = \$171.25$

31. $T(x) = 3750 + 0.28(x - 30,000); T(32,450) = 4436$

33. $f(x) = -0.21x + 73.7;$ Life expectancy decreases by 0.21 years per cigarette.

35. $F(x) = \dfrac{9}{5}x + 32; F(45) = 113°$ F.

37. $A(t) = 750(1.042)^t; \$848.52; \$921.30$

39. \$2612.91

41. 1.94 million; 2.03 million; 2.12 million

43. 7.3 billion; 7.8 billion; 8.2 billion

45. 121.4 million; 130.2 million

47. Uganda: 20.6 years; Mexico: 50.0 years; Finland: 350 years

49. Only 1 cupcake is left at 5:00. $C(x) = 512 \cdot (0.5)^x$

51. $V(t) = 22000(0.9)^t; V(12) = \$6213.45$

53. $A(5) = 15.3$ gm; $A(8) = 13.1$ gm

55. About $60,300

57. $f(x) = 5(x - 4)^2$

59. $f(x) = \sqrt{\dfrac{x + 3}{8}}$

61. $f(x) = \dfrac{\sqrt{(x + 6)^4 + 7}}{x - 1}$

Section 1.3

1.

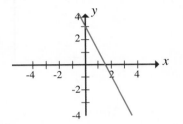

3.

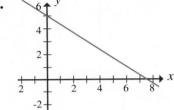

5.

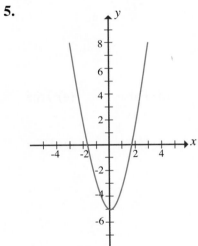

7.

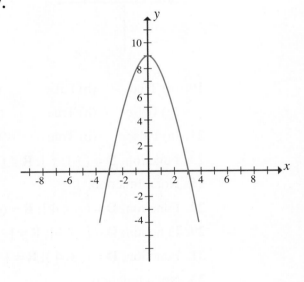

9.

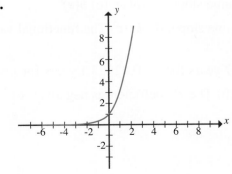

11.

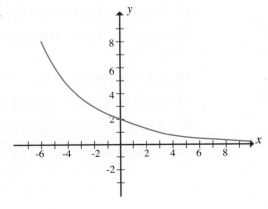

13.

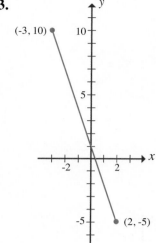

15.

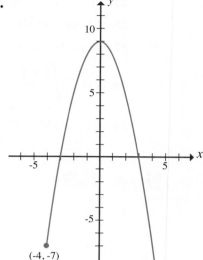

17.

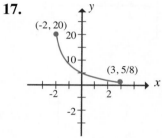

19. (a) True   (b) False   (c) True   (d) False   (e) True   (f) True
(g) False   (h) True

21. (a) False   (b) True   (c) True   (d) False   (e) False   (f) True

23. Function; $\mathbf{D} = [-1, 5]$; $\mathbf{R} = [-4, 2]$

25. Not a function.

27. Function; $\mathbf{D} = (-\infty, 4]$; $\mathbf{R} = (-\infty, 2]$

29. Function; $\mathbf{D} = [-3, 4]$; $\mathbf{R} = [-1, 2]$

31. Function; $\mathbf{D} = [-3, 4)$; $\mathbf{R} = \{-3, -2, -1, 0, 1, 2, 3\}$

33. Not a function.

35. (a) $f(x)$ and $g(x)$; They have the same slope; (b) $h(x)$; (c) $h(x)$

37. (a) $f(x)$ and $g(x)$; They have the same slope; (b) k; (c) The functional values are always positive.

39. (a) 5000; (b) $f(t) = 5\,e^{0.03t}$; (c) 34.7 years for $r = 0.02$; 23.1 years for $r = 0.03$

41. (a) It opens down instead of up; (b) The x^2 coefficient is negative; (c) -3, 2, and 4

43. It rotates counterclockwise.

45. (a) $f(x) = x^2 - 3$; (b) $f(x) = -2x + 3$

Section 1.4

1. **D** = $(-\infty, \infty)$, **R** = $[-1, \infty)$; decreasing over $(-\infty, 1]$; increasing over $[1, \infty)$; absolute minimum of -1 at $x = 1$; no maximum; zeroes at $x = 0$ and 2.

3. **D** = $[-4, 4]$, **R** = $[-2, 2]$; increasing over $[-4, 0]$; decreasing over $[0, 4]$; absolute minimum of -2 at $x = -4$ and 4; absolute maximum of 2 at $x = 0$; zeroes at x = -2 and 2.

5. **D** = $(-\infty, 4]$, **R** = $[-2, \infty)$; decreasing over $(-\infty, -2]$ and $[2, 4]$; increasing over $[-2, 2]$; strictly local minimum of 1 at $x = -2$; strictly local maximum of 3 at $x = 2$; absolute minimum of -2 at $x = 4$; no absolute maximum; zero at $x = 3$.

7. **D** = $[-4, \infty)$, **R** = $[-4, \infty)$; increasing over $[-4, -2]$ and $[2, \infty)$; decreasing over $[-2, 2]$; absolute minimum of -4 at $x = -4$; strictly local maximum of 4 at $x = -2$; strictly local minimum of 1 at $x = 2$; no absolute maximum; zero at $x = -3$.

9. **D** = $[-5, 4]$, **R** = $[-3, 1]$; increasing over $[-5, -3]$ and $[0, 2]$; decreasing over $[-3, 0]$ and $[2, 4]$; strictly local minima of -1 and -2 at $x = -5$ and 0 respectively; strictly local maximum of 0 at $x = 2$; absolute minimum of -3 at $x = 4$; absolute maximum of 1 at $x = -3$; zeroes at $x = -4, -2$, and 2.

11. Maximum of 6 at $x = 1.5$; minimum of -1 at $x = -2$; increasing over $[-2, 1.5]$; zero at $x = -1.5$

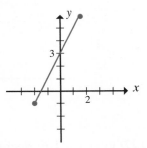

13. Maximum of 3.8 at $x = 2$; minimum of 2.2 at $x = 3$; decreasing over $[2, 3]$.

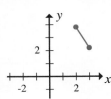

15. Maximum of 4 at $x = 0$; increasing over $(-\infty, 0]$; decreasing over $[0, \infty)$; zeroes at $x = -2$ and 2.

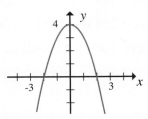

17. Maximum of 1 at $x = -1$; increasing over $(-\infty, 1]$; decreasing over $[1, \infty)$; zeroes at $x = -2$ and 0.

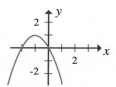

19. Minimum of 0.75 at $x = 2.5$; decreasing over $(-\infty, 2.5]$; increasing over $[2.5, \infty)$; no zeroes.

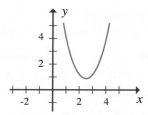

21. Maximum of 9 at $x = -1$; increasing over $(-\infty, -1]$; decreasing over $[-1, \infty)$; zeroes at $x = -4$ and 2.

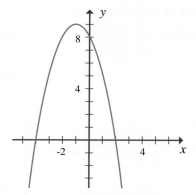

23. Minimum of $-147/36$ at $x = 5/6$; decreasing over $(-\infty, 5/6]$; increasing over $[5/6, \infty)$; zeroes at $x = -1/3$ and 2.

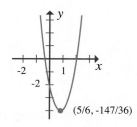

25. Maximum height $= 144$ ft at $t = 1$ sec; strikes the ground at $t = 4$ sec.

27. Maximum height $= 256$ ft at $t = 3$ sec; strikes the ground at $t = 7$ sec.

29. Local maximum of 3 at $x = 0$; absolute minimum of 2 at $x = -1$ and 1.

31. Local maximum of 2 at $x = 0$; local minimum ≈ -2.6 at $x \approx 1.7$; zeroes at $x \approx -0.6$, 0.7, 2.3

33. Local minimum ≈ -4.3 at $x \approx -1.7$; local minimum ≈ 0.1 at $x \approx 0.4$; local maximum ≈ 1.9 at $x \approx -0.5$; local maximum ≈ 6.3 at $x \approx 1.7$; zeroes at $x = -2, -1$, and 2.

35. $P(d) = 0.03d + 1$; $P(1500) = 46$ atm.

37. Maximum revenue is $1012.50 obtained for $x = 225$ donuts.

39. $A(x) = x(300 - x)$; Maximum area of 22,500 ft² for length $= 150$ ft and width $= 150$ ft.

41. (a) The supply function; the demand function

(b) Equilibrium price = $6 per bushel; equilibrium quantity = 4000 bushels per week.

(c) price greater than $6 per bushel; price less than $6 per bushel

43.

T (°**Kelvin**)	λ_{max} (**microns**)
1250	2.4
1500	2.0
1750	1.7
2000	1.5

λ_{max} is a decreasing function of temperature.

45. Displacement: graph #2; Velocity: graph #1

Chapter 1 Review Test

1. No. Some consonants are midway between two vowels and so there would be a choice for the assignment.

2. Domain = {People in the room}; Range = {eye colors}

3. The probability of heart disease for a person increases as the amount of fat consumed by the person increases.

4. Rene Descartes

5. Acceleration is a decreasing function of distance.

6. 393.4; 381; 366.4; decreasing.

7. $V(t) = -300t + 7200$; $V(10) = 4200$.

8. (a) $f(x) = 3x - 4$ (b) $g(x) = 5 \cdot 4^x$

9. (a) and (c)

10. (a) 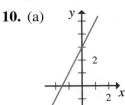   (b)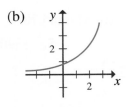

11. (a) $\mathbf{D} = [-3, 4]$; $\mathbf{R} = [-4, 2]$; Max = 2; Min = −4

(b) $\mathbf{D} = [-4, 2]$; $\mathbf{R} = [0, 3]$; Max = 3; Min = 0

12. The maximum height is 144 ft.; It takes 5 seconds for the object to hit the ground.

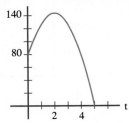

13. (a) True   (b) False   (c) False   (d) True   (e) True

14. (a) $[-2, 0]$; $[3, \infty)$ (b) 1; 3   (c) 3; 0   (d) −5 and −1

15. $A(t) = 3200(1.047)^t$; $A(2) = \$3507.87$; $A(5) = \$4026.09$

16. $A(t) = 217e^{0.016t}$; $A(4) = 231$ thousand; $A(8) = 246$ thousand

17. (a) $f(x) = 6(x + 3)^2$ (b) $f(x) = \sqrt[3]{7x - 4}$

Chapter 2

Section 2.1

1. (a) \$735.30	(b) \$5890.63	(c) \$554.79	
3. (a) \$5035.30	(b) \$70,890.63	(c) \$23,054.79	
5. (a) \$7526.43	(b) \$48,845.58	(c) \$121,811.22	
7. \$4.11			
9. (a) \$9063.83	(b) \$10,249.19	(c) \$12,408.57	(d) \$16,084.53
11. (a) \$2528.83	(b) \$2544.75	(c) \$2546.19	(d) \$17.36
13. (a) \$7338.51	(b) \$26,900.97	(c) \$76,769.62	
15. (a) 6.3%	(b) 6.45%	(c) 6.49%	(d) 6.50%

17. *Bank of Oz*: 6.18%; *Auntie Em*: 6.24%; *Auntie Em*

19. \$7238.95; \$7220.94; \$7212.17

21. \$4317.90   **23.** \$13,219.68   **25.** \$55,514.29

27. (a) \$4190.26   (b) \$235,575.22   (c) \$119,311.85

Section 2.2

1. (a) \$100,451.50   (b) \$17,765.78   (c) \$6741.70

3. \$68,427.34   **5.** \$340,884.78

7. $F(t) = 75\left(\dfrac{1.005^{12t} - 1}{0.005}\right)$

9. $F(5) = \$5232.75$; $F(10) = \$12,290.95$; $F(20) = \$34,653.07$

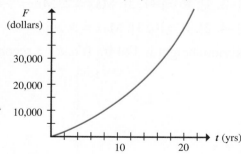

11. \$1615.61; \$1614.77

13. (a) \$216.43   (b) \$729.11   (c) \$73,333.60

15. \$117.51   **17.** \$11,101.37   **19.** \$126.56

21. $R = \$235.04$, Annual cost $= \$2820.48$; $R = \$54.85$, Annual cost $= \$2852.27$

23. $R = \$178.91$, Annual cost $= \$4651.66$; $R = \$386.99$, Annual cost $= \$4643.88$

25. (a) $\$100,451.50 - \$36,000 = \$64,451.50$

    (b) $\$17,765.78 - \$9000 = \$8765.78$

    (c) $\$6741.70 - \$4900 = \$1841.70$

27. (a) $\$100,000 - \$51,943.20 = \$48,056.80$

    (b) $\$62,000 - \$43,746.60 = \$18,253.40$

    (c) $\$3,500,000 - \$1,833,340 = \$1,666,660$

29. (a) 3906   (b) 3280   (c) 329,554,457   (d) $1\left(\dfrac{63}{64}\right)$

Section 2.3

1. (a) $\$16,679.16$ (b) $\$5094.56$ (c) $\$3520.57$

3. $\$86,117.35$

5. $\$2501.26$

7. (a) $R = \$560.43$, Total interest $= \$5350.96$

    (b) $R = \$6628.16$, Total interest $= \$445,816.00$

    (c) $R = \$16,630.72$, Total interest $= \$139,352.96$

9. $R = \$498.21$, Total interest $= \$2935.56$; Three-year term, two-year term

11. $R = \$1877.92$, Total interest $= \$5046.72$

13.

Number	Payment	Interest	Applied to Principal	Balance
1	734.62	644.00	90.62	91,909.38
2	734.62	643.37	91.25	91,818.13
3	734.62	642.73	91.89	91,726.24

15. (a)

Number	Payment	Interest	Applied to Principal	Balance
1	558.41	170.33	388.08	27,611.92
2	558.41	167.97	390.44	27,221.48

    (b) $B_{12} = 30,113.79 - 6929.73 = \$23,184.06$

    (c) First year interest $= 12(558.41) - 4815.94 = \$1884.98$

17. (a)

Number	Payment	Interest	Applied to Principal	Balance
1	553.96	440.94	113.02	82,886.98
2	553.96	440.34	113.62	82,773.36
3	553.96	439.73	114.23	82,659.13

    (b) $B_{60} = \$75,037.95$

19. $R = \$1119.61$, $B_6 = \$23,181.86$, Interest $= \$1899.52$

21. $\$42,299.90$

23. $\$354,203.32$

25. $P = R(1 + i)^{-1} + R(1 + i)^{-2} + \ldots + R(1 + i)^{-n}$

$= R(1 + i)^{-1} [\, 1 + R(1 + i)^{-1} + R(1 + i)^{-2} + \ldots + R(1 + i)^{-(n-1)} \,]$

$= R(1 + i)^{-1} \left(\dfrac{(1+i)^{-n} - 1}{(1+i)^{-n} - 1} \right)$

$= R \left(\dfrac{(1+i)^{-n} - 1}{1 - (1+i)} \right)$

$= R \left(\dfrac{(1+i)^{-n} - 1}{-i} \right)$

$= R \left(\dfrac{1 - (1+i)^{-n}}{i} \right)$

27. (a) $R = \$331.50$, Total cost $= 60(331.50) - 1200 = \$18,690$

(b) $R = \$306.91$, Total cost $= 60(306.91) = \$18,414.79$; Option (b).

Chapter 2 Review Test

1. $2.88

2. Annually: \$3345.45; Monthly: \$3365.31; Daily: \$3367.13

3. Central Bank: 6.67%; National Trust: 6.45%; Central Bank

4. \$69,299.40; \$70,757.10

5. $R = \$371,346.85$; Aggregate Contribution $= \$3,713,468.50$

6. $1615.21

7. (a)

Number	Payment	Interest	Applied to Principal	Balance
1	551.90	117.50	434.40	23,065.60
2	551.90	115.33	436.57	22,629.03
3	551.90	113.15	438.75	22,190.28

(b) $B_{12} = 24,949.43 - 6808.00 = \$18,141.43$

(c) $1264.23

Chapter 3

Section 3.1

1. (a), (d), and (e) are statements.

3.

x	p	L(p)
0	$40 \le 50$	True
3	$46 \le 50$	True
6	$52 \le 50$	False
9	$58 \le 50$	False
12	$64 \le 50$	False

5.

m	n	p	$L(p)$
0	10	$100 = 100$	True
3	5	$34 = 100$	False
6	6	$72 = 100$	False
6	8	$100 = 100$	True
7	7	$98 = 100$	False
8	6	$100 = 100$	True
9	1	$82 = 100$	False
10	0	$100 = 100$	True

7. (a) I do not ride my bike everyday.

(b) I ride my bike everyday or I weigh less than 200 pounds.

(c) If I ride my bike everyday, then I weigh less than 200 pounds.

(d) If I weigh less than 200 pounds, then I ride my bike everyday.

(e) I do not ride my bike everyday and I weigh less than 200 pounds.

(f) It is not the case that I ride my bike everyday and I weigh less than 200 pounds.

(g) I ride my bike everyday if and only if I do not weigh less than 200 pounds.

(h) It is not the case that I ride my bike everyday if and only if I weigh less than 200 pounds.

9. (a) $q \Rightarrow r$ (b) $q \wedge \sim p \wedge \sim r$ (c) $\sim p \wedge \sim q$ (d) $\sim r \Rightarrow \sim q$

(e) $\sim p \Rightarrow (q \vee r)$ (f) $p \Leftrightarrow \sim q$ (g) $r \Leftrightarrow (\sim p \wedge q)$ (h) $\sim (p \vee (q \wedge r))$

11. (a) "A polygon is a square," is the antecedent. "A polygon is a rectangle," is the consequent.

(b) "A number is rational," is the antecedent. "A number is real," is the consequent.

(c) "An electric current flows through a wire," is the antecedent . "A magnetic field is created," is the consequent.

(d) "His movies gross $50 million or are seen by 10 million people," is the antecedent. "His movies make a profit," is the consequent.

(e) "The topic is gun control or welfare," is the antecedent. "Much debate occurs in Congress," is the consequent.

(f) "It doesn't rain or snow," is the antecedent. "We go fishing," is the consequent.

13. (a) $p \vee \sim r$ (b) $(\sim p \wedge q) \vee r$ (c) $p \wedge \sim q \wedge \sim r$

(d) $\sim p \vee \sim (q \wedge r)$ (e) $(p \wedge q) \vee (\sim q \wedge r)$

15. No. It cannot be assigned a consistent logical value.

17. Yes.

Section 3.2

1. The statements, "Nick enjoys baseball," and "Nick enjoys spinach," must both be true.

3. False; True.

5. True; False.

7.

x	p	q	$L(p)$	$L(q)$	$L(p \wedge q)$	$L(p \vee q)$
500	$500 > 1000$	$500 < 1500$	F	T	F	T
1000	$1000 > 1000$	$1000 < 1500$	F	T	F	T
1250	$1250 > 1000$	$1250 < 1500$	T	T	T	T
1499	$1499 > 1000$	$1499 < 1500$	T	T	T	T
1500	$1500 > 1000$	$1500 < 1500$	T	F	F	T
1600	$1600 > 1000$	$1600 < 1500$	T	F	F	T

9.

x	y	p	q	$L(p)$	$L(q)$	$L(p \wedge q)$	$L(p \vee q)$
62	15	$62 \geq 75$	$77 = 100$	F	F	F	F
70	30	$70 \geq 75$	$100 = 100$	F	T	F	T
75	25	$75 \geq 75$	$100 = 100$	T	T	T	T
80	20	$80 \geq 75$	$100 = 100$	T	T	T	T
90	11	$90 \geq 75$	$101 = 100$	T	F	F	T

11. T   **13.** F   **15.** T

17.

p	q	$\sim p \Rightarrow q$
T	T	F **T** T
T	F	F **T** F
F	T	T **T** T
F	F	T **F** F

19.

p	q	$\sim(p \wedge q)$
T	T	**F** T T T
T	F	**T** T F F
F	T	**T** F F T
F	F	**T** F F F

21.

p	$p \wedge \sim p$
T	T **F** F
F	F **F** T

23.

p	q	$(p \Rightarrow q) \Rightarrow (q \Rightarrow p)$
T	T	T T T **T** T T T
T	F	T F F **F** T T T
F	T	F T T **F** T F F
F	F	F T F **T** F T F

25.

p	q	r	$(p \wedge q) \vee r$
T	T	T	T T T **T** T
T	T	F	T T T **T** F
T	F	T	T F F **T** T
T	F	F	T F F **F** F

p	q	r	$(p \wedge q) \vee r$
F	T	T	F F T T
F	T	F	F F T F F
F	F	T	F F F T T
F	F	F	F F F F F

27.

p	q	r	$p \Rightarrow (q \vee \sim r)$
T	T	T	T T T T F
T	T	F	T T T T T
T	F	T	T F F F F
T	F	F	T T F T T
F	T	T	F T T T F
F	T	F	F T T T T
F	F	T	F T F F F
F	F	F	F T F T T

29. T

31. Either logical value is possible.

33. F

35. T

37. F

39.

p	q	p<=>q			(p=>q) ∧ (q=>p)							
T	T	T	T	T	T	T	T	T	T	T	T
T	F	T	F	F	T	F	F	F	F	T	T
F	T	F	F	T	F	T	T	F	T	F	F
F	F	F	T	F	F	T	F	T	F	T	F

41.

p	q	p=>q			~pvq		
T	T	T	T	T	F	T	T
T	F	T	F	F	F	F	F
F	T	F	T	T	T	T	T
F	F	F	T	F	T	T	F

43. Either Senator Farnsworth did not vote for the tax cut or he did not vote for the new spending bill.

45. Erika did not receive a B in Math and she did not receive a C in English.

47. $x \leq 2$ or $y \neq 35$

49. $x + y \neq 4$ and $x \geq 13.5$

51. $y \geq 85$ and $y \leq 90$

53. $(x \geq 10 \text{ or } y \geq 10)$ and $2x + 3y \neq 50$

55. Converse: If the temperature of the earth's core is between 4000 and 5000° C., then it is made principally of iron.

Contrapositive: If the temperature of the earth's core is not between 4000 and 5000° C., then it is not made principally of iron.; The contrapositive.

57. Converse: If $f(2) = 19$, then $f(x) = 8x + 3$.

Contrapositive: If $f(2) \neq 19$, then $f(x) \neq 8x + 3$.; The contrapositive.

🏆**59.** $p \Rightarrow q \equiv \sim(p \wedge \sim q)$

🏆**61.** $p \wedge q \equiv \mathbf{NOR}(\mathbf{NOR}(p, q), \mathbf{NOR}(p, q))$

Section 3.3

1. Yes   **3.** No   **5.** No   **7.** Yes

9. No   **11.** Yes   **13.** Yes   **15.** Yes

17. True   **19.** True   **21.** False   **23.** False

***25.** True   **27.** Valid. *MP*   **29.** Invalid.   **31.** Valid. *MT*

33. Invalid.   **35.** Invalid.

37. Valid.   **39.** Invalid.

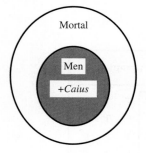   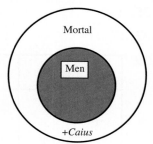

41. Valid.   **43.** Invalid.

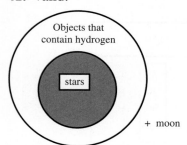   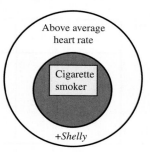

45. Valid.   **47.** Invalid.

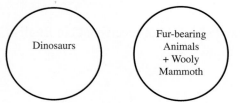   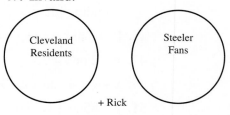

49. Invalid.

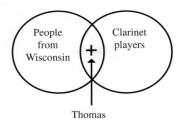

Thomas

51. Invalid.

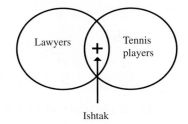

Ishtak

53. Alexander will pass his math class.

55. No conclusion is possible.

57. The cork has a density less than or equal to 1.0 gm/cm^3.

59. No conclusion is possible.

61. Popeye does not eat spinach.

63. Khantana is not a history major.

Section 3.4

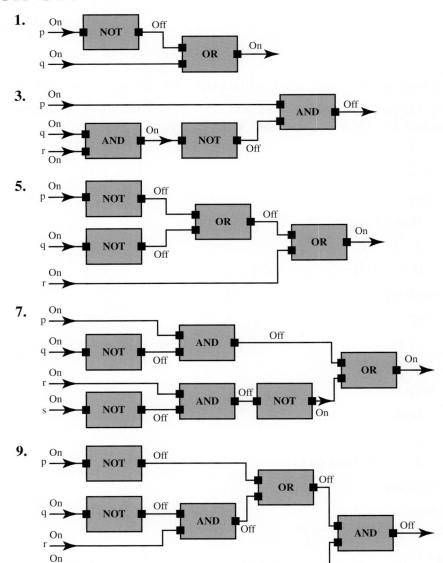

11. if $(x < y)$ **then**

   $y + 60 \rightarrow m$

   else

   $x + 60 \rightarrow m$

   endif

13. if $(r < 75$ or $r > 100)$ **then**

   $3 * r + 1 \rightarrow s$

   else

   $2 * r + 1 \rightarrow s$

   endif

15. if $(M = N$ and $M + N \leq 1000)$ **then**

   $sqrt(M + N) \rightarrow P$

   else

   $sqrt(M * N) \rightarrow P$

   endif

17. (a) 78   (b) 136   (c) 12

19. 2   **21.** 16

   4   4

   6   2

   8

   10

   12

23. The boolean expression, $n > 1$, will remain false and so the loop cannot be exited.

25. The boolean expression, $S > 40$ and $m < 12$, will remain false because $m < 12$ will remain false. Therefore, the loop cannot be exited.

27. 1 $3 \rightarrow x$

   2   $5 \rightarrow y$

   3   **loop**

   4   **flip**(x, y)

   5       $x + 1 \rightarrow x$

   6   **if** $(x = 9)$ **then go to** 8 **endif**

   7   **endloop**

   8   **stop**

29. 1 $7 \rightarrow x$

   2   $1 \rightarrow y$

   3   **loop**

   4   **flip**(x, y)

   5       $y + 1 \rightarrow y$

   6   **if** $(y = 13)$ **then go to** 8 **endif**

   7   **endloop**

   8   **stop**

31. 1 $2 \rightarrow x$

   2   $4 \rightarrow y$

3   **loop**

4   **flip**(x, y)

5     $x + 1 \rightarrow x$

6     $y + 1 \rightarrow y$

7   **if** $(x = 10)$ **then go to 9endif**

8   **endloop**

9   **stop**

33. 1 $1 \rightarrow x$

2   **loop**

3   **flip**(x, x)

4     $x + 1 \rightarrow x$

5   **if** $(x = 13)$ **then go to 7endif**

6   **endloop**

7   **stop**

35. 1   **print** "Input property value."

2   **read** v

3   **if** $(v < 80{,}000)$ **then**

4     $0.01 * v \rightarrow t$

5   **else**

6     $800 + 0.02 * (v - 80{,}000) \rightarrow t$

7   **endif**

8   **print** "Your property tax is ", t

9   **stop**

37. 1   **print** "Input purchase cost."

2   **read** x

3   **if** $(x < 100)$ **then**

4     $x - 0.1 * x \rightarrow s$

5   **else**

6     $x - 0.15 * x \rightarrow s$

7   **endif**

8   $s + 0.06 * s \rightarrow b$

8   **print** "Your final bill is ", b

9   **stop**

39. 1   **loop**

2   **print** "Input the length and width of a rectangle."

3   **read** l, w

4   **if** $(l > 0$ and $w > 0)$ **then**

5       $2 * l + 2 * w \rightarrow p$

6   **else**

7   **go to** 10

8   **endif**

9   **print** "The perimeter of the rectangle is ", p

10   **endloop**

11   **stop**

41. 1   **print** "Input the coefficients (in the right order) of a quadratic equation."

2   **read** a, b, c

3   $b * b - 4 * a * c \rightarrow D$

4   **if** $(D \geq 0)$ **then**

5   **print** "This equation has real roots."

6   **else**

7   **print** "This equation has complex roots which are not real."

8   **endif**

9   **stop**

43. 1   **print** "Please input a letter and a Caesar shift constant."

2   **read** a, k

3   $ord(a) + k \rightarrow x$

4   x **mod** $26 \rightarrow y$

5   $chr(y) \rightarrow b$

6   **print** "The letter shifted by ", k, " positions is", b

7   **stop**

45. : Disp "Input property value."

: Prompt V

: If (V < 80000): Then

: 0.01*V $\rightarrow$ T

: Else

: 800 + 0.02*(V − 80000) $\rightarrow$ T

: End

: Disp "Your prop. tax is", T

: Stop

(a) \$610;   (b) \$1116;   (c) \$2130

47. : Disp "Input purchase cost."

: Prompt X

: If (X < 100): Then

: X − 0.1*X $\rightarrow$ Y

: Else

```
: X - 0.15*X → Y
: End
: Y + 0.06*Y → B
: Disp "Your bill is", B
: Stop
```

(a) \$74.41; (b) \$91.73;   (c) \$123.66;   (d) \$190.09

49.
```
: Disp "Input a,b,c coefficients."
: Prompt A,B,C
: B^2 - 4*A*C → D
: If (D ≥ 0) : Then
: Disp "Real roots"
: Else
: Disp "Roots are not real."
: End
: Stop
```

(a) Real roots; (b) Roots are not real; (c) Real roots; (d) Real roots

51. Yes; The additional goto statements reduces program efficiency and does not adhere to a top-down approach. It also makes the code harder to understand.

53. 1 $0 \rightarrow S$

    2 $1 \rightarrow n$

    3 **loop**

    4   **print** "Input next value."

    5   **read** x

    6     $S + x^2 \rightarrow S$

    7     $n + 1 \rightarrow n$

    8   **if** $(n \leq 10)$ **then go to** 10 **endif**

    9 **endloop**

   10 $sqrt(S/10) \rightarrow R$

   11 **print** "The result is ", R

   12 **stop**

```
: 0 → S : 1 → N
: While (N • 10)
   : Disp "Input next value."
   : Prompt X
   : S + X^2 → S
   : N + 1 → N
: End
```

: $\sqrt{(S/10)} \rightarrow \text{R}$

: Disp "The result is ", R

: Stop

(a) 12.40967365; (b) 62.04836823

55. 1 **print** "Input the first term and the common ratio."

2 **read** x, r

3 $x \rightarrow S$

4 $1 \rightarrow n$

5 **loop**

6    $x * r \rightarrow x$

7    $S + x \rightarrow S$

8    $n + 1 \rightarrow n$

9   **if** $(n = 50)$ **then go to** 11 **endif**

10 **endloop**

11 **print** "The sum is", S

12 **stop**

: Disp "Input first term and ratio."

: Prompt A, R

: A $\rightarrow$ S : 1 $\rightarrow$ N

: While (N < 50)

  : A*R $\rightarrow$ A

  : S + A $\rightarrow$ S

  : N + 1 $\rightarrow$ N

: End

: Disp "The sum is", S

: Stop

(a) 14;   (b) 39.99997735;   (c) 3,825,728,995

57. 1 $1 \rightarrow a$

2 $1 \rightarrow b$

3 $2 \rightarrow n$

4 **print** "The first 30 Fibonacci numbers are: "

5 **print** a, b

6 **loop**

7    $n + 1 \rightarrow n$

8    $a + b \rightarrow c$

9   **print** c

10   **if** $(n = 30)$ **then go to** 14 **endif**

11   $b \to a$

12   $c \to b$

13 **endloop**

14 **stop**

```
:1 → A : 1 → B : 2 → N
:Disp "The first 30 Fibo No. are:"
:Disp A, B
:While (N < 30)
   :N + 1 → N
   :A + B → C
   :Disp C
   :B → A : C → B
:End
```

The 30th term is 832,040.

Chapter 3 Review Test

1.   (b), (e), (f), and (g) are statements.

2.   (a) $p \wedge q$ (b) $\sim(p \vee q)$ (c) $\sim q \Rightarrow \sim p$

3.   (a) T;   (b) F;   (c) F;   (d) T;   (e) T

4.   (a) Michael studies 30 hours per week or he does not get good grades.

    (b) If Michael studies 30 hours per week, then he gets good grades and he goes to a party on Saturday.

    (c) It is not the case that Michael studies 30 hours per week and he goes to a party on Saturday.

5.

x	y	p	q	$L(p)$	$L(q)$	$L(p \wedge q)$	$L(p \vee q)$
−8	10	−8 > 0	2 = 10	F	F	F	F
−1	11	−1 > 0	10 = 10	F	T	F	T
1	7	1 > 0	8 = 10	T	F	F	T
5	5	5 > 0	10 = 10	T	T	T	T
25	−15	25 > 0	10 = 10	T	T	T	T

6.

p	q	p ∧ q ⇔ ~(p ∨ q)
T	T	T T T **F** F T T T
T	F	T F F **T** F T T F
F	T	F F T **T** F F T T
F	F	F F F **F** T F F F

7.   Yes, it is a tautology.

8.   (a) T   (b) F

9.   (a) $x + y \neq 17$ or $5z \leq 21$ (b) $A = 100$ and $B > 200$

10. $[(p \Rightarrow q) \wedge \sim q] \Rightarrow \sim p$. Valid *MT* syllogism.

11. Invalid.

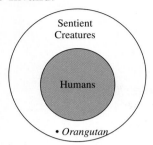

12. (a) Mercury orbits the sun in an elliptical path.   (b) No conclusion is possible.

13.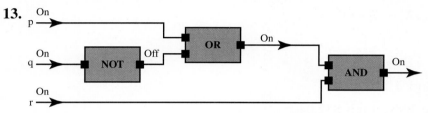

14.   2

5

8

11

14

17

15. 1   **print** "Please input your purchase cost."

2   **read** x

3   **if** $(x < 50)$ **then**

4       $x + 0.02 * x \rightarrow A$

5   **else**

6       $x + 0.03 * x \rightarrow A$

7   **endif**

8   **print** "The amount owed is ", A

9   **stop**

Chapter 4

Section 4.1

1. The rotation of Earth on its axis. Yes, because Earth rotates $360°/24$ hr $= 15°$/hr.

3. No parallax apparent to the naked eye.

5. Earth is the stationary center of the universe. Every celestial body travels in a circular path.

7. $15°/hr = 0.25°/min = 15'/min = 0.25'/sec = 15"/sec.$

9. 19,500 light-seconds

11. 0.5° per hour

13. 5.9×10^8 km

15. $0.013° = 47"$

17. 6.3 ly

19. $0.0023° = 0.14'$; $0.007° = 0.44'$; Yes.

21. 0.26 AU = 39 million km; $0.018° = 65"$

23. 150 million km

25. 1.4 million km

27. 30,000 miles

29. Chicago, Illinois

Section 4.2

1. The Ptolemaic system used epicycles and deferents. In the Copernican system, the outer planets appear to reverse as Earth passes them on its smaller orbit around the sun.

3. $x = 18.0, y = 31.2$

5. $x = 69.7, y = 29.6$

7. 29.9 km/sec

9. 13.1 km/sec

11. 5 pc = 16.3 ly = 1.03 million AU

13. 0.14 arc-seconds

15. 31 trillion km.

17. The ellipse with A and B separated by 1 inch; A and B separated by 1 inch; A and B separated by 6 inches; A and B separated by 1 inch.

19. 0.0068

21. 0.0484

23. 29.6 AU; Sometimes Neptune is farther from the sun than Pluto.

25. 4.09 AU

27. 0.055

29. $v = \dfrac{9.52\pi}{\sqrt[3]{p}}$; Uranus: 6.83 km/sec; Neptune: 5.46 km/sec; Pluto: 4.76 km/sec

31. 440 AU.

33. diameter = 220 million − 78 million = 142 million km

35. 209 million km; Roughly, yes.

37. 77.2 AU; 678 yr

🏆**39.** At aphelion, $\theta = \sin^{-1}(0.47) = 28°$; at perihelion, $\theta = \sin^{-1}(0.31) = 18°$

🏆**41.** Area $= \pi a^2 \sqrt{1 - e^2}$; No.

Section 4.3

1. 58.8 m/sec; 176 m

3. $v = 4.4\sqrt{d}$; 44 m/sec; 98.4 m/sec; 139 m/sec; 278 m/sec; Yes.

5. Mercury: $t = \sqrt{500/1.9} = 16.2$ sec, impact velocity $= 61.6$ m/sec;

Earth: $t = \sqrt{500/4.9} = 10.1$ sec, impact velocity $= 99.0$ m/sec;

Saturn: $t = \sqrt{500/5.2} = 9.8$ sec, impact velocity $= 102$ m/sec.

Saturn; Mercury.

7. For $d = 4$ m, $v = \sqrt{2(9.8)(4)} = 8.9$ m/sec on Earth. Assuming a maximum safe impact velocity of 8.9 m/sec, on Pluto you could jump from the height d where $8.9 = \sqrt{2(0.7)d}$. Solving this for d gives $d = 56.6$ m.

9. 54.5 kg; 46.5 lb; 305 lb; 108 lb.

11. 82.0 kg; 804 N; 181 lb.

13. Converse: If the television is on, then someone is watching television. False.

Contrapositive: If the television is off, then no one is watching television.

15. Converse: If a polygon is a rhombus, then it is a square. False.

Contrapositive: If a polygon is not a rhombus, then it is not a square. True.

17. Converse: If a house costs at least $180,000, then it is on Paradise Lane. False.

Contrapositive: If a house costs less than $180,000, then it is not on Paradise Lane. True.

19. Converse: If $2x + 15 = 39$, then $x = 12$. True.

Contrapositive: If $2x + 15 \neq 39$, then $x \neq 12$. True.

21. Converse: If my shirt is not blue, then it is solid red. False.

Contrapositive: If my shirt is blue, then it is not solid red. True.

23. True.

Converse: If $f(a) < f(b)$, then $f(x)$ is an increasing function over $[a, b]$. False.

Contrapositive: If $f(a) \geq f(b)$, then $f(x)$ is not an increasing function over $[a, b]$. True.

25. False.

Converse: If $f(x)$ is a linear function with positive slope, then $f(x)$ is an increasing function. True

Contrapositive: If $f(x)$ is not a linear function with positive slope, then $f(x)$ is not an increasing function. False.

27. True.

Converse: If $f(2) = 23$, then $f(x) = 5x + 13$. False.

Contrapositive: If $f(2) \neq 23$, then $f(x) \neq 5x + 13$. True.

29. The Copernican system is not the only system in which a full series of the phases of Venus would be observed. In the Ptolemaic system, Venus is always located between the sun and Earth and therefore can only exhibit a crescent phase.

31. *Dialogue Concerning the Two Chief World Systems.* (Usually just referred to as *The Dialogue.*)

33. It is impossible for the velocity of $m + M$ to be both less than and greater than the velocity of M. Therefore, Aristotle is wrong. The velocity of a falling object does not depend on mass.

Section 4.4

1. Woolsthorpe, England; the bubonic plague.

3. 45 pc; 5 pc.

5. 16

7. The force would be doubled.

9. 1.913 GU; 0.431 GU; 0.00272 GU; 0.000643 GU; 0.130 GU

11. 0.925 m/sec²; 0.148 m/sec²; 0.058 m/sec²

13. The radius of Uranus is greater than the radius of Venus; The radius of Venus = 6,040,000 m = 6,040 km; The radius of Uranus = 25,600,000 m = 25,600 km; Yes. Note that $(25,600/6040)^2 = 18 \approx 20$, the difference due to round-off error.

15. 5.99×10^{24} kg; This answer matches our previous value to two significant digits.

17. (a) Io   (b) Callisto   (c) 1.37×10^4 m/sec

19. 42,000 km

21. 0.0005 newtons

23. Newton's laws: unified terrestrial and celestial motion, provided simple axioms from which many observable phenomena could be deduced, and they were predictive.

🏆**25.** By knowing the radius r and period p of the moon, we can compute its circular velocity by $v_c = \dfrac{2\pi r}{p}$. Then we set this value equal to $\sqrt{\dfrac{GM}{r}}$ and solve for M.

🏆**27.** Since $F_p < F_s$, the gravitational force of the sun on the moon is greater than that exerted by Earth. A body primarily in orbit around the sun, but whose path is strongly influenced by Earth. Tug-of-war ratio = 0.4563.

Chapter 4 Review Test

1. He placed the sun at the center of the orbits for Mercury and Venus.

2. An axiom is a statement universally accepted as true. The model is built from observations of the phenomenon assuming the axioms to be true.

3. Polaris; Scranton, PA.

4. About 500 light-seconds.

5. 30°

6. 9.55 AU; 9.68 km/sec.

7. 3.57 pc = 11.6 ly.

8. 0.970; cigar

9. 0.308 AU.

10. Converse: If the temperature is over 100°, then it is a July day in Phoenix.

    Contrapositive: If it is not over 100°, then it is not a July day in Phoenix. No; Yes.

11. He observed the moons of Jupiter revolving around Jupiter. He also observed that Venus underwent a complete sequence of phases.

12. 26 sec.; 96.2 m/sec.

13. 350 pc.

14. Mars also has a greater mass than Mercury.

15. 0.0028 GU.

16. Yes. The imparted acceleration depends only on the distance. $a = 0.00077$ m/sec².

17. 7.26×10^{22} kg.

Chapter 5

Section 5.1

1. The vertex set is $V(G) = \{B, G, H, P\}$.

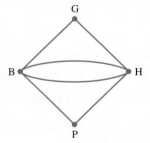

3. Yes. 10 connections.

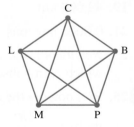

5. The vertex set is $V(G) = \{A, CR, S1, H, R, S2, BH, C, S3\}$.

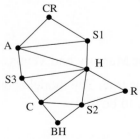

7. Not isomorphic.

9. Not isomorphic. The graph on the left contains a bridge.

11. Isomorphic.

13. (a) 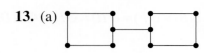   **(b)**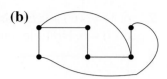

15. $\delta(a) = 3$, $\delta(b) = 2$, $\delta(c) = 3$, $\delta(d) = 3$, $\delta(e) = 3$, $\delta(f) = 3$, $\delta(g) = 2$, $\delta(h) = 1$, $\delta(k) = 4$,

Vertices a, c, d, e, f, and h are odd. Vertices b, g, and k are even.

17. Two simple circuits based at a are $abca$ and $abcda$. Two simple paths from d to f are dcf and $dabcf$. Four simple circuits based at h are $hjnh$, $hjkgh$, $hjkmnh$, and $hjnmkgh$. Five simple paths from n to k are nmk, njk, $nhjk$, $nhgk$, and $njhgk$.

19. $G_1 \cong G_4$, $G_3 \cong G_5 \cong G_6$

21.

Graph	T_1	T_2	T_3	T_4	T_5	T_6	T_7
# edges	1	2	3	4	5	6	7
Deg sum	2	4	6	8	10	12	14

The degree sum of any graph is equal to twice the number of edges.

23. (a) 6; (b) 12; (c) 12

25. 21, 55, 190

27.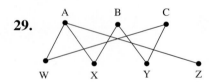

29.

A   B   C

W   X   Y   Z

31.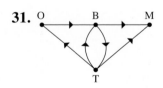

O   B   M

T

33. False

35. False

37. False

39. False

41. True

43. True

Section 5.2

1. $\delta(A) = \delta(B) = \delta(C) = \delta(D) = 2$. $\delta(E) = 4$. One Euler circuit is $ABECDEA$.

3. $\delta(a) = \delta(e) = 3$. $\delta(b) = \delta(d) = \delta(f) = 4$. $\delta(c) = 2$. No Euler circuit exists.

5. $\delta(A) = \delta(C) = \delta(E) = \delta(G) = \delta(I) = \delta(J) = \delta(L) = 2$. $\delta(B) = \delta(D) = \delta(F) = \delta(H) = \delta(K) = 4$. One Euler circuit is $ABCDEFDBLKHFGHIJKA$.

7. $\delta(A) = \delta(C) = \delta(E) = \delta(G) = \delta(I) = \delta(K) = 2$. $\delta(B) = \delta(F) = \delta(H) = \delta(L) = 4$. $\delta(D) = \delta(J) = 3$. $\delta(O) = 6$. No Euler circuit exists.

9.

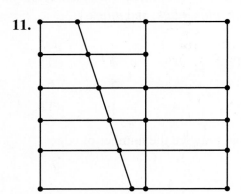

11.

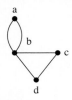

13. An Euler circuit is *abcdba*.

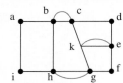

15. *abcdcedfebgfaga* is one circuit.

17. *ABCODCDEFOAFA* is one circuit.

19. *abcbedcdfghjefghija* is one circuit.

23. *abcbhgkcdekefghia* is one circuit.

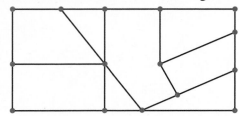

25. There are 10 vertices of odd degree.

27. (a) $\delta(A) = \delta(C) = \delta(D) = 2$. $\delta(B) = \delta(E) = 4$.

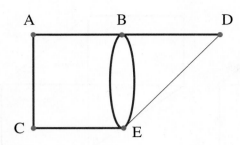

29. (a) $\delta(A) = \delta(B) = \delta(F) = 2$. $\delta(D) = 4$. $\delta(C) = \delta(E) = 6$.

(b)

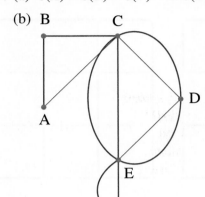

31. False   **33.** True   **35.** True   **37.** True

Section 5.3

1. False   **3.** False   **5.** True   **7.** False

9. True   **11.** True

13.

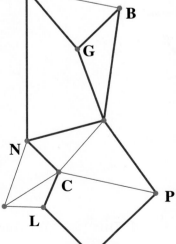

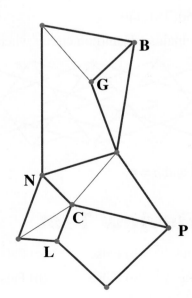

15. (a)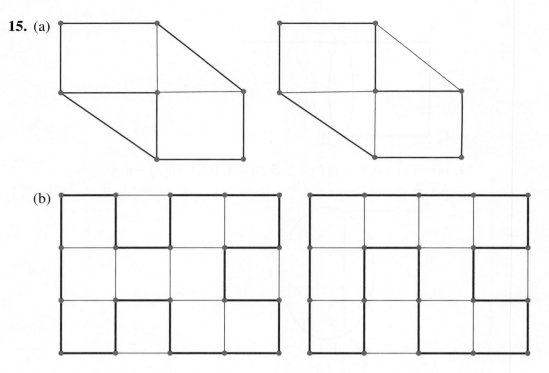

(b)

17. No. A graph having a vertex of degree one will not have a Hamilton circuit.

19. (a) Using the nearest neighbor algorithm, the minimum weight Hamilton circuit is *ABCFDEA* with weight 20.5. Using the brute force algorithm, the minimum weight Hamilton circuit is *ABCDFEA* with weight 16.8.

(b) Using either the nearest neighbor or brute force algorithm, the minimum weight Hamilton circuit is *AGFEDCBA* with weight 34.5.

21. (a) *DNABD* with total distance 4194 miles. (b) Same solution of *DNABD*.

23. Both methods yield the optimal solution of *BANMB* with total distance 3519 miles.

25. 12; 360; 181,440

29. B_{mn} contains a Hamilton circuit when $m = n$.

31.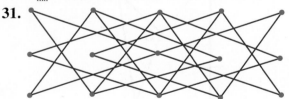

33. $m = 5$ and $n = 6$.

Chapter 5 Review Test

1. (a) True   (b) False   (c) False   (d) True   (e) True

(f) True   (g) False   (h) False   (i) True   (j) False

2.

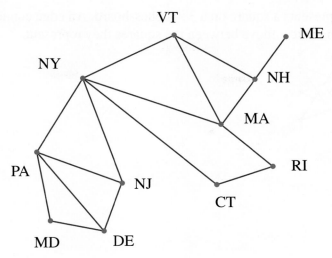

3. (a) 66; (b) 2520; (c) 300; (d) 2000.

4.

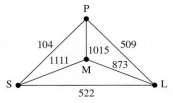

5. (d)

6. Yes. *abcdbgdegaefa*

7. $\delta(A) = \delta(D) = \delta(E) = 2$. $\delta(B) = \delta(C) = 4$.

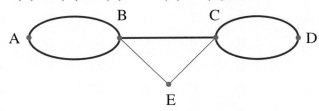

8. The optimal solution is *SPMLS* with a total weight of 2514 miles.

9.

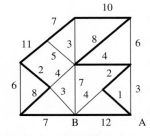 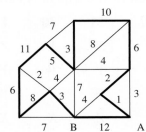

10. A bipartite graph has two sets of vertices S_1 and S_2 such that every edge connects a vertex in S_1 to a vertex in S_2.

11. Every vertex represents a square on a 3 × 4 chessboard. An edge connects two vertices if the knight can make a move between the squares they represent.

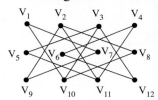

Chapter 6

Section 6.1

1. (a)

Number of Ballots	3	3	1	2	1
First	D	Y	Z	Z	D
Second	Y	Z	D	Y	Z
Third	Z	D	Y	D	Y

(b) No.

(c) Yes. Denali.

3. (a) 24

(b)

Number of Ballots	3	2	3	2	1	1
First	B	B	C	A	D	D
Second	C	A	B	D	C	A
Third	D	C	A	B	A	B
Fourth	A	D	D	C	B	C

(c) No

(d) Yes. Betelgeuse.

5. 5 points for 1st place, 4 points for 2nd place, 3 points for 3rd place, 2 points for 4th place, and 1 point for 5th place.

7. The Borda count totals for each candidate are 204 for Perot, 196 for Clinton, and 194 for Bush. Therefore, the winner is Perot.

9.

Number of Ballots	51	30	11	9
First	X	Y	Z	W
Second	Y	W	Y	Y
Third	W	Z	W	Z
Fourth	Z	X	X	X

In this table, X has a Borda count total of 254 and Y has a total of 333. So Y wins the election even though X received a majority of first-place votes.

11.

Player, Team	1st	2nd	3rd	Total
Coghlan, Marlins	17	6	2	105

Happ, Phillies	13	11	8	106
Hanson, Braves	2	6	9	37
McCutchen, Pirates	0	5	0	15
McGehee, Brewers	0	3	4	13

13. 251

15.

Percentage	35	5	10	8	13	21	8
First	L	L	B	D	J	D	D
Second	D	D	L	B	B	J	J
Third	B	J	D	J	L	L	B
Fourth	J	B	J	L	D	B	L

Borda count for Lincoln: $40(4) + 10(3) + 34(2) + 16(1) = 270$

Borda count for Douglas: $37(4) + 40(3) + 10(2) + 13(1) = 301$

Section 6.2

1. Plurality: San Diego

Borda count: San Diego

Hare method: Miami

Yes, they are different.

3. Plurality: Yoshi

Borda count: Xavier

Hare method: Yoshi

Yes, they are different.

5. Winner: *A*

7. Rosborough was eliminated after the first round because he received the fewest first preference votes (1636). Of the 1636 votes that listed him first, 716 listed Rattenbury second, 399 listed Ellis second, and 484 listed Smith second. An Exh. (exhausted) ballot is one that has no candidate listed after eliminating the last candidate. There were 37 such ballots after Rosborough was eliminated.

9. 3560; 1506

11. Winner using plurality method: Borda method

Winner using Borda method: Hare method

Winner using Hare method: Plurality

No, there is no Condorcet candidate in this preference table.

13.

Percentage of Ballots	28	12	27	9	8	16
First	B	B	P	P	H	H
Second	H	P	H	B	P	B
Third	P	H	B	H	B	P

Note: This table is not unique. Many other tables would satisfy the required conditions.

15. No, the Hare system is not independent of irrelevant alternatives. For example, given the following preference table, B will be eliminated in the first round and then A wins the election in the second round with 65% of votes.

Percentage of Ballots	40	35	25
First	A	C	B
Second	C	B	A
Third	B	A	C

However, if we change the table to the following:

Percentage of Ballots	40	35	25
First	A	B	B
Second	C	C	A
Third	B	A	C

then, B wins in the first round with 60% of votes. The only change occurred in the middle column where B and C are switched, but the relative ranking between B and A was not altered.

17. Yes, the Condorcet candidate is C. The plurality winner is A, which demonstrates that the plurality method does not satisfy the Condorcet criterion.

19. (a) The total points accumulated by X would increase. (b) The points of no other candidates would increase. (c) X would remain the winner. (d) The Borda count method satisfies the monotonicity criterion.

21. (a) Arizona would be ranked first with Borda count.

(b) No, this table does not have a Condorcet candidate.

(c) One example of a 4-cycle in this table: $A > I > D > P > A$.

(d) Yes. $A > I > K > D > P > A$.

23.

Percentage of Ballots	40	40	20
First	A	B	E
Second	B	C	A
Third	C	D	B
Fourth	D	E	C
Fifth	E	A	D

25. No. 94 voted in the first round, 95 voted in the second round, and 98 voted in the third round. This cannot happen with a preference ballot since only one round of voting would occur.

27. Yes, the Coombs method satisfies the majority criterion.

Section 6.3

1. No. A is not a Condorcet candidate.

3. No. B is not a Condorcet candidate. Also, C received a majority of votes and did not win.

5. (a) Harrison; (b) Harrison: 71.4%, Ajuin: 60.7%, Elizabeth: 39.3%; (c) Yes.

7. $2^3 - 1 = 7$; $2^4 - 1 = 15$; $2^n - 1$

9. Probably $\{B\}$. It may be possible to get some voters in $\{B\}$ to switch to $\{A, B\}$.

11. B would win using the plurality method. D wins according to the approval table.

Number of Ballots	6	3	2	2
A		X		
B		X		X
C			X	X
D		X	X	X

13. independent, B, C, one-on-one, Condorcet, perfectly fair

15.

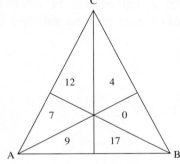

17.

	A	B	C
Preference	31	34	**35**
Approval	**76**	62	62
Borda	**107**	96	97

$A > B$ by 52 to 48 and $A > C$ by 55 to 45. So A is a Condorcet winner.

19.

	A	B	C
Preference	**20**	14	11
Approval	22	33	**35**
Borda	42	**47**	46

$C > B$ by 23 to 22 and $C > A$ by 23 to 22. So C is a Condorcet winner.

Chapter 6 Review Test

1. (a)

Number of Ballots	2	3	4	3
First	D	B	C	A
Second	A	A	D	D
Third	B	C	A	B
Fourth	C	D	B	C

(b) No

(c) C

2. Borda count method; No

3.
Player	1st	2nd	3rd	Total
Angel Berroa	12	7	7	88
Hideki Matsui	10	14	2	94
Rocco Baldelli	5	0	16	41
Jody Geru	0	6	2	20
Mark Texeira	1	1	1	9

Hideki Matsui; Independence of Irrelevant Alternatives

4. Plurality: Z; Borda: X; Hare: Y

5. 5 points for first place, 3 points for second place, and 1 point for third place.

6. The four fairness criteria are Majority, Independence of Irrelevant Alternatives, Monotonicity, and Condorcet. The Borda Count method satisfies the Monotonicity Criterion.

7. The Approval Method

8. (a) Yes. $P > C > I > X > P$

    (b) Yes, there are two 3-cycles. $X > C > I > X$; $P > I > X > P$

    (c) No.

9.
Number of Ballots	23	15	7
First	A	B	C
Second	B	C	B
Third	C	A	A

10. (a) Hunter Mountain; (b) No.

11. irrelevant; C; C; one-on-one; Condorcet; impossible

12.
	A	B	C
Plurality	10	13	12
Approval	27	17	26
Borda	37	30	38

$A > C$ by 19 to 16 and $A > B$ by 18 to 17. so A is a Condorcet winner.

Chapter 7

Section 7.1

1. (a) $S = \{2, 3, 4, 5, 6, 7, 8, 9, 10, 11, 12, 13, 14, 15, 16, 17, 18\}$

    (b) $E = \{2, 4, 6, 8, 10, 12, 14, 16, 18\}$

    (c) $A = \{10, 11, 12, 13, 14, 15, 16, 17, 18\}$

    (d) $L = \{2, 3, 4, 5\}$

    (e) $P = \{2, 3, 5, 7, 11, 13, 17\}$

3. (a) $\{(1, 1), (1, 2), (1, 3), (1, 4), (1, 5), (1, 6),$

$(2, 1), (2, 2), (2, 3), (2, 4), (2, 5), (2, 6),$

$(3, 1), (3, 2), (3, 3), (3, 4), (3, 5), (3, 6),$

$(4, 1), (4, 2), (4, 3), (4, 4), (4, 5), (4, 6)\}$ (Red die listed first in each pair.)

(b) $F = \{(1, 1), (1, 3), (1, 5), (3, 1), (3, 3), (3, 5)\}$

(c) $B = \{(3, 6), (4, 5), (4, 6)\}$

(d) $N = \{((2, 1), (3, 1), (3, 2), (4, 1), (4, 2), (4, 3)\}$

(e) $T = \{(1, 3), (2, 4), (3, 5), (4, 6)\}$

(f) $A = \{(1, 3), (1, 4), (1, 5), (1, 6), (2, 4), (2, 5), (2, 6), (3, 5), (3, 6), (4, 6)\}$

5. (a) $S = \{0, 1, 2, 3, \ldots, 98, 99, 100\}$

(b) $S = \{10000, 10001, 10002, \ldots 99997, 99998, 99999\}$

(c) $S = \{0, 1, 2, 3\}$

(d) $S = \{HHH, HHT, HTH, HTT, THH, THT, TTH, TTT\}$

7. (a) $S = \{HHH, HHT, HTH, HTT, THH, THT, TTH, TTT\}$

(b) $A = \{HHH, HHT, HTH, HTT, THH, THT, TTH\}$

(c) $B = \{HHT, HTH, THH\}$

(d) $C = \{HTT, THT, TTH\}$

(e) $D = \{TTT\}$

9. (a) $S = \{0, 00, 1, 2, 3, \ldots, 34, 35, 36\}$

(b) $R = \{9, 32, 30, 7, 5, 34, 3, 36, 1, 27, 25, 12, 19, 18, 21, 16, 23, 14\}$ (Ordered clockwise)

(c) $B = \{20, 28, 11, 33, 17, 22, 15, 24, 13, 10, 29, 8, 31, 6, 26, 4, 35, 2\}$ (Ordered clockwise)

(d) $M = \{19, 21, 23, 25, 27, 30, 32, 34\}$

(e) $G = \{0, 00\}$

(f) $E = \{2, 4, 6, 8, 10, 12, 14, 16, 18, 20, 22, 24, 26, 28, 30, 32, 34, 36, 0, 00\}$

Section 7.2

1. (a) False; (b) True; (c) True; (d) True; (e) False; (f) True; (g) False; (h) True

3. (a) People who like basketball or hockey; (b) People who do not like football; (c) People who do not like basketball but do like football; (d) People who like neither hockey nor football.

5. (a) $\overline{B} = \{-3, -2, 0, 3\}$; (b) $A \cap B = \{-1, 1\}$;

(c) $\overline{A} \cup B = \{-3, 2, 3\}$; (d) $\overline{A \cup B} = \{-3, 3\}$; (e) $\overline{A \cup \overline{B}} = \{2\}$

7. (a) $\overline{A} \cap B$

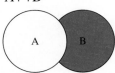

(b) $A \cup \overline{B}$

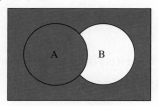

(c) $\overline{\overline{A} \cap B} = A \cup \overline{B}$. Same diagram as (b).

(d) $\overline{\overline{A} \cup B} = A \cap \overline{B}$

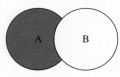

9. (a) $\overline{F \cap \overline{S}} = \{1, 3, 4, 6, 8\}$ (b) $\overline{F} \cup \overline{S} = \{1, 2, 3, 5, 6, 7\}$

(c) $F \cap S = \{4, 8\}$ and $F \cap \overline{S} = \{2, 5, 7\}$. So $(F \cap S) \cup (F \cap \overline{S}) = F$;

(d) $S \cap F = \{4, 8\}$ and $S \cap \overline{F} = \{3\}$. So $(S \cap F) \cup (S \cap \overline{F}) = S$.

11. For example, partition them by height: less than 5 ft, 6 in and greater than or equal to 5 ft, 6 in.

15. (a) $A - B$

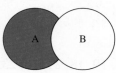

(b) $B - A$

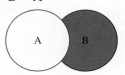

(c) $A - \overline{B} = A \cap B$

(d) $\overline{B - A} = \overline{B \cap \overline{A}} = \overline{B} \cup A$

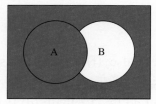

17. 19

19. $n(A \cap B) = 7$, $n(B) = 60$, and $n(\overline{B}) = 65$.

21. $g(A) = 2^{n(A)}$; 2^{52}

23. (a) $f(L) = \{\beta, \delta, \gamma\}$ (b) $f(L \cap M) = \emptyset$ (c) $n(L) = 4$ (d) $n(f(L)) = 3$
(e) $n(f(L \cap M)) = 0$

25. (a) 845; (b) 205; (c) 155

27. (a) 525; (b) 270; (c) 75

29. (a) 6; (b) 16; (c) 73; (d) 39; (e) 50

31. (a) $A \cap B \cap C$; (b) $A \cap (\overline{B \cup C})$; (c) $B \cap (\overline{A \cup C})$; (d) $(A \cup B) \cap \overline{C}$; (e) $(B \cap C) \cap \overline{A}$

Section 7.3

1. (a) $S = \{HH, HT, TH, TT\}$; (b) 3/4; (c) 1/4; (d) 1/2

3. (a) $\{(1, 1), (1, 2), (1, 3), \ldots, (6, 4), (6, 5), (6, 6)\}$;

(b) 5/12; (c) 1/6; (d) 1/9; (e) 5/18

5. If A and B are mutually exclusive, then $p(A \cap B) = 0$. Therefore, $p(A \cup B) = p(A) + p(B)$.

7. (a) $S = \{10, 11, 12, 13, 14, 15, 16, 17, 18, 19, 20\}$

(b) $\dfrac{6}{11}$ (c) $\dfrac{5}{11}$ (d) $\dfrac{4}{11}$ (e) $\dfrac{8}{11}$

9. (a) $\dfrac{1}{2}$ (b) $\dfrac{1}{4}$ (c) $\dfrac{7}{8}$ (d) $\dfrac{1}{8}$ (e) $\dfrac{5}{8}$

11. $p(A \cup B \cup C) = 1$ because $U = A \cup B \cup C$. Since A and $B \cup C$ are mutually exclusive, $p(A \cup B \cup C) = p(A) + p(B \cup C)$. Since B and C are mutually exclusive, $p(B \cup C) = p(B) + p(C)$. Therefore, $1 = p(A \cup B \cup C) = p(A) + p(B) + p(C)$.

13. (a) $\dfrac{1}{10}$ (b) $\dfrac{1}{100}$ (c) $\dfrac{1}{2}$ (d) $\dfrac{9}{10}$

15. (a) $\dfrac{1}{4}$ (b) $\dfrac{3}{4}$ (c) $\dfrac{13}{16}$ (d) $\dfrac{3}{8}$

17. (a) $\dfrac{21}{32}$ (b) $\dfrac{27}{32}$ (c) $\dfrac{7}{32}$ (d) $\dfrac{3}{16}$

19. (a) 0.845 (b) 0.205 (c) 0.155

21. (a) $\dfrac{9}{19}$ (b) $\dfrac{1}{19}$ (c) $\dfrac{11}{38}$ (d) $\dfrac{4}{19}$ (e) $\dfrac{10}{19}$

23. −$1.90

25. 0.5

	w	w
R	Rw	Rw
w	ww	ww

27. Yes. 0.25

	B	b
B	BB	Bb
b	Bb	bb

29. Yes. 0.25

	D	d
D	DD	Dd
d	Dd	dd

31. Six spots. The probability is 16/36.

33. 10:3; 1:1; 2:7

Chapter 7 Review Test

1. (a) True; (b) True; (c) False; (d) False; (e) True; (f) False; (g) True; (h) False

2. No. The given set does not contain all the possible outcomes.

3. (a) $\{t, w, x, z\}$; (b) $\{u, y\}$; (c) $\{s, v\}$; (d) $\{t, u, w, x, y, z\}$; (e) 5; (f) 3; (g) 1

4. See Figure 7.2.4.

5. 25

6. 250

7. (a) 1/2; (b) 1/4; (c) 7/8; (d) 1/8; (e) 5/8

8. (a) 840; (b) 137; (c) 160

9. 0.77

10. (a) 9/19; (b) 10/19; (c) 12/19

11. (a) 1/2; (b) 1/4

12. −$1.75

Chapter 8

Section 8.1

1. $\overline{X} = 38.4$, $M = 37$, range $= 44$. When oldest retires: $\overline{X}$ changes to 37.8, M doesn't change.

3. $\overline{X} = 9.32$, $M = 9.45$. Throw out scores: $\overline{X} = 9.39$, $M = 9.45$. The mean is closer to the median, so this is probably a good practice to follow.

5. $\overline{X} = 85.3$, $M = 85.5$, $\overline{X}g = 86.0$, $Mg = 86.4$. Yes, this distribution is approximately symmetric.

7. $\overline{X}g = 14.25$, $Mg = 12.67$

Goals	Frequency	Rel. frequency
4 - 9	5	5/29 = 0.17
9 - 14	13	13/29 = 0.45
14 - 19	6	6/29 = 0.21
19 - 24	2	2/29 = 0.07
24 - 29	1	1/29 = 0.03
29 - 34	2	2/29 = 0.07

Ages	Freq.	ρ
40 - 44	2	.05
44 - 48	3	.07
48 - 52	10	.24
52 - 56	10	.24
56 - 60	8	.19
60 - 64	5	.12
64 - 68	3	.07
68 - 72	1	.02

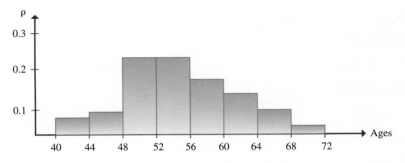

9. $\overline{X}g = 54.8$, $Mg = 54.3$

11. $\overline{X}g = 67.4$, $Mg = 67.4$. Yes, this distribution is approximately symmetric.

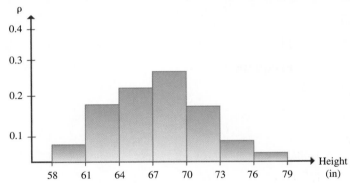

13.

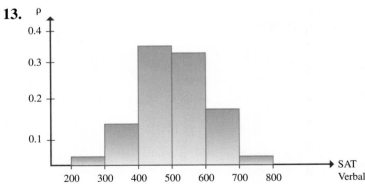

$\overline{X}g = 494$, $Mg = 492$. Yes, this distribution is approximately symmetric.

15. (a) 26.5%   (b) 41.5%   (c) 62.75%   (d) 13.5%

17. 63%

19. 41%

21. 61%

23. 100%

25. 25–34 : 43.5%, 35–44 : 37.5%, 45–54 : 32.5%

27. Increasing; decreasing

29. (a) Skewed: $\overline{X} > M$ (b) Symmetric: $\overline{X} \approx M$

    (c) Skewed: $\overline{X} < M$ (d) Symmetric: $\overline{X} \approx M$

31. $Q_1 = 31{,}079$; $Q_3 = 38{,}700$; Q_2

33.

Class	Frequency
25000–30,000	10
30,000–35,000	20
35,000–40,000	9
40,000–45,000	6
45,000–50,000	4
50,000–55,000	1

35. (a)

Class	Frequency
20.5–21.5	5
21.5–22.5	10
22.5–23.5	14
23.5–24.5	11

(b)

Class	Frequency
20.5–21.3	5
21.3–22.1	3
22.1–22.9	10
22.9–23.7	14
23.7–24.5	8

(c)

Class	Frequency
20.5–21.0	2
21.0–21.5	3
21.5–22.0	2
22.0–22.5	8
22.5–23.0	5
23.0–23.5	9
23.5–24.0	8
24.0–24.5	3

37. $\overline{X} = 74.28$, $M = 74.4$

39. $\overline{X} = \$924$, $Q_1 = \$860$, $M = \$930$, $Q_3 = \$1030$

41. Team runs: $\overline{X} = 733.8$, $Q_1 = 702$, $M = 741$, $Q_3 = 774.5$

43. Passing yards: $\overline{X} = 3676.4$, $Q_1 = 3266.5$, $M = 3546$, $Q_3 = 4005.5$

Section 8.2

1. $\sigma = 13.5$. Oldest retires: σ decreases to 11.3, because 40 is closer to the mean than 62.

3. $\sigma = 0.44$. Throw out scores: σ is smaller, new value is 0.20.

5. (a) $\sigma g = 12.8$, (b) $\sigma = 12.4$; They are fairly close to each other.

7. $\sigma g = 6.61$

9. $\sigma g = 6.4$

11. $\sigma g = 4.0$

13. $\sigma g = 101$

15. 242, 413, 615, 736

17. $z_{21.7} = 3.19$

19. $z_{99.3} = 0.43$

21. $z_{300} = 0.87$

23. 156.37

25. 4 (or less)

27. $\overline{X} = 70.78$, $\sigma = 7.44$

29. 63.34; 70.78; 78.22

31. $\overline{X}g = 3.03$, $\sigma g = 0.57$

33. $\overline{X} = \$924$, $\sigma = \$166$; Saturn, $z = -1.59$; Chevrolet, $z = -0.66$; Dodge, $z = 0.91$

35. $\overline{X} = \$35,039$, $\sigma = \$6206$; $\$25,730$; $\$35,039$; $\$44,348$

37. Team averages: $\overline{x} = .2616$, $\sigma = 0.1201$; Home Runs: $\overline{x} = 160.3$, $\sigma = 33.2$

39. $\overline{X} = 3676.4$, $\sigma = 552.6$

Section 8.3

1. (a) 0.25   (b) 0.81   (c) 0.19   (d) 0.5

3. (a) 0.25   (b) 0.9375   (c) 0.36   (d) 0.345

5. (a) 0.2116   (b) 0.2944   (c) 0.57   (d) 0.4

7. (a) 0.5   (b) 0.5   (c) 0.125   (d) 0.125   (e) 0.75

9. (a) 1/4   (b) 1/4   (c) 3/8   (d) 5/8

11. 1.77

13. $M = \sqrt{\dfrac{1}{m}}$

15. $\mu = 1.67$; $M = 1.77$; Yes.

17. (a) A (b) $1 - A$ (c) $0.5 - A$

19. (a) 0.0174   (b) 0.4032   (c) 0.9484   (d) 0.8578   (e) 0.3037   (f) 0.9893

21. (a) 0.1359   (b) 0.3336   (c) 0.3446   (d) 0.4796

23. (a) 0.2033   (b) 0.2578   (c) 0.5159   (d) 0.6293   (e) 0   (f) 0.3232

25. (a) A(98.4–100), B(95.2–98.3), C(88.8–95.1), D(85.6–88.7)

    (b) Same as in Example 4.

27. (a) 9.2%   (b) 2.3%   (c) 88.5%

29. (a) 5.94%   (b) 50%   (c) 32.62%   (d) 68.92%

31. (a) 0.18 to 0.34 grams   (b) 68.26%

(c) more than 0.42 or less than 0.10 grams (d) 4.56%

33. 1.65 standard deviations

35. (b) $p(10 < x < 20) = 0.1776$ and $p(30 < x < 40) = 0.2823$

37. (a) 0.053   (b) 0.947   (c) 0.406   (d) 0.248

39. (a) 0.125   (b) 0.229   (c) 0.296   (d) 0.350

Section 8.4

1. No. First, you are personally involved in the selection process and so you may avoid talking to certain types of people. Second, you are talking to a distinct subgroup of the college population – those students who actually go to the library. Third, those who are present during 5:00–7:00 p.m. may be primarily part-time students who are employed full time during the day. The bulk of full-time day students have a smaller chance of being chosen.

3. Cameron, Lon, Kam, David, Rodolfo, Maurice, Rachel, Alma,

Chan, Maria, Stephanie, Enrico, Marilyn, Gabriel, Kadhri

5. [23.38, 25.88]   **7.** [8.13, 8.95]

9. [0.0594, 0.0656]   **11.** [23.58, 25.68]

13. [8.1, 8.9]   **15.** 95%; 95%

17. [35, 57]; m.o.e. = 11%

19. [53.0, 65.4]; m.o.e. = 6.2%

21. [31.1, 38.5]; m.o.e. = 3.7%

23. Increase n; margin of error $= 1.96(\sigma/\sqrt{n})$ which is a decreasing function of n.

25. [2.36, 2.54]; m. o. e. = 0.09

27. 3.5%

29. [65.7, 72.3]

31. $2.58 \left(\dfrac{\sigma}{\sqrt{n}} \right)$

33. 0.1469

35. For $\hat{p} = 53\%$, m.o.e. = 2.2%; For $\hat{p} = 85\%$, m.o.e. = 1.6% ($n = 2000$ in each case.)

37. Acceptance interval is [6.7, 7.3]; 7.4 is significant.

39. Acceptance interval is [27.2, 28.8]; 27.7 is not significant.

41. No, not necessarily.

Chapter 8 Review Test

1. *A*   **2.** *B*   **3.** *A*

4. *C*   **5.** 0.65

6. (a) 16% (b) 13% (c) 41.5%

7. $\overline{X} = 748.3$; $M = 717.5$; Yes; No

9. Carl F. Gauss

10. $\overline{X}_g = 408$; $M_g = 354$; $\sigma_g = 260$

11. 79.67%

12. [13.2, 13.6]; m.o.e. = 0.2

Chapter 9

Section 9.1

1. $[\,2\,]_5 = \{\, \ldots, -8, -3, 2, 7, 12, 17, \ldots \,\}$

$[\,3\,]_5 = \{\, \ldots, -7, -2, 3, 8, 13, 18, \ldots \,\}$

$[\,4\,]_5 = \{\, \ldots, -6, -1, 4, 9, 14, 19, \ldots \,\}$

3. $[\,2\,]_6, [\,6\,]_7, [\,1\,]_2, [\,0\,]_{11}, [\,13\,]_{18}$

5. $[\,18\,]_{39}, [\,48\,]_{74}, [\,472\,]_{528}, [\,176\,]_{463}$

7. 4, 8, 13, 1, 0, 12, 81, 239, 39, 9

9. Thursday

11. 38, 122, 206

13. Hand a number from 1 to 42 to each guest as they arrive. Seat everyone in residue class $[0]_7$ at the first table, everyone in class $[1]_7$ at the second table, etc.

15. $[0]_4$ is a subset of $[0]_2$; $[1]_4$ is a subset of $[1]_2$; $[0]_2 = [0]_4 \cup [2]_4$ where $\cup$ means the union (combination) of the two classes.

17. $[0]_{12} = C,$ $\qquad [4]_{12} = E,$ $\qquad [7]_{12} = G,$

$[3]_{12} = D\# \text{ or } Eb,$ $\quad [8]_{12} = G\# \text{ or } Ab,$ $\quad [13]_{12} = C\# \text{ or } Db,$

$[19]_{12} = G,$ $\qquad [24]_{12} = C,$ $\qquad [30]_{12} = F\# \text{ or } Gb$

19. $C = [\,0\,]_{12},$ $\qquad D\# = [\,3\,]_{12},$ $\qquad F = [\,5\,]_{12},$ $\qquad Gb = [\,6\,]_{12},$

$A\# = [\,10\,]_{12},$ $\quad Bb = [\,10\,]_{12},$ $\quad B\# = [\,0\,]_{12}$

21. C#, Gb, Bb, B, E#

23. 3-5-10, 4-6-11, E-Gb-B

25.

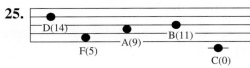

27.

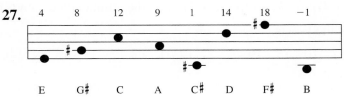

29. $[0]_{12} = C$; $[2]_{12} = D$; $[4]_{12} = E$; $[5]_{12} = F$

31.

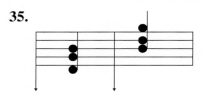

7-11-13 1-7-11   (C#-G-B)

33.

13-17-20 1-5-8   (C#-F-G#)

35.

2-7-11 11-14-19   (B-D-G)

37. 2-5-10 (D-F-Bb)

39. 0.74 AU; Amazingly close.

41. 4.76 AU; Quite close.

Section 9.2

1. Horizontal translation

3. Vertical translation

5. Reflection across the x-axis

7. Reflection across the y-axis

9. Horizontal translation of 10 units

11. Rotation of 180°

13. Reflection across the x-axis

15. $(-12, 2)$

17. $(-26, 4)$

19. $(7, -3)$

21. Rotation of 90° followed by reflection across the y-axis. Or a reflection across the x-axis followed by a rotation of 90°.

23. $(a, b) \rightarrow (a, -(b + 4))$; $(a, b) \rightarrow (a, -b + 4)$; Yes, the composite depends on the order.

25. Reflection across the y-axis.

27. 8 - 12 - 15 - 20 - 7 - 10 - 15 - 19 - 5 - 8 - 12 - 17 - 3 - 7 - 12 - 15

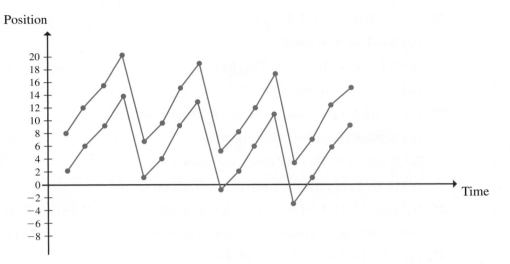

29. (a) Position

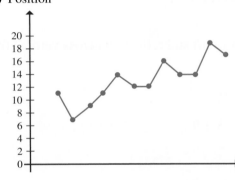

(b) 6 - 2 - 4 - 6 - 9 - 7 - 7 - 11 - 9 - 9 - 14 – 12

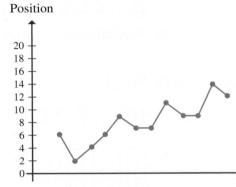

(c) (−17) - (−19) - (−14) - (−14) - (−16) - (−12) - (−12) - (−14) - (−11) - (−9) - (−7) - (−11)

(d) 17 - 19 - 14 - 14 - 16 - 12 - 12 - 14 - 11 - 9 - 7 - 11

Position

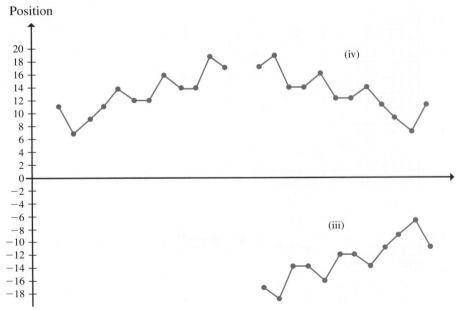

31. (a) 5-9-10-12-8-7-2-3

(b) 0-(−1)-4-5-9-7-6-2

(c) 0-1-(−4)-(−5)-(−9)-(−7)-(−6)-(−2)

(d) 2-(−2)-(−3)-(−5)-(−1)-0-5-4

33. (a) Vertical translation of −4 notes. (Transposition)

(b) Vertical reflection. (Retrograde)

(c) Rotation of 180°. (Retrograde-inversion)

(d) Horizontal reflection across $p = 7$. (Inversion)

35. Subject: 13-11-14-13-11-21; Interval lengths: (−2)-(+3)-(−1)-(−2)-(+10)

Inversion: 20-21-18-20-21-11; Interval lengths: (+1)-(−3)-(+2)-(+1)-(−10)

37. (a) 7-12-14-17-14-12-7-12-14-14-12

(b) 12-17-19-22-19-17-12-17-19-19-17

(c) 11-6-4-1-4-6-11-6-4-4-6

39. Vertical translation (transposition) and reflection across vertical (retrograde).

Section 9.3

1. $[\,6\,]_7$ **3.** $[\,2\,]_9$ **5.** $[\,9\,]_{10}$ **7.** $[\,16\,]_{25}$

9.

+	0	1	2	3
0	0	1	2	3
1	1	2	3	0
2	2	3	0	1
3	3	0	1	2

3; 2

11.

+	0	1	2	3	4	5	6	7	8	9
0	0	1	2	3	4	5	6	7	8	9
1	1	2	3	4	5	6	7	8	9	0
2	2	3	4	5	6	7	8	9	0	1
3	3	4	5	6	7	8	9	0	1	2
4	4	5	6	7	8	9	0	1	2	3
5	5	6	7	8	9	0	1	2	3	4
6	6	7	8	9	0	1	2	3	4	5
7	7	8	9	0	1	2	3	4	5	6
8	8	9	0	1	2	3	4	5	6	7
9	9	0	1	2	3	4	5	6	7	8

7; 4; 2

13.

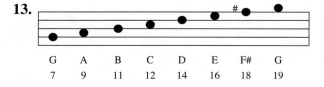

G	A	B	C	D	E	F#	G
7	9	11	12	14	16	18	19

15.

Bb	C	D	Eb	F	G	A	Bb
−2	0	2	3	5	7	9	10

17.

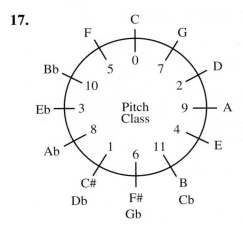

19. D; E

21. B = Cb

23. $k^{\#}(2) = [\,2 \cdot 7\,]_{12} = [\,14\,]_{12} = [\,2\,]_{12} = D$

25. $k^{b}(1) = [\,5\,]_{12} = F$

27. $k^{b}(5) = [\,5 \cdot 5\,]_{12} = [\,25\,]_{12} = [\,1\,]_{12} = Db$

29. A; B

31. Yes; Yes; No

33. $97 \equiv 7 \pmod 9$; $18 \equiv 0 \pmod 9$; $10 \equiv 1 \pmod 9$; $87 \equiv 6 \pmod 9$; $-25 \equiv 2 \pmod 9$; $4 \equiv 4 \pmod 9$; $35 \equiv 8 \pmod 9$; $-31 \equiv 5 \pmod 9$; $48 \equiv 3 \pmod 9$

35. $0 \equiv 0 \pmod 8$; $7 \equiv 7 \pmod 8$; $14 \equiv 6 \pmod 8$; $21 \equiv 5 \pmod 8$; $28 \equiv 4 \pmod 8$; $35 \equiv 3 \pmod 8$; $42 \equiv 2 \pmod 8$; $49 \equiv 1 \pmod 8$

Section 9.4

1. 5 **3.** 4 **5.** 6 **7.** 3, 7

9. 7 **11.** 1 **13.** 4 **15.** 9

17. 6 **19.** 0 **21.** 8 **23.** 1

25. 5 **27.** 4 **29.** 8 **31.** 5

33. 7 **35.** 2 **37.** 5 **39.** 1, 3, 5, 7

41. B W M Z Z   Q A P C U   I V

43. T H E   C O D E   I S   U N B R E A K A B L E

45. The decrypting function is $P = 21(C - 2) \bmod 26$.

T H E   E A G L E   H A S   L A N D E D

47. A L L   Y O U   N E E D   I S   L O V E

49. $8^2 = 64 \equiv 1 \pmod 9$, $7^2 = 49 \equiv 4 = 2^2 \pmod 9$, $36 \equiv 9 \pmod 9$, $25 \equiv 16 \pmod 9$

51. $a^3 - a = a(a^2 - 1) = a(a - 1)(a + 1)$. Since $a - 1$, a, $a + 1$ are three consecutive integers, one of them must be divisible by 3. Therefore, $a^3 \equiv a \pmod 3$.

53. If p is a prime, then $a^p \equiv a \pmod p$ for every integer a.

Chapter 9 Review Test

1. $[\,2\,]_3 = \{\ldots, -4, -1, 2, 5, 8, 11, \ldots\}$

$[\,4\,]_7 = \{\ldots, -10, -3, 4, 11, 18, 25, \ldots\}$

$[\,11\,]_{10} = \{\ldots, -19, -9, 1, 11, 21, 31, \ldots\}$; 11

2. 263

3. $[\,0\,]_{14}$ is a subset of $[\,0\,]_7$.

4. (a) $[\,1\,]_7$ (b) $[\,0\,]_4$ (c) $[\,13\,]_{15}$ (d) $[\,6\,]_{10}$

5. 3; 1

+	0	1	2	3	4	5
0	0	1	2	3	4	5
1	1	2	3	4	5	0
2	2	3	4	5	0	1
3	3	4	5	0	1	2
4	4	5	0	1	2	3
5	5	0	1	2	3	4

6. D; G; A#; C#

7.

Eb   F#   A   C   G#

8.

9–14–17   2–5–9

9. (a) Rotation of 180°   (b) Reflection across x-axis   (c) Horizontal translation of k units.

10. (a) $(-5, -1)$ (b) $(-6, 6)$

11. Reflection across the y-axis.

12. (a) 12-11-7-5-7-2-0-11-7-5-7   (b) 3-1-3-7-(−4)-(−2)-3-1-3-7-8

(c) (−12)-(−11)-(−7)-(−5)-(−7)-(−2)-0-(−11)-(−7)-(−5)-(−7)

13. B; C#

14. $[\,21\,]_{12} = [\,9\,]_{12} = $ A; $[\,10\,]_{12} = $ Bb

15. $0 \equiv 0 \pmod 6$; $11 \equiv 5 \pmod 6$; $22 \equiv 4 \pmod 6$; $33 \equiv 3 \pmod 6$; $44 \equiv 2 \pmod 6$;

$55 \equiv 1 \pmod 6$

16. (a) 5   (b) 12   (c) 10   (d) 4

17. 4

18. 6

19. H A P P Y   D A Y   Y O U   A R E   D O N E

Chapter 10

Section 10.1

1. 3.5 million; 3.9 million

3. 1.8 feet

5. 2.3 feet

7. 3.7 feet

9. 1650 km/hr; 1430 km/hr; 427 km/hr

The graph of this table of values reveals a function that is far from being linear. The most reliable answer is the one that is bracketed by the closest stakes: 1650 km/hr.

11. 832 minutes

13. 782.3 minutes

15. Supply: $L(x) = 75(x - 2.05) + 1015$; $L(2.10) = 1018.75 \approx 1019$ thousand;

Demand: $L(x) = -81.25(x - 2.05) + 1002$; $L(2.10) \approx 998$ thousand.

17. $1.94 and $2.05

19. 0.245 Atm.

21. 26.13 seconds

23. 0.3317; Since the ratios for these dates are 0.3119 and 0.3544, the interpolated value must be between these numbers. However, these are much farther apart in time than the stakes used in Exercise 22.

25. 2.368; Using a calculator, $\sqrt[3]{15} \approx 2.466$ giving an error $= |2.466 - 2.368| = 0.098$

27. 3.622; Using a calculator, $\sqrt[3]{50} \approx 3.684$ giving an error $= |3.684 - 3.622| = 0.062$

29. $L(x) = 100(x - 2) + 25$; $L(2.4) = 65$; Calculator value $5^{2.4} \approx 47.59$. Error $= 17.41$.

31. $L(x) = 0.0923(x - 16) + 6$; $L(30) = 7.292$; Calculator value $\sqrt[4]{30} + \sqrt{30} \approx 7.818$. Error $= 0.526$.

Section 10.2

1. $w(x) = 1.99x + 4.73$; Domain $= [10.5, 25.0]$

3. $\bar{x} = 17.6$; $\bar{y} = 39.8$; $w(17.6) = 39.8$

5. $w(625) = 3.5$; absolute error $= 0.1$; relative error $= 2.9\%$

9. Positive slope; (a) 1920: -0.2; 1930: -0.1; 1940: 0.0 (Answers may vary slightly.)

(b) 1990: 0.3; 2000: 0.38; 2010: 0.42 (Answers may vary slightly.)

(c) 1990: 0.4; 2000: 0.55; 2010: 0.7 (Answers may vary slightly.)

11. $w(x) = 0.02(x - 5) + 0.32$; $w(22) = 0.66$.

13. $w(x) = 433x + 115$

15. Supply prediction: $w(1.80) = 894$; Demand prediction: $w(1.80) = 1110$.

17. $w(7.2) = 39$; 0, 0%; 3, 7.1%; 6, 13.3%

19. $w(x) = 0.95 + 4.1$; $w(78) = 78.2 \approx 78$

21. $w(x) = 0.43x + 153$

23. $w(635) = 427$; The total mass of the other two books must be less than 273 gm.

25. $w(x) = 1.5x + 0.46$

27. $w(x) = 0.0013x - 0.25$

Section 10.3

1. Power   **3.** None of these.   **5.** Hyperbolic.

7. Power.   **9.** Exponential.   **11.** $y = 25\sqrt{x}$

13. $y = 0.005x^2$ **15.** $y = 4.56 \cdot (1.99)$ **17.** $\lambda_{max} = \dfrac{2{,}920{,}000}{T}$

19. $A(t) = 0.01t^2$

21. $d = \dfrac{3.2}{p}$

23. $y = 216 \cdot (1.02)^t$. In 2020, $y = 216 \cdot (1.02)^{40} = 477$; In 2025, $y = 216 \cdot (1.02)^{45} = 527$

25. $f(x) = 34.5 \cdot (0.951)^x$; $f(12) = 18.9$

Chapter 10 Review Test

1. 210 lbs.

2. $w(x) = 7.00x - 323$

3. $w(84) = 265$; absolute error = 25; relative error = 8.6%

4. $L(5.3) = 41.6$; calculator value = 39.4; absolute error = 2.2

5. No. $w(70.1) = 240 \neq 98.8$ which it should if $w(x)$ were the Wald function.

6. (a) D; (b) B; (c) A; (d) E; (e) C; (f) B

7. $y = 81.14 \cdot (1.004)^t$; When $t = 65$, $y = 105$

Glossary

Absolute error The absolute value of the difference between a predicted value and the actual value.

Absolute maximum The largest value a function attains in its range.

Absolute minimum The smallest value a function attains in its range.

Acceleration Rate of change of velocity with respect to time. It has both a numerical value and a direction.

Acceptance interval An interval centered on an assumed population mean that is used to test for whether it contains a mean computed from a sample.

Accidental Symbol preceding a note or note name, indicating the pitch to be played higher or lower by one half-step.

Action The steps taken by a function to process an element x in its domain.

Addition principle For any sets A and B, $n(A \cup B) = n(A) + n(B) - n(A \cap B)$.

Adjacent edges Two edges that meet at a common vertex.

Adjacent vertices Two vertices joined by an edge.

Aggregate contribution The total amount paid by the investor in an annuity. It is equal to the sum of the rent payments.

Algebraic statement Statement containing one or more variables.

Algorithm Sequence of specific steps to compute a value or solve a problem.

Amortization A schedule of regular payments for the purpose of repaying a loan.

Angular diameter The angular separation of opposite sides of an observed object.

Angular separation The measure of the angle formed at the observer's position by the two lines of sight to two separate objects.

Annuity A sequence of equal payments paid or received at regular intervals of time.

Annuity due Annuity in which the rent is paid at the beginning of the period.

Answer The segment of notes altered by a transformation of the subject that appears later in a fugue or canon.

Antecedent In the conditional statement $p \Rightarrow q$, p is the antecedent.

Aphelion The point in the orbit of an object in our solar system where the object is at a maximum distance from the sun.

Approval voting The voting method in which each person votes for as many candidates as he or she finds acceptable. The winner is the candidate with the most approval votes.

Approximating curve A curve that comes arbitrarily close to a set of data points and can be obtained as the graph of a modeling function.

Arcminute One-sixtieth of a degree. Used as a unit for angular separation.

Arcsecond One-sixtieth of a minute of arc. Used as a unit for angular separation.

Argument Collection of a sequence of statements.

Arrow's impossibility theorem It is mathematically impossible for a democratic voting method to satisfy all the fairness criteria.

Astronomical unit The mean distance from Earth to the sun, equal to about 93 million mi or 150 million km. Abbreviated AU.

Astronomy The science of the observation and study of the universe.

Asymptote A line used as a guide in graphing a function. The graph approaches it forever but does not intersect with it.

Axiom A statement universally accepted as true.

Base The factor by which an exponential function is consistently multiplied per unit change in the independent variable. It is denoted by b in the general formula for the function. It is necessary for $b > 0$ and $b \neq 1$. Also, the initial and terminal vertex of a circuit.

Bass clef Clef identifying the staff for those notes that are to be played primarily by the left hand on the piano.

Biased sample A sample gathered by a process in which the computed statistic is consistently too high or too low.

Biconditional The connective *if and only if*.

Binary Having exactly two possible values.

Bipartite graph A graph whose only edges consist of those that join every vertex in one group of vertices to every vertex in another distinct group.

Bit The smallest computer memory unit. It is binary valued.

Boolean expression Computer programming name for an algebraic statement. It is either true or false.

Borda count method The voting method using preference ballots in which points are awarded for each rank and the total number of points is computed for each candidate. The winner is the candidate with the highest point total.

Boundary markers The numbers that form the left and right endpoints of the interval classes.

Bridge An edge whose removal changes a connected graph into a disconnected graph.

Brute-force algorithm A problem-solving procedure in which the total weight of every Hamilton circuit is computed in order to determine an optimal circuit.

Byte Computer memory storage unit, usually composed of 8 magnetized bits.

Caesar cipher An encryption technique in which the ciphertext is created from the plaintext according to $C = (P + s)$ **mod** 26.

Cardinality The cardinality $n(S)$ of a finite set S is the number of elements contained in S.

Cartesian coordinate system A system of perpendicular number lines called axes used to graph functions. Also referred to as rectangular coordinate system.

Cell Computer memory storage unit composed of 1 or more bytes.

Central force The force on a moving body directed toward a central fixed point.

Central Limit Theorem If each value in a population is the result of summing over the same independent factors, then the distribution of that population approaches a normal distribution as the number of factors increases.

Central processing unit The brain of a computer that performs the arithmetic and logical operations. Abbreviated CPU.

Centripetal acceleration The acceleration on a moving body directed toward the center of its path of motion that results from a central force.

Chord A set of three or more notes to be played simultaneously.

Ciphertext The letters in an encrypted message.

Circle of fifths A geometric scheme for identification of the major keys with the corresponding number of accidentals.

Circuit A path in a graph whose terminal vertex is the same as its initial vertex.

Circular velocity Constant speed at which a body under the influence of a central gravitational force must move to maintain a circular path.

Clef Symbol preceding a staff that identifies the correspondence between lines and spaces and the pitches they represent.

Complement The complement $\overline{A}$ of a set A consists of the elements in the universal set that are not elements of A.

Complete bipartite graph A bipartite graph in which every vertex in one vertex group is joined by an edge to every vertex in the second group.

Complete graph A graph in which every pair of vertices is joined by an edge. The complete graph on n vertices is denoted by K_n.

Complete residue system modulo m A set of m integers, no two of which are congruent modulo m.

Composite The transformation resulting from the consecutive application of two or more transformations.

Compound interest Interest that is repeatedly computed at the end of regular time intervals and added to the accrued amount.

Compound statement Statement consisting of two or more prime statements.

Computer Machine that performs numerical computations and processes information.

Conclusion Final statement in an argument.

Conditional The connective *if ... then*.

Condorcet candidate A candidate associated with a preference table who is preferred by a majority of voters in one-on-one comparisons with each of the other candidates.

Condorcet criterion If one candidate defeats the other candidates in one-on-one comparisons, then that candidate should be the winner of the election.

Condorcet's paradox An election involving n candidates in which an n-cycle occurs.

Confidence interval An interval centered on a sample statistic that contains a population parameter with a specified level of confidence.

Congruence modulo m Two integers a and b are congruent modulo m if their difference is divisible by m.

Conjunction The connective *and*.

Connected graph A graph in which every vertex is connected to at least one other vertex.

Connective Word used to combine two statements into a compound statement.

Consequent In the conditional statement $p \Rightarrow q$, the consequent is q.

Continuous compounding Compounding of money according to the function $A(t) = Pe^{rt}$.

Contrapositive For the conditional statement "if P, then Q," the contrapostive is "if not Q, then not P." Its truth value is the same as that of the original statement.

Converse For the conditional statement "if P, then Q," the converse is "if Q, then P." Its truth value is not related to that of the original statement.

Copernican system Model of the planetary system with the sun at the center and Earth rotating on its own axis.

Copernican velocity Speed of a planet assuming it traveled in a circular orbit.

Crescent A phase of an illuminated body in which less than one-half of the illuminated side is visible from Earth.

Crossing An intersection of two edges that is not considered to be a vertex.

Cryptology The study of the techniques used for creating secret codes.

Decision structure Programming method for making a decision.

Decreasing A function f is said to be decreasing over an interval if $f(a) > f(b)$ whenever $a < b$ for any pair of numbers a, b in the interval.

Deduction Process used to reach a conclusion from a given set of premises.

Deferent Imaginary circle centered on Earth invented by Hipparchus, around which the center of a planet's epicycle moved.

Degree of a vertex The number of edges that meet at vertex v.

De Morgan's laws For any sets A and B, $\overline{A \cup B} = \overline{A} \cap \overline{B}$ and $\overline{A \cap B} = \overline{A} \cup \overline{B}$.

Density function A function whose graph specifies how a population is distributed. The area under the graph and over the domain of the function must equal 1.

Dependent variable The variable in a two-variable relationship whose value depends on the value assigned to the independent variable.

Deviation The difference between a specific data value and the mean of the collection of which the value is a member.

Dialogue The book written by Galileo, which defends the Copernican system through an extended conversation among three men.

Dichotomous preferences A division of candidates by a voter into two subsets of preferred and nonpreferred choices. The voter considers the members within each particular subset to be equally qualified.

Disconnected graph A graph that is not connected.

Disjoint sets Two sets whose intersection is empty; i.e., two sets A and B for which $A \cap B = \varnothing$.

Disjunction The connective *or*.

Dispersion The spread of any collection of data values.

Distribution The manner in which a set of numbers is apportioned or arranged between the extremes.

Domain The set of elements that are the inputs to a particular function.

Eccentricity Numerical value between 0 and 1 indicating the extent to which an ellipse departs from a circle.

Edge A line segment drawn between two vertices that represents some relationship between them.

Effective rate or yield The percentage increase of an investment attained in 1 year.

Ellipse The set of all points whose sum of distances from two fixed foci is constant.

Elongation The angular separation between the sun and a planet.

Empty set The set containing no elements. It is denoted by the symbol $\varnothing$.

Enharmonic spellings Different names for the same pitch.

Epicycle Imaginary circle around which a planet moved in the Ptolemaic system. The center of the epicycle moved along the deferent.

Equilibrium price The price for which the supply quantity is equal to the demand quantity.

Equilibrium quantity The quantity attained by both the supply and demand functions at the equilibrium price.

Equivalent Descriptive of two statement forms whose truth tables have identical final columns. Also called *logically* equivalent.

Euler circuits A closed path through a graph that contains every edge exactly once.

Eulerization Procedure in which edges are duplicated in order to create a new graph containing no odd vertices.

Even vertex A vertex whose degree is an even integer.

Event A subset of outcomes from a sample space.

Expected value The average value per repetition of a numerically valued experiment after a large number of repetitions. Signified as $E(X)$.

Experiment An action or a process for which the outcome cannot be predicted with certainty.

Exponential averaging The technique for finding an exponential function whose graph forms an approximating curve to a set of data points.

Exponential function A function of the form $f(x) = kb^x$.

Extended graphs The graphs obtained by extending a smooth curve between points on the graph of a function whose domain is only a subset of the interval displayed on the x-axis.

First quartile A number that is greater than 25% of the values in a collection and less than 75% of the values.

Flat Accidental indicating a note to be played lower by a half-step, signified by $\flat$.

Focus One of the two fixed points inside an ellipse, from which the distances to any point on the ellipse total a constant value.

Force An entity applied to an object, which changes its velocity. It is equal to the product of the mass of the object and its acceleration.

Formula The precise expression for a function that describes its action in terms of the independent variable.

Free (or tonal) inversion A melodic inversion in which some notes may be slightly altered to maintain the tonality of the music.

Frequency The number of data values in an interval class. Also, the number of sound waves or vibrations per second reaching the ear that produce a particular pitch.

Function A relationship between two sets that assigns to every element in the first set (domain) exactly one element in the second set (range).

Future value The amount of an annuity or some current sum of money that will accumulate with interest over some future period of time.

Gaussian (normal) distribution A distribution discovered by Carl Gauss whose z-scores form a population having density function $f(z) = \dfrac{1}{\sqrt{2\pi}} e^{-(z^2/2)}$.

Generalized Caesar cipher An encryption technique in which the ciphertext is created from the plaintext according to $C = (rP + s) \bmod 26$.

Geocentric Earth-centered.

Geometric mean The nth principal root of the product of n values.

Gibbous A phase of an illuminated body in which more than one-half of the illuminated side is visible from Earth.

Gram The basic measure of mass in the metric system.

Graph A collection of vertices and edges.

Graph of a function f The set of all ordered pairs (x, y) plotted on a Cartesian coordinate system in which the y-coordinate is equal to $f(x)$ and x is a member of the domain of f.

Graph theory The study of the properties and applications of graphs.

Group estimate of the mean An estimate of the mean made by summing the products of the midpoints and the corresponding relative frequencies of the interval classes of a relative frequency distribution.

Group estimate of the median An estimate of the median made by locating the median interval and approximating its location in the interval. Denoted by M_g.

Group estimate of the standard deviations An estimate of the standard deviation from a relative frequency distribution made by taking the square root of a weighted average of the squared deviations.

Half-step Interval between two adjacent piano keys.

Hamilton circuit A closed path through a graph that contains every vertex exactly once.

Hare system The voting method using preference ballots such that if no candidate has a majority of first-place votes, then the candidate with the fewest first-place votes is eliminated, and all the ballots are adjusted accordingly. This process continues until one candidate receives a majority. Also called *instant runoff*.

Heliocentric Sun-centered.

Heuristic algorithm A problem-solving procedure that produces a solution quickly. However, this solution may not be optimal.

Histogram A pictorial display of a relative frequency distribution with the interval classes represented on a horizontal axis and the relative frequencies represented by areas of rectangles situated over the intervals.

Horizontal translation Horizontal movement of a point by k units in the Cartesian plane and having the functional representation $(a, b) \rightarrow (a + k, b)$.

Hypothetico-deductive method A method of inquiry in which a theory gains increasing acceptance as more evidence for its validity is observed. This is *not* the same as a deductive method, by which a definitive conclusion is reached.

Image The specific functional value in the range of a function associated with a particular value in the domain of the function.

Increasing A function f is said to be increasing over an interval if $f(a) < f(b)$ whenever $a < b$ for any pair of numbers a, b in the interval.

Incrementation Repeated increase of the value of a variable by a fixed amount.

Independence of irrelevant alternatives (IIA) In an election that declares candidate A to be the winner, the order on any preference ballot can be changed if it does not affect the ranking of candidate A with respect to candidate B, and B still does not win.

Independent variable The variable in a two-variable relationship or a function that varies freely among the allowed values.

Inertia The ability of a body to resist a change in its state of motion.

Initialization Assignment of an initial value to a variable.

Input device Computer component that transmits information to the memory.

Insincere voting The act of voting in a manner that does not reflect the true wishes of the voter.

Instant runoff Another name for the Hare system of voting.

Intercepts Points on the coordinate axes where a graph intersects them.

Interest An amount of money earned in an investment or a fee charged for the lending of money.

Interest period Period of time over which the interest rate is applied.

Interest rate The percentage to be applied to the accrued amount in an investment or the remaining balance in a loan when determining the interest.

Interpretation A substitution of English words for the objects and symbols in a statement form.

Intersection of two sets The intersection $A \cap B$ of two sets A and B is the set of elements belonging to both A and B.

Interval classes Adjacent intervals of real numbers used for sorting a large collection of data values.

Inverse-square law A general principle by which the effect of a phenomenon on an object varies inversely as the square of its distance to the source.

Isomorphism A one-to-one correspondence between the vertices and edges of two graphs. The degrees of corresponding vertices must be equal.

Kepler's laws of planetary motion:

1. The orbit of each planet is an ellipse, with the sun at one focus.

2. The line joining a planet and the sun sweeps out equal areas in equal amounts of time.

3. The square of the period of a planet's revolution is proportional to the cube of its mean distance from the sun.

Key The name of a major scale according to the first note. Also the name of an ordered pair of integers that provides the information needed to decipher an encrypted message.

Key signature A group of one or more accidentals at the beginning of a composition to identify the key.

Kilogram 1,000 grams.

Law of large numbers If p represents the probability that an event A occurs and r represents the relative frequency of A, then r approaches p as the number of repetitions increases.

Law of universal gravitation Any body attracts another body with a force directed along the line joining their centers, which is proportional to the product of the masses of the bodies and inversely proportional to the square of the distance between them.

Laws of motion The three basic principles of motion, which Newton states in the beginning of the *Principia* and from which he derives the causes for many celestial and terrestrial phenomena. See Figure 4.4.5.

Least residue The smallest nonnegative integer in a residue class.

Least-residue system modulo m A complete residue system consisting entirely of least residues.

Level of confidence A percentage that gives the portion of the time a confidence interval will contain the population parameter.

Light-year Distance traveled by light in 1 yr, equal to about 9.5 trillion km.

Linear congruence The congruence equation $ax \equiv b \pmod{m}$.

Linear fit A straight line that comes as close as possible to the points in a scatter plot.

Linear function A function of the form $f(x) = mx + k$.

Linear interpolation Prediction of a functional value based on the equation of the straight line connecting two adjacent ordered pairs.

Logic The study of the principles of deductive reasoning.

Logic circuit Combination of logic gates (i.e., electrical circuits and connective gates) that simulates a statement form.

Logic gate Computer mechanism that simulates a logical connective.

Logical value True or false.

Logically equivalent See *equivalent*.

Looping Program structure for repeating a set of steps.

m-modularization Grouping of the integers according to residue classes modulo m.

Machine language The program instructions in a special language that the computer understands.

Major scale A specific arrangement of eight notes separated by steps according to whole-whole-half-whole-whole-whole-half.

Majority criterion If a candidate receives the majority of the first-place votes in an election, then that candidate is the winner.

Margin of error A number that is added and subtracted from an estimate ($\bar{x}$ or $\hat{p}$) to create a confidence interval for a parameter (μ or p).

Mass Property of an object that is a measure of the amount of matter in the object. It can be thought of as a measure of its inertia.

Mathematical modeling Simulating a real-world phenomenon through the creation of a mathematical structure.

Mean The sum of the values in a collection of data, divided by the number of values. Signified by $\bar{x}$.

Measure of central tendency The typical or average number of a collection of data.

Median A number that is greater than 50% of the values in a collection and less than 50% of the values. Signified by M.

Median interval The interval class in a histogram or relative frequency distribution that contains the median.

Melodic inversion A reflection of a segment of notes across a horizontal line through a pivot note.

Memory The component of a computer where information is electronically stored in locations called cells .

Minimal Hamilton circuit A Hamilton circuit in a weighted graph whose total weight is less than that of any other Hamilton circuit.

Minimum Eulerization Eulerization of a graph that uses a minimum number of duplicated edges.

Mod A function often included in programming languages and calculators. x mod m is the least residue of $[x]_m$.

Model Function, system of equations, computer program, or other vehicle capable of processing quantitative information to make predictions.

Modeling type A function chosen from the choices: linear ($y = mx + k$), power ($y = kx^p$), hyperbolic ($y = k/x$), or exponential ($y = kb^x$).

Modular arithmetic Arithmetic done with residue classes.

Modulus The positive integer m used in defining congruence modulo m.

Modus ponens Valid reasoning scheme with form $[(p \Rightarrow q) \wedge p] \Rightarrow q$. Abbreviated as MP. Also called *direct reasoning*.

Modus tollens Valid reasoning scheme with form $[(p \Rightarrow q) \wedge \sim q] \Rightarrow \sim p$. Abbreviated as MT. Also called *indirect reasoning*.

Monotonicity criterion If candidate X is the winner in an election before her or his ranking is elevated one place on any ballot, then X remains the winner.

Mortgage Loan used to purchase real estate.

Multiplicative inverse The number that, when multiplied by a, yields 1 modulo m is the multiplicative inverse of a modulo m.

Mutually exclusive event Events that cannot occur at the same time.

n-cycle A pattern of circular rankings of n candidates in an election.

Natural exponential function An exponential function with base given by a power of e.

Natural laws Foundational principles that provide a basis of explanation of other relationships found in nature.

Natural philosophy A set of principles or ideas concerning the workings of nature. The addition of mathematics to formalize such ideas led to the recasting of this phrase as *science*.

Nearest-neighbor algorithm A heuristic algorithm for forming a Hamilton circuit that proceeds from one vertex to an adjacent vertex along the lowest-weight edge that does not complete a circuit until it returns to the initial vertex.

Negation Reversal of the logical value of a statement with some form of *not*.

Network Another name for a graph.

Newton Unit of force in the metric system.

Node Another name for a vertex.

Nonlinear function A function whose graph is not a straight line.

Normal curve The bell-shaped graph of a Gaussian or normal distribution.

Note Symbol written on the music staff whose location indicates to the musician the pitch to play and the length of time the pitch is to be maintained. Often used interchangeably with the word *pitch*.

Note graph The graph of a position sequence as a function of time.

Note of inversion The note on the musical staff about which a segment of notes is rotated 180°.

Number theory The study of integers and associated properties.

Octave Interval between pitches that are 12 half-steps apart. This implies that the frequency of the higher pitch is twice that of the lower pitch.

Odd vertex A vertex whose degree is an odd integer.

Odds against an event The ratio of the probability that the event does not occur to the probability that it does occur.

Odds for an event The ratio of the probability that the event does occur to the probability that it does not occur.

Ordinary annuity Annuity in which the rent is paid at the end of the period.

Output device Computer component that displays results from the performance of a computer program.

Parabola The graph of a quadratic function.

Paradigm shift An altering of the standard model for a phenomenon.

Paradox Sentence or set of conditions leading to a sentence with an undecidable logical value.

Parallax The change of position of a close object with respect to a more distant background when viewed from two different locations.

Parameter The constants k, p, and b used in a nonlinear modeling function such as $y = kx^p$, $y = \dfrac{k}{x}$, or $y = kb^x$.

Parameters Measurements of central tendency and dispersion associated with a population.

Parsec The distance to an object at which the stellar parallax has a measure of 1 arcsecond (1′). It is equal to 3.26 light-years (ly).

Partition For a set S, pair of subsets A and B such that $A \cap B = \varnothing$ and $A \cup B = S$.

Path A sequence of consecutive adjacent vertices.

Payment period or interval Period of time between rent or annuity payments.

Perfect fifth An interval between two notes that spans 7 half-steps. This corresponds to 5 degrees on the staff.

Perfect fourth An interval between two notes that spans 5 half-steps. This corresponds to 4 degrees on the staff.

Perihelion The point in the orbit of an object in our solar system where the object is at a minimum distance from the sun.

Period Time between repetitions of a repeating phenomenon.

Pitch Property of sound that identifies it as high or low compared to other sounds. It is a function of the frequency of the sound waves producing it.

Pitch class Residue class modulo 12 containing the key positions of all pitches that are one or more octaves apart.

Pivot note The note on the staff across which an inversion is applied.

Pixel Smallest element of a television or computer screen.

Plaintext The letters in a message before it is encrypted.

Planar graph A graph that is isomorphic to a graph without crossings that is contained in a plane surface.

Plurality method The voting method in which the candidate with the most votes wins.

Population An infinite (or very large) collection of data.

Position sequence A sequence of positions on the piano keyboard corresponding to a segment or sequence of musical notes.

Preference ballot A ballot in which the voter ranks the candidates from most favored to least favored.

Preference table A table that shows the number of voters who have cast a preference ballot of each particular ranking of candidates.

Premise Any of the statements of an argument except for the conclusion.

Present value The original or current amount of money in an interest-bearing account or annuity.

Primary chord Chord whose notes lie between key positions 0 and 11 inclusive.

Primary note A note whose key position is between 0 and 11 inclusive.

Prime A positive integer having only itself and 1 as factors.

Prime statement Statement that does not contain a connective.

Principal The original amount of money deposited in an investment or borrowed from a lender.

Principia The grand treatise of Isaac Newton that established the foundations of physics and astronomy. Its explanations for natural phenomena are mathematically derived from a core set of three laws of motion.

Probability The value approached by the relative frequency of the event as the number of repetitions of the experiment increases.

Probability of an event A (in a uniform sample space S) A function p with domain consisting of the subsets (events) of S and range consisting of the unit interval $[0, 1]$. For any event A, $p(A) = \dfrac{n(A)}{n(S)}$.

Program List of instructions for a computer to translate to machine language and then perform.

Pseudocode Description of an algorithm that the programmer designs before writing the program.

Ptolemaic system Model of the planetary system with Earth at the center and employing epicycles and deferents to explain retrograde motion of the planets.

Quadratic function Any function of the form $f(x) = ax^2 + bx + c$.

Quadrivium A group of studies in medieval universities consisting of arithmetic, music, geometry, and astronomy.

Random sample A sample gathered from a population in such a way that every member of the population has an equal chance of being selected.

Range The set of elements that are the outputs of a particular function. Also, the difference between the largest and the smallest values in a collection.

Rate The change in the functional value of a linear function per unit change in the independent variable.

Reflection across a horizontal A reflection across any line parallel to the x-axis in the Cartesian plane. The reflection across the x-axis has the functional representation $(a, b) \rightarrow (a, -b)$.

Reflection across a vertical A reflection across any line parallel to the y-axis in the Cartesian plane. The reflection across the y-axis has the functional representation $(a, b) \rightarrow (-a, b)$.

Relative error The result of dividing an absolute error by the observed value or actual value of a function and converting to a percentage.

Relative frequency The fractional portion of a collection of data values that lie in a particular class. It is obtained by dividing the frequency of the class by the total number of values in the collection.

Relative frequency distribution A table displaying interval classes and their corresponding relative frequencies.

Rent Payment made or received in an annuity.

Residue (or congruence) class Set of numbers such that the difference between any two members of the set is a multiple of a common integer m. Denoted by $[x]_m$.

Retrograde Musical terminology for the sequence of notes obtained by reflecting the note graph of an initial sequence across a vertical.

Retrograde inversion Musical terminology for the sequence of notes obtained by the rotation of the note graph of an initial sequence through an angle of $180°$.

Retrograde motion The apparent reversal of the movement of a planet.

Rotation A rotation of $90°$ in the Cartesian plane has the functional representation $(a, b) \rightarrow (-b, a)$. A rotation of $180°$ in the Cartesian plane has the functional representation $(a, b) \rightarrow (-a, -b)$.

Sample A subset of a larger collection or population.

Sample space The set of all distinct possible outcomes of an experiment.

Sampling distribution A population of means of samples, all of the same size.

Scatter plot The graph of a set of data points obtained from an experiment or observation of a phenomenon.

Scientific method A formalized procedure for exploring natural phenomena involving:

1. Careful observation and the recording of data.

2. Objective analysis of the data and the creation of a model to fit the data.

3. Use of the model to make predictions that can be tested against new observations.

Set A collection of objects.

Set difference The set $A - B$ consisting of those elements of A that are not also in B. It must necessarily be equal to the set $A \cap \overline{B}$.

Sharp Accidental indicating a note to be played higher by a half-step, signified by #.

Significance level The percentage of time that a randomly collected sample mean would lie outside a prescribed acceptance interval.

Significance test A test involving a sample mean $\overline{x}$. If $\overline{x}$ lies outside the acceptance interval, it is said to be significant.

Simple annuity Annuity in which the interest period is equal to the payment period.

Simple circuit A circuit in which the base is the only vertex used more than once.

Simple interest The interest earned on the principal P at an annual rate r over t years and given by Prt.

Simple path A path in which no vertex occurs more than once.

Single transferable voting systems (STVS) A voting method in which votes are transferred after a candidate has been eliminated.

Sinking fund Annuity established for the purpose of accumulating a specific amount of money in the future.

Skewed distribution A distribution in which the main body of the values is clustered at one end. The mean is not equal to the median in a skewed distribution.

Slope Characteristic of a linear graph computed by $y_1 - y_0 / x_1 - x_0$ for any two points (x_0, y_0) and (x_1, y_1) on the line. It is numerically equal to the rate.

Staff Set of five parallel lines forming the grid on which notes are written to compose music.

Stakes The two data points used in linear interpolation whose first coordinates bracket the value used as the input for the prediction.

Standard deviation The primary measure of dispersion of any collection of data. It is the square root of the mean of the squared deviations.

Standard normal distribution The distribution whose density function is given by $f(z) = \dfrac{1}{\sqrt{2\pi}} e^{-(z^2/2)}$. It has mean 0 and standard deviation 1.

Statement Declarative sentence that has exactly one logical value.

Statement form Any expression of statement symbols and connectives.

Statistics The science of collecting, organizing, and analyzing large amounts of data in a cohesive manner for the purpose of reaching meaningful conclusions.

Stellar parallax One-half of the angular displacement in the apparent position of a star when observed from points on Earth's orbit that are separated by 180°.

Strategic manipulation Submission of a ballot or block of ballots in which voters have voted insincerely.

Strict (or real) inversion A melodic inversion in which the intervals between consecutive notes are preserved but in the opposite direction.

Strictly local maximum The largest value of a function for an interval of "nearby" domain values. It is not an absolute maximum. It appears on the graph as the y-coordinate of a point at the top of a minor hill.

Strictly local minimum The smallest value of a function for an interval of "nearby" domain values. It is not an absolute minimum. It appears on the graph as the y-coordinate of a point at the bottom of a minor valley.

Subject Fundamental theme of notes in a fugue or canon that appears near the beginning and is transformed later into altered segments called answers.

Subset Set A is a subset of set B if every element of A is also an element of B.

Syllogism An argument consisting of two premises and a conclusion.

Symbolic logic The science of using formal abstractions and procedures to systemize the principles of deduction and valid reasoning. Also called *logical calculus*.

Symmetric distribution A distribution where a vertical line through the median divides the histogram into two mirror images. The mean is equal to the median in a symmetric distribution.

Tautology A statement form (or its interpretation) that is true for every possible combination of logical values for its prime statements.

Term Time interval over which the parameters of an annuity are in effect.

Tonic The first note of a major scale.

Transformation A change in location of the points in the Cartesian coordinate system resulting from a well-defined function.

Transforming the data Altering either all the first coordinates or all the second coordinates in a set of data points, which produces a new set of points whose graph is approximately linear.

Transposition The changing of the key of a piece of music. It is the musical analog of a vertical translation of a note graph.

Traveling Salesperson Problem (TSP) The problem of finding a minimum-weight Hamilton circuit in a complete weighted graph.

Treble clef Clef identifying the staff for those notes that are to be played primarily by the right hand on the piano.

Truth table A table listing the logical value of a statement form for every combination of values for the prime statements.

Uniform sample space A finite sample space in which each outcome is equally likely to occur.

Union of two sets The union $A \cup B$ of two sets A and B is the set of objects belonging to either A or B.

Universal set A set of all possible elements associated with a certain phenomenon or experiment.

Valid Descriptive of an argument whose truth table corresponding to its statement form is a tautology.

Variable A letter or symbol that represents any one of a variety of quantities or objects.

Velocity The rate of change of position of a moving body, i.e., its speed and its direction of motion.

Venn diagrams Pictorial diagram using circles for sets, initiated by John Venn and used to analyze the validity of arguments.

Vertex The point on a parabola at which the curve turns around. The x-coordinate of the vertex is given by $x = -b/2a$, where $f(x) = ax^2 + bx + c$ is the associated quadratic function. Also, one of a set of points that all represent a similar type of object. The vertex set of a graph G is denoted by $V(G)$.

Vertical line test A graph represents a function of the variable associated with the horizontal axis if no vertical line intersects the graph more than once.

Vertical translation Vertical movement of a point by k units in the Cartesian plane and having the functional representation $(a, b) \rightarrow (a, b + k)$.

Wald function The linear function created by using two average ordered pairs of points from a set of ordered pairs observed from some experiment or phenomenon.

Wald line The graph of the Wald function.

Weight Amount of force that a body exerts on the surface of a planet or moon as a result of gravity. Also, a numerical value assigned to an edge.

Weighted graph A graph in which every edge has a weight.

Whole step Interval equivalent of two consecutive half-steps.

z-score A measure of distance of any data value from the mean. It is equal to the deviation of the value divided by the standard deviation.

Zeroes of the function The value r of a function f such that $f(r) = 0$.

Zodiac A narrow, beltlike region around the sky containing the apparent paths of the sun and planets.

Chapter 1 The Concept of Function

Berresford, Geoffrey C. and Andrew M. Rockett. 2010. *Applied Calculus*. 5th ed. Belmont, CA: Brooks/Cole.

Dubinsky, Ed and Harel Guershon, ed. 1992. *The Concept of Function.: Aspects of Epistemology and Pedagogy*, MAA Notes, Number 25. Washington, DC: Mathematical Association of America.

Thomas Jr., George B., Maurice D. Weir, and Joel R. Hass. 2013. *Thomas' Calculus*, 13th ed. New Jersey: Pearson.

White, Alvin M. ed. 1993. *Humanistic Mathematics*. MAA Notes, Number 32. Washington, DC: Mathematical Association of America.

Chapter 3 Logic and Computer Science

Agostini, F. 1986. *Math and logic games*. New York: Harper-Collins.

Bell, E. T. 2008. *Men of mathematics*. Clermont, FL: Paw Prints.

Bondi, C., ed. 1991. *New applications of mathematics*. London: Penguin Books.

Cooper, D. 1993. *Oh! Pascal!* 3rd ed. New York: W.W. Norton.

Epstein, R. L. and W. A. Carnielli. 1989. *Computability, computable functions, logic, and the foundations of mathematics*. Belmont, CA: Wadsworth.

Kelly, J. J. 1996. *The essence of logic.* Upper Saddle River, NJ: Prentice Hall.

Mendelson, E. 2009. *Introduction to mathematical logic.* 5th ed. London: Chapman & Hall.

Monroe, J. and R. Wicander. 2005. *The changing earth.* 4th ed. Belmont, CA: Brooks-Cole.

Newman, J. R. 2003. *The world of mathematics.* Mineola, NY: Dover Publications.

Russell, B. 1924. *Introduction to mathematical philosophy.* London: George Allen & Unwin.

Stoll, R. R. 1979. *Sets, logic, and axiomatic theories.* San Francisco: W. H. Freeman.

Van Dyke, F. 1995. A visual approach to deductive reasoning. *Mathematics Teacher*, NCTM 88, no. 6 (September).

Chapter 4 Astronomy and the Methods of Science

Cohen, I. B. 1981. Newton's discovery of gravity. *Scientific American* (March) 166–179.

Giancoli, D. C. 2010. *Physics.* 6th ed. Boston: Addison Wesley.

Gingerich, O. 1982. The Galileo Affair. *Scientific American* 133–143 (August).

—. 1983. From Aristarchus to Copernicus. *Sky & Telescope* 410–412 (November).

—. 1994. Circles of the gods: Copernicus, Kepler, and the ellipse. *Bulletin of the American Academy of Arts and Sciences* 47, no. 4 (January).

Greene, B. 2000. *The elegant universe.* New York: Vintage Books.

Hutchins, R. M., ed. 1990. Ptolemy: The almagest; Nicholas Copernicus: On the revolutions of the heavenly spheres; Johannes Kepler: Epitome of Copernican astronomy. *Great Books of the Western World: Encyclopedia Britannica*, 2nd ed.

Koupelis, T. 2014. *In quest of the universe.* Burlington, MA: Jones & Bartlett.

Lerner, L. and E. Gosselin. 1973. Giordano Bruno. *Scientific American* 86–94 (April).

Pasachoff, J. and A. Filippenko. 2007. *The cosmos.* Belmont, CA: Brooks/Cole.

Newman, J. R. 2003. *The world of mathematics.* Mineola, NY: Dover Publications.

Peebles, P. J. E. and D. Wilkinson. 1967. The primeval fireball. *Scientific American* (June) 28–37.

Rickey, V. F. 1987. Isaac Newton: Man, myth, and mathematics. *The College Mathematics Journal* 18, no. 5 (November) 362–389.

de Santillana, G. 1955. *The crime of Galileo.* Chicago: University of Chicago Press.

Seeds, M. A. 2011. *Horizons.* Belmont, CA: Wadsworth.

Shawl, S. J., R. R. Robbins, and W. H. Jefferys. 1999. *Discovering astronomy.* 3rd ed. Dubuque: Kendall Hunt.

Simmons, G. F. 1992. *Calculus gems.* New York: McGraw-Hill.

Chapter 5 Graph Theory

Chartrand, Gary, and Ping Zhang. 2012. *A First Course in Graph Theory.* Dover. Originally Boston: McGraw Hill.

COMAP. Solomon Garfunkel et. al. 2011. *For All Practical Purposes*, 9th ed. New York: Freeman.

Gross, Jonathan L. and Jay Yellen. 2005. *Graph Theory and Its Applications.* 2nd ed. Boca Raton, FL: Chapman and Hall/CRC.

Ore, Oystein. 1963. *Graphs and Their Uses.* New York: Random House.

Chapter 6 Voting Methods

Black, D. 1986. *The theory of committees and elections.* Dordrecht: Kluwer.

Brams, S. J., and D. R. Herschbach. 2001. The science of elections. *Science* 292 (May).

Brams, S. J., and P. C. Fishburn. 2002. Voting procedures. *Handbook of social choice and welfare*, vol. 1. K. J. Arrow, A. K. Sen, and K. Suzumura, eds. Philadelphia: Elsevier.

Brams, S. J. 2007. *The presidential election game.* 2nd ed. Wellesley, MA: A. K. Peters/ CRC Press.

——. 2008. *Mathematics and democracy.* Princeton, NJ: Princeton University Press.

Center for Election Science, http://www.approvalvoting.org/.

Saari, D. G. 2003. *Basic geometry of voting.* Berlin: Springer-Verlag.

Truchon, M. 1998. *Figure skating and the theory of social choice*, cahier 9814, cahier 98-16, du Centre de Recherche en Economie et Finance Appliquées (CREFA) (October).

Chapter 7 Probability

Denny, Mark and Steven Gaines. 2000. *Chance in Biology: Using Probability to Explore Nature.* Princeton, NJ: Princeton University Press.

Packel, Edward. 2006. *The Mathematics of Games and Gambling*, 2nd ed. Washington, DC: Mathematical Association of America.

Ross, Sheldon. 2009. *Introduction to Probability Models*, 10th ed. Burlington, MA: Academic Press (Elsevier).

Chapter 8 Statistics

Brockett, P. and A. Levine. 1984. *Statistics and probability.* Philadelphia: Saunders College Publishing.

Cleveland, W. S. 1985. *The elements of graphing data.* Monterey, CA: Wadsworth.

Congressional Quarterly, Inc. 1990. *Editorial Research Reports* 1:16 (April).

Consortium for Mathematics and Its Applications (CoMap). 2011. *For all practical purposes.* New York: W. H. Freeman.

La Liga Football League. 2009–2010. http://soccernet.espn.go.com/stats/topscorers.

Triola, M. F. 2011. *Elementary statistics.* 11th ed. Upper Saddle River, NJ: Pearson.

U.S. Bureau of Labor Statistics 2010 (April), http://www.bls.gov.

U.S. Geological Survey, 2009. Earthquakes Hazard Program, http://neic.usgs.gov/neis/ eqlists/.

Chapter 9 Mathematics in Music and Cryptology

Beutelspacher, A. 1994. *Cryptology.* Washington, DC: Mathematical Association of America.

James, J. 1993. *The music of the spheres: Music, science and the natural order of the universe.* New York: Springer-Verlag.

Johnson, C. M. 2001. Functions of number theory in music. *Mathematics Teacher* 94 no. 8 (November) 700–707.

—. 2009. Introducing group theory through music. *Mathematics Teacher* 103, no. 2 (September) 116–123.

Long, C. T. 1995. *Elementary introduction to number theory*. 3rd ed. Long Grove, IL: Waveland Press.

Maor, E. 1979. What is there so mathematical about music? *Mathematics Teacher* (September) 414–422.

Ore, O. 1967. *Invitation to number theory*. Washington, DC: The Mathematical Association of America.

O'Shea, T. 1979. Geometric transformations and musical composition. *Mathematics Teacher* (October) 523–528.

Ottman, R. W., and F. Mainous. 2003. *Rudiments of music*. 4th ed. Upper Saddle River, NJ: Prentice-Hall.

Sinkov, A. 1966. *Elementary cryptanalysis: A mathematical approach,* 6th printing. Washington, DC: Mathematical Association of America.

Stephenson, B. 1994. *The music of the heavens: Kepler's harmonic astronomy.* Princeton, NJ: Princeton University Press.

Chapter 10 Creating a Model from Data

Duffin, R. J. 1991. Lines of best fit by graphics and the Wald line, *American Mathematical Monthly* 98, no. 9 (November) 835–839.

Francis, J. C. 1993. *Management of investments*. New York: McGraw-Hill.

Giordano, F., W. Fox, S. Horton, and M. Weir. 2008. *A first course in mathematical modeling.* Belmont, CA: Brooks/Cole.

Hoaglin, D., F. Mosteller, and J. Tukey. 1983. *Understanding robust and exploratory data analysis.* Hoboken, NJ: John Wiley.

NASA Goddard Institute for Space Studies. 2012. (May 10, 2012). http:/www.giss.nasa.gov.

$f(0)=0$

Index

Note: Page numbers followed by *f*, indicate materials in figures.

A

absolute error, 516
absolute maximum, 54
absolute minimum, 54
acceleration, 209
acceptance intervals, 431, 432
accidental, 448
action, 23
actuaries, 370
addition principle, 344
adjacent edges, 250
adjacent vertices, 250
aggregate contribution, 93
algebraic statement, 112
algorithms, 15, 154
alleles, 357
Almagest (Ptolemy), 192
Alpha centauri, 180, 187
amortization, 97
    of loans, 95–102
angular diameter, 183, 183*f*
angular separation, 181, 181*f*
annuity, 85–90, 86*f*
annuity due, 85
annular eclipse, 184
answer (in a fugue), 468
antecedent, 115
aphelion, 199

apogee, 203
apparent magnitudes, 13
approval voting, 315
approximately symmetric distribution, 381, 381*f*
approximating curve, 539
arcminutes, 181
arcseconds, 181
argument, 136
Aristarchus, 184, 185
Aristotle, 190, 190*f*
Arrow's impossibility theorem, 314
Art of the Fugue, 469*f*
Astronomia Nova (The New Astronomy) (Kepler), 199
astronomical unit, 180
astronomy, 176, 176*f*
asymptote, 543
atomic number, 7
axioms, 178, 179

B

base, 250
bass clef, 449
Beethoven's *Symphony No. 4*, retrograde in, 466, 467*f*
Bessel, Friedrich Wilhelm, 195
biased sample, 424, 424*f*
biconditional, 113
Big Dipper constellation, 180*f*, 187
binary, 151
binary systems, 187
bipartite graph, 258
bits, 151

black hole, 239
Boole, George, 122
Boolean expressions, 155
Borda count method, 295
boundary markers, 371
Boyle's law, 542*f*
bridge, 251
Bruno, Giordano, 208
brute-force algorithm, 275
byte, 151

C

Caesar cipher, 496
Cannaboro county, 436
Cardano, Girolamo, 350
cardinality, 343
Cartesian coordinate system, 39, 510
Cartesian plane, 460*f*
Cavalieri, Bonaventura, 206
cells, 153
central force, 224
central force law, 224, 238
central limit theorem, 413, 427
central processing unit (CPU), 153
centripetal acceleration, 224
chord, 450, 451*f*
ciphertext, 496
circle of fifths, 478–484, 482*f*
circuit, 250
circular velocity, 231
clef, 449
clock arithmetic, 489
common ratio, 86
complement, 339
complete bipartite graph, 283
complete graph, 252
complete residue system modulo *m*, 487
composite, 462, 462*f*
compound interest, 75
compound statements, 112
computers, 151
conclusion, 136
conditional, 113
Condorcet candidate, 306
Condorcet criterion, 306
Condorcet, Marquis de, 305
Condorcet's paradox, 307
confidence intervals, 416, 423–432
congruence, 443
congruence class, 443
conjunction, 113
connected graph, 251
connectives, 111
consequent, 115
continuous compounding, 84
contrapositive, 130, 215
converses, 130, 214
Copernican system, 193–197, 194*f*
Copernican velocity, 194
Copernicanism, defense of, 208–220
Copernicus, Nicholas, 193
CPU. *See* central processing unit
crescent phase, 214
crossing, 248
cryptology, 14, 495
    modular arithmetic and, 489–499

current flow
    AND gate, 151, 152*f*
    OR gate and NOT gate, 151, 152*f*
cycloid, 191, 192*f*

D

data
    dispersion of, 381, 391–397
    representation of, 371–381
De Morgan, Augustus, 342, 342*f*
De Morgan's laws, 342
De Revolutionibus Orbium Caelestium (Copernicus), 193
decision-making process, 246
decision structure, 155
decreasing function, 11
decrypting (deciphering), 496
deduction, 110
deferent, 191, 192
degree of a vertex, 249
DeMoivre, Abraham, 410
density function, 405
dependent variable, 5
Descartes, Rene, 38, 208
deterministic phenomena, 330
deviation, 391
Dialogue Concerning the Two Chief World Systems (Galileo), 216
dichotomous preference, 316
digraph, 258
disconnected graph, 251
disjoint sets, 340
disjunction, 113
dispersion of data, 381, 391–397
distribution
    of ages, in Kingsbridge County, 377, 378*f*
    of data, 377
    of errors, 409
    function of, 391
    of sample means, 426
domain, 6

E

eccentricity, 199
edges, 247
effective rate or yield, 78, 79*f*
ellipse, 198
elongation of planet, 189
empty set, 338
encrypting (enciphering), 496
enharmonic spellings, 449
enumeration, 424
epicycle, 191, 192
equilibrium point, 57
equilibrium price, 57
equilibrium quantity, 57
equivalent, 129
Eratosthenes, 188
errors, distribution of, 409
escape velocity, 237
Eudoxus, 191
Euler circuit, 252, 260–270
Euler, Leonhard, 15, 260
Eulerization, 262
    of Manhattan graph, 263
even vertex, 250
event, 332
expected value, 357

experiments, 330
exponential averaging, 544
extended graphs, 55

F

federal income tax amounts, histogram of, 372*f*
Fermat, Pierre de, 39, 498
Fermat's last theorem, 498
flat, 449
focus, 198
force, 226
formula, 6
free (or tonal) inversion, 468
frequency, 372, 442
frequency distribution, with amended last class, 373*f*
function of distribution, 391
functional expressions, creation of, 23
    exponential functions, 27–31
    linear functions, 23–27
functions, 4–17
  defined, 6
  graph of, 38–43
future value, 72

G

Galilei, Galileo, 16, 208–217, 208*f*, 217*f*
Gauss, Carl Friedrich, 408–409, 408*f*
Gaussian (normal) distribution, 409
generalized Caesar cipher, 497
genotypes, 357–358
geocentric, 191
geocentric system, 179, 179*f*
geocentric universe, 179
geometric mean, 544
geometric sum, 86
geometry, 11
geostationary satellite, 237
gibbous phase, 213–214
Gourmet Club election, 294
grams, 211
graph of a function *f*, 24, 38
graph theory, 246, 262
    Euler circuits, 260–270
    graphs and paths, 246–259
    Hamilton circuits, 271–286
graphing calculators, 44
graphs
    of function, 38–44
    interpretation of, 50–58
    and paths, 246–259
gravitational constant, 229
group estimate of the mean, 376, 376*f*, 396
group estimate of the median, 379, 379*f*
group estimate of the standard deviations, 396
grouped data, 372

H

half-step, 442
Halley, Edmond, 221, 233
Halley's Comet, 233*f*
Hamilton circuit, 252, 271–286
Hamilton, William Rowan, 273
Hare system, 303
Harmonice Mundi (The Harmony of the World)
    (Kepler), 200
heliocentric system, 185

Heracleides, 182, 182*f*
heuristic algorithms, 275
Hipparchus, 12, 191
histogram, 372, 372*f*, 391*f*
    of data collection, 393*f*
      to estimate percentages, 377, 377*f*
      same standard deviation but different
        means, 395*f*
horizontal translation, 459, 459*f*
hypothetico-deductive method, 215

I

IIA. *See* Independence of Irrelevant Alternatives
image, 8
increasing function, 11
incrementation, 158
Independence of Irrelevant Alternatives (IIA), 297
independent variable, 5
inertia, 226
inference, 423
infinite sample space, 405
initialization, 158
input device, 153
insincere voting, 297
instant runoff, 303
intercepts, 41
interest, 72
    and effective yield, 72–80
interest period, 75
interest rate, 72
interpolate, 509
interpretation, 125
intersection of two sets, 339
interval classes, 371
inverse-square law, 222, 223*f*
Ironwheel Tire Company, 374
isomorphism, 247

K

Kepler, Johannes, 197–201, 197*f*
Kepler's first law, 198, 198*f*
Kepler's laws of planetary motion, 198
Kepler's second law, 199, 199*f*
Kepler's third law, 200
key, 479, 497
key signature, 480–481, 481*f*
keyboard, positions on, 448*f*
kilogram, 211
Kingsbridge County
    distribution of ages in, 377, 378*f*
    percentage of people, 378, 378*f*
knowledge, quadrivium of, 178–179, 450*f*
Kuratowski, Kazimierz, 252

L

latitude angle, 189
Law of Demand, 57
law of inertia, 226
law of large numbers, 350, 406
Law of Supply, 57
law of universal gravitation, 228
laws of motion, 226, 227*f*
least residue, 444
least-residue system modulo *m*, 479
level of confidence, 427, 428
light-second, 186

light-year, 180
linear congruence, 494
linear diameter, 183, 183*f*
linear fit, 509
linear functions, 23–27
    defined, 24
linear interpolation, 509, 511
linear regression, 539
literacy, 370
loans, amortization of, 95–102
logic circuit, 152
logic error, 138
logic gate, 135, 151
logical value, 111
logically equivalent, 129
looping, 155
lunar eclipses, observations of, 191*f*

M

m-modularization, 443
machine language, 154
Maestlin, Michael, 198
major scale, 479, 480*f*
major triads, 450
majority criterion, 295
Manhattan graph, Eulerization of, 263
margin of error, 423, 428
Mars
    position of, 182*f*
    surface gravity on, 211
mass, 211
matching problems, 258
mathematical modeling, 508
    linear interpolation, 509–517
    modeling process, 508
    nonlinear modeling functions, 537–546
    Wald function, linear prediction with, 522–529
mean, 374
    estimates for, 397
measure of central tendency, 374
median, 374
median interval, 379
melodic inversion, 468
memory, 153
Mendel, Gregor, 357
meteorology, 9–10
middle C, 448, 449
minimal Hamilton circuit, 274
minimum Eulerization, 263
mod, 446–447, 496–498
model, 508
modeling type, 537
modular arithmetic, 489, 490*f*
    and cryptology, 489–499
modulus, 443
modus ponens, 139
modus tollens, 140
monotonicity criterion, 305
moon, orbit of, 227*f*
mortgage, 98
multiplicative inverse, 495
music, 442
"music of the spheres," 178, 179
mutually exclusive events, 340

N

n-cycle, 307
natural laws, 232
natural philosophy, 176
nearest-neighbor algorithm, 277
negation, 113
network, 247
neutrons, 9
Newton, Isaac, 15, 129*f*, 211, 221–234, 408
Night Sky, 176*f*
nodes, 247
nonlinear functions, 27, 537
nonlinear modeling functions, 537–546
normal curve, 409, 409*f*, 410*f*
normal distribution, 405–416, 416*f*, 423
note graph, 464, 464*f*
note of inversion, 466
notes, 447, 447*f*
    residue classes and, 442–451
    sample pattern of, 464*f*
number theory, 14–15, 443
numeracy, 370
numerical analysis, 410

O

octave, 442
odd vertex, 250
odds against an event, 364
odds for an event, 363
optimal circuit, 252
ordinary annuity, 85
output device, 153
outstanding principal, 99

P

parabola, 41, 52
paradigm shift, 178, 195
paradox, 112
parallax, 179
parameters, 405, 537
parsec, 196
particular distribution, 394
partition, 340
paths, 250
    graphs and, 246–259
payment period or interval, 85
perfect fifth, 469
    interval, 483, 483*f*
perfect fourth, 483
perigee, 203
perihelion, 199
period, 446
Periodic Table of the elements, 7–8, 8*f*
Philosophiae Naturalis Principia Mathematica (Newton), 226
piano key positions, 463*f*
piano keyboard, part of, 442*f*
pitch, 442
    names of, 449*f*
pitch class, 448
pivot note, 469
pixels, 158
plaintext, 496
planar graph, 248

plane, 459
planets, 177
    distances and periods of, 193–194, 194*f*
    elongation of, 189
Plato, 190, 191
plurality method, 295
Polaris star, 177
Pomp and Circumstance (Edward), 451*f*
population, 405
position sequence, 464
preference ballots, 293
preference table, 294
premise, 136
present value, 72
primary chord, 450, 450*f*
prime, 494
prime statements, 112
principal, 72
Principia, 226, 226*f*
probability, 350–354
    properties of, 354–358
    sample spaces and events, 330–334
    sets, 338–345
probability of event *A*, 351
program, 153
pseudocode, 154
Ptolemaic system, 182, 190–193, 191*f*
Ptolemy, Claudius, 191
Punnett square, 358
Pythagoras, 7, 178, 179, 179*f*

Q

quadratic function, 41
quadrivium of knowledge, 178–179, 450*f*

R

random digit table, 424
random sample, 424
range, 6, 371
rate, 24
real inversion, 468
reflection, 459, 461*f*, 466*f*
reflection across a horizontal, 461
reflection across a vertical, 461
relative error, 525
relative frequency, 350, 372
relative frequency distribution, 372, 372*f*, 376
    standard deviation estimation from, 396*f*
rent, 85
residue (or congruence) class, 443
residue classes and notes, 442–451
retrograde, 466, 466*f*
retrograde inversion, 466
retrograde motion, 182, 195, 195*f*
rotation, 459, 460, 467*f*
routing problem, 260
Russell, Bertrand, 110
Russian peasant method for power residues, 492

S

SADD election. *See* Students Against Drunk Driving election
sample, 423
sample means, distribution of, 426
sample space, 330

sampling distribution, 426, 427
Saros cycle, 446
scatter plot, 522
science
    ancient milestones, 177–189
    defense of Copernicanism, 208–220
    defined as, 176
scientific method, 178
selection bias, 425
set, 6, 330, 338
set difference, 347
sharp, 448
Sidereus Nuncius (Galileo), 213
significance level, 431
significance tests, 431
simple annuity, 85
simple circuit, 250
simple interest, 72
    credit cards, 74
simple path, 250
single transferable voting systems (STVS), 304
sinking fund, 88
skewed distribution, 381
slope, 24
small-angle formula, 183
solar eclipse, 188
staff, 449
stakes, 512
standard deviation, 391, 392*f*
    estimates for, 397
    from relative frequency distribution, estimation, 396*f*
standard normal curve, 410*f*
standard normal distribution, 410
standardized normal distribution table, 410*f*, 412
stars, in Big Dipper constellation, 180*f*
statement, 111
statement forms, 125
statistics, 370
statistics capability, using calculator with, 387–390, 402–404
stellar parallax, 195–196
strategic manipulation, 297
strict (or real) inversion, 468
strictly local maximum, 54
strictly local minimum, 54
Students Against Drunk Driving (SADD) election, 303*f*, 304*f*
STVS. *See* single transferable voting systems
subset, 338
substitution alphabet, 496
survey, results of, 430, 430*f*
syllogism, 138
symbolic logic, 122
symmetric distribution, 381

T

tautology, 136
term, 72
Texas Instruments (TI), 160
Titius–Bode rule, 205, 409
tonal inversion, 468
tonic, 479
transformation, 459–471
transforming the data, 539
translations, 459
transposition, 464

Traveling Salesperson Problem (TSP), 276
treble clef, 449, 449f, 450f
truth table, 121
    biconditional, 124, 124f
    conditional statement, 123, 124f
    for conjunction, 121, 121f
    negation and disjunction, 122, 122f
    prime statement symbols, 127, 128f
TSP. See Traveling Salesperson Problem

U

uniform sample space, 351
union of two sets, 339
universal set, 338

V

valid, 136
variable, 5
velocity, 209
Venn diagrams, 141, 340
Venn, John, 340
Venus, phases of, 214f
vertex, 52
vertex-oriented circuit, 271
vertical line test, 43
vertical translation, 459, 460f
    of 2 notes, 465f
vertices, 247

von Neumann, John, 153
voting methods
    approval voting, 314–319
    Borda count method, 295–296
    first-place votes, 294–295

W

Wald, Abraham, 522
Wald function, 523
    linear prediction, 523
Wald line, 523
wanderer, 177
web-generated histograms, 390
weight, 211, 274
weighted average, 396
weighted graph, 274
Weyl, Hermann, 215
Whitehead, Alfred North, 110
whole step, 442
Wiles, Andrew J., 498

Z

z -score, 393
zeroes of the function, 53
Zetonia, annual income of people, 380, 380f
Zimmer, Carl, 279
zodiac, 182